Lecture Notes in Computer Science 11472

Commenced Publication in 1973
Founding and Former Series Editors:
Gerhard Goos, Juris Hartmanis, and Jan van Leeuwen

More information about this series at http://www.springer.com/series/7409

Jan Mazal (Ed.)

Modelling and Simulation for Autonomous Systems

5th International Conference, MESAS 2018
Prague, Czech Republic, October 17–19, 2018
Revised Selected papers

 Springer

Editor
Jan Mazal (ID)
NATO Modelling and Simulation Centre
of Excellence
Rome, Italy

ISSN 0302-9743   ISSN 1611-3349   (electronic)
Lecture Notes in Computer Science
ISBN 978-3-030-14983-3   ISBN 978-3-030-14984-0   (eBook)
https://doi.org/10.1007/978-3-030-14984-0

Library of Congress Control Number: 2019933337

LNCS Sublibrary: SL3 – Information Systems and Applications, incl. Internet/Web, and HCI

This Springer imprint is published by the registered company Springer Nature Switzerland AG
The registered company address is: Gewerbestrasse 11, 6330 Cham, Switzerland

Preface

This volume contains the full papers presented at the MESAS 2018 Conference: Modelling and Simulation for Autonomous Systems, held during October 17–19, 2018, in Prague. The initial idea to launch the MESAS project was introduced by the Concept Development and Experimentation Branch at the NATO Modelling and Simulation Centre of Excellence in 2013. From that time, the event gathers together—in keynote, regular, poster, and way ahead sessions—fully recognized experts from different technical communities in the military, academia, and industry. The main tracks of the 2018 edition of MESAS were "Future Challenges of Advanced M&S Technology," "Swarming—R&D and Application," "M&S of Intelligent Systems—AI, R&D and Application," and "Analysis of Tensor-Based Image Segmentation Using Echo State Networks AxS in Context of Future Warfare and Security Environment (Concepts, Applications, Training, Interoperability, etc.)." The community of interest submitted 66 papers for consideration. Each submission was reviewed by two Program Committee members or selected independent reviewers. After the review process the committee decided to accept 46 papers for presentation (in seven sessions) and these papers were also accepted to be included in the conference proceedings.

December 2018   Jan Mazal

MESAS 2018 Organizer

NATO Modelling and Simulation Centre of Excellence
(NATO M&S COE)

The NATO M&S COE is a recognized international military organization activated by the North Atlantic Council in 2012, and does not fall under the NATO Command Structure. Partnering nations provide funding and personnel for the center through a memorandum of understanding. The Czech Republic, Italy, and the USA are the contributing nations, as of this publication. The NATO M&S COE supports NATO transformation by improving the networking of NATO and nationally owned M&S systems, promoting cooperation between nations and organizations through the sharing of M&S information, and serving as an international source of expertise.

The NATO M&S COE seeks to be a leading world-class organization, providing the best military expertise in modelling and simulation technology, methodologies, and the development of M&S professionals. Its state-of-the-art facilities can support a wide range of M&S activities including but not limited to: education and training of NATO M&S professionals on M&S concepts and technology with hands-on courses that expose students to the latest simulation software currently used across the alliance; concept development and experimentation using a wide array of software capability and network connections to test and evaluate military doctrinal concepts as well as new simulation interoperability verification; and the same network connectivity that enables the COE to become the focal point for NATO's future distributed simulation environment and services.

https://www.mscoe.org/

Organization

General Chair

Adriano Fagiolini   University of Palermo, Italy

Technical Program Committee Chair

Petr Stodola   University of Defence, Czech Republic

Technical Program Committee

Ronald C. Arkin	Georgia Institute of Technology, VA, USA
Ozkan Atan	University of Van Yüzüncü Yil, Turkey
Richard Balogh	Slovak University of Technology in Bratislava, Slovakia
Luca Bascetta	Politecnico di Milano, Italy
Marco Biagini	Modelling and Simulation Centre of Excellence, Italy
Antonio Bicchi	University of Pisa, Italy
Dalibor Biolek	University of Defence, Czech Republic
Agostino Bruzzone	University of Genoa, Italy
Erdal Cayirci	University of Stavanger, Norway
Fabio Corona	Modelling and Simulation Centre of Excellence, Italy
Andrea D'Ambrogio	University of Rome Tor Vergata, Italy
Riccardo Di Matteo	University of Genoa, Italy
Radek Doskočil	University of Defence, Czech Republic
Michal Dub	University of Defence, Czech Republic
Jan Faigl	Czech Technical University in Prague, Czech Republic
Jan Farlik	University of Defence, Czech Republic
Pavel Foltin	University of Defence, Czech Republic
Petr Frantis	University of Defence, Czech Republic
Corrado Guarino Lo Bianco	University of Parma, Italy
Karel Hajek	University of Defence, Czech Republic
Kamila Hasilova	University of Defence, Czech Republic
Václav Hlaváč	Czech Technical University in Prague, Czech Republic
Jan Hodicky	Centre for the Security and Military Strategic Studies University of Defense in Brno, Czech Republic
Jam Holub	FEE CTU Prague, Czech Republic
Jaroslav Hrdina	Brno University of Technology, Czech Republic
Thomas C. Irwin	Joint Force Development, DoD, USA
Shafagh Jafer	Embry-Riddle Aeronautical University, Florida, USA
Sebastian Jahnen	Universität der Bundeswehr, Germany
Jason Jones	Modelling and Simulation Centre of Excellence, Italy

Lukáš Kopečný	Brno University of Technology, Czech Republic
Piotr Kosiuczenko	Military University of Technology in Warsaw, Poland
Tomas Krajnik	Czech Technical University, Czech Republic
Tobias Kuhn	NATO M&S COE, Italy
Miroslav Kulich	Czech Technical University, Czech Republic
Václav Křivánek	University of Defence in Brno
Jan Leuchter	University of Defence, Czech Republic
Pavel Manas	University of Defence, Czech Republic
Marina Massei	University of Genoa, Italy
Jan Mazal (Ed.)	Modelling and Simulation Centre of Excellence, Italy
Vladimir Mostyn	VSB - Technical University of Ostrava, Czech Republic
Pierpaolo Murrieri	Selex ES, a Finmeccanica company, Italy
Andrzej Najgebauer	Military University of Technology in Warsaw, Poland
Petr Novak	VSB-TUO, Czech Republic
Josef Prochazka	University of Defence, Czech Republic
Lucia Pallottino	University of Pisa, Italy
Stefan Pickl	Universität der Bundeswehr, Germany
Vaclav Prenosil	Masaryk University, Czech Republic
Libor Preucil	Czech Technical University in Prague - CIIRC, Czech Republic
Dalibor Procházka	University of Defence, Czech Republic
Paolo Proietti	MIMOS, Italy
Jan Rohac	Czech Technical University in Prague, Czech Republic
Milan Rollo	Czech Technical University in Prague, Czech Republic
Marian Rybansky	University of Defence, Czech Republic
Martin Saska	Czech Technical University in Prague, Czech Republic
Marc Schmitt	Universität der Bundeswehr, Germany
Vaclav Skala	University of West Bohemia, Czech Republic
Marcin Sosnowski	Jan Dlugosz University in Czestochowa, Poland
Julie M. Stark	Office of Naval Research Global, USA
Andreas Tolk	The MITRE Corporation, USA
Petr Vasik	Brno University of Technology, Czech Republic
Jiří Vokřínek	Czech Technical University in Prague, Czech Republic
Premysl Volf	Czech Technical University in Prague, Czech Republic
Ludek Zalud	Brno University of Technology, Czech Republic
Fumin Zhang	Georgia Institute of Technology, VA, USA
Radomír Ščurek	VSB - Technical University of Ostrava, Czech Republic

NATO M&S CoE Director

Vincenzo Milano	Modelling and Simulation Centre of Excellence, Italy

MESAS 2018 Event Director

Jason Jones   Modelling and Simulation Centre of Excellence, Italy

MESAS 2018 Event Manager and Proceedings Chair

Jan Mazal (Ed.)   Modelling and Simulation Centre of Excellence, Italy

MESAS 2018 Organizing Committee

Felice Daiello   Modelling and Simulation Centre of Excellence, Italy
Fulvio Morghese   Modelling and Simulation Centre of Excellence, Italy
Paolo Proietti   MIMOS, Italy

Contents

Swarming - R&D and Application

M&S of Intelligent Systems - AI, R&D and Application

AxS in Context of Future Warfare and Security Environment (Concepts, Applications, Training, Interoperability, etc.)

Future Challenges of Advanced M&S Technology

Modelling, Simulation, and Planning for the MoleMOD System

Michaela Brejchová[1], Miroslav Kulich[2]([✉])[iD], Jan Petrš[3], and Libor Přeučil[2]

[1] Faculty of Electrical Engineering, Czech Technical University in Prague,
Prague, Czech Republic
brejcmi3@fel.cvut.cz
[2] Czech Institute of Informatics, Robotics, and Cybernetics,
Czech Technical University in Prague, Prague, Czech Republic
{kulich,preucil}@cvut.cz
[3] Faculty of Architecture, Czech Technical University in Prague,
Prague, Czech Republic
petrjan@fa.cvut.cz
http://imr.ciir.cvut.cz

Abstract. MoleMOD is a heterogeneous self-reconfigurable modular robotic system to be employed in architecture and civil engineering. In this paper we present two components of the MoleMOD infrastructure - a test environment and a planning algorithm. The test environment for simulation and visualization of active parts as well as passive blocks of MoleMOD is based on Gazebo - a powerful general-purpose robotic simulator. The key effort has been put into preparation of realistic models of passive and active components taking into account their physical characteristics. Moreover, given a starting configuration of the MoleMOD system and a final configuration an approach to plan collision-free trajectories for a fleet of active parts is introduced.

Keywords: Modelling · Simulation · Planning · Self-reconfiguration · Modular robotic systems

1 Introduction

MoleMOD is a unique self-reconfigurable modular robotic system developed at Czech Technical University in Prague [3,4]. The system is heterogeneous as it consists of active robots and passive modular building blocks and it is inspired by colonies like termites, ants or bees, which permanently rebuild and adapt their "houses" to surroundings and current conditions (Fig. 1).

The robots are flexible, can rotate, and are able to connect to the passive blocks as well as pick up and carry them along a given trajectory. This approach offers extensive possibilities of reconfiguration and adaption due to separated mobile and passive units. The two parts separation gave the system its acronym Mole (animal) + MOD (module). The passive part can be imagined as units

© Springer Nature Switzerland AG 2019
J. Mazal (Ed.): MESAS 2018, LNCS 11472, pp. 3–15, 2019.
https://doi.org/10.1007/978-3-030-14984-0_1

Fig. 1. The MoleMOD system.

of regular 3D lattice, just like individual crystals or voxels, respectively, in a crystalline lattice or virtual digitized volume.

The active part, as it is now, is further decomposed to three essential parts: soft/flexible body, revolving heads and a rotator in the centre. The head is primarily used for screwing the passive units to hold together and the secondary function is to ride over the construction site1. The flexible body allows for the peristaltic movement, through the trajectories within the passive block conglomerate. A secondary function of the body, not least important, is the manipulation with the blocks, so picking up and carrying. Finally, the rotator allows rotation of the blocks as is typical for majority of modular robotic systems (Fig. 2).

The MoleMOD system is very adaptable and can be used in many various situations, especially at locations that are not safe for people, or there is a problem to build – places like deserts, mountains or polar regions, which cannot be inhabited, but it can be necessary to build there. MoleMOD may not be used only for building houses, but also for bridges, pylons or research stations.

The system does not need cranes or other external construction machines. Therefore it is quite easy to transport it to the building site. It will even be possible to transfer only robots; building modules will be created by a 3D printer from local materials. This way the system could be used for the colonization of another planet.

Other uses may be for example temporary constructions. Tribunes for sports events, such as the Olympic Games or the World Cup races, markets, exhibitions, festivals, events that last only a few days or weeks. No less important is the possibility of using the system in case of a disaster. It can be building of bridges after a flood, shelters for people who have lost their homes due to a catastrophe and so on. Also, it can be used after a nuclear disaster, when the presence of humans is not possible because of radiation.

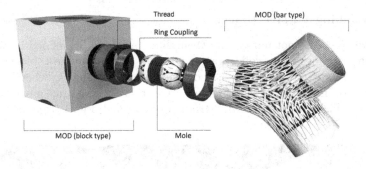

Fig. 2. Structure of the system.

The rest of the paper is organized as follows. Sect. 2 is dedicated to the description of the simulation environment for MoleMOD based on the Gazebo simulator, while the planning approach is presented in Sect. 3. The final comments and future work are described in Sect. 4.

2   Simulation Environment

This section provides a description of the Gazebo simulator and design of models (both passive blocks and active parts) for testing the MoleMOD system in Gazebo. Gazebo is an open-source robotics simulator, which can be used to design robots, test algorithms and artificial intelligence systems using realistic scenarios. The simulator offers indoor and outdoor environments with the possibility of setting several properties, such as wind, gravitation, friction and so on. Gazebo includes multiple physics engines (ODE,Bullet, Simbody and DART), a library of robot models and environments, several types of sensors and functional graphical and programmatic interfaces [1].

A simulation environment in Gazebo is described in so called world files, which include specification of elements such as robots, lights, sensors or static objects. The files use SDF (Simulation Description Format), an XML format originally developed for Gazebo.

Model files are similar to world files but contain only specifications for a model. The model created by this file can be included in a world file, so it is possible to use one model several times without rewriting the entire code. Also, there is the online model database.

SDF models can be just simple shapes but also complex robots. Basically, a model consists of links, joints, sensors, collision objects, visuals and plugins [2]. A **link** contains the physical properties. It is a body of the model or its part. It may have many collision and visual elements. A **collision** element is a geometry that is used to check collisions. A link can contain many collisions. A **visual** element visualize parts of a link. A link can have many visuals or none. A **joint** connects two links. Each joint has a parent and a child, an axis of rotation and some other properties. A **sensor** collects data from the world and these are then used by plugins.

2.1   Modelling Passive Blocks

As passive blocks do not move, only their shape needs to be defined in the form of a triangular mesh – a collection of triangles. Even simplest blocks, cubes with two straight circular tunnels, consist of many triangles, when a precise approximation is needed, see Fig. 3 (left). We, therefore, assume square-shaped tunnels for which only a fraction of triangles is needed, Fig. 3 (right).

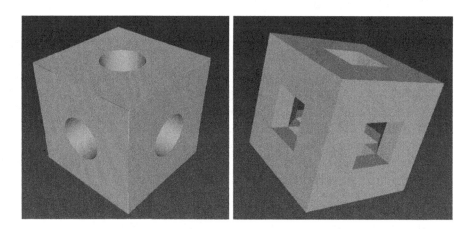

Fig. 3. A cube-shaped block with two tunnels. Precise approximation of a cube-shaped block with circular tunnels contains 6174 triangular faces (left), while a description of a block with squared tunnels contains 352 faces.

Triangular meshes can be defined in Collada (Collaborative Design Activity) files (.dae), which can be created either manually or using some modelling tool like MeshLab (www.meshlab.net/) or Blender (www.blender.org). This can be imported into a world file similarly to Listing 1.1.

```
<sdf version="1.4">
    <world name="default">
        <include>
            <uri>model://sun</uri>
        </include>
        <include>
            <uri>model://ground_plane</uri>
        </include>
        <model name="my_mesh">
            <pose>0 0 .2 0 0 0</pose>
            <link name='tunnel'>
                <visual name='visual'>
                    <transparency>0.5</transparency>
                    <geometry>
                        <mesh>
                            <uri>file://1.dae</uri>
```

```
                              <scale>1 1 1</scale>
                          </mesh>
                      </geometry>
                  </visual>
                  <collision name='collision'>
                      <geometry>
                          <mesh>
                              <uri>file://1.dae</uri>
                              <scale>1 1 1</scale>
                          </mesh>
                      </geometry>
                  </collision>
              </link>
          </model>
      </world>
  </sdf>
```

Listing 1.1. A world file with mesh importing

2.2 Modelling Active Parts

Active parts – robots – consist of three components: a soft body, revolving heads and a rotator. Precise modeling of these will be unnecessarily complex which will significantly slow down simulation of the whole MoleMOD system. Gazebo has, moreover, no ways how to simulate soft components. We thus model the robots making use of components available in Gazebo – links and joints. To describe the modelling process, we start with a simple model with limited functionality and make the model more complex and powerfull subsequently in several steps.

The Basic Model. The basic model is constructed from two cubes connected by a prismatic joint, see Fig. 4. The model is limited to move only in one direction forwards or backwards by expanding and contracting the joint and changing the frictions of the links.

The forward motion of the model consists of four parts:

1. setting frictions
2. the joint expansion
3. setting frictions
4. the joint contraction

The frictions of the links are set in a model plugin. Before moving itself, it is necessary to lower the friction of the front cube (the first cube in the direction of the movement). For expanding the joint, a positive velocity is set to the joint in the plugin which leads to movement of the front cube.

The simulator updates in fixed intervals. When a predefined number of iterations is done, the plugin changes the sign of the velocity and swaps frictions – higher for the front cube and lower for the back cube. After repeating the same number of iterations, the joint position is set to the initial (contracted) state.

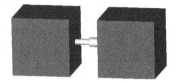

Fig. 4. The basic model

The Rotating Model. For turning, it is needed to add a revolute joint in the middle of the model. Because the simulator cannot connect two joints, it is necessary to add two links and one more prismatic joint. Now the model is made of two head cubes, two small middle cubes, two prismatic joints and one revolute joint, Fig. 5.

The first prismatic joint connects the first head cube with the first middle cube. This cube is joined with the second middle cube by the revolute joint. These both cubes are immaterial and they are placed on themselves. The second middle cube is connected to the second cube with the second prismatic joint.

After these changes, the code of the plugin has to be adapted. For translation, both prismatic joints expand and contract at the same time.

Rotation is similar to translation. The four parts of motion are:

1. setting frictions
2. the joint rotation to the direction, where we want to turn
3. setting frictions
4. the joint rotation to the different direction

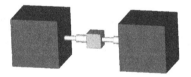

Fig. 5. The rotating model

At first, the friction of the cube to be shifted is reduced. The plugin then sets an input velocity to the revolute joint. This velocity can be positive or negative depending on the direction of the turn. Finally, frictions are swapped and velocity with an opposite sign is set to the joint. In contrast to the moving ahead, this motion stops after one step and will never be repeated, because the model would spin on the same place.

Translation/Rotation Controller. Unfortunately, the movements of the previous models are not precise enough. The joints move for the same time, but it does not guarantee that their final position is the same. To make the motion more accurate, a simple controller has been designed. The input to the controller is a position of the joint - the length of the prismatic joint or the angle of the rotation of the revolute joint.

```
1   if (position < required_value - accuracy) {
2       setVel(vel);
3   } else if (position > required_value + accuracy) {
4       setVel(-vel);
5   } else {
6       setVel(0.0);
7   }
```

Listing 1.2. A joint controller

Listing 1.2 shows a primitive controller that sets a positive velocity to the joint, if its position is lower than the required one, negative if larger or zero if it is in the interval determined by the deviation.

For better control, we can divide the possible positions into more intervals. The result will be similar to the example above; it will only contain more conditions. The controller used in the simulator is divided into five intervals. At the beginning of the motion, the joint is set to an initial speed. When the joint position is close to the required position, the speed is decreased to half the value.

The controller accuracy is 50 μm. In this range, the joint is set to zero velocity. If the position is larger or smaller (depending on the direction of the motion) the plugin sets a negative speed to the joint. Thanks to this condition there is no need to set the reverse speed for the contracting, the controller will solve it. Also, it is not necessary to count updates; one step consists of expanding to input length and contracting back to the initial state.

The second input for ahead motion is a number of iterations, but in this case, it is not the number of updates, but the number of calls of the controller (the total sum of all expanding and contracting). The distance the model moves is equal to the product of one step length and the number of iterations.

The Final Model. The model with just one revolute joint is not sufficient. To turn in a tunnel or lift a building unit, at least two revolute joints are needed. It is also not convenient to use a simple revolute joint, because it can rotate only around one axis. With the joints, the robot could rotate sideways or up and down, but could not do both of these operations. A solution to this problem is to use another joint. The joint is called universal in the simulator and it can rotate around two axes.

The final model consists of three main cubes, four prismatic joints, two universal joints and four little immaterial cubes that are between joints as depicted in Figs. 6 and 7.

Fig. 6. Moving in a tunnel - the final model

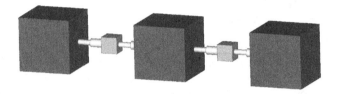

Fig. 7. The final model

2.3 Experiments

The models for the simulator have been made; the next step is to test them. In the previous chapter, the model of the MoleMOD robot was introduced and two simple moves (forward moving and rotation) were described.

Forward movement works on the same principle that has been described previously, see Fig. 8. The only difference is that the final model has four prismatic joints instead of two. The model could use all four joints when moving, but it is simpler to use only two, for example, the first and the fourth joint. That will spare us larger changes in the code and also the motion will be more accurate.

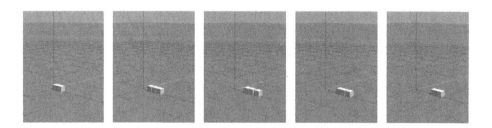

Fig. 8. Forward moving

The rotation remains practically the same with only two minor changes. The last model contains two universal joints, so it is needed to decide which joint will rotate. Then the rotation axis has to be set because the universal joint can rotate around two axes. An example of rotation is depicted in Fig. 9.

However, for simulating the work of the system, these two motions are not sufficient. For the basic version of the planning, it is essential to add lifting

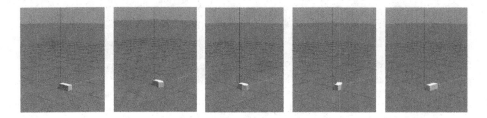

Fig. 9. Turning right

(putting a module on a next block or lifting a module just up), shifting modules and moves of robots in tunnels.

Robot movement in a straight tunnel is exactly the same as the forward moving. The length of the motion is adjusted according to the size of the block, see Fig. 10.

Fig. 10. Moving in the straight tunnel

Turning in a tunnel is more complicated. Because the robot is only a little bit smaller than the tunnel, it is impossible to turn around at once. It is needed to combine both types of moves - rotation and translation. This can be achieved by using a joint controller, see an example in Fig. 11

Fig. 11. Turning in the module

Lifting is, similarly to turning, a combination of translational and rotational motion, see Fig. 12.

Fig. 12. Lifting

3 Planning

Planning in MoleMOD is based on the A* algorithm – an informed method for state-space search. A state is represented by an arrangement of passive blocks and robots positions. We assuming two-dimensional space, which can be described as a two-dimensional matrix, where each cell stores the information whether the corresponding space is empty, it contains a block, or a block with a robot.

An action is performed by movement of robots, which can relocate some blocks. The simplest action is moving from one block to another. The natural condition for this action is that the position, where the robot moves to, is neighboring to the current robot position, lies inside a given area and also it contains an empty block. The robot can move to the right, to the left, up and down this way.

Another type of movement is lifting of a block. In reality, the model expands and partially inserts into the blocks beside. After that, the robot lifts one block up and place it on the top of the second one. Finally, the model retracts into one of the two blocks. To reduce the number of possibilities, we assume that the robot picks up the block, where it originally was, and remains in the block after the movement. To the motion can be executed, there has to be a free space around the moving block.

Putting down is similar to lifting. The robot expands into the block under its position and then contracts with the top one. For simplicity, the model starts and finishes again in the moved block.

The last motion that is possible with only one robot is moving a block to the right or the left. The model is in the block that we want to shift. If the place next to the block is free and under it, there is another block, the robot expands to this block. Then it moves the block to the empty position and contracts. The moving finishes again in the shifted block.

All movements, which were mentioned above, are valid also for the case with more robots. The advantage is that they can be performed in parallel, so the entire construction is done faster.

Besides, robots can work on a movement together. A block is lifted by one robot to a certain position, where a second robot takes it and completes the move as shown in Fig. 13.

The planning algorithm applies one action, where cooperation of two robots is used. When the block is lifted by one robot, it is like stairs, the block moves

Fig. 13. Cooperation of two robots

not only up but also to a side. In some cases, however, it is necessary to move just upwards. The movement starts identically as lifting. The robot in the block expands to the block beside and then elevates the block one position up. The second robot has to be in the block that lies on the block where the robot stretched to before. The second robot expands and "catches" the block and the first robot contracts to the underlying block.

3.1   Cost Calculation

The evaluation function of A* has a standard form $f(x) = g(x) + h(x)$, where $g(x)$ is the cost of the path from the start node to x, and $h(x)$ is a heuristic function that estimates the cost of the cheapest path from x to the goal.

One possibility of computing a value of the cost g is based on distance of the start and current states: the distance of a robot or a block moving is equal to $|i' - i| + |j' - j|$, where (i, j) are coordinates of the robot or the robot in the first state and (i', j') are its coordinates in the second state. However, computing the distance in every step means to find the robot or block that has just been moved, and calculate how far has shifted. That is senselessly complicated. Because the number of movements is limited and each has a specific distance that never changes, it is much simpler to assign a value d to every movement m:

$$g(x_n) = g(x_{n-1}) + d(m).$$

While the case of one robot is simple as the robot can perform only one movement in one step, the case with more robots is more complicated. For example, two robots can shift two blocks at the same time or gradually in two steps. The second option does not have any advantages, it only prolongs the building, so it is necessary to obviate it. That can be done by adding 1 to the distance in every step:

$$g(x_n) = g(x_{n-1}) + d(m) + 1,$$

where $d(m)$ is the sum of costs of the motions that have been made to get from the state x_{n-1} to the state x_n.

Counting the distance to the goal state is more difficult. A robot moves to a block, shifts it, moves to another block, shifts it and so over and over again until the goal state is reached. It is almost impossible to calculate the real cost.

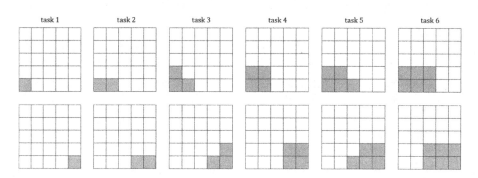

Fig. 14. Planning tasks for time measurement

When a block moves between two states, the cost is equal to $|i' - i| + |j' - j|$. If more blocks are moved, the cost is $\sum_{n=1}^{N} |i'_n - i_n| + |j'_n - j_n|$, where N is the number of blocks that have been moved, and $[i_n, j_n]$ and $[i'_n, j'_n]$ are coordinates of n-th block before and after the movement.

Costs of robots movement have to be added to the total cost, but it is hard to reckon them in advance. It depends on the number of robots and blocks and their positions. Every time a robot moves a block to a correct location, it has to move to the position next to the block. The minimal cost of the path from a block to another is one and the total cost of the way between blocks is equal to n_{wrong}, where n_{wrong} is the number of the blocks in the wrong positions which do not contain an robot. Therefore, the estimation of the cost of the path from the current state x to the goal state x_G is:

$$h(x) = \sum_{n=1}^{n_{wrong}} (|i_{Gn} - i_n| + |j_{Gn} - j_n|) + n_{wrong} - n_{robot},$$

where $[i_{Gn}, j_{Gn}]$ is a goal position of n-th block.

3.2 Experiments

Several experiments have been performed to demonstrate feasibility and time complexity of the planning algorithm. Six planning tasks depicted in Fig. 14 were run. In each pair, the upper state represents the initial layout of the blocks and the state below the goal arrangement. All tasks have been successively solved for different numbers of robots and for all cases the time of the planning was measured.

Table 1 shows that the time needed to find a solution, if the system contains more robots, is significantly higher than the case of one robot. However, the ratio of blocks and robots is also important. Robots need sufficient amount of space to move, so in the case that most blocks are occupied, the number of options is reduced. The other case is, if the quantity of blocks is quite higher than the number of robots, so the robots have plenty of space for moving. Each robot can perform some movements, the number of moves depends on the specific conditions. One more robot adds its motions and combinations of its moves and moves of others. The larger space and the more blocks it contains, the more movements will be possible.

Table 1. Time of planning

Time [ms]						
	1 robot	2 robots	3 robots	4 robots	5 robots	6 robots
Task 1	0.24					
Task 2	1.87	2.02				
Task 3	3.08	18.23	5.10			
Task 4	9.00	75.60	117.95	9.91		
Task 5	15.50	224.10	614.93	332.15	18.63	
Task 6	35.64	1212.75	4390.14	4000.69	706.35	22.48

4   Conclusion

In this paper we presented MoleMOD – heterogeneous self-reconfigurable modular robotic system and two software parts of it – the simulation environment and the planning algorithm. The system is still in the first phase of development and thus only first results were presented which can be further improved. Especially the planning algorithm has a potential for next improvements. First, better heuristics, which more precisely estimates the cost to the goal, needs to be designed. This will guide the A* algorithm to search the state space more effectively by expanding less states. Secondly, we will investigate more advanced planning algorithms (e.g., hierarchical) which will further improve computational complexity of the planning. Finally, extension of the algorithm into the 3D case will be done. Regarding the simulation environment, we would like to equip robots with sensors to accelerate development of control algorithms for robots.

Acknowledgement. This work has been supported by the European Union's Horizon 2020 research and innovation programme under grant agreement No 688117, and by the European Regional Development Fund under the project Robotics for Industry 4.0 (reg. no. CZ.02.1.01/0.0/0.0/15 003/0000470).

References

1. Open Source Robotics Foundation: Gazebo (2018). http://gazebo.org/
2. Open Source Robotics Foundation: SDF (2018). http://sdformat.org/
3. Petrš, J.: MoleMOD (2017). http://www.studioflorian.com/projekty/347-jan-petrs-molemod
4. Petrš, J., Havelka, J., Florián, M., Novák, J.: MoleMOD - on design specification and applications of a self-reconfigurable constructional robotic system. In: Fioravanti, A., et al. (eds.) ShoCK! - Sharing Computational Knowledge! - Proceedings of the 35th eCAADe Conference, vol. 2, pp. 159–166 (2017)

MUAVET – An Experimental Test-Bed for Autonomous Multi-rotor Applications

Jan Chudoba$^{(\boxtimes)}$, Viktor Kozák, and Libor Přeučil

Czech Institute of Informatics, Robotics and Cybernetics,
Czech Technical University in Prague,
Jugoslávských Partyzánu 1580/3, Prague, Czech Republic
jan.chudoba@cvut.cz
http://ciirc.cvut.cz

Abstract. Multi-rotor flying vehicles (referred as UAVs here) are well suited for many applications, including patrolling, inspection, reconnaissance or mapping and they are already used in many cases. In most cases they are used in manual mode, mainly due to legal issues, but autonomous methods and algorithms for UAV navigation are under extensive development nowadays.

This article describes a test-bed system for development of the fully autonomous multi-UAV systems called MUAVET (Multi-UAV Experimental Test-bed). The system incorporates several UAVs, communicating with the base station, able to be controlled either locally from the on-board-running control application, centrally from the base station, or by any combined approach. The system provides basic safety features and autonomous/automatic functions which significantly reduce the risks of crash or loosing the UAV, even in case of wrong command from the user application. The function and the performance of the MUAVET system is demonstrated in the search-and-rescue scenario of multiple UAVs searching for the visually-marked ground objects placed in the designated area and cooperating with the autonomous ground vehicle supposed to reach the found targets on ground.

1 Introduction

Unmanned aerial vehicles (UAVs) are widely and commonly used nowadays for the tasks, where manned airplane or helicopter utilization is not efficient, economic, or possible. Majority of these tasks are focused on the inspection or monitoring of the specified area of facility from the air. The mapping of the area is another significant area of the UAV utilization. Mainly due to the actual legislative restrictions, most of the operations are remote controlled. This is demanding for the need of sufficient number of trained pilots, especially in multi-UAV tasks, as well as stress laid on individual pilots during the flight.

Current level of UAV development is prepared for the fully autonomous operations, which can significantly simplify the mission setup and reduces the necessary personnel requirements. As a result, the cost of the mission may be

J. Mazal (Ed.): MESAS 2018, LNCS 11472, pp. 16–26, 2019.
https://doi.org/10.1007/978-3-030-14984-0_2

decreased. A fully autonomous UAV system has to be safe in the first place. This means mainly that risk of accidental crash of the vehicle or flight in the unintended direction is minimized as much as possible.

This article is focused on multi-rotor unmanned helicopters, able to take-off and land vertically and stay on the position in the air. These vehicles have typically lower cruise speed and significantly lower flight time on single charged battery than fixed-wing vehicles. Multi-rotor UAVs are optimal for smaller area operations, where total flight trajectory length does not exceed few kilometers. The significant advantage of multi-rotors is that they only need sufficiently flat terrain for take-off and landing, but no extra equipment or runway.

The topic of this article is a description of a test-bed system for evaluation of methods and algorithms for the control of multiple UAVs. Similar systems were built in the past, like the Multi-vehicle Experimental Platform for Distributed Coordination and Control built by H. How's team, used e.g. in [1]. This test-bed consists of several fixed wing planes, ground rovers and blimps. The system addresses the issues of distributed command and control algorithms, network topologies, resource allocation, fleet autonomy and human-in-the-loop operation control. Most past or existing UAV test-beds with multi-rotor UAVs are small-scale and indoor. The test-bed used at the GRASP laboratory [2] uses several AscTec Hummingbird quad-rotors controlled under Vicon [3] motion capture system used as a ground-truth localization system. The size of the test-bed arena is approximately $5 \times 4 \times 3.5$ m. Basic UAV control regulators are provided on different levels, allowing UAV control by variously skilled users and different methods. System provides basic functions like initial positioning of UAVs or collision avoidance. There is also a simulator as a part of the system. Another example of the similar test-bed system is mentioned in [4]. There are also software-based simulated testbeds available, like the OpenUAV system presented in [5], providing cloud support and allowing users to use it without difficult initial setup. The OpenUAV project is provided as an open-source. As the simulated environments can not cover the richness and all possible complications of the real world, they can significantly help to solve the basic problems before transfer to the hardware platform.

This article describes a design and development of the test-bed system called MUAVET (Multi-UAV Experimentation Test-bed), intended for testing of autonomous methods and algorithms for control and coordination of multiple UAVs fulfilling the common mission. It is an outdoor system with the expected operation area up to $1\,\mathrm{km}^2$. The main requirements which affected the system design were

- low cost and usage of off-the-shelf components,
- maximal safety (automatic detection of dangerous situations and response to prevent damage),
- scalability for up to tens of UAVs and
- simple but flexible API for user application development.

The basic navigation of the UAVs is Global Navigation Satellite System (e.g. GPS) based.

The designed system is indented for experimental and evaluation purposes only. It is not expected to be used in real situations.

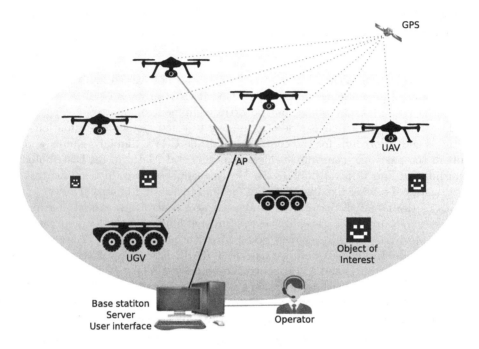

Fig. 1. MUAVET system overview (from [6]).

2   System Description

The overview of the MUAVET system is depicted in Fig. 1. There is a base station containing server computer, wireless communication access point and an user interface. Several UAVs and ground vehicles (UGVs) are connected to the server through the access point.

The MUAVET system consists of hardware and software part. The software part is designed as independent on the hardware, so any UAV (or possibly other robot) may be integrated after proper interfacing.

2.1   Test-Bed Hardware

The hardware used for the demonstration and evaluation was selected with regard to minimal cost and sufficient necessary abilities. One of the most important factors is possibility of controlling the UAV flight controller from the installed on-board computer.

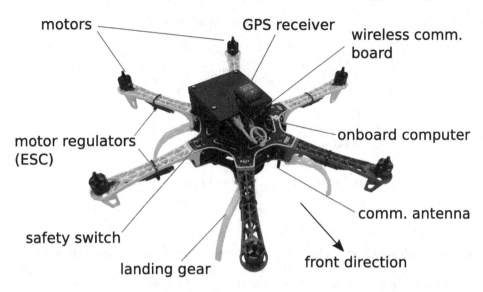

motors

GPS receiver

wireless comm. board

motor regulators (ESC)

onboard computer

safety switch

comm. antenna

landing gear

front direction

Fig. 2. MUAVET test-bed hexa-copter with propellers dismounted (from [6]).

The test-bed hardware is based on the DJI F550 hexa-copter frame [7] with motors installed (see Fig. 2). The frame is further equipped with the 3DR Pixhawk mini flight controller, providing good stability, flight performance and sufficient adjustability and configurability. This flight controller is running on the open-source Pixhawk firmware [8].

The PC-level computer (Intel NUC) is installed as the UAV on-board computer. This brings sufficient power on-board, suitable even for camera image processing applications. Additionally, the on-board code update and development is easy and fast. The presence of PC on-board also allows easy connection of various devices or sensors using standard interfaces.

Since the longer-range communication with the base station is required, an extra wireless communication module was installed on-board. This module is based on Mikrotik Routerboard IEEE802.11/bg WiFi with external antenna. The communication range in the open space was successfully tested up to 1 km, which is over expected operation range of the UAV.

The main sensor installed on-board the UAV is a camera. The global-shutter camera was installed due to possible vibrations, which usually significantly distort image of rolling-shutter cameras, rendering their output unusable for machine processing. For the demonstration purposes, the visual pattern detector method based on AprilTag [9] library was installed, giving the UAV ability to autonomously recognize artificial marks placed on ground.

The Table 1 summarizes the basic parameters of the UAV.

The second hardware part of the MUAVET system is a base station. Main components of the base station include a server computer and a wireless access point with an antenna. For field experiments, battery operation of the base

Table 1. UAV basic parameters

Parameter	Value
Frame weight	480 g
Total take-off weight	1200 g–2400 g
Diameter incl. propellers	800 mm
Max. ascend velocity	8.0 m/s
Max. cruise velocity	20.0 m/s
Max. horizontal acceleration	15 m/s^2
Primary battery	4-cell (14.8 V), 6750 mAh
Max. flight time	approx. 15 min

station is advantageous. The possible extension of the base station includes a RTK GNSS base station for centimeter-level positioning of UAVs. A computer with the display and input device is typically part of the base, forming the system user interface and serving as a platform for executing user applications.

2.2 Software Architecture

The MUAVET system is designed as a centralized system with the base station computer (server) as a central element providing connection between UAVs and user control applications, as seen in Fig. 3.

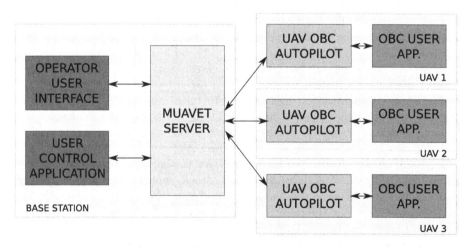

Fig. 3. MUAVET software structure (from [6]).

This approach was selected regarding the following advantages:

- central computer can permanently monitor states of all UAVs,
- user application connects to the single fixed point (server socket) and
- server can filter messages sent by user to UAVs in order to prevent communication channel overload.

Even though the system is designed as centralized, UAVs do not need permanent connection to the server for the safe operation. The control system of the UAV is designed as semi-autonomous, which means an UAV can fulfill the assigned task even without the active server connection. Only consequence of the lost connection is the actual inability to send new command and monitor actual state from the base station. However, since the commands are re-transmitted until acknowledgment, the UAV will get the command as soon as the connection is re-gained.

Table 2. API functions used to control an UAV from user application. Functions marked with (*) are provided for increased safety of the system and prevention of unintended UAV action.

API function	Description
Arm	Enable activation of UAV motors (*)
Disarm	Disable UAV motors (*)
Takeoff	Start the motors and take-off to pre-defined height
LandOnPosition	Fly to specified position and land there
Land	Land on the current position
HoldPosition	Stop moving and hold current position
FlyTo	Fly to specified position and stop there
FlyTrajectory	Fly through specified list of way-points
TrajectoryPause	Stop while flying through the waypoints
TrajectoryContinue	Continue paused flight
EnableLocalControl	Enable/disable control from the on-board computer (*)
SetFlyVelocity	Set a cruise velocity of the flight
SetMaxTiltAngle	Set the limit for the allowed tilt angles
RequestCameraImage	Start camera image capturing

The UAVs in the MUAVET system are controlled by the user applications, which may be running on base station or on-board the UAV. The on-board application can provide fully autonomous control without permanent server connection. Base-station application is advantageous in situations, where more UAVs need to be controlled simultaneously and connection losses are not expected. The application programming interface used by the user application is same on the server as well as on-board, so the same application may transferred from base to

UAV or vice-versa with minimal or no modification needed. From the obvious reason, controlling of another UAV from on-board computer of second UAV is not possible. The user API allows to control the UAV using the functions listed in the Table 2. The UAV measured data and telemetry is provided to the user application using the callback function mechanism.

Communication between UAVs is possible through the base station access point, as long as both communicating UAVs have an active connection. An advanced mode of direct communication between UAVs is possible by configuring the wireless modules in mesh mode.

Data sent between the UAV and the base station are handled differently, based on the data nature. These data may be basically divided into commands (sent to UAV) and sensor data (sent from UAV). The sensor data are sent continuously with the period given by the speed of measurement or defined requested period of sending. Typically, only the most recent data are important. This means, that in case of short loss of connection, some data may be lost without much trouble, as long as most recent data are received when connection is re-gained. For this reason, the connection between base station and the UAV is based on UDP protocol, which does not provide secure delivery. The advantage of this approach is communication overload prevention in such situations.

The UAV commands, on the other hand, has to be safely delivered, since the command is sent only once by the user application. Since the commands are sent over the same UDP protocol, the acknowledge and re-transmit mechanism is implemented in the base station server, ensuring the secure command delivery. Moreover, this mechanism is improved in a way to prevent overloading of the communication when user tries to send commands faster than what is possible to actually send. When a new command is sent by the user, it cancels the previous command of the same type, so for example the pending command to "fly to requested position" may be overridden by a newer "hold position" command, so the previous command is not delivered to UAV since it is already obsolete.

Several safety features and mechanisms are implemented on different levels of control. These levels include mainly the firmware in the flight controller HW module and the autopilot software running on-board the UAV PC. Since the on-board PC or its control application may fail, some basic safety functionalities are included in the flight controller. These include dangerous discharge of the battery, leaving an operation area in defined radius from the take-off position, or loss of the control commands from the PC, which may indicate PC or application failure. Since the activation of the safety mechanisms on the flight controller is not expected under normal conditions, the reaction results in imminent landing.

Safety mechanisms implemented in the autopilot control application running on the on-board PC should be activated sooner than the corresponding mechanisms in the flight controller. When a battery is discharged under a defined safe level, the UAV is ordered to return to the take-off position and land there. During this return flight, user can not override the UAV operation by a different command, with the exception of imminent landing. The short-term loss of the connection between the base and the UAV is allowable. When the connection is lost for longer period, the UAV interrupts the current operation and returns to the take-off position. In this case, the user may send new command to the UAV and interrupt the return when connection is re-gained.

Additionally a basic collision avoidance is implemented, preventing collision between two UAVs. When an UAV presence is detected in a expected trajectory of a second UAV, the second UAV flight is paused, holding the UAV on the current place until the first one clears the way. The collision avoidance mechanism is designed only as a last chance to prevent accident and may cause deadlock between two UAVs. Therefore, the flight trajectories should be planned as collision-free by the user applications in the first place. Since the UAVs do not have sensors for detecting obstacles physically, the collision avoidance is realized only by evaluating UAV positions and velocities. Collisions avoidance with fixed obstacles is possible by adding positions of the known obstacle boundaries to the collision avoidance subsystem, so UAVs are not allowed to fly into these areas.

3   Experimental Evaluation

For the purpose of the system functionality and performance evaluation, the search & rescue - type of experiment was performed. In this experiment, the ground vehicle (UGV, see Fig. 4a) was integrated into the system to demonstrate the scalability and possibility to integrate different types of robots. The UAVs divided the whole operation area into several sections depending on the number of UAVs and selected method of coverage. The algorithm for the area division and the trajectory planning was provided by our partners from King Abdulaziz University [10]. Each of the UAVs then executed a trajectory providing full coverage of the assigned area, considering the height above the ground and camera field of view. There were several visual marks placed in the area (see Fig. 4), representing the objects of interest.

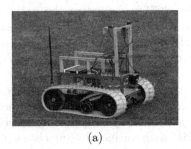 

(a)   (b)

Fig. 4. (a) Cameleon UGV used in the experiment. (b) Visual target used as an object of interest.

The mission execution was controlled by a central application commanding all UAVs and the UGV through the system user API. This setup allowed easy time synchronization of all controlled vehicles. The detected and localized position of the installed objects were reported to the mission control application, which ordered the UGV to visit the places of the object positions in the order of detections. The resulting trajectories are visible in the Fig. 5.

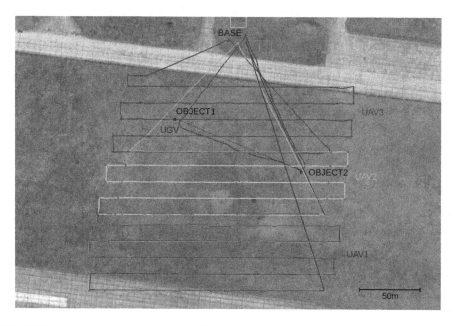

Fig. 5. Search and rescue experiment UAV and UGV trajectories (from [11], modified).

The 3 used UAVs flew total distance of 4271 m, including the paths from the base to the starting position and return to land. The total covered area was about 200×200 m and the time of search mission was about 640 s. The positions of all robots in time are plotted in the Fig. 6. The cruise speed of all UAVs was set to 3 m/s.

This experiment successfully demonstrated ability to simultaneously control several UAVs and coordination between air and ground robots. The precision of the real flight trajectory error from the planned trajectory and the precision of the ground object detection corresponds to the achievable precision of the common GNSS system (typ. error 1–3 m [12]).

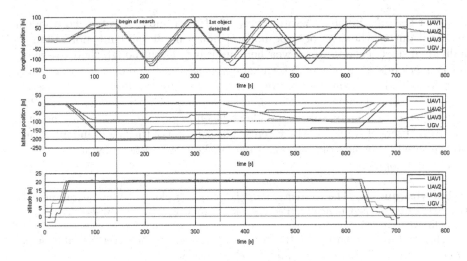

Fig. 6. Position of all robots in time.

4   Conclusion

In this article, the multi-UAV experimental test-bed MUAVET was presented. The test-bed is able to execute experiments and tests with multiple coordinated UAVs. A skilled pilot is not needed for the usage of the system. The UAVs are controlled using a simple API either from the centralized application running at the base station computer, or by autonomous control application which may run on-board. Alternatively, the both control approaches may be combined, so the UAV may be autonomous and commanded by higher-level command from the base. The UAV may be controlled on different levels of control, while the basic mode is the control by GPS way-points which UAV flies through using the defined velocity.

The UAV telemetry is transmitted to the base in a way which does not overload the communication channel in case of worse wireless connection. When the communication to the UAV is lost for short period, actual data transferred to the base may be lost, but the UAV may still continue the mission since it is designed as autonomous. Commands sent from the base to the UAV are acknowledged and re-transmitted if necessary, so they can not be lost permanently.

The system safety is a primary objective and also one of the biggest challenges. Dangerous situations and common faults are permanently automatically evaluated on-board the UAV. These include mainly discharged battery state or loss of the connection to the base for the longer period. In such dangerous cases, the UAV selects one of the possible options, depending on the actual level of danger, typically return to the take-off position and land or land immediately on the current place. These safety mechanisms are implemented on different HW/SW levels, so in case the software running on-board the UAV PC fails, the low level flight controller is able to handle the situation.

After the individual UAVs were properly tested and calibrated, no accident occurred during the tens of experimental hours in the air and more than 30 individual starts even in worse weather conditions. Even though there are many safety mechanisms implemented, the system has still to be considered possibly dangerous and constant supervision is needed during the flight. One reason is dependence of the navigation system on the available GNSS systems (e.g. GPS), which may occasionally fail to provide sufficiently precise position, as reported by several users of similar UAV systems. The solution of the better independence on the satellite navigation systems is one of the future directions of the further system development.

Acknowledgments. The system MUAVET was developed in co-operation with the King Abdulaziz University, Faculty of Computing and IT, Research group of Dr. Ahmed Barnawi.

This work was supported by the Technology Agency of the Czech Republic under project TE01020197 Center for Applied Cybernetics 3.

References

1. Richards, A., Bellingham, J., Tillerson, M., How, J.: Coordination and control of multiple UAVs. In: Proceedings of the AIAA Guidance, Navigation and Control Conference, AIAA 2002, p. 4588 (2002)
2. Michael, N., Mellinger, D., Lindsey, Q., Kumar, V.: The GRASP multiple micro-UAV testbed. IEEE Robot. Autom. Mag. **17**(3), 56–65 (2010)
3. VICON optical motion capture cameras. https://vicon.com/products/camera-systems/
4. Palacios, F.M., Quesada, E.S.E., Sanahuja, G., Salazar, S., Salazar, O.G., Carrillo, L.R.G.: Test bed for applications of heterogeneous unmanned vehicles. Int. J. Adv. Robot. Syst. **14**(1), 1729881416687111 (2017)
5. Schmittle, M., et al.: OpenUAV: a UAV testbed for the cps and robotics community. In: Proceedings of the 9th ACM/IEEE International Conference on Cyber-Physical Systems, ICCPS 2018, pp. 130–139. IEEE Press, Piscataway (2018)
6. Chudoba, J.: Muavet - system specifications. Technical report, Czech Technical University in Prague, Czech Institute of Informatics, Robotics and Cybernetics (2018)
7. DJI. https://www.dji.com/
8. Pixhawk project. https://pixhawk.org/
9. Wang, J., Olson, E.: AprilTag 2: efficient and robust fiducial detection. In: 2016 IEEE/RSJ International Conference on Intelligent Robots and Systems (IROS), pp. 4193–4198. IEEE (2016)
10. Barnawi, A., Al-Barakati, A.: Design and implantation of a search and find application on a heterogeneous robotic platform. J. Eng. Technol. **6**(Special Issue on Technology Innovations and Applications), 381–391 (2017)
11. Chudoba, J.: Muavet final experiment. Technical report, Czech Technical University in Prague, Czech Institute of Informatics, Robotics and Cybernetics (2018)
12. WAAS T&E Team William J. Hughes Technical Center. Global positioning system (GPS), standard positioning service (SPS), performance analysis report. Technical report 96, Federal Aviation Administration, 1284 Maryland Avenue SW, Washington, DC 20024, January 2017. http://www.nstb.tc.faa.gov/reports/PAN96_0117.pdf

Trident Snake Robot Motion Simulation in V-Rep

Roman Byrtus and Jana Vechetová[✉]

Faculty of Mechanical Engineering, Brno University of Technology,
Brno, Czech Republic
{171155,161790}@vutbr.cz

Abstract. We present a simulation of a trident snake robot motion in two local control models, the original one and its nilpotent approximation. More precisely, we derive the control system from the kinematics of a trident snake robot, calculate its nilpotent approximation and compare these two models by simulating their local motion planning in software V-Rep.

Keywords: Non–holonomic kinematics · Local control · Snake robot · Nilpotent approximation

1 Introduction

The locomotion of snake–like robots is currently widely studied for its interesting property that the forward movement is realized just by bending in joints, i. e. the wheels are not driven, which guarantees adaptability to various surfaces, see [7,9] for a classical or [4] for an algebraic approach. In this paper, we elaborate a simplified model of a trident snake robot introduced by Ishikawa in [2] and further elaborated algebraically in [5]. We derive kinematics based on non–slip and non–slide sideways condition for the passive wheels. Then we provide a description of the control model based on controlling vector fields and we show that the mechanism is locally controllable.

Consequently, we determine an approximation of the controlling distribution which comes from a sub–Riemannian distance which we call a nilpotent approximation, see [6] for details.

Finally, simple motion planning algorithms are assembled based on geometric control theory, see [8,12]. We compare the original control model with the approximated one by a simulation in V-Rep software [1].

2 Robot Model

We deal with a three-headed snake robot moving on the planar surface which is called a trident snake robot. The mechanism was introduced in [2] and combines a snake–like motion control with a capability of carrying loads on its triangular

© Springer Nature Switzerland AG 2019
J. Mazal (Ed.): MESAS 2018, LNCS 11472, pp. 27–42, 2019.
https://doi.org/10.1007/978-3-030-14984-0_3

platform, such as sensors, cameras etc. Indeed, the robot is composed of a platform in the shape of an equilateral triangle with circumscribed circle of radius r and three branches of serial links (also called legs) which are connected to the root block via actuated joints at the vertices of the triangle. In the general case of multiple links in each branch, there is also an actuator between each pair of links. Each link has a pair of passive wheels at the centre, which are assumed not to slip, nor slide sideways. It provides an important snake-like property that the ground friction in the direction perpendicular to the link is considerably higher than the friction of a simple forward move.

In our case we assume that each leg has only one link of the length $2l$, therefore the distance between a wheel and the adjacent joint is the same as the radius $r = l = 1$. Thus we consider the model with legs of length $l = 1$ and a pair of wheels at their ends, see Fig. 1, [6].

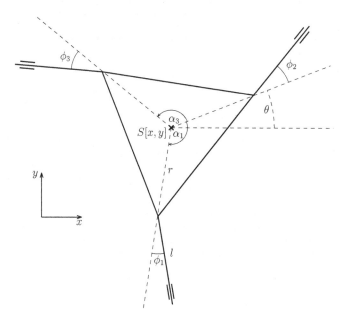

Fig. 1. Trident snake model.

To describe the actual position and space orientation of a trident snake robot we need the set of 6 generalized coordinates

$$q = (x, y, \theta, \phi_1, \phi_2, \phi_3),$$

where (x, y) represent the position of the centre point S in 2D space with respect to a fixed coordinate system. The absolute orientation of the mechanism is represented by the angle θ and the last three coordinates represent the rotation of the appropriate leg ϕ_i relatively to the platform. Thus we have the configuration vector $\mathbf{w} := (x, y, \theta)^T$ and the shape vector $\phi := (\phi_1, \phi_2, \phi_3)^T$.

3 Kinematics and Control

To derive the kinematics description of the trident snake robot, let x_i, y_i, $i = 1, 2, 3$, denote the wheel positions. Then

$$\begin{pmatrix} x_i \\ y_i \end{pmatrix} = \begin{pmatrix} x + \cos(\alpha_i + \theta) + \cos(\alpha_i + \theta + \phi_i) \\ y + \sin(\alpha_i + \theta) + \sin(\alpha_i + \theta + \phi_i) \end{pmatrix}, \tag{1}$$

where α_i is the i–th central angle within the triangle. The order of angles is described in Fig. 1, namely

$$\alpha_1 = -\frac{2}{3}\pi, \qquad \alpha_2 = 0, \qquad \alpha_3 = \frac{2}{3}\pi.$$

The following three constraints follow from the non–slip and non–slide assumption on the wheels meaning that the actual velocity vector is always parallel to the link:

$$\dot{x}_i \sin(\alpha_i + \theta + \phi_i) = \dot{y}_i \cos(\alpha_i + \theta + \phi_i), \quad i = 1, 2, 3. \tag{2}$$

Differentiating the equations (1) and substituting into (2) we obtain the dynamical system

$$\begin{pmatrix} \dot{\phi}_1 \\ \dot{\phi}_2 \\ \dot{\phi}_3 \end{pmatrix} = \begin{pmatrix} \sin(\theta + \alpha_1 + \phi_1) & -\cos(\theta + \alpha_1 + \phi_1) & -(1 + \cos\phi_1) \\ \sin(\theta + \alpha_2 + \phi_2) & -\cos(\theta + \alpha_2 + \phi_2) & -(1 + \cos\phi_2) \\ \sin(\theta + \alpha_3 + \phi_3) & -\cos(\theta + \alpha_3 + \phi_3) & -(1 + \cos\phi_3) \end{pmatrix} \begin{pmatrix} \dot{x} \\ \dot{y} \\ \dot{\theta} \end{pmatrix}. \tag{3}$$

Note that this system can be transformed easily (extracting the rotation matrix) to the form in which parameter θ is eliminated and thus it corresponds to non–inertial frame of reference, [2].

To define the control system of a trident snake robot we transform (3) into the system of ODEs where the vector $(u_1, u_2, u_3) = (\dot{x}, \dot{y}, \dot{\theta}) = \dot{\mathbf{w}}$ is the vector of controlling parameters. Hence we have the following control system in the form $\dot{q} = Gu$:

$$\begin{pmatrix} \dot{x} \\ \dot{y} \\ \dot{\theta} \\ \dot{\phi}_1 \\ \dot{\phi}_2 \\ \dot{\phi}_3 \end{pmatrix} = \begin{pmatrix} \cos\theta & -\sin\theta & 0 \\ \sin\theta & \cos\theta & 0 \\ 0 & 0 & 1 \\ \sin(\alpha_1 + \phi_1) & -\cos(\alpha_1 + \phi_1) & -1 - \cos(\phi_1) \\ \sin(\alpha_2 + \phi_2) & -\cos(\alpha_2 + \phi_2) & -1 - \cos(\phi_2) \\ \sin(\alpha_3 + \phi_3) & -\cos(\alpha_3 + \phi_3) & -1 - \cos(\phi_3) \end{pmatrix} \begin{pmatrix} u_1 \\ u_2 \\ u_3 \end{pmatrix}. \tag{4}$$

where the control matrix G is a matrix of dimension 6×3 whose columns are considered as the controlling vector fields g_1, g_2, g_3.

To check local controllability and, essentially, to define the remaining controlling vector fields, we compute Lie brackets of g_1, g_2, g_3. For example, if we assume a coordinate change $(x, y, \theta, \phi_1, \phi_2, \phi_3) =: (x_1, x_2, x_3, x_4, x_5, x_6)$, the vector fields g_1 and g_2 are of the form

$$g_1 = \cos(\theta)\partial_{x_1} + \sin(\theta)\partial_{x_2} + \sin(\phi_1 - \tfrac{2}{3}\pi)\partial_{x_4} + \sin(\phi_2)\partial_{x_5} + \sin(\phi_3 + \tfrac{2}{3}\pi)\partial_{x_6},$$

$$g_2 = -\sin(\theta)\partial_{x_1} + \cos(\theta)\partial_{x_2} - \cos(\phi_1 - \tfrac{2}{3}\pi)\partial_{x_4} - \cos(\phi_2)\partial_{x_5} - \cos(\phi_3 + \tfrac{2}{3}\pi)\partial_{x_6},$$

and the computation of Lie brackets in a matrix form is realized in the following way

$$[g_1, g_2] = \frac{\partial g_1}{\partial x^i} g_2 - \frac{\partial g_2}{\partial x^i} g_1$$

$$
=
\begin{pmatrix}
\frac{\partial \cos\theta}{\partial x} & \frac{\partial \cos\theta}{\partial y} & \cdots & \frac{\partial \cos\theta}{\partial \phi_2} & \frac{\partial \cos\theta}{\partial \phi_3} \\
\frac{\partial \sin\theta}{\partial x} & \frac{\partial \sin\theta}{\partial y} & \cdots & \frac{\partial \sin\theta}{\partial \phi_2} & \frac{\partial \sin\theta}{\partial \phi_3} \\
\frac{\partial 0}{\partial x} & \frac{\partial 0}{\partial y} & \cdots & \frac{\partial 0}{\partial \phi_2} & \frac{\partial 0}{\partial \phi_3} \\
\frac{\partial(\sin(\phi_1-\frac{2}{3}\pi))}{\partial x} & \frac{\partial(\sin(\phi_1-\frac{2}{3}\pi))}{\partial y} & \cdots & \frac{\partial(\sin(\phi_1-\frac{2}{3}\pi))}{\partial \phi_2} & \frac{\partial(\sin(\phi_1-\frac{2}{3}\pi))}{\partial \phi_3} \\
\frac{\partial(\sin(\phi_2))}{\partial x} & \frac{\partial(\sin(\phi_2))}{\partial y} & \cdots & \frac{\partial(\sin(\phi_2))}{\partial \phi_2} & \frac{\partial(\sin(\phi_2))}{\partial \phi_3} \\
\frac{\partial(\sin(\phi_3+\frac{2}{3}\pi))}{\partial x} & \frac{\partial(\sin(\phi_3+\frac{2}{3}\pi))}{\partial y} & \cdots & \frac{\partial(\sin(\phi_3+\frac{2}{3}\pi))}{\partial \phi_2} & \frac{\partial(\sin(\phi_3+\frac{2}{3}\pi))}{\partial \phi_3}
\end{pmatrix}
\begin{pmatrix}
-\sin\theta \\
\cos\theta \\
0 \\
-\cos(\phi_1-\frac{2}{3}\pi) \\
-\cos(\phi_2) \\
-\cos(\phi_3+\frac{2}{3}\pi)
\end{pmatrix}
$$

$$
-
\begin{pmatrix}
\frac{\partial(-\sin\theta)}{\partial x} & \frac{\partial(-\sin\theta)}{\partial y} & \cdots & \frac{\partial(-\sin\theta)}{\partial \phi_2} & \frac{\partial(-\sin\theta)}{\partial \phi_3} \\
\frac{\partial(\cos\theta)}{\partial x} & \frac{\partial(\cos\theta)}{\partial y} & \cdots & \frac{\partial(\cos\theta)}{\partial \phi_2} & \frac{\partial(\cos\theta)}{\partial \phi_3} \\
\frac{\partial 0}{\partial x} & \frac{\partial 0}{\partial y} & \cdots & \frac{\partial 0}{\partial \phi_2} & \frac{\partial 0}{\partial \phi_3} \\
\frac{\partial(-\cos(\phi_1-\frac{2}{3}\pi))}{\partial x} & \frac{\partial(-\cos(\phi_1-\frac{2}{3}\pi))}{\partial y} & \cdots & \frac{\partial(-\cos(\phi_1-\frac{2}{3}\pi))}{\partial \phi_2} & \frac{\partial(-\cos(\phi_1-\frac{2}{3}\pi))}{\partial \phi_3} \\
\frac{\partial(-\cos(\phi_2))}{\partial x} & \frac{\partial(-\cos(\phi_2))}{\partial y} & \cdots & \frac{\partial(-\cos(\phi_2))}{\partial \phi_2} & \frac{\partial(-\cos(\phi_2))}{\partial \phi_3} \\
\frac{\partial(-\cos(\phi_3+\frac{2}{3}\pi))}{\partial x} & \frac{\partial(-\cos(\phi_3+\frac{2}{3}\pi))}{\partial y} & \cdots & \frac{\partial(-\cos(\phi_3+\frac{2}{3}\pi))}{\partial \phi_2} & \frac{\partial(-\cos(\phi_3+\frac{2}{3}\pi))}{\partial \phi_3}
\end{pmatrix}
\begin{pmatrix}
\cos\theta \\
\sin\theta \\
0 \\
\sin(\phi_1-\frac{2}{3}\pi) \\
\sin(\phi_2) \\
\sin(\phi_3+\frac{2}{3}\pi)
\end{pmatrix}
=
\begin{pmatrix}
0 \\
0 \\
1 \\
1 \\
1
\end{pmatrix}
$$

In the same way we compute all Lie brackets of vector fields g_1, g_2, g_3. Denoting $g_{12} = [g_1, g_2], g_{13} = [g_1, g_3]$ and $g_{23} = [g_2, g_3]$, we have

$$
g_{12} =
\begin{pmatrix}
0 \\
0 \\
0 \\
1 \\
1 \\
1
\end{pmatrix},
\quad
g_{13} =
\begin{pmatrix}
\sin\theta \\
\cos\theta \\
0 \\
\frac{1}{2} - \cos(\phi_1 - \frac{2}{3}\pi) \\
-1 - \cos(\phi_2) \\
\frac{1}{2} - \cos(\phi_3 + \frac{2}{3}\pi)
\end{pmatrix},
\quad
g_{23} =
\begin{pmatrix}
\cos\theta \\
\sin\theta \\
0 \\
-\frac{\sqrt{3}}{2} + \sin(\phi_1 - \frac{2}{3}\pi) \\
\sin(\phi_2) \\
\frac{\sqrt{3}}{2} + \sin(\phi_3 + \frac{2}{3}\pi)
\end{pmatrix}.
$$

Note that the local controllability of the trident snake robot may be now easily checked by Chow – Rachevsky theorem, see e.g. [8].

4 Approximated Model

To simplify our model locally, we construct privileged coordinates according to Bellache's algorithm, see e.g. [3]. For the model (4) we first exclude the rotation matrix which leads to elimination of θ from the controlling matrix and thus the vector fields are modified as follows:

$$g_1 = \partial_{x_1} + \sin\left(x_4 - \frac{2}{3}\pi\right)\partial_{x_4} + \sin(x_5)\partial_{x_5} + \sin\left(x_6 + \frac{2}{3}\pi\right)\partial_{x_6},$$

$$g_2 = \partial_{x_2} - \cos\left(x_4 - \frac{2}{3}\pi\right)\partial_{x_4} - \cos(x_5)\partial_{x_5} - \cos\left(x_6 + \frac{2}{3}\pi\right)\partial_{x_6},$$

$$g_3 = \partial_{x_3} - (1 + \cos x_4)\partial_{x_4} - (1 + \cos x_5)\partial_{x_5} - (1 + \cos x_6)\partial_{x_6}.$$

We compute Lie brackets again but we evaluate the result at the point $q = (x_1, x_2, x_3, x_4, x_5, x_6) = (0, 0, 0, 0, 0, 0)$. Therefore the approximation is local in the neighbourhood of q. The computation leads to the result:

$$g_4 = [g_1, g_2] = \partial_{x_4} + \partial_{x_5} + \partial_{x_6},$$
$$g_5 = [g_1, g_3] = \partial_{x_4} - 2\partial_{x_5} + \partial_{x_6},$$
$$g_6 = [g_2, g_3] = -\sqrt{3}\partial_{x_4} + \sqrt{3}\partial_{x_6}.$$

We now determine the so–called adapted frame which, in general, is a coordinate change. Therefore we denote the basis $\{x_1, x_2, x_3, x_4, x_5, x_6\}$ by x and the new basis $\{y_1, y_2, y_3, y_4, y_5, y_6\}$ as y. If we denote by $[g_k^i]_x$ the i-th coordinate of vector field g_k in the basis x $(i, k = 1, \ldots, 6)$ then we have for the first one:

$$[g_1^1]_x = 1, \qquad [g_1^2]_x = 0, \qquad [g_1^3]_x = 0,$$
$$[g_1^4]_x = \sin\left(x_4 - \frac{2}{3}\pi\right), [g_1^5]_x = \sin(x_5), [g_1^6]_x = \sin\left(x_6 + \frac{2}{3}\pi\right).$$

The essential condition on the adapted frame for any point p of the configuration space is the following:

$$\frac{\partial}{\partial y_i}\Big|_p = g_i|_p, \qquad i = 1, 2, \ldots, 6. \tag{5}$$

In our particular case, the condition (5) transforms as

$$[g_k^i]_y = \sum_{j=1}^{6} \frac{\partial y_i}{\partial x_j}[g_k^j]_x,$$

for $i, k = 1, \ldots, 6$. Hence for each vector field evaluated in $p = (0, 0, 0, 0, 0, 0)$ we have in a matrix form

$$\begin{pmatrix} 1 \\ 0 \\ 0 \\ 0 \\ 0 \\ 0 \end{pmatrix} = [g_1]_y = \begin{pmatrix} \frac{\partial y_1}{\partial x_1} - \frac{\sqrt{3}}{2}\frac{\partial y_1}{\partial x_4} + \frac{\sqrt{3}}{2}\frac{\partial y_1}{\partial x_6} \\ \frac{\partial y_2}{\partial x_1} - \frac{\sqrt{3}}{2}\frac{\partial y_2}{\partial x_4} + \frac{\sqrt{3}}{2}\frac{\partial y_2}{\partial x_6} \\ \frac{\partial y_3}{\partial x_1} - \frac{\sqrt{3}}{2}\frac{\partial y_3}{\partial x_4} + \frac{\sqrt{3}}{2}\frac{\partial y_3}{\partial x_6} \\ \frac{\partial y_4}{\partial x_1} - \frac{\sqrt{3}}{2}\frac{\partial y_4}{\partial x_4} + \frac{\sqrt{3}}{2}\frac{\partial y_4}{\partial x_6} \\ \frac{\partial y_5}{\partial x_1} - \frac{\sqrt{3}}{2}\frac{\partial y_5}{\partial x_4} + \frac{\sqrt{3}}{2}\frac{\partial y_5}{\partial x_6} \\ \frac{\partial y_6}{\partial x_1} - \frac{\sqrt{3}}{2}\frac{\partial y_6}{\partial x_4} + \frac{\sqrt{3}}{2}\frac{\partial y_6}{\partial x_6} \end{pmatrix},$$

$$\begin{pmatrix} 0 \\ 1 \\ 0 \\ 0 \\ 0 \\ 0 \end{pmatrix} = [g_2]_y = \begin{pmatrix} \frac{\partial y_1}{\partial x_2} + \frac{1}{2}\frac{\partial y_1}{\partial x_4} - \frac{\partial y_1}{\partial x_5} + \frac{1}{2}\frac{\partial y_1}{\partial x_6} \\ \frac{\partial y_2}{\partial x_2} + \frac{1}{2}\frac{\partial y_2}{\partial x_4} - \frac{\partial y_2}{\partial x_5} + \frac{1}{2}\frac{\partial y_2}{\partial x_6} \\ \frac{\partial y_3}{\partial x_2} + \frac{1}{2}\frac{\partial y_3}{\partial x_4} - \frac{\partial y_3}{\partial x_5} + \frac{1}{2}\frac{\partial y_3}{\partial x_6} \\ \frac{\partial y_4}{\partial x_2} + \frac{1}{2}\frac{\partial y_4}{\partial x_4} - \frac{\partial y_4}{\partial x_5} + \frac{1}{2}\frac{\partial y_4}{\partial x_6} \\ \frac{\partial y_5}{\partial x_2} + \frac{1}{2}\frac{\partial y_5}{\partial x_4} - \frac{\partial y_5}{\partial x_5} + \frac{1}{2}\frac{\partial y_5}{\partial x_6} \\ \frac{\partial y_6}{\partial x_2} + \frac{1}{2}\frac{\partial y_6}{\partial x_4} - \frac{\partial y_6}{\partial x_5} + \frac{1}{2}\frac{\partial y_6}{\partial x_6} \end{pmatrix},$$

$$\begin{pmatrix} 0 \\ 0 \\ 1 \\ 0 \\ 0 \\ 0 \end{pmatrix} = [g_3]_y = \begin{pmatrix} \frac{\partial y_1}{\partial x_3} - 2\frac{\partial y_1}{\partial x_4} - 2\frac{\partial y_1}{\partial x_5} - 2\frac{\partial y_1}{\partial x_6} \\ \frac{\partial y_2}{\partial x_3} - 2\frac{\partial y_2}{\partial x_4} - 2\frac{\partial y_2}{\partial x_5} - 2\frac{\partial^2 1}{\partial x_6} \\ \frac{\partial y_3}{\partial x_3} - 2\frac{\partial y_3}{\partial x_4} - 2\frac{\partial y_3}{\partial x_5} - 2\frac{\partial y_3}{\partial x_6} \\ \frac{\partial y_4}{\partial x_3} - 2\frac{\partial y_4}{\partial x_4} - 2\frac{\partial y_4}{\partial x_5} - 2\frac{\partial y_4}{\partial x_6} \\ \frac{\partial y_5}{\partial x_3} - 2\frac{\partial y_5}{\partial x_4} - 2\frac{\partial y_5}{\partial x_5} - 2\frac{\partial y_5}{\partial x_6} \\ \frac{\partial y_6}{\partial x_3} - 2\frac{\partial y_6}{\partial x_4} - 2\frac{\partial y_6}{\partial x_5} - 2\frac{\partial y_6}{\partial x_6} \end{pmatrix},$$

$$\begin{pmatrix} 0 \\ 0 \\ 0 \\ 1 \\ 0 \\ 0 \end{pmatrix} = [g_4]_y = \begin{pmatrix} \frac{\partial y_1}{\partial x_4} + \frac{\partial y_1}{\partial x_5} + \frac{\partial y_1}{\partial x_6} \\ \frac{\partial y_2}{\partial x_4} + \frac{\partial y_2}{\partial x_5} + \frac{\partial y_2}{\partial x_6} \\ \frac{\partial y_3}{\partial x_4} + \frac{\partial y_3}{\partial x_5} + \frac{\partial y_3}{\partial x_6} \\ \frac{\partial y_4}{\partial x_4} + \frac{\partial y_4}{\partial x_5} + \frac{\partial y_4}{\partial x_6} \\ \frac{\partial y_5}{\partial x_4} + \frac{\partial y_5}{\partial x_5} + \frac{\partial y_5}{\partial x_6} \\ \frac{\partial y_6}{\partial x_4} + \frac{\partial y_6}{\partial x_5} + \frac{\partial y_6}{\partial x_6} \end{pmatrix},$$

$$\begin{pmatrix} 0 \\ 0 \\ 0 \\ 0 \\ 1 \\ 0 \end{pmatrix} = [g_5]_y = \begin{pmatrix} \frac{\partial y_1}{\partial x_4} - 2\frac{\partial y_1}{\partial x_5} + \frac{\partial y_1}{\partial x_6} \\ \frac{\partial y_2}{\partial x_4} - 2\frac{\partial y_2}{\partial x_5} + \frac{\partial y_2}{\partial x_6} \\ \frac{\partial y_3}{\partial x_4} - 2\frac{\partial y_3}{\partial x_5} + \frac{\partial y_3}{\partial x_6} \\ \frac{\partial y_4}{\partial x_4} - 2\frac{\partial y_4}{\partial x_5} + \frac{\partial y_4}{\partial x_6} \\ \frac{\partial y_5}{\partial x_4} - 2\frac{\partial y_5}{\partial x_5} + \frac{\partial y_5}{\partial x_6} \\ \frac{\partial y_6}{\partial x_4} - 2\frac{\partial y_6}{\partial x_5} + \frac{\partial y_6}{\partial x_6} \end{pmatrix},$$

$$\begin{pmatrix} 0 \\ 0 \\ 0 \\ 0 \\ 0 \\ 1 \end{pmatrix} = [g_6]_y = \begin{pmatrix} -\sqrt{3}\frac{\partial y_1}{\partial x_4} + \sqrt{3}\frac{\partial y_1}{\partial x_6} \\ -\sqrt{3}\frac{\partial y_2}{\partial x_4} + \sqrt{3}\frac{\partial y_2}{\partial x_6} \\ -\sqrt{3}\frac{\partial y_3}{\partial x_4} + \sqrt{3}\frac{\partial y_3}{\partial x_6} \\ -\sqrt{3}\frac{\partial y_4}{\partial x_4} + \sqrt{3}\frac{\partial y_4}{\partial x_6} \\ -\sqrt{3}\frac{\partial y_5}{\partial x_4} + \sqrt{3}\frac{\partial y_5}{\partial x_6} \\ -\sqrt{3}\frac{\partial y_6}{\partial x_4} + \sqrt{3}\frac{\partial y_6}{\partial x_6} \end{pmatrix}.$$

We get a system of 36 PDEs. To solve them, we group together 6, each containing a particular y_i. We demonstrate the computation of y_1 which is composed of the first rows of the above matrices.

This system of PDEs is in the matrix form

$$
\begin{pmatrix}
1 & 0 & 0 & -\frac{\sqrt{3}}{2} & 0 & \frac{\sqrt{3}}{2} \\
0 & 1 & 0 & \frac{1}{2} & -1 & \frac{1}{2} \\
0 & 0 & 1 & -2 & -2 & -2 \\
0 & 0 & 0 & 1 & 1 & 1 \\
0 & 0 & 0 & 1 & -2 & 1 \\
0 & 0 & 0 & -\sqrt{3} & 0 & \sqrt{3}
\end{pmatrix}
\begin{pmatrix}
\frac{\partial y_1}{\partial x_1} \\
\frac{\partial y_1}{\partial x_2} \\
\frac{\partial y_1}{\partial x_3} \\
\frac{\partial y_1}{\partial x_4} \\
\frac{\partial y_1}{\partial x_5} \\
\frac{\partial y_1}{\partial x_6}
\end{pmatrix}
=
\begin{pmatrix}
1 \\ 0 \\ 0 \\ 0 \\ 0 \\ 0
\end{pmatrix}.
\tag{6}
$$

By inverting the matrix and multiplying (6) from the left we get

$$
\begin{pmatrix}
\frac{\partial y_1}{\partial x_1} \\
\frac{\partial y_1}{\partial x_2} \\
\frac{\partial y_1}{\partial x_3} \\
\frac{\partial y_1}{\partial x_4} \\
\frac{\partial y_1}{\partial x_5} \\
\frac{\partial y_1}{\partial x_6}
\end{pmatrix}
=
\begin{pmatrix}
1 & 0 & 0 & 0 & 0 & -\frac{1}{2} \\
0 & 1 & 0 & 0 & -\frac{1}{2} & 0 \\
0 & 0 & 1 & 2 & 0 & 0 \\
0 & 0 & 0 & \frac{1}{3} & \frac{1}{6} & -\frac{\sqrt{3}}{6} \\
0 & 0 & 0 & \frac{1}{3} & -\frac{1}{3} & 0 \\
0 & 0 & 0 & \frac{1}{3} & \frac{1}{6} & \frac{\sqrt{3}}{6}
\end{pmatrix}
\begin{pmatrix}
1 \\ 0 \\ 0 \\ 0 \\ 0 \\ 0
\end{pmatrix}.
\tag{7}
$$

Final step to find y_1 is to integrate the following equation which we obtain from (7):

$$
\frac{\partial y_1}{x_1} = 1.
$$

Hence the result is

$$
y_1 = x_1 + c(x_1).
$$

We use the same procedure to find the rest of y_i. And finally we get a linear transformation (at 0) of the following form

$$
\begin{pmatrix}
y_1 \\ y_2 \\ y_3 \\ y_4 \\ y_5 \\ y_6
\end{pmatrix}
=
\begin{pmatrix}
1 & 0 & 0 & 0 & 0 & 0 \\
0 & 1 & 0 & 0 & 0 & 0 \\
0 & 0 & 1 & 0 & 0 & 0 \\
0 & 0 & 2 & \frac{1}{3} & \frac{1}{3} & \frac{1}{3} \\
0 & -\frac{1}{2} & 0 & \frac{1}{6} & -\frac{1}{3} & \frac{1}{6} \\
-\frac{1}{2} & 0 & 0 & -\frac{\sqrt{3}}{6} & 0 & \frac{\sqrt{3}}{6}
\end{pmatrix}
\begin{pmatrix}
x_1 \\ x_2 \\ x_3 \\ x_4 \\ x_5 \\ x_6
\end{pmatrix}.
$$

With this adapted frame we continue to the nilpotent approximation.

5   Nilpotent Approximation

Let us note that the theory necessary to deriving the nilpotent approximation may be found in [3]. For the trident snake robot, it was elaborated in [6]. The reasons, why the nilpotent approximation is so outstanding, may be found in [8] and [12]. We just note that it was used to derive the periodic input necessary for any motion planning and it is suitable for the so–called piecewise constant input motion model. For the purposes of this paper we just show the result in the form:

$$\hat{g}_1 = \partial_{y_1} - \frac{1}{2}y_2\partial_{y_4} + \left(-\frac{1}{2}y_2 - y_3\right)\partial_{y_5} - \frac{1}{2}y_1\partial_{y_6},$$

$$\hat{g}_2 = \partial_{y_2} + \frac{1}{2}y_1\partial_{y_4} - \frac{1}{2}y_1\partial_{y_5} + \left(\frac{1}{2}y_2 - y_3\right)\partial_{y_6},$$

$$\hat{g}_3 = \partial_{y_3}.$$

The family $(\hat{g}_1, \hat{g}_2, \hat{g}_3)$ is called the *nilpotent approximation* of (g_1, g_2, g_3) at p associated with the coordinates y. It follows that every bracket of the vector fields $\hat{g}_1, \hat{g}_2, \hat{g}_3$ is zero if the length is greater than 1. Note that the $\partial_{y_1}, \partial_{y_2}$ and ∂_{y_3} coordinate functions of g_1, g_2 and g_3 are formed by constants and the $\partial_{y_4}, \partial_{y_5}$ and ∂_{y_6} coordinate functions are linear polynomials in y_1, y_2, y_3.

We obtain the remaining three vector fields by taking Lie brackets of $(\hat{g}_1, \hat{g}_2, \hat{g}_3)$. Hence we get

$$\hat{g}_4 = [\hat{g}_1, \hat{g}_2] = \partial_{y_4},$$
$$\hat{g}_5 = [\hat{g}_1, \hat{g}_3] = \partial_{y_5},$$
$$\hat{g}_5 = [\hat{g}_2, \hat{g}_3] = \partial_{y_6}.$$

6   Motion Planning

Basically, the idea is that the controlling vector fields determine locally the allowed direction of an infinitesimal motion in spatial coordinates, i.e. they generate translations and rotations. Lie brackets then form a reconfiguration of the robot so that the initial position is restored and the motion according to controlling vector field may be applied again.

We present the motion planning algorithm with an initial point

$$q_0 = (x, y, \theta, \phi_1, \phi_2, \phi_3) = (0, 0, 0, 0, 0, 0).$$

Evaluating the controlling vector fields and their Lie brackets at q, we receive

$$\bar{G}(\mathbf{0}) = \begin{pmatrix} 1 & 0 & 0 & 0 & 0 & 1 \\ 0 & 1 & 0 & 0 & 1 & 0 \\ 0 & 0 & 1 & 0 & 0 & 0 \\ -\frac{\sqrt{3}}{2} & \frac{1}{2} & -2 & 1 & 1 & -\sqrt{3} \\ 0 & -1 & -2 & 1 & -2 & 0 \\ \frac{\sqrt{3}}{2} & \frac{1}{2} & -2 & 1 & 1 & \sqrt{3}. \end{pmatrix}$$

Let us briefly describe the effect of each vector field motion. The original description may be found in [2]. The effect of g_1 motion is the following. The body of the snake moves along the x–axis and leg's angles change proportionally to $-1:0:1$. It means that the first leg moves in the anti–clockwise direction and the third vice versa. The analysis of the effect of g_2 and g_3 motions is left to the reader and we just provide the list of the appropriate vector fields:

$$g_1(\mathbf{0}) = \left(1, 0, 0, -\frac{\sqrt{3}}{2}, 0, \frac{\sqrt{3}}{2}\right), \quad g_2(\mathbf{0}) = \left(0, 1, 0, \frac{1}{2}, -1, \frac{1}{2}\right)$$

and

$$g_3(\mathbf{0}) = (0, 0, 1, -2, -2, -2).$$

Next step is to describe the realization of Lie bracket motions which in fact are certain combination of two controlling vector fields. Their evaluation at q is of the form

$$g_{12}(\mathbf{0}) = (0, 0, 0, 1, 1, 1), \quad g_{13}(\mathbf{0}) = (0, 1, 0, 1, -2, 1)$$

and

$$g_{23}(\mathbf{0}) = \left(1, 0, 0, -\sqrt{3}, 0, \sqrt{3}\right).$$

For example, with g_{12} motion the configuration of a robot is not changed, only legs are rotated in the clockwise direction proportionally to $1:1:1$. Similarly, the effect of g_{13} and g_{23} motions can be described.

Consequently, it is clear that any local motion is given as a linear combination of the controlling vector fields together with an additional linear combination of Lie brackets. Therefore the question of the Lie bracket motion realization arises as it can not be realized directly. Indeed, it may be found in [2] and [8] that the periodic input plays the role of these motions realizer, i. e. the input $u(t) = \begin{pmatrix} u_1(t) \\ u_2(t) \\ u_3(t) \end{pmatrix}$ in the form

$$\begin{aligned} u(t) &= (-A\omega \sin(\omega t), A\omega \cos(\omega t), 0)^T && \text{for } g_{12}, \\ u(t) &= (-A\omega \sin(\omega t), 0, A\omega \cos(\omega t))^T && \text{for } g_{13}, \\ u(t) &= (0, -A\omega \sin(\omega t), A\omega \cos(\omega t))^T && \text{for } g_{23}, \end{aligned}$$

where $A \in \mathbb{R}$ is a positive amplitude and $\omega \in \mathbb{N}$ is a frequency. Particularly, we set the following values for our Lie bracket motions:

$$\begin{aligned} A &= 0.3, & \omega &= 3 \text{ rad/s} && \text{for } g_{12}, \\ A &= 0.2, & \omega &= 3 \text{ rad/s} && \text{for } g_{13}, \\ A &= 0.25, & \omega &= 2 \text{ rad/s} && \text{for } g_{23} \end{aligned}$$

and visualize the evolution of the leg's angles in Fig. 2.

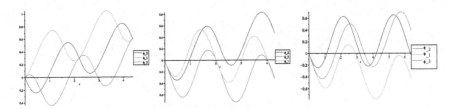

Fig. 2. Evolution of ϕ_i for Lie bracket motions in MAPLE.

Note that each graph visualizes two cycles (periods) of the motion and simulation values of A, ω were derived experimentally in [2].

Finally, motion of the robot in the axis x direction, i.e. movement from $q_{init} = (0,0,0,0,0,0)^T$ to $q_{final} = (1,0,0,0,0,0)^T$, is achieved by a combination of vector field g_1 and Lie bracket g_{23} in the following form:

$$2g_{23}(0) - g_1(0) = (1,0,0,0,0,0)^T.$$

The y axis motion and rotation with respect to a fixed coordinate system is then realized by these combinations:

$$2g_2(0) - g_{13}(0) = (0,1,0,0,0,0)^T, \quad 2g_{12}(0) + g_3(0) = (0,0,1,0,0,0)^T.$$

For more advanced motion models of a trident snake robot, see [10,11].

7   Simulation in V-REP

For the software simulation of basic motions we use V-REP (Virtual Robot Experimentation Platform), see [1]. This environment includes physical properties and influences of surrounding environment.

The model of a trident snake robot is composed of several basic objects (body, legs, wheels and joints), which were created in Autodesk Inventor Professional 2018 and then imported into V-REP. The simulation is controlled by scripts (written in the LUA language), which are attached to objects. Furthermore, the objects, scripts and settings of the simulation together form so-called scenes in V-REP (showed in Fig. 3). Before running the simulation, we select the desired physical engine (in our case, we used the Vortex engine for our simulations) and also the time step.

In V-REP, script is divided into four main parts. In the *init* section we initialize values before the main part of the simulation begins. The *actuation* section serves to control the robot, where, in our case, we control the target positions of the joints connecting the body of the robot and its legs. The *sensing* section is reserved for code managing sensors in the scene. Finally, the *cleanup* section is used to run the code while the simulation ends.

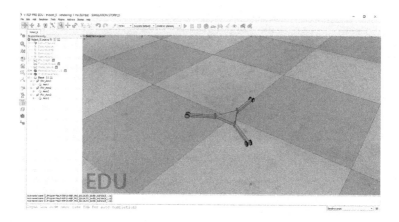

Fig. 3. A scene in V-REP.

Let us introduce the structure of our control script (for motion g_{12}) before presenting a chosen few simulated motions.

```
--getting V-REP handles to specific simulation objects
base = sim.getObjectHandle('Base');
cJoint_A = sim.getObjectHandle('RV_Arm1');
cJoint_B = sim.getObjectHandle('RV_Arm2');
cJoint_C = sim.getObjectHandle('RV_Arm3');

--parameters of the construction of the robot
alpha1 = -2/3 * math.pi;
alpha2 = 0;
alpha3 = 2/3 * math.pi;

--parameters of the motion
A = 0.3;                    --amplitude
w = 3;                      --angular frequency
sim.setObjectFloatParameter(cJoint_A,
sim.jointfloatparam_pid_p, 0.550);

T = 2.5*math.pi/w; --time of motion for periodic input
```

Here we present the main part of our init section. We declare variables pointing to objects in the scene, to which we refer later. At first, we initialize variables that characterize our situation, these are parameters describing the construction of the robot and then parameters which are specific for each motion. Furthermore, we configure the position control (PID) parameters, different for every motion.

```
t = sim.getSimulationTime();
step = sim.getSimulationTimeStep();

th = sim.getObjectOrientation(base, -1)[3];

if (t < 2*T) then

    --get the orientations of the arms
    phi1 = sim.getJointPosition(cJoint_A);
    phi2 = sim.getJointPosition(cJoint_B);
    phi3 = sim.getJointPosition(cJoint_C);

    --calculate inputs for specific motions
    u1 = -A*w*math.sin(w*t);
    u2 = A*w*math.cos(w*t);
    u3 = 0;

    --calculating the change of the angles
    dphi1 = math.sin(th+phi1+alpha1)*u1-math.cos(th+phi1+alpha1)*
    u2-(1+math.cos(phi1))*u3;
    dphi2 = math.sin(th+phi2+alpha2)*u1-math.cos(th+phi2+alpha2)*
    u2-(1+math.cos(phi2))*u3;
    dphi3 = math.sin(th+phi3+alpha3)*u1-math.cos(th+phi3+alpha3)*
    u2-(1+math.cos(phi3))*u3;

    --calculating the final angle of the joint, Euler method
    PHI_1 = step*dphi1 + phi1;
    PHI_2 = step*dphi2 + phi2;
    PHI_3 = step*dphi3 + phi3;

    --turn the servomotors
    sim.setJointTargetPosition(cJoint_A, PHI_1);
    sim.setJointTargetPosition(cJoint_B, PHI_2);
    sim.setJointTargetPosition(cJoint_C, PHI_3);
end
```

Here we introduce the main part of our actuation script. In this part of code, we get the current orientation of the base and also the angular position of our joints. Then we use them to calculate the target position of each joint for the current time step, using Euler method.

After presenting the code, let us introduce some results of our work in V-REP. The sequence od positions of the trident snake, during the simulation of motion in the flow of the basic vector fields g_1, g_3 from time $t = 0$ to $t = 1$, can be seen in Figs. 5 and 7, respectively. Note that in the simulation graphs included, see Figs. 4, 6 and 8, all the values are displayed w.r.t. the scene's coordinate system.

In this section we present three corresponding graphs for every vector field motion. In Figs. 4, 6, 8, 9, 12 and 13, the first graph represents the evolution of the rotation for each leg ϕ_1, ϕ_2, ϕ_3 during the simulation time. The second graph shows the evolution of the absolute orientation of the mechanism θ. And the third one shows the evolution of the position of the centre point S of the mechanism in the $2D$ space.

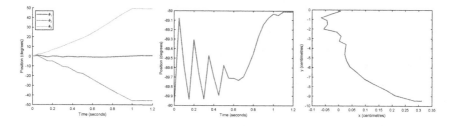

Fig. 4. Realization of g_1 motion.

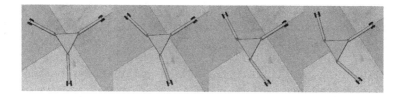

Fig. 5. Realization of g_1 motion.

As we can see, the motion in the direction of the vector field g_1 results in the movement of the entire mechanism along one axis, with symmetric rotation of two of the snake's legs. As the movement g_2 is of the similar behaviour, we only present the graphs of the related simulation in V-REP.

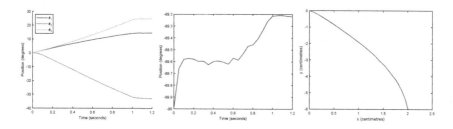

Fig. 6. Realization of g_2 motion.

The last of the three basic field motions g_3 causes the rotation of the robot. In other words, the final position (x, y) of the center of the robot's body is the same as before the simulation, whereas the absolute orientation θ has changed together with rotations of the legs. Note that all three legs have changed in the same direction with the same value.

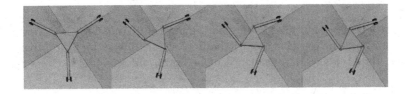

Fig. 7. Realization of g_3 motion.

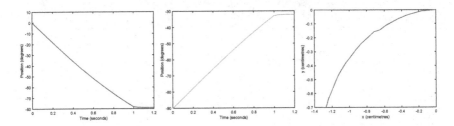

Fig. 8. Realization of g_3 motion.

After the analysis of basic vector filed motions let us move to more complicated ones. Thus, here we introduce all three Lie bracket motions realized in V-REP.

The first is the vector field motion g_{12} during time $t \in (0, 2T)$ with T as a period of the appropriate periodic input.

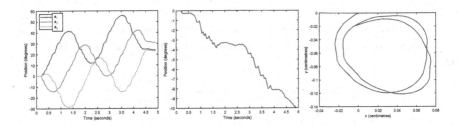

Fig. 9. Realization of g_{12} motion.

Note that g_{12} causes rotation in the anti–clockwise direction, see Fig. 9, and the body follows small circular path anti–clockwise. In Fig. 10, the process is shown from the initial position to the final one.

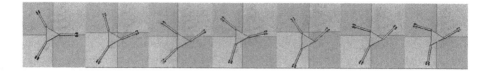

Fig. 10. Realization of g_{12} motion.

Simulation of g_{13} motion has the following result. Trident snake robot moves along y–axis and the rotation θ of the body in the final position is approximately zero, see Fig. 11 for V-Rep sequence of position and Fig. 12 for characteristic graphs.

Fig. 11. Realization of g_{13} motion.

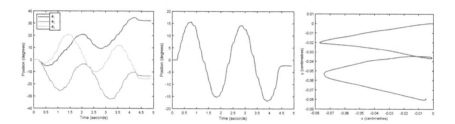

Fig. 12. Realization of g_{13} motion.

For the last one g_{23} motion, we can see that it causes translation in the x-axis direction, rotation angle of the body is approximately zero as well as the y coordinate of the centre, see Figs. 13 and 14. However, the body moves along x–axis. Applying g_{23} motion, the final position of a trident snake can be seen in Fig. 14.

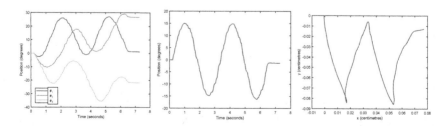

Fig. 13. Realization of g_{23} motion.

We also analyze the behaviour of the approximated model. In the following graphs in Fig. 15, we present the difference between the input coming to V-REP simulations for the original model and the approximated one. This comparison is presented for the g_{13} motion.

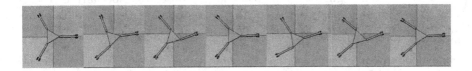

Fig. 14. Realization of g_{23} motion.

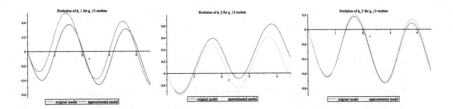

Fig. 15. Comparison of models for g_{13} motion in MAPLE.

Note that the difference between the trident snake position after the motion controlled by the original distribution and the nilpotent approximation is in order 10^{-2} and thus may be considered as negligible.

8   Conclusion

We described a non–holonomic system called a trident snake robot. We derived kinematic equations corresponding with constrains on friction forces on passive wheels.

To simplify our model we use a nilpotent approximation of the controlling distribution, so that the Lie algebra generated by approximated vector fields $\hat{g}_1, \hat{g}_2, \hat{g}_3$ is nilpotent, i.e. all Lie brackets of length greater than 1 vanish.

We have also presented an overview of motions caused by controlling vector fields and their combinations to obtain a pure translation (in x an y direction) and rotation around z-axis. By combining these elementary motions we can reach arbitrary point in our state space which is well known consequence of Chow–Rashevsky Theorem. More precisely, the dimension of a state space of the system is the same as the dimension of a controlling distribution of this system we came to the fact that trident snake robot is locally controllable.

We presented the results of simulations in an environment called V–REP. We simulated also more complicated motions appropriate to the Lie brackets of vector fields which were realized by periodic input which brought certain inaccuracy to this simulation.

Finally, we compared our models in V-Rep and Maple. The difference between the original and the approximated model is in order 10^{-2} which, compared to the robot size and further inaccuracies caused by dynamics and construction, may be considered negligible. This fact may be observed especially in Fig. 15 where the evolution of particular angles for g_{13} motion are depicted. This proves that, for the robot control, nilpotent approximation is convenient.

Acknowledgement. The first author was supported by a grant of the Czech Science Foundation (GAČR) no. 17–21360S. The second author was supported by a grant no. FSI-S-17-4464.

References

1. Coppelia Robotics: V-REP User Manual. http://www.coppeliarobotics.com/helpFiles/
2. Ishikawa, M.: Trident snake robot: locomotion analysis and control. In: IFAC Proceeding Volumes, pp. 895–900 (2004)
3. Jean, F.: Control of Nonholonomic Systems: From Sub-Riemannian Geometry to Motion Planning. Springer, New York (2014). https://doi.org/10.1007/978-3-319-08690-3
4. Hrdina, J., Návrat, A., Vašík, P., Matoušek, R.: CGA-based robotic snake control. Adv. Appl. Clifford Algebr. **27**(1), 621–632 (2017). https://doi.org/10.1007/s00006-016-0695-5
5. Hrdina, J., Návrat, A., Vašík, P., Matoušek, R.: Geometric control of the trident snake robot based on CGA. Adv. Appl. Clifford Algebr. **27**(1), 633–645 (2017). https://doi.org/10.1007/s00006-016-0693-7
6. Hrdina, J., Návrat, A., Vašík, P., Matoušek, R.: Nilpotent approximation of a trident snake robot controlling distribution. Kybernetika **53**, 1118–1130 (2017). https://doi.org/10.14736/kyb-2017-6-1118
7. Liljebäck, P., Pettersen, K.Y., Stavdahl, O., Gravdahl, J.T.: Snake Robots, Modelling, Mechatronics, and Control. Springer, London (2013). https://doi.org/10.1007/978-1-4471-2996-7
8. Murray, R.M., Zexiang, L., Sastry, S.S.: A Mathematical Introduction to Robotic Manipulation. CRC Press, Boca Raton (1994)
9. Návrat, A., Vašík, P.: On geometric control models of a robotic snake. Note di Matematica **37**, 119–129 (2017). https://doi.org/10.1285/i15900932v37suppl1p119
10. Paszuk, D., Tchoń, K., Pietrowska, Z.: Motion planning of the trident snake robot equipped with passive or active wheels. Bull. Polish Acad. Sci. Techn. Sci. **60**(3), 547–555 (2009). https://doi.org/10.2478/v10175-012-0067-9
11. Pietrowska, Z., Tchoń, K.: Dynamics and motion planning of trident snake robot. J. Intell. Robotic Syst. **75**, 17–28 (2014). https://doi.org/10.1007/s10846-013-9858-y
12. Selig, J.M.: Geometric Fundamentals of Robotics. Monographs in Computer Science. Springer, New York (2004). https://doi.org/10.1007/b138859

Modelling and Optimization of the Air Operational Manoeuvre

Agostino G. Bruzzone[1]([⊠]), Josef Procházka[2], Libor Kutěj[2],
Dalibor Procházka[2], Jaroslav Kozůbek[2], and Radomir Scurek[3]

[1] University of Genoa, Genoa, Italy
agostino@itim.unige.it
[2] University of Defence, Brno, Czech Republic
{jan.mazal,josef.prochazka,libor.kutej,
dalibor.prochazka,jaroslav.kozubek}@unob.cz
[3] WSB Uniwersity Dąbrowa Górnicza, Dąbrowa Górnicza, Poland
radomir.scurek@gmail.com

Abstract. Increasing complexity of the operational environment and advanced technology implementation in combat will probably lead to a serious limitation of human performance in all operational domains and activities in the future. With except of the clear indications, that tactical robotics will outperform human soldiers in many routine tasks on the battlefield, the area of operational decision making (resistible for decades to some automation) seems to be slowly approaching to the same stage. Presented article discusses the fundamental theory of optimization of the air operational maneuver and present the approach to the solution. The solution is highly theoretical and uses a modelling and simulation as an experimental platform to the visualization and evaluation of solution. The problem of air operational maneuver is specific in this case by many variables imposed on initial parametrization of the task (starting and destination point could not be known at the beginning, only "air operational" area should be selected) and very wide search of possible courses of action and the best "multi criteria" choice identification.

Keywords: UAV · Safety maneuver modelling · ISR · Optimization · Air maneuverer

1 Introduction

Contemporary highly dynamic military operational environment brings many changes and new challenges which were not significant or apparently visible before [9]. One of the today's significant trends in military is continuous pursue for the effectiveness and its improvement in context of lack of qualified personnel. This gives us the motivation for search of the solution of selected operational problems, which could be further partially or fully automated and save the human effort and increase effectiveness in mission execution [7].

© Springer Nature Switzerland AG 2019
J. Mazal (Ed.): MESAS 2018, LNCS 11472, pp. 43–53, 2019.
https://doi.org/10.1007/978-3-030-14984-0_4

2 State of the Art

Even thought a lot of publications dedicated to the area of Air/UAV maneuver optimization were found, it is still an actual topic.

After a publication analyses, before this paper was written, it could be mentioned, that majority of the papers dedicated to the Air maneuverer optimization area fell to the several sub-topical segments, mainly swarming and formation optimization, many of them are close to the air traffic control tasks, problems related to the traveling salesman problem (TSP) or optimization in collision avoidance, for instance: [1–6]. Very few papers were found, which are close to the operational optimization like: [10–12], especially in more complex and multi criteria context, when the operational situation is considered.

3 Approach to the Solution

Following solution is an continuum or more "operational" extension to the solution of "Modelling of the UAV safety maneuver for the air insertion operations" published in 2016 [11], what was more "tactically oriented" and initial parameters like destination point and UAV take-off point was known.

In this case we were interested in the best (safeties) and stable 3D flight path through the restricted flight corridor from the north to the south. It means that we search the entry and exit point where airplane (UAV) penetrate the "north" and "south" plane of the 3D corridor and path between these points fulfilling following optimization condition:

$$AIR_{path} = \min \rightarrow \sum_{i=1}^{M} A_{I_i,J_i,K_i};\tag{1}$$

where:

$A_{x,y,z}$– 3D safety matrix of operational area, derived from the set of analyses (3)
I_i, J_i, K_i– are the mathematical progressions coding the individual components/axes of the 3D path.

The condition:

$$\forall i \in (1 \ldots M) = \; > (|I_{i+1} - I_i| + |J_{i+1} - J_i| + |K_{i+1} - K_i|) < 3,\tag{2}$$

means that that two following elements of A matrix are adjacent. Also, because of the initial condition, were entry and exit points are not specified, the solution search for the best:

$A_{I_i,J_i,K_i}; i = 1$– entry point of the flight, must lay on the "north" plane of the cuboid representing a 3D air operation area.
$A_{I_i,J_i,K_i}; i = M$– exit point of the flight, must lay on the "south" plane of the cuboid representing a 3D air operation area.

Schema presenting the placement of the operational areas is depicted on following Fig. 1. Were ELOR shows the operating area for enemy forces with "light" weapons (affecting 3D AOA in a different way than enemy with "heavy" weapons), and EHOR present the operating area for the enemy forces with "heavy" weapons, limited with their maneuver to the roads (indicated by the green crosses), contrary to the enemy forces with "light" weapons (they are not limited to the roads with their ground movement).

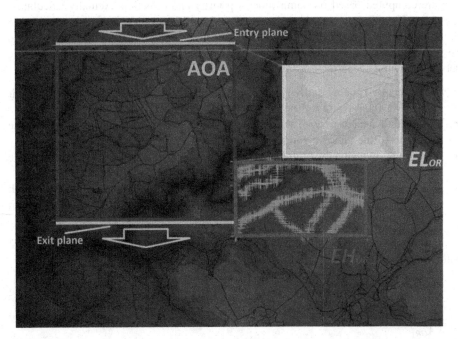

Fig. 1. Schema presenting the placement of the operational areas, AOA – Area of Operational Air Maneuver, EH_{OR} represents the operating area for the enemy forces with "heavy" weapons, EL_{OR} shows the operating area for enemy forces with "light" weapons, from the computer application developed by the authors.

Calculation of the 3D safety matrix includes set of analyses linked to set selected criteria in the Air Operational Area – AOA. Set to of these analyses could be fully or partially automated in context of updated operational situation, the latest data are usually available in C4ISTAR systems. For the mentioned example was selected following approach in calculation of 3D safety matrix:

$$A_{x,y,z} = \sum_{m=1}^{N} \left(SV_m \cdot GOA_m(x, y, z) \right) \tag{3}$$

$A_{x,y,z}$ - 3D safety matrix of operational area

$GOA_m(x, y, z)$ - Geo-operational analyses matrix (storing a partial results)

SV_m - safety weights defining the priorities in analyses in particular case of operational task solution.

The count of analyses affecting coefficients of the safety matrix should be linked to the desired purpose and information available from the Operational Area. Security analyses usually represent the partial security aspects to the global security situation and are computed based on commander's priority and selection. Actually calculating the A_i coefficients is a separate operational task and it could be based on other additional requirements and expected danger. Also single analyses should be normalized to be weighted and summarized with the others. To experiment with mentioned case, we set up following scenario:

- We expect two enemy entity types with two various weapons that can endanger Air manoeuvre (heavy and light).
- The ground maneuverer of the enemy with the heavy weapons is limited to the roads and to the area we suppose them to operate.
- The ground maneuverer of the enemy with the light weapons is limited only to the area we suppose them to operate.
- Due to the mentioned conditions, we could calculate all possible positions for both of the enemy entities and calculate safety coefficient for any (ground and air) point of Air Operational Area.

For the purpose of simulation experiments we define some terms and calculate following analyses:

- Air Operational Area as a space for the 3D safety matrix, AOA see Fig. 1.
- Area of enemy operation (with heavy and light weapons) as EH_{OR} and EL_{OR}.
- Visibility analyses of the all points of AOA from the as EH_{OR} and EL_{OR}, named VA_{OR}.
- Appearance analysis of the enemy with "light" weapons in the EL_{OR} and threat analyses for the AOA from the EL_{OR} area.
- Appearance analysis of the enemy with "heavy" weapons in the EH_{OR} and threat analyses for the AOA from the EH_{OR} area.
- Specification of the Air manoeuvre constraints.

Whole solution is rather complicated, consists of several thousand lines of C++ code, so the detailed description fells out of the frame of this article, in any case following scheme on Fig. 2, and generally describes the core procedures/parts of the solution algorithm:

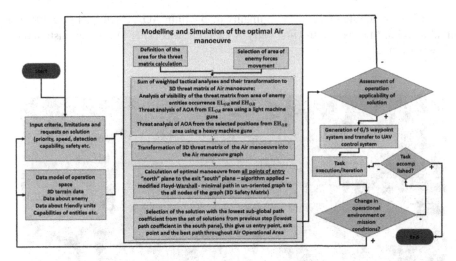

Fig. 2. Generic algorithm of the solution.

For example, of application of individual analyses were used following Geo-operational analyses, similarly as in the reference published in 2016 [11, 13]:

Analysis of "light" weapons enemy threat from EL_{OR} operational area:

$$GOA_1 = C_{alk} \cdot F_v(S, D) \cdot P_{zlk}(S, D), \tag{4}$$

Analysis of "heavy" weapons enemy threat from EH_{OR} operational area:

$$GOA_2 = C_{atk} \cdot F_v(S, D) \cdot P_{ztk}(S, D) \tag{5}$$

where:

- C_{alk} is a operational coefficient of a multi-criteria evaluation defined for GOA_1 analysis.
- C_{atk} is a operational coefficient of a multi-criteria evaluation defined for GOA_2,
- $F_v(S, D)$ is a visibility function from the source - S point to the destination - D point on a (digital) terrain model, $0 \le F_v(S, D) \le 1$.
- $P_{zlk}(S, D)$ is a probability hit of a target at the position of $D(x, y, z)$ using a "light weapon" from the point of $S(x, y, z)$ on terrain model.
- $P_{ztk}(S, D)$ is a probability hit of a target at the position of $D(x, y, z)$ using a "heavy weapon" from the source point of $S(x, y, z)$ on terrain model.
- $S(x, y, z)$ is the source point on a (digital) terrain model.
- $D(x, y, z)$ is the destination point on a (digital) terrain model.

To demonstrate solution of the problem, a computer application was developed (C++), where mentioned approaches and algorithms were implemented. The input data for the operating environment were taken from the highly detailed terrain database of the selected region in the Czech Republic provided by GEO service of the Czech Armed Forces. General overview of the map and various operating areas is presented on the Fig. 1, example of Air path optimality map within AOA calculated from the first initial iteration (means that the first entry XYZ point was (1, 1, 1)), is demonstrated on following Fig. 3:

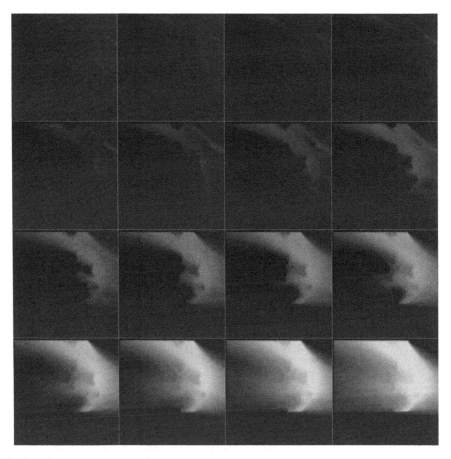

Fig. 3. 16 cuts (from the ground to top, starting from the top left corner to the bottom right) of 3D Air optimality path map within AOA, darker point is better in terms of the path safety (application developed by the authors).

On the 3D safety matrix, there were investigated/calculated huge number of optimal paths coming from the each possible entry point on the "north" pane of the matrix (1, y, z), (5000 m × 1600 m). For presentation purpose, the sampling step was degraded up to the 100 m vertically and 715 m horizontally. Final Path optimality map from the "south" matrix pane, based on the entry point is demonstrated on the following Fig. 4, with the red-white dot indicating the best exit point from the AOA/3D safety matrix.

Fig. 4. The figure is presenting $15 \times 7 = 105$ final Path Optimality Maps taken from the "south" matrix pane (500, y, z), with the 10 m resolution, scale is from black to white (darker is better), red dots indicate the best position for the exit of the AOA within its entry point (Color figure online)

The results of calculations and number of simulated flights is presented on the next Fig. 5, on right graph, there are presented exit point from the 3D matrix in the "south" pane, on the right graph, there are showed initial sampled entry points in the "north" pane on the 3D matrix (bigger circles) with exit point (same as the right graph) and connection lines represent correspondence between entry and exit point within a single sub-solution/iteration.

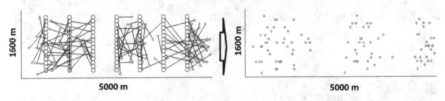

Fig. 5. Graphs of calculated exit points (right graph) and entry and exit points (left graph) from and to the AOA/3D matrix, left graph also show the linkage between entry and exit point within a single iteration.

Based on a previous search process for the best coefficient from the "exit" pane, consolidating the lowest total sum of possible threat, the optimal solution was easily found from all simulations, searching the lowest coefficient at the exit pane after the Floyd-Warshall algorithm execution. When the best candidate (exit point) was identified, the back search to the entry point was executed and this is depicted on the Fig. 6, were the 16 cuts of 3D path map are demonstrated with appearance of the optimal path in each layer indicated by the red dots. Path search was applied to the non-oriented weighted graph (topology of 26 connecting neighboring cells, totally $512 \times 512 \times 160$ nodes).

Based on the selected sampling of input (entry) point in the AOA, there were calculated 105 optimal paths. All these paths are visualised in random colour on the Fig. 7, including its altitude profile and overlay with the map of AOA. It is apparent at the first look, that majority of the paths converge to the three air corridors indicating the alternative and the most safety areas. If we are searching for the best path and we are flexible in the entry and exit point selection, the bold red path was calculated as

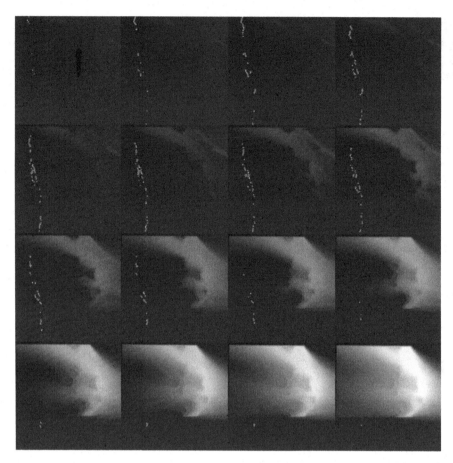

Fig. 6. 16 cuts of the 3D path map, with the best Air path indicated in each layer (red dots), integration of all points, creates a continuous path, it is demonstrated on the next Fig. 6. (red path), individual cuts represent the minimal sum of the safety coefficients to each point of the area in each altitude layer, taken from the application developed by the authors (Color figure online)

absolutely best option from the all possible candidates. This path includes also two alternative paths approximately in the middle third, which possess the same safety ratio.

More detailed altitude profile of the Air optimal path is presented on the Fig. 8. Looking at the altitude profile of all calculated paths, the altitude also indicates slight shape or limitation within its scope (close to the "north" plane, higher altitude is preferred), see Fig. 7, in the middle.

Explained approach was chosen to demonstrate a one of the possible way to the operational problem solution. It also shows the opportunity to operational task automation and its close relation with real time operational decision support in context of C4ISTAR systems or potentially UAV's operational autonomous/adaptive path planning [14].

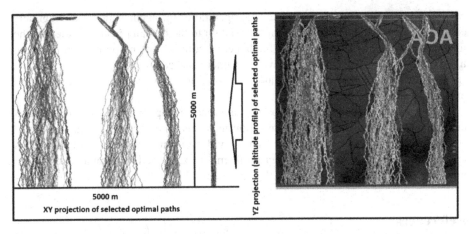

Fig. 7. Illustration of 105 sub optimal Air paths, based on the entry of the "north" pane, the best path with the lowest sum of safety coefficients is highlighted by thick red dots, correspondence with AOA is illustrated on the right. (Color figure online)

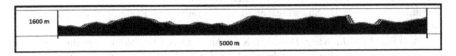

Fig. 8. Altitude profile of the final Air optimal path.

The computer application, demonstrating the possible approach to the solution was executed on PC with INTEL Core-5 (1.8 GHz) processor and the solution was found approximately within a 30 min, contemporary application also offers a large area for optimization and highly parallel processing architecture implementation (GPU, CUDA and so.).

4   Conclusion

As it was mentioned, the importance of effective automation of operational planning aspects in various areas (as a decision support component) is constantly on increase. It is necessary to say, that there appears a great potential in automation and optimization of operational tasks, which are closer to the human high-level reasoning instead of low-level engineering problems and Modelling and Simulation methods could be successfully applied [8]. From this point of view, a crucial aspect to evaluate an operational planning dealing with OPFOR and other players (e.g. civilians, neutral units, suspect ones) is strongly related to the ability to evaluate their behaviours; it is fundamental to create some effective models and behaviour that reproduce the actions/reactions based on the different boundary conditions. It is evident that use of Intelligent Agents driving objects during simulation enhance largely the effectiveness of simulation approach in this context as well as in other joint scenarios [15, 16]. Also we have to understand, that almost any operational problem follows a pragmatic concept, what means the rationality in the relation to the human or certain side and the final achievement. Mainly it fulfils a fundamental criteria's of an optimization problem like maximization of profit or achievement under the condition of limited resources spending, like minimization of the task cost, time for execution, effort, danger areas explosion and so.

Presented solution shows the possible approach in operational problem solution dedicated to the air manoeuvre optimization in operational conditions with undefined starting and destination point, what means new dimension of options and calculations leading to a higher decision area then problems with selected constraints and known initial inputs.

References

1. Geiger, B.: Unmanned aerial vehicle trajectory planning with direct methods. A Dissertation in Aerospace Engineering. The Pennsylvania State University, Pennsylvania, USA (2009)
2. Waseem, A.K.: Safe trajectory planning techniques for autonomous air vehicles. A dissertation work. University of Leicester, United Kingdom (2005)
3. Tsourdos, A., White, B., Shanmugavel, M.: Cooperative Path Planning of Unmanned Aerial Vehicles, p. 214. Wiley, Hoboken (2010). ISBN: 978-0-470-74129-0
4. Duan, H.B., Ma, G.J., Wang, D.B., Yu, X.F.: An improved ant colony algorithm for solving continuous space optimization problems. J. Syst. Simul. **19**(5), 974–977 (2007)
5. Qu, Y.H., Pan, Q., Yan, Y.G.: Flight path planning of UAV based on heuristically search and genetic algorithms. In: Proceedings of the IEEE 32nd Annual Conference, pp. 45–50 (2005)
6. Liu, C.A., Li, W.J., Wang, H.P.: Path planning for UAVs based on ant colony. J. Air Force Eng. Univ. **2**(5), 9–12 (2004)
7. Kress, M.: Operational Logistics: The Art and Science of Sustaining Military Operations. Springer, Heidelberg (2002). https://doi.org/10.1007/978-1-4615-1085-7
8. Rybar, M.: Modelovanie a simulacia vo vojenstve. Ministerstvo obrany Slovenskej republiky, Bratislava (2000)
9. Washburn, A., Kress, M.: Combat Modeling: International Series in Operations Research & Management Science. Springer, Heidelberg (2009)
10. Mokrá, I.: Modelový přístup k rozhodovacím aktivitám velitelů jednotek v bojových operacích. Disertační práce. Brno: Univerzita obrany v Brně, Fakulta ekonomiky a managementu. 120 s (2012)
11. Mazal, J., Stodola, P., Procházka, D., Kutěj, L., Ščurek, R., Procházka, J.: Modelling of the UAV safety manoeuvre for the air insertion operations. In: Hodicky, J. (ed.) MESAS 2016. LNCS, vol. 9991, pp. 337–346. Springer, Cham (2016). https://doi.org/10.1007/978-3-319-47605-6_27
12. Rybansky, M.: Modelling of the optimal vehicle route in terrain in emergency situations using GIS data. In: 8th International Symposium of the Digital Earth (ISDE8) 2013, Kuching, Sarawak, Malaysia 2014 IOP Conference Series: Earth Environmental Science, vol. 18, p. 012071 (2014). https://doi.org/10.1088/1755-1315/18/1/012131. ISSN 1755-1307
13. Rybanský, M., Vala, M.: Relief impact on transport. In.: ICMT 2009 - International Conference on Military Technologies 2009, Brno, Czech Republic, 9 p. (2009). ISBN 978-80-7231-649-6 (978-80-7231-648-9 CD)

14. Drozd, J., Stodola, P., Křišťálová, D., Kozůbek, J.: Experiments with the uas reconnaissance model in the real environment. In: Mazal, J. (ed.) MESAS 2017. LNCS, vol. 10756, pp. 340–349. Springer, Cham (2018). https://doi.org/10.1007/978-3-319-76072-8_24
15. Bruzzone, A.G.: MS2G as Pillar for Developing Strategic Engineering as a New Discipline for Complex Problem Solving. Keynote Speech at I3 M, Budapest, September 2018
16. Bruzzone, A.G., Massei, M.: Simulation-based military training. In: Mittal, S., Durak, U., Ören, T. (eds.) Guide to Simulation-Based Disciplines. SFMA, pp. 315–361. Springer, Cham (2017). https://doi.org/10.1007/978-3-319-61264-5_14

Spatiotemporal Models of Human Activity for Robotic Patrolling

Tomáš Vintr[1]([✉]) [ID], Kerem Eyisoy[2] [ID], Vanda Vintrová[3] [ID], Zhi Yan[4] [ID],
Yassine Ruichek[4] [ID], and Tomáš Krajník[1] [ID]

[1] Artificial Intelligence Center, Faculty of Electrical Engineering,
Czech Technical University, Prague, Czech Republic
`tomas.vintr@fel.cvut.cz`
[2] Department of Computer Engineering, Faculty of Engineering,
Marmara University, Istanbul, Turkey
[3] Faculty of Informatics and Statistics, University of Economics,
Prague, Czech Republic
[4] EPAN Research Group, LE2I-CNRS,
University of Technology of Belfort-Montbliard (UTBM), Belfort, France

Abstract. We present a method that allows autonomous systems to
detect anomalous events in human-populated environments through
understating of their structure and how they change over time. We represent the environment by temporary warped space-hypertime continuous models derived from patterns of changes driven by human activities
within the observed space. The ability of the method to detect anomalies
is evaluated on real-world datasets gathered by robots over the course of
several weeks. An earlier version of this approach was already applied to
robots that patrolled offices of a global security company (G4S).

1 Introduction

Autonomous patrolling robots or monitoring sensor networks in general should
be able to correctly detect anomalous behaviour of agents. This goal is usually
achieved by using complicated rule base created by an expert, but we believe
that this goal should be achieved through the self learning of an autonomous
system. We present a method, that can create a model of changes in human
populated environments based on the past observations, and decide, what is
anomalous compared to its previous experience.

As the robots gradually attain the ability of long-term operation, they not
only encounter new challenges due to the fact that the environment changes
over time [12,15], but by interpreting these changes, they also get new opportunities to learn not only the environment structure, but also its dynamics [8,17].

The work has been supported by the Czech Science Foundation project 17-27006Y,
MŠMT project FR-8J18FR018 and PHC Barrande project 40682ZH (3L4AV). We
thank the School of Computer Science, University of Lincoln, UK for providing us
with the data, namely Dr Grzegorz Cielniak for being the long-suffering, but tolerant
experimental human subject.

© Springer Nature Switzerland AG 2019
J. Mazal (Ed.): MESAS 2018, LNCS 11472, pp. 54–64, 2019.
https://doi.org/10.1007/978-3-030-14984-0_5

However, a robot, which continuously operates withing a given environment, can learn the environment dynamics within a few days and after that, new observations typically conform to the already-created model and do not bring significant additional information [3, 13]. Thus, to refine the environment model, a robot should be able to determine (or even predict), that a given observation is or will be novel, unusual or anomalous. The ability to detect and predict anomalous situations thus significantly contributes to the quality of the robot environment model and therefore to the efficiency of its operation [7, 22, 23]. Furthermore, detecting anomalous situations is often required in cases that the robot is deployed in security-related scenarios [8]. Since the environmental changes observed indoors are typically caused by people, detection of anomalous situations is closely tied to modelling the human behaviour.

The aim of our method is to create a model of human presence in the patrolling robot operational area. Such a model has to predict the regular behaviour and incorporate the detection of anomalous events. Anomalous events, outliers in data, and novelties in streams of data are defined only vaguely as measurements, that are somehow strange – unexpected, inconsistent, suspicious, or rare – compared to the previous experience of a human observer [6], [1] and there are many approaches to solve different scenarios in different scientific fields [11]. In our case, we use the outlier detection based on the model [21]. Our model is a function that estimates the Bernoulli distribution [4] of occurrences over the time line – similarly to FreMEn model [3, 13, 14].

For the time dependent events prediction we usually use methods for time series forecasting. The time series forecasting understands time-dependent events as a combination of three components that can be analysed separately - a trend, seasonal and cyclic patterns [9]. The trend describes long-term increase or decrease in data, seasonal patterns describe periodical changes in data and cyclic patterns are fluctuations in data with unknown periodicity. In the proposed method we assume that it is possible to neglect the trend and cyclic patterns due to the nature of the studied problem, where the environmental changes are induced directly by human routines and habits. Therefore, our model is derived from periodical patterns, that explain the majority of events in human populated environment. Using the time series forecasting with neglected trend to analyse the time dependent human behaviour patterns was firstly introduced in [13] and it demonstrated its efficiency in different robotic applications and scenarios [16, 20, 23].

Contrary to usual approach of modelling multiple seasonalities, where individual patterns are analysed separately [5], we create a special vector space, that reflects every selected seasonality, project the time series into it, and then analyse the periodical features of human behaviour in that space, see Fig. 1. For every seasonal pattern we create extra two dimensions and project the time series into a circle on that plane in the way that every measurement from the same position relative to the periodicity is on the same position on the circle. The proposed projection then reflects not only the seasonality derived from the structure of the vector space, but also the continuity derived from a circularity of the projection.

2 Method Description

Our method supposes that there are some patterns – functions derived from the human behaviour – over the time line. To create models of different features in data, we need to estimate parameters of their distributions. But the time line, understood as domain of some time dependent function, is not suitable to estimate the distribution of features that occur rarely, because we cannot repeatedly observe the same situation. As we assume the periodical nature of these features, we can transform the time line into the different, constrained vector space. The idea behind this transformation is based on the following idea [25]: the human behaviour is very similar during every morning as opposed to the difference in behaviour during morning and afternoon of one randomly chosen day. Similarly, human behaviour five minutes before midnight and five minutes after midnight could be probably very similar, although in fact we compare behaviour in two different days. Contrary to that, human behaviour during Sunday afternoon is probably different from behaviour on Monday afternoon, etc. Therefore, our method transforms time line into the set of circles based on chosen periodicities detected in data, see Fig. 1. For example, when it chooses periodicity of one day, every measurement from the dataset is projected into this circle in a way that measurements from the same time in different days are at the same position. Than we can study periodical features of human behaviour with chosen periodicity and estimate distribution parameters, as illustrated in Fig. 1. It is possible to extend this vector space with new dimensions and project data into the corresponding circles. We call this vector space with such projection of data the *warped hypertime*. When the model of mixture of distributions over warped hypertime is created, we can define regular and anomalous human behaviour in the robot's operational area.

The detection of the most influencing periodicities in the data can be done using the Spectral analysis [2], parameters of the mixture of distributions can be gathered using the Expectation-Maximisation model for Gaussian mixtures [26] and parameters of distribution of residuals between model and measured data for anomaly detection can be obtained using the robust descriptive statistic [18]. The whole process can be described as follows:

- identification of the most influencing periodicities in data,
- creation of warped hypertime,
- estimation of parameters of mixture of distributions,
- calculation of the parameters of the residuals between model and data.

2.1 Identification of Periodicities in the Training Data

Let us have time series $R(t_i)$, $i = 1 \ldots n$, where $R(t_i) = 1$ when the sensor detects an occurrence of the studied phenomenon and $R(t_i) = 0$ otherwise. Than we use the spectral analysis to find periodicities in the data, specifically

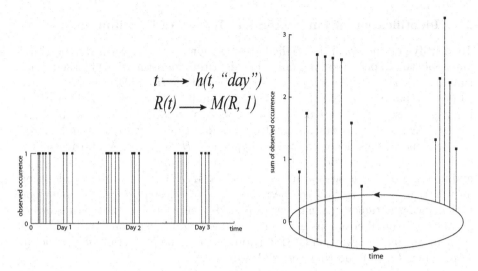

Fig. 1. Example of the warped hypertime projection. Positive detections during three days are projected into a warped hypertime with a one day period. The parameters of the distribution of a random time dependent phenomenon that exhibit a periodic behaviour can be easily estimated.

we will use the method derived from the Frequency Map Enhancement [13]. This method is suitable to analyse time series with binary data and it is robust to missing values. For every considered period T_τ, $\tau = 1 \ldots \Upsilon$, corresponding to the frequency $f_\tau = 1/T_\tau$ we calculate a component of the frequency spectrum

$$\gamma_\tau = \frac{1}{n} \sum_{i=1}^{n} (R(t_i) - \mu) e^{(-1j)t_i f_\tau 2\pi}, \tag{1}$$

where $\mu = \frac{1}{n} \sum_{i=1}^{n} R(t_i)$, sort them in descending order and return corresponding periods ordered accordingly.

2.2   Building the Warped Hypertime

For every chosen period T_τ we can create two $2d$ warped hypertimes – projections of $R(t_i)$ and $1 - R(t_i)$ into $\{t|h(t, T_\tau)\}$ (Fig. 1), where the function h is defined as follows:

$$h(t, T_\tau) = \left(\cos\left(\frac{2\pi t}{T_\tau}\right), \sin\left(\frac{2\pi t}{T_\tau}\right) \right). \tag{2}$$

The time series $R(t_i)$ reflect detected occurrences of a studied phenomenon and time series $1 - R(t_i)$ reflect complement phenomenon, 'non-occurrences'. The warped hypertimes with the projection of the same time series can be combined into a warped hypertime of higher dimension

$$R(t_i) \rightarrow \{t| (h(t, T_{\tau_1}), h(t, T_{\tau_2}), \ldots)\}. \tag{3}$$

2.3   Identification of Parameters of Mixture of Distributions

To identify parameters of the studied phenomenon, we create warped hypertimes and apply the Expectation Maximisation algorithm for estimating Gaussian Mixture Models (EM GMM). For both, warped hypertime H and EM GMM, we have to provide parameters. For EM GMM it is number of clusters c and for H we need to choose proper periods T. As the periods are ordered by their influence to $R(t_i)$ (Subsect. 2.1), we can build models in an iterative manner and choose the best one. Let us denote a warped hypertime H, which is build using first p of most influencing periods and with projection of time series $R(t_i)$ as $H(R, p)$ and a model over this space with c clusters as $M(R, p, c)$, then we can describe an iterative process to find the best model as two nested loops, the inner one with growing parameter p (number of periods) and the outer one with growing parameter c (number of clusters). The obtained models are compared using the Root-mean-square deviation (RMSD) [10] between the projection of a model to a time line $\hat{R}_{c,p}(t_i)$ and time series $R(t_i)$:

$$E_{c,p} = \sqrt{\frac{1}{n} \sum_{i=1}^{n} \left(\hat{R}_{c,p}(t_i) - R(t_i) \right)^2}, \tag{4}$$

where

$$\hat{R}_{c,p}(t_i) = \hat{R}^*_{c,p}(t_i) \frac{\sum_{i=1}^{n} R_{c,p}(t_i)}{\sum_{i=1}^{n} \hat{R}^*_{c,p}(t_i)}, \tag{5}$$

where

$$\hat{R}^*_{c,p}(t_i) = \frac{\hat{R}^1_{c,p}(t_i)}{\hat{R}^1_{c,p}(t_i) + \hat{R}^0_{c,p}(t_i)} \tag{6}$$

where $\hat{R}^1_{c,p}(t_i)$ is a modelled value of occurrence $M(R, p, c)$ at time t_i and $\hat{R}^0_{c,p}(t_i)$ is a modelled value of non-occurrence $M(1 - R, p, c)$ at the same time. $\hat{R}^*_{c,p}(t_i)$ is the estimation of relative frequency of occurrences in time t_i, and after application of normalisation (5) we can think of $\hat{R}_{c,p}(t)$ as about the function that estimates probability value of the Bernoulli distribution [4] of a studied phenomenon in time. The model with the lowest value of $E_{c,p}$ is chosen.

The structure of the iterative process is as follows:

1. $c = 1$, $p = 1$, $E_{1,0} = \inf$ and define $\max(c)$
2. create $H(R, p)$, build $M(R, p, c)$, calculate $E_{c,p}$,
3. if $E_{c,p-1} < E_{c,p}$ then $c := c + 1$, $p := 1$ else $p := p + 1$,
4. if $c \neq \max(c)$ then goto 2,
5. return $M(R, p, c)$ corresponding to minimal $E_{c,p}$.

2.4   Analysis of Residuals

Our model estimates the probability of occurrence of the studied phenomenon over the time line. The distribution of occurrence in specific time match the

Bernoulli distribution. Therefore, to test hypothesis that the detected occurrence is an outlier, we only need to subtract the detected value (0 or 1) from estimated probability and compare that with chosen significance level α.

Using this assumption about the distribution and the model, we can choose some level of the "extreme" difference between the model and the measurement $L = 1 - \alpha$ and label absolute values of differences between $\hat{R}_{c,p}(t)$ and $R(t)$ that are larger than L as outliers and vice versa:

$$\forall t : o(t) = \begin{cases} 1 \ldots \left| R(t) - \hat{R}_{c,p}(t) \right| <= L \\ -1 \ldots \left| R(t) - \hat{R}_{c,p}(t) \right| > L \end{cases}. \tag{7}$$

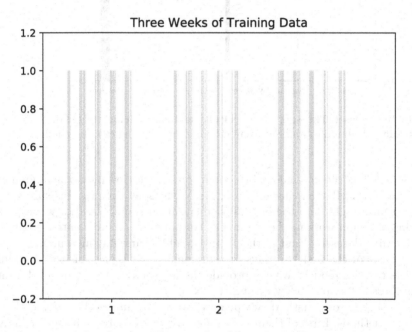

Fig. 2. Longer training dataset. Three weeks of detections of one person in his office. Time series values acquire 1 when the person is detected and 0 otherwise.

3 Experiments

We evaluated our method on a real world dataset from Lincoln university. It consists of twenty weeks of detection of one person in his office. We divided this dataset into the training dataset that consists of first three weeks (Fig. 2) and the test dataset that consists of the eighteenth week (Fig. 3). The eighteenth week was chosen based on the unusual amount of the anomalous behaviour. We also manually labelled the outliers in the test dataset to be able to evaluate the correctness of our outlier detection method. There are three situations that

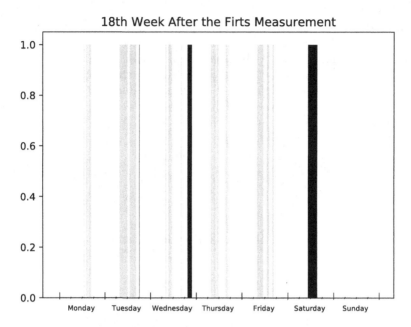

Fig. 3. Test dataset. The eighteenth week of detections with highlighted outliers (black) on Tuesday, Wednesday, and Saturday. The outliers were labelled manually.

were labelled as the anomalous behaviour: on Tuesday evening, the human subject came back after leaving his office, on Wednesday evening this subject was working until night, and the last labelled situation is on Saturday, when he was working on a non-working day. From the training dataset we created twenty one new training datasets. First of them consists of the first day measurements, second consists of the first two days measurements, etc. Using this set of training datasets to train models, we can provide the information about the evolution of the outliers detection ability during the learning.

We compare our method, warped hypertime technique *WHyTe*, with four different methods. First of them is *Prophet*, the open source time series analysis tool created by Facebook [24], second is *FreMEn* as defined in [13] using five most influencing periodicities, and the last two are histograms *Hist24* and *Hist168*. Hist24 calculates the mean of occurrences for every hour in a day and Hist168 calculates one hour means over the whole week. We choose basic periodicities for histograms using our knowledge of most prominent periodicities form WHyTe. Contrary to histograms, WHyTe and FreMEn calculate the most prominent periodicities automatically, no prior assumptions about the dataset are required.

To quantify the outlier detection power of the tested methods, we chose the Matthews correlation coefficient [19], which is suitable to measure the quality of binary classifications, in our case inliers and outliers. We created models for all of 21 training datasets consisting of data form the first day, the first two days, ..., the first twenty one days of measurements (Fig. 2), and detected outliers in the 18th

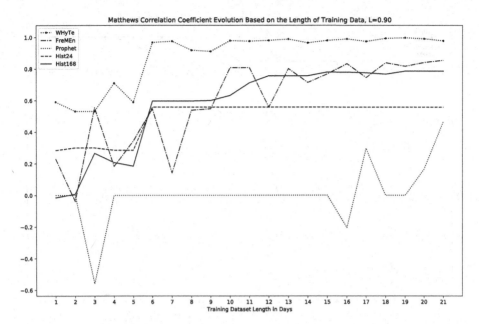

Fig. 4. Evolution of Matthews correlation coefficient using $L = 0.90$. On the x axis there are numbers of days used to train model, on the y axes are values of Matthews correlation coefficient $\langle -1; 1 \rangle$. A coefficient of value 1 means correct labelling of outliers by the corresponding method.

week of measurements (Fig. 3) using the models's prediction of the behaviour of the studied subject. The values of Matthews correlation coefficients for different methods and different training datasets ordered by the length of these datasets for $L = 0.90$ and $L = 0.99$ are shown in Figs. 4 and 5 respectively.

We can see in Figs. 4 and 5 that WHyTe needed shorter time to learn features of the studied subject behaviour than the original FreMEn. The ability of WHyTe to detect the anomalous behaviour on our dataset is significantly better than the ability of FreMEn. Compared to the other methods, WHyTe exhibits robustness to the choice of significance level. Moreover, WHyTe is better compared to the popular technique in robotics, the histograms, even if these histograms use correctly predefined periodicities. Hist168 with $L = 0.99$ is able to maximise its ability to predict outliers, but it needed two and half weeks to fully train. WHyTe was properly trained during the first ten days. Prophet with its default setting, representing commercial tools for time series analysis, was unable to learn the model as fast as the others. It should be noted that it was able to predict outliers similarly to FreMEn when the models were created over five and more weeks. But it also predicts a trend, contrary to the other tested methods, and therefore it takes longer time to train. Based on our experiments, the assumption that the trend should be neglected for the human behaviour analysis over several weeks was correct.

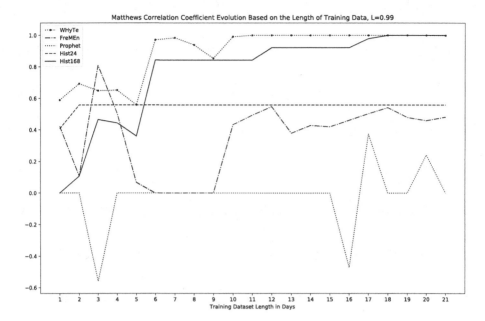

Fig. 5. Evolution of Matthews correlation coefficient using $L = 0.99$. On the x axis there are numbers of days used to train model, on the y axes are values of Matthews correlation coefficient $\langle -1; 1 \rangle$. A coefficient of value 1 means correct labelling of outliers by the corresponding method.

4   Conclusion

We proposed a new method to detect the anomalous behaviour of humans in their natural environment. This method projects the time series that describe the human behaviour into the special vector space. Such projection reflects seasonality and continuity of the studied phenomenon and features of this phenomenon can be analysed and described using usual statistical methods. The experiments performed on a real-world dataset demonstrate that the method outperforms the other state of the art, both in terms of predictive accuracy and ability to detect anomalies. Furthermore, it needs shorter time to learn, it is more precise in detecting outliers, and it is more robust to the choice of significance level for division between inliers and outliers. Moreover, it uses usual statistical tools that are part of common analytic libraries. In the future work, we will extend this method to be able to model phenomenons described by a general vector and we will evaluate this method on a wider range of datasets with different scenarios.

References

1. Barnett, V., Lewis, T.: Outliers in Statistical Data. Wiley, Hoboken (1974)
2. Brockwell, P.J., Davis, R.A.: Introduction to Time Series and Forecasting. Springer, Cham (2016). https://doi.org/10.1007/978-3-319-29854-2

3. Coppola, C., Krajnık, T., Duckett, T., Bellotto, N.: Learning temporal context for activity recognition. In: European Conference on Artificial Intelligence (2016)
4. Evans, M., Hastings, N., Peacock, B.: Bernoulli distribution. In: Statistical Distributions, 3rd edn, pp. 31–33. Wiley, Hoboken (2000)
5. Gould, P.G., Koehler, A.B., Ord, J.K., Snyder, R.D., Hyndman, R.J., Vahid-Araghi, F.: Forecasting time series with multiple seasonal patterns. Eur. J. Oper. Res. **191**(1), 207–222 (2008)
6. Grubbs, F.E.: Procedures for detecting outlying observations in samples. Technometrics **11**(1), 1–21 (1969)
7. Hanheide, M., Hebesberger, D., Krajník, T.: The when, where, and how: an adaptive robotic info-terminal for care home residents. In: Proceedings of the 2017 ACM/IEEE International Conference on Human-Robot Interaction, HRI 2017, pp. 341–349. ACM, New York (2017). https://doi.org/10.1145/2909824.3020228
8. Hawes, N., et al.: The strands project: long-term autonomy in everyday environments. IEEE Robot. Autom. Mag. **24**(3), 146–156 (2017). https://doi.org/10.1109/MRA.2016.2636359
9. Hyndman, R.J., Athanasopoulos, G.: Forecasting: principles and practice. OTexts (2018)
10. Hyndman, R.J., Koehler, A.B.: Another look at measures of forecast accuracy. Int. J. Forecast. **22**(4), 679–688 (2006)
11. Ilango, V., Subramanian, R., Vasudevan, V.: A five step procedure for outlier analysis in data mining. Eur. J. Sci. Res. **75**(3), 327–339 (2012)
12. Krajník, T., Fentanes, J.P., Mozos, O.M., Duckett, T., Ekekrantz, J., Hanheide, M.: Long-term topological localisation for service robots in dynamic environments using spectral maps. In: IEEE/RSJ International Conference on Intelligent Robots and Systems (IROS), pp. 4537–4542. IEEE (2014)
13. Krajník, T., Fentanes, J.P., Santos, J.M., Duckett, T.: FreMEn: frequency map enhancement for long-term mobile robot autonomy in changing environments. IEEE Trans. Robot. **33**(4), 964–977 (2017)
14. Krajnik, T., Fentanes, J.P., Cielniak, G., Dondrup, C., Duckett, T.: Spectral analysis for long-term robotic mapping. In: 2014 IEEE International Conference on Robotics and Automation (ICRA), pp. 3706–3711. IEEE (2014)
15. Krajník, T., Fentanes, J.P., Hanheide, M., Duckett, T.: Persistent localization and life-long mapping in changing environments using the frequency map enhancement. In: 2016 IEEE/RSJ International Conference on Intelligent Robots and Systems (IROS), pp. 4558–4563. IEEE (2016)
16. Krajník, T., Kulich, M., Mudrová, L., Ambrus, R., Duckett, T.: Where's waldo at time t? Using spatio-temporal models for mobile robot search. In: IEEE International Conference on Robotics and Automation (ICRA) (2015)
17. Kunze, L., Hawes, N., Duckett, T., Hanheide, M., Krajnik, T.: Artificial intelligence for long-term robot autonomy: a survey. In: IEEE Robotics and Automation Letters, p. 1 (2018). https://doi.org/10.1109/LRA.2018.2860628
18. Maronna, R., Martin, R.D., Yohai, V.: Robust Statistics, vol. 1. Wiley, Chichester (2006)
19. Matthews, B.W.: Comparison of the predicted and observed secondary structure of T4 phage lysozyme. Biochimica et Biophysica Acta (BBA)-Protein Struct. **405**(2), 442–451 (1975)
20. Fentanes, J.P., Lacerda, B., Krajník, T., Hawes, N., Hanheide, M.: Now or later? Predicting and maximising success of navigation actions from long-term experience. In: IEEE International Conference on Robotics and Automation (ICRA) (2015)

21. Rousseeuw, P.J., Leroy, A.M.: Robust Regression and Outlier Detection, vol. 589. Wiley, New York (2005)
22. Santos, J.M., Krajník, T., Duckett, T.: Spatio-temporal exploration strategies for long-term autonomy of mobile robots. Robot. Autonom. Syst. **88**, 116–126 (2017)
23. Santos, J.M., Krajnik, T., Pulido Fentanes, J., Duckett, T.: Lifelong information-driven exploration to complete and refine 4D spatio-temporal maps. Robot. Autom. Lett. **1**(2), 684–691 (2016)
24. Taylor, S.J., Letham, B.: Forecasting at scale. Am. Stat. **72**(1), 37–45 (2018)
25. Vintr, T., Molina, S., Fentanes, J., Cielniak, G., Duckett, T., Krajník, T.: Spatio-temporal models for motion planning in human populated environments. In: Student Conference on Planning in Artificial Intelligence and Robotics (2017)
26. Xu, L., Jordan, M.I.: On convergence properties of the EM algorithm for Gaussian mixtures. Neural Comput. **8**(1), 129–151 (1996)

Distributed Simulation Environment of Unmanned Aerial Systems for a Search Problem

Stanisław Skrzypecki[(✉)], Dariusz Pierzchała, and Zbigniew Tarapata

Cybernetics Faculty, Military University of Technology, Warsaw, Poland
{stanislaw.skrzypecki, dariusz.pierzchala,
zbigniew.tarapata}@wat.edu.pl

Abstract. In the paper the programmable simulation environment for Unmanned Aerial Systems (UAS) and the preliminary research of UAV's swarm applied for a search problem in a large-scale terrain are presented. Proposed approach is based on distributed simulation, multiagent systems and multiresolution modelling in order to perform studies on UAV modelled as a swarm with determined, autonomous, combined behaviours. Some software agents have been simulated in constructive component while others have been controlled by virtual simulator (VBS3) with interoperability provided by DIS or HLA protocols. Furthermore, the biological inspired algorithms (e.g. PSO algorithm and other modifications) have been used to model UAVs' actions. The preliminary results lead to conclusion of usability of the environment in solving search problem and modelling UAV's movements and behaviours.

Keywords: UAS · UAV's swarm · Distributed simulation ·
Autonomous systems · Biological inspired algorithms · PSO algorithm

1 Introduction

Regarding rapid development of teleinformatics and mechatronics the technical parameters of Unmanned Aerial Vehicle (UAV) are significantly improving. The same process consequence with rising number of opportunities to use them in many others areas, including transport of small-sized goods [6], monitoring and information gathering [16] or entertainment [18]. The wide spectrum of usage also includes military operations: from ISR[1] to combat missions[2]. The development of UAV has led to the concept of Unmanned Aerial Systems (UAS). The definition of UAS means a system consisting of three components: UAV, a control system (operated by human or in autonomous way) and C3 system (communication, command and control) [22]. Such a system, composed of many UAVs, acting jointly to achieve a goal, is often referred to as a drone swarm. The simulations of drone swarm have demonstrated very high effectiveness of the solution [7]. The concepts of using that type of configuration of

---

[1] Intelligence, surveillance and reconnaissance operations [19].

[2] E.g. killing of the leader of al-Qaida Anwar al-Awlaki by a U.S. drone attack in 2011.

© Springer Nature Switzerland AG 2019
J. Mazal (Ed.): MESAS 2018, LNCS 11472, pp. 65–81, 2019.
https://doi.org/10.1007/978-3-030-14984-0_6

aerial vehicles is not a most recent idea. Herein, it is worth to mention the Office of the Secretary of Defense's technical report "Unmanned Aircraft Systems Roadmap: 2005–2030". The report assumed continuous development of UAVs - starting from a remotely controlled single aircraft, through those adapting to variable flight conditions and operating in tactical groups (controlled in a distributed manner), ending with fully autonomous swarms of aircrafts [9]. In recent years projects as LOCUST[3] and PER-DIX[4] have shown a high level of advancement in the pursuit of these concepts.

The paper presents a programmable simulation environment for modelling Unmanned Aerial Systems and studies confirming the functionality of this environment. As a case study, the problem of using UAVs swarms to search for signal sources in a large-scale terrain is presented. However, development of new algorithms was not the aim of the paper. The simulation environment is based on the original DisSim simulation engine [2, 8], which uses distributed discrete simulation techniques, multiagent systems and multiresolution modelling. In the current version of the environment, the possibility of collision of aircrafts with each other and real limitations of communication between UAVs have been omitted. Regardless of the above, there is wide range of possibilities to determine a lot of model parameters such as the maximum distance between aircrafts (communication range) or the minimum number of aircrafts remaining in the communication range (communication redundancy). However, the most important assumption of the environment is the ability to prepare models in which individual UAVs are characterized by autonomy in making decisions, deterministic models (predetermined behaviours, movement and arrangement of a whole group) or mixed models. An autonomism is ensured by the use of the biological inspired algorithms, including the Particle Swarm Optimization (PSO) algorithm [3] and its modifications. Likewise, the occurrence of interactions between objects, including those from other simulators, is important in the prepared environment. Interacting simulators may have different command levels (single measures, tactical, operational or strategic) as well as types of simulation (constructive simulation or virtual simulation). For this reason, the environment is required to maintain objects states at various levels and to process states between those levels with interoperability provided with the DIS or HLA protocols.

The paper is organized as follows. In the Sect. 2 the programmable UAS simulation environment was described. It includes discrete event-driven simulation, DisSim as simulation engine, concept of environment and description of particular scenarios implementation. In the Sect. 3 as the case study of environment functionality testing, the problem of using UAVs swarms to search for signal sources in a given area was introduced. Moreover three algorithms solving this problem and results of comparison were presented. The Sect. 4 presents some paper summary and conclusions.

---

[3] Project LOCUST (Low-Cost UAV Swarming Technology) – the group of drones launched in 2016 from the ground launchers (up to 30), each aircraft is capable to execute self-conducting programmed actions [20].

[4] Project PERDIX – micro-drone swarm consisted of 103 Perdix drones launched in 2017 from three F/A-18 Super Hornets fighters and demonstrated advanced swarm behaviours such as collective decision-making, adaptive formation flying, and self-healing [17]; Perdix drone was initially developed at the MIT University in 2010–2011 [21].

2 Programmable UAS Simulation Environment

2.1 Discrete Event-Driven Simulation

The basis of the programable UAS simulation environment is the discrete event simulation (DES). DES is characterized by discrete timestep changes resulting from sequentially planned moments of event realizations. The event e from finite set of events E is the algorithmically planned state change of the object (system) at a given simulation time t:

$$e = <t,f_e^S> , e \in E, t \in T \tag{1}$$

A set of attributes values of the modelled system object at any time t of the simulation time is called the system state S. It is defined as the following ordered four:

$$S = <o,a,v,t> , o \in O, a \in A_c, v \in V_a^c, t \in T \tag{2}$$

$O = \{o = \{id, c\}, c \in C^0\}$ – a set of c class objects identified by id with a unique value in the collection of objects,
C^0 – nonempty set of modelled objects classes,
A_c – nonempty set of attributes defined for the object class $c \in C^0$,
V_a^c – set of acceptable values for attribute $a \in A_c$ of the object class $c \in C^0$,
$t \in T$ – simulation time which is variable from countable subset $T \subset R_+ \cup \{0\}$; it means that $(\exists t_i, t_k) (\neg \exists t_j) (t_i < t_j < t_k)$ so there are two such moments that there is no other between them - it indicates a discrete time passage.

The function of state change f_e^S determines the state $s \in S$ in which the system will be located at the moment t after the realisation of event e:

$$f_e^S : T \times S \rightarrow S \tag{3}$$

2.2 Simulation Engine

In the presented simulation environment the discrete event-driven simulation is being executed through the DisSim package. It is a simulation engine implemented in Java that supports sequentially events (*BasicSimStateChange*) realisation planned for many simulation objects (*BasicSimEntity*) and stored in a shared event calendar (*BasicSimCalendar*). The software offers a dynamic configuration of objects system (plug-in system), a number of auxiliary classes for generating pseudo-random numbers, monitoring and gathering variable states or calculation of statistics on accumulated time series [2, 8]. The DisSim package has been enhanced to include: (a) distributed protocols DIS and HLA; (b) multiagent systems; (c) multiresolution modelling [11]. It made it possible to build programmable environment for constructive simulation allowing studies on behavioural, movement and groups models of UAVs.

As a result of layered environmental concept presented in Sect. 2.3, the DisSim has the following packages model classifying interfaces and classes for various purposes:

- **simspace:**
  - **core** – discrete event simulation execution;
  - **process** – continuous processes modelled with the DES;
- **connector:**
  - **agent** – adaptation of a single agent to participation in distributed simulation;
  - **aggregates** – adaptation of simulation objects to the requirements of multiresolution modelling;
  - **dis** – communication via the DIS protocol with other simulators;
  - **hla** – communication via the HLA protocol with other simulators;
  - **statechanges** – includes classes of messages sent between agents and plugins (statechanges), interfaces of statechanges and enumerations of message types;
- **broker** – implementation of broker and observer design patterns;
- **gui** – graphical user interface;
- **monitors** – variables monitoring and statistical analysis;
- **scripts** – enables plugins system;
- **config** – importing configuration files and managing simulation parameters;
- **random** – pseudo-random variables generation.

2.3   Concept of Environment

The concept of the environment (shown in Fig. 1) is based on the layer structure. The following layers are responsible for different functions: (a) GUI layer; (b) the analytical layer; (c) the aggregation layer; (d) agents layer; (e) distribution layer. The components of each layer has been designed based on interfaces and abstract classes that provide the modularity and extensibility of the solution. Each simulation object is a simulation agent and it can be aggregated at any resolution level at the same simulation time. The prepared agent mechanisms allow the exchange of simulation events (so-called statechanges) with DIS or HLA plugins. These plugins are responsible for further messages exchange.

GUI layer (shown in Fig. 2) is intended to manage the simulation execution, to create and manage UAVs and to observe objects from interoperating simulators. All objects are visualised on the two-dimensional map of a terrain and specified in the internal list of objects. For each object, appropriate attributes such as object type, aggregation level and position or level of damage can be checked. GUI also specifies the visibility of objects with regarding to their resolution level and provides the user with information about the messages transmitted between the simulators (communication event log).

The analytics layer is responsible for gathering information and parameters' values of simulated objects. It allows to examine the characteristics of processes and preparing specific measures. Examples of such measures are as follows: task completion status, task's execution time, lost aircrafts' count, total travelled distance, average distance between aircrafts or energy consumption.

SIMULATION ENVIRONMENT

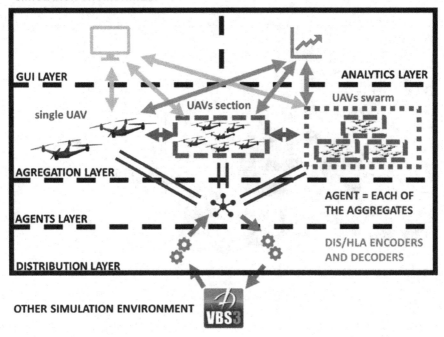

Fig. 1. Concept of UAVs swarm simulation environment

The analytics and the GUI layers interact with the aggregates. The aggregates are the objects immersed in the aggregation layer and characterized by different levels of resolution - from individual UAVs, through sections, to UAV swarms. According to relationships between them, the state change of the object at one level propagates (basing on events or intervals) and determines the state of the aggregate that is in the relationship. Processes of transition to the lower or higher resolution using the specified rules are called, respectively, aggregation and disaggregation. They are presented in the Fig. 3. The proposed approach of maintaining different details' levels of simulated objects and defining rules for processing states into an aggregated or disaggregated models is the field of study of multiresolution modelling. That solution enables management of individual UAVs, entire group or as part of the cooperation of this UAV group with other combat measures (as support for ground operations or air assault measures). Moreover, interoperability with simulators on different levels of command will be realized by passing the state of objects at the appropriate level of resolution. High-resolution entity (HRE-type objects) are very detailed – these can be single UAV, and an object with a high aggregation level (LRE-type object which means low-resolution entity) can be UAV swarm [1]. Further, the environment allows the management of aggregates at abstraction levels going beyond the cooperating simulators, which would be represented only in the GUI layer or in the analytical layer for research purposes.

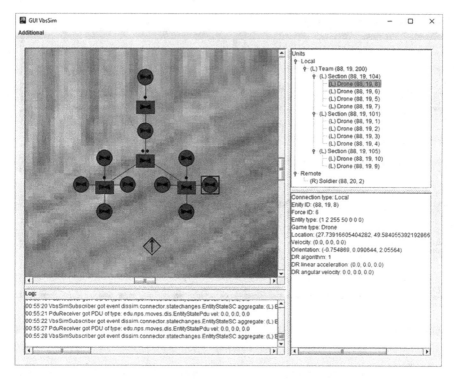

Fig. 2. GUI Layer showing the UAV swarm divided into sections and individual aircrafts

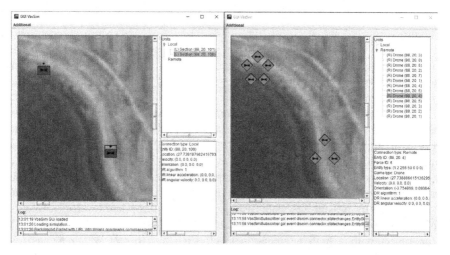

Fig. 3. The process of disaggregation in the aggregation layer – the movement of two sections of the BSP results with movement of individual aircrafts

Each of aggregates from the aggregation layer can be a simulation agent in the agent layer simultaneously. The agent, which is the basis of a multi-agent approach, represents every object participating in the simulation. It has appropriate data structures to store its state and knowledge both about the environment and other agents. In order of using the self-preservation rules and the rules of reacting to external stimuli, there are determined the actions to be taken by the given object. Presented in Fig. 4 reaction to the event of firing is one example of these rules appliance. The multi-agent approach allows for simulation of objects autonomy and modelling the behaviour of individual aggregates from single aircrafts to whole groups. The agent layer interoperates with other simulators by sending events via the event bus (and receiving respectively).

The distribution layer is responsible for the cooperation of the presented environment with other simulators using distributed simulation protocols. These can be simulators controlling different scopes of simulation, however working in one simulation experiment. The most commonly used distributed simulation protocols are *Distributed Interactive Simulation* (DIS) and *High Level Architecture* (HLA)[5]. The main difference between them is the way of exchanging messages: in the case of DIS – directly between simulators using structured PDUs (Packet Data Unit), in the case of HLA – using a central component called RTI (run-time infrastructure), which manages their transmission only to specific simulators.

Fig. 4. Reaction to simulation events in the agent layer – group of 5 UAV is fling in compact order, after firing the aircrafts are being dispersed.

Distributed simulation gives possibility to model UAV behaviours in a wide range of scenarios. Moreover, the approach avoids the necessity of implementation full behavioural models of objects which the UAVs would interact. In the tests of environment, the Virtual Battlespace 3 system (VBS3) has been used for this purpose. Due to the possibility of performing high-resolution virtual simulation characterized by high quality of graphical imaging, VBS3 allows visualization of the UAV group cooperation model. On the other hand, advanced simulation algorithms and interactivity allows to perform UAV interactions with other objects [14]. However, this would not be possible without the

---

[5] Commonly used distributed simulation protocols: DIS (35%), HLA (35%), TENA (16%), CTIA (3%), other (7%) [4].

software encoders and decoders located inside the distribution layer of the environment. At first, events sent from the agent layer are received from the event bus, and then the encoders are responsible for sending them to another simulator in the format corresponding to the chosen protocol - the decoders are responsible for the reverse process. If the DIS protocol is used, the layer is responsible for sending the PDU to the IP addresses specified in the options and receiving incoming packets to the port which the listener is being set on. In the case of the HLA protocol, the layer is responsible for connection with RTI, as well as for subscribing and publishing selected methods and objects of the federation in accordance with the FOM model. In the presented solution, to ensure compatibility with the DIS protocol, we have used FOM model RPR 2.0. Figure 5 shows the state of two cooperating simulators: VBS3 and DisSim working in a common simulation experiment. The right side of the graphic presents the section of five objects held locally in DisSim (blue colour) and two objects from another simulator (red colour). On the left is VBS3 system with the same objects - five UAVs are in the air, while the two UAVs maintained by the VBS3 simulation are on the surface.

Fig. 5. The state of two simulators cooperating with the use of the distribution layer (Color figure online)

2.4   Particular Scenarios Implementation

The DisSim package prepared for performing a distributed event-driven simulation with multiresolution and multiagent modelling approach results with possibility of implementing appropriate scenarios in order to conduct research on UAS. Such a scenario requires: the preparation of objects extending *SimObject* class, defining aggregation rules by implementing the interfaces *IAggreagate* and *IAggregateRule*, using encoders and decoders for the proper PDUs (in the case of DIS Protocol) or interaction and objects handles (in the case of the HLA protocol). It is also necessary to extend the abstract classes for the subscriber and event broker as well as the abstract *EntityManager* class by mapping the prepared *SimObject* classes to the *EntityType* attribute compliant with the DIS protocol. Section 3 describes the test scenarios implemented in simulation environment we have prepared in such a way. The presented research problem is served as a case study in order to present functionality of the concept. A class diagram of prepared scenarios is presented in Fig. 6.

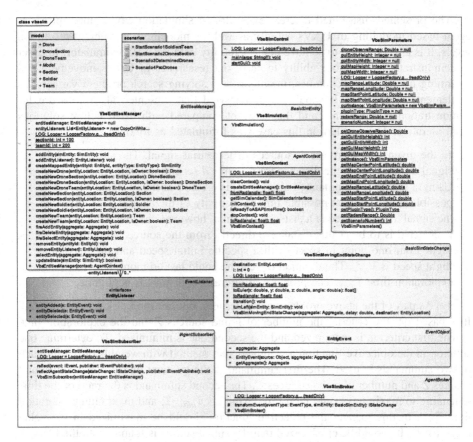

Fig. 6. Class model for simulation scenarios build on DisSim package

3 Biological Inspired Algorithm for Search Problem – Case Study

The search problem is well known problem and has been already used for UAVs missions. The most apparent problem basing on searching is so called Multi-UAV Cooperative Reconnaissance Problem. The UAVs usage in search mission and reconnaissance was formulated also as multiple travelling salesman problem(TSP) solved by tabu search algorithm [12]. It was also solved by simulated annealing problem with finding the shortest path connecting each target and UAV [5] or with A* used to find the shortest path between two points and then task assignment problem [10]. Alike PSO algorithms and its modifications have been already used for UAVs [15]. Moreover, Voronoi diagrams or quadtrees can be used for solving search problem.

The presented distributed discrete event-driven simulation environment provides agent-based modelling of UAVs activity as well as modelling at different resolution levels. It is also prepared to achieve interoperability with the VBS3 simulator. In order to demonstrate the solutions ability to carry out researches of UAVs movement and swarming models and techniques, the algorithms solving search problem variant was implemented. However, It should be emphasized that presented algorithms are for demonstration purpose only and they are not compared with listed before solutions from literature. The problem in this case was formulated as follows:

> In the given area there are K sources of terrestrial signal. The strength of the signal decreases with the distance from source d according to the function $G(d)$. Using a group of L unmanned aircraft vehicles, find as many signal sources as possible in the shortest possible time. Sensitivity of the aircraft sensors allows the signal to be detected at a distance c from it, however, the moment the source is detected is to be at this distance c directly from the source. Aircrafts start a flight from one point in space with a specific initial speed and the maximum flight speed is V_{max}. The duration of search should last no more than maximum exploration time T_{max}. The problem of aircrafts collisions is negligible.

As part of the above problem, a multicriteria optimization task was formulated. It is based on two criteria: the number of found sources Z and the exploration time T. The first criterion is maximized and the second one is minimized. To determine one solution in the objective function a weighted sum method with weights of 0.5 for each of two summands is used. The first summand is the quotient of the number of found sources Z and number of signal sources K. The second summand is the ratio of the time remaining to maximum exploration time, which is $T_{max}-T$, and maximum exploration time T_{max}. In this manner the objective function should be maximized in the value range [0, 1]. It should be emphasized that the signal source is found when the UAV is located at a distance d not larger than the range of the aircraft sensors c rather than when the sensors detect only the signal itself in the area defined by the maximum range of the signal source f (assumed that $c < f$). It means that the maximum value of the function determining the signal strength $G(d)$ is within the range of the aircraft sensors.

In order to solve the problem, three algorithms has been implemented:

- Algorithm I – exact algorithm,
- Algorithm II – PSO algorithm with maximum speed and search area restrictions,
- Algorithm III – PSO algorithm with maximum speed and search area restrictions modified with maximal remembered signal value resetting for given aircraft and for all aircrafts in case of source being detected.

Algorithm I – Exact Algorithm. The exact algorithm is a full overview of the search area by all aircraft. The area is divided into the number of aircraft. Each area segment is checked by one aircraft to check the signal level entirely. Algorithm I has been illustrated in Fig. 7.

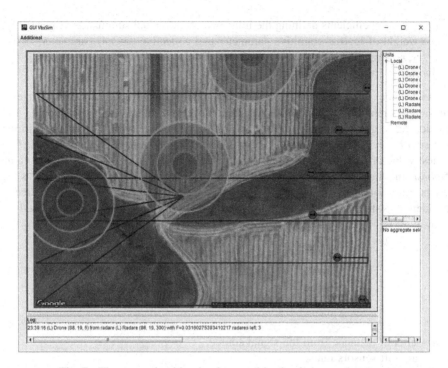

Fig. 7. The exact algorithm implemented in simulation environment

Algorithm II – PSO Algorithm with Maximum Speed and Search Area Restrictions. As a second solution, a modified version of the PSO algorithm with the global neighbourhood was used. As modification the maximum UAV (agent) speed was determined, start point of all UAV was established and space of exploration was limited. In addition, if the designated new position of UAV is beyond the search space, the new velocity (understood as position vector change in time unit) is generated randomly to be in acceptable solutions set.

The formula for the new velocity of the i-th agent in the next step $t + 1$ (in case the new designated position is in the acceptable solutions set) is as follow:

$$v_{t+1}^i = \frac{c_1 r_1 v_t^i + c_2 r_2 \left(x_{pbest}^i - x_t^i\right) + c_3 r_3 \left(x_{gbest} - x_t^i\right)}{\left\|c_1 r_1 v_t^i + c_2 r_2 \left(x_{pbest}^i - x_t^i\right) + c_3 r_3 \left(x_{gbest} - x_t^i\right)\right\|} V_{max} \tag{4}$$

and for new location of the i-th agent in the next step $t + 1$:

$$x_{t+1}^i = x_t^i + v_{t+1}^i \tag{5}$$

x_{pbest}^i – the best position found by the i-th agent,
x_{gbest} – the best position found by a set of neighbours (here: all the aircrafts),

c_1, c_2, c_3 – coefficients defining the influence of individual summands, respectively, the weight of inertia, cognitive parameter and collective (social) parameter, r_1, r_2, r_3 – random values from a uniform distribution U(0, 1). $\|vector\|$ – norm of the vector.

Assessment the aircraft position in the simulation step is conducted with usage of the signal strength function $G(d)$ according to the distance from the source. The function has a great influence on formula for the new velocity of the i-th agent in the next step $t + 1$. Signal strength function is used to determine the value of *pbest* (the highest value of $G(d)$ found by the i-th aircraft and its best-known position) and value of *gbest* (the highest value of the $G(d)$ function found by a set of neighbours and the best position found by them). Moreover, it determines when a given signal source is considered as found (function value equal to 1). Function $G(d)$ is formulated as follows:

$$G(d) = \begin{cases} 1 & ,d \leq c \\ 0 & ,d > f + c \\ 1 - \frac{d-c}{f} & ,c < d \leq f + c \end{cases} \tag{6}$$

d – aircraft distance from the signal source,
c – aircraft sensors range,
f – maximum source signal range.

Figure 8 presents an interpretation of the function $G(d)$ conditions:

(1) Function value is equal to 1, when the source of the signal S is directly in the range of the sensors of the aircraft D (source is found);
(2) Function value is equal to 0, when the range of the aircraft sensors D is beyond the coverage area of the S signal source;
(3) Function values are in the range (0, 1), when the range of sensors of the aircraft D is in the coverage area of the signal source S, but not directly over this source.

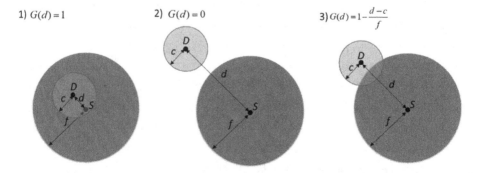

1) $G(d) = 1$ 2) $G(d) = 0$ 3) $G(d) = 1 - \frac{d-c}{f}$

Fig. 8. Interpretation of the function $G(d)$ conditions for PSO algorithms

Algorithm III – Modified PSO Algorithm with Maximal Signal Value Resetting.
As a third solution, the PSO algorithm based on the Algorithm II modifications was
reused. Modification includes limitations on the maximum velocity and the search area,
moreover additionally variable for each signal source determining whether it has
already been found has been applied. This variable determines new conditions, namely,
in each iteration:

- if the new signal source was found by any aircraft, the value of *gbest* and its
  location is zeroed for each aircraft,
- if any aircraft is above the already found source, the value of *pbest* and its location
  is zeroed only for that aircraft.

Owing to such a solution, if any radar is found, the UAV swarm focuses on finding
another one. Figure 9 presents the environment during experiments for a scenario
implementing the modified PSO algorithm.

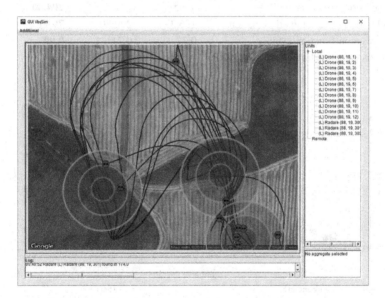

Fig. 9. Twelve aircrafts and three signal sources during experiments for a scenario implementing the modified PSO algorithm.

Algorithms Comparison. In order to compare quality of algorithms, basing on
established model parameters, the criteria values and the value of the target function for
the experiments were calculated. Table 1 presents the results obtained in a series of 10
experiments. It can be noticed that the heuristic algorithm II does not show better
results than the exact Algorithm I, what could have been expected. However, the
applying of further modifications to the heuristic algorithm has conducted in Algorithm III to best results. Nevertheless, it should be noted that such experiments ought to
be carried out for a much larger number of parameters in order to be able to draw more

precise conclusions. Taking into account the results of the PSO algorithm, once in the case of applying only limitations to the search area and maximum speed, and the second time with the above restrictions by adding resetting the values of the local target functions after finding the signal source, the heuristic algorithm have to be well adjusted to the problem model. Otherwise, it can product much worse results than the exact algorithm.

Table 1. Comparison of algorithm quality for the following model parameters: 10 measurements, 4 signal sources, 15 UAVs, maximum exploration time 1000 s, signal range 80 m, range of aircraft sensors 5 m, maximum speed 8 m/s, search area 400×600 m, PSO coefficient weight: $c_1 = 10$, $c_2 = 4$, $c_3 = 1$.

	Algorithm I	Algorithm II	Algorithm III
Found signal sources average	**4,00**	3,80	**4,00**
Found signal sources standard deviation	**0,00**	0,42	**0,00**
Exploration time average	455,98	534,70	**204,80**
Exploration time standard deviation	136,33	290,80	**79,71**
Optimisation function average	0,77	0,71	**0,90**
Optimisation function standard deviation	0,07	0,19	**0,04**

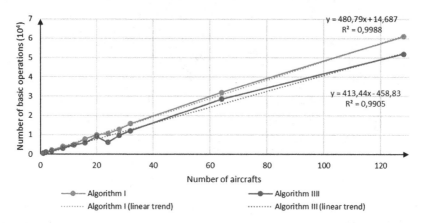

Fig. 10. Graph of the computational complexity of Algorithm I (exact) and Algorithm III (with maximal signal value resetting) at a constant number of signal sources equal to 4 and increasing number of aircrafts.

During the examination of the presented algorithms in the simulation environment, in addition to the comparison of the quality of the algorithms, the computational and memory complexity has been also analysed. Complexity has been calculated in depend of increasing number of UAV, signal sources or maximum exploration time. Considering the concept of iterative calculation of subsequent aircraft positions, the differences in complexity (in general) are not significant and relative to certain dependencies asymptotically aspire to the linear functions such as shown in Fig. 10. However, in each case, biologically inspired Algorithm III has showed better coefficients of these functions than Algorithm I.

4 Conclusion

In the paper the simulation environment was presented as a platform for UAS testing. Moreover, it can be used to study autonomous vehicles, boats (including submarines) or other robots used in an automation of production processes in factories or even processes in agriculture. The environment is a practical consequence of both agent-based modelling of UAV behaviours and modelling them at diverse resolution levels. It also managed to achieve interoperability with the VBS3 simulator with support of distributed simulation protocols. The proposed solution has been used to conduct research on the use of the UAV group to search for sources of terrestrial signal in a given area. The described problem was a case study in order to present the functionality and usefulness of environment. We defined a multi-criteria optimization task based on two criteria: number of found sources, time of exploration. Thereafter, there were proposed three algorithms to solve the problem: one exact algorithm and two heuristic algorithms basing on the PSO algorithm. In order the algorithms were used for demonstration purposes, there were not compared to other solutions. Considering the results of experiments, Algorithm III (involving maximum speed and search area restrictions and modified with maximal remembered signal value resetting for given aircraft and for all aircrafts in case of source being detected) seems to be the best. In case of this algorithm, the lowest value of the average exploration time were obtained as well as the highest value of the optimization function. Moreover, each of experiments ended with finding all signal sources.

The present state of work is actually a promising beginning to further wider research. The presented results show the usability and functionality of the prepared environment for simulation of UAS and problem solving. Further steps and directions of development envisage extension of models and methods (both for the problem presented and for the new problems), strike a balance with maintaining the autonomy of individual objects, and controlling and hierarchizing a group of objects. As part of considering centralized and hierarchical solutions, it should be possible to define a partially determined model (e.g. specifying only the preferred number of aircrafts in the sections) or strongly determined (e.g. a strict determination of the UAVs sections as a part of larger group with provided autonomy of individuals aircrafts or with autonomy of sections with strict determination of position arrangement of UAVs within this section) [13]. In addition, it also seems important to take into account collisions between aircrafts as well as between aircrafts and other objects, so as to explore more heuristic algorithms for UAS appliance (mainly in the field of biologically inspired algorithms). These algorithms shall be examined in a larger range of test cases and parameters. Especially, the interactions with external objects, e.g. air defense measures maintained in VBS3 simulation, should be provided.

References

1. Davis, P.K., Bigelow, J.H.: Experiment in Multiresolution Modeling (MRM). RAND (1998)
2. Dyk, M., Najgebauer, A., Pierzchała, D.: SenseSim: an agent-based and discrete event simulator for wireless sensor networks and the internet of things. In: 2015 IEEE 2nd World Forum on Internet of Things (WF-IoT) Proceedings (2015). ISBN 9781509003655
3. Eberhart, R., Kennedy, K.: A new optimizer using particle swarm theory. In: Proceedings of the Sixth International Symposium on Micro Machine and Human Science, MHS 1995, Nagoya, pp. 39–43 (1995)
4. Gustavsson, P.M., Björkman, U., Wemmergard, J.: LVC aspects and integration of live simulation 09F-SIW-090. In: SISO Fall Simulation Interoperability Workshop (2009)
5. Hutchison, M.G.: A method for estimating range requirements of tactical reconnaissance UAVs. In: AIAA's 1st Technical Conference and Workshop on Unmanned Aerospace Vehicles, Portsmouth, Virginia, pp. 120–124 (2002). https://doi.org/10.2514/6.2002-3438
6. Kückelhaus, M.: Unmanned aerial vehicle in logistics. A DHL perspective on implications and Use cases for the logistics industry, DHL customer solutions & innovation (2014)
7. Munoz, M.F.: Agent-based simulation and analysis of a defensive UAV swarm against an enemy UAV swarm, Naval Postgraduate School, Monterey, California (2011)
8. Najgebauer, A., et al.: The qualitative and quantitative support method for capability based planning of armed forces development. In: Nguyen, N.T., Trawiński, B., Kosala, R. (eds.) ACIIDS 2015. LNCS (LNAI), vol. 9012, pp. 212–223. Springer, Cham (2015). https://doi.org/10.1007/978-3-319-15705-4_21
9. Office of the Secretary of Defense, Unmanned Aircraft Systems Roadmap: 2005–2030, Technical report, Department of Defense (2005)
10. Ousingsawat, J., Campbell, M.E.: Establishing trajectories for multi-vehicle reconnaissance. In: AIAA Guidance, Navigation, and Control Conference and Exhibit, Providence, Rhode Island (2004)
11. Pierzchała, D., Skrzypecki, S.: Multi-agent and multi-resolution distributed simulation DisSim – VBS, Simulation in research and development (in Polish: Symulacja w badaniach i rozwoju), vol. 7 no. 1-2/2016, pp. 25–34 (2016)
12. Ryan, J.L., Bailey, T.G., Moore, J.T., Carlton, W.B.: Reactive tabu search in unmanned aerial reconnaissance simulations. In: Proceedings of the 1998 Winter Simulation Conference, Grand Hyatt, Washington D.C., pp. 873–879 (1998)
13. Tarapata, Z.: Movement simulation and management of cooperating objects in CGF systems: a case study. In: Jędrzejowicz, P., Nguyen, N.T., Howlet, R.J., Jain, L.C. (eds.) KES-AMSTA 2010. LNCS (LNAI), vol. 6070, pp. 293–304. Springer, Heidelberg (2010). https://doi.org/10.1007/978-3-642-13480-7_31
14. Wantoch-Rekowski, R. (Red.): Technologies for designing live and virtual simulators in a programmable virtual simulation environment VBS3 (in Polish: Technologie projektowania trenażerów i symulatorów w programowalnym środowisku symulacji wirtualnej VBS3). Wydawnictwo Naukowe PWN, Warszawa (2016)
15. Xie, R., Wang, X., Wang, X., Wei, H.: UAV flight performance optimization based on improved particle swarm algorithm. In: Su, C.-Y., Rakheja, S., Liu, H. (eds.) ICIRA 2012. LNCS (LNAI), vol. 7506, pp. 396–403. Springer, Heidelberg (2012). https://doi.org/10.1007/978-3-642-33509-9_39
16. Aerial Surveillance & Monitoring, observation with UAV/drone. https://www.microdrones.com/en/industry-experts/security/monitoring/

17. Department of Defense Announces Successful Micro-Drone Demonstration > U.S. DEPARTMENT OF DEFENSE > News Release View. https://www.defense.gov/News/News-Releases/News-Release-View/Article/1044811/department-of-defense-announces-successful-micro-drone-demonstration/
18. DroneBase - Drones and the Future of Entertainment. https://blog.dronebase.com/2017/07/10/drones-and-the-future-of-entertainment/
19. ISR Unmanned Aerial Vehicles: Intelligence, Surveillance, Reconnaissance (ISR). https://www.militaryfactory.com/aircraft/uav-intelligence-surveillance-reconnaissance.asp
20. LOCUST: Autonomous, Swarming UAVs Fly into the Future. http://www.navy.mil/submit/display.asp?story_id=86558
21. Project Perdix | Beaver Works. https://beaverworks.ll.mit.edu/CMS/bw/projectperdixcapstone/
22. Unmanned Aerial Systems (UAS) - SKYbrary Aviation Safety. https://www.skybrary.aero/index.php/Unmanned_Aerial_Systems_(UAS)

Note on Signature of Trident Mechanisms with Distribution Growth Vector (4,7)

Stanislav Frolík[✉]

Faculty of Mechanical Engineering, Brno University of Technology,
Brno, Czech Republic
stanislav.frolik@vutbr.cz

Abstract. We recall some basic concepts of differential geometry and control theory and their application in robotics, namely we describe so called generalized trident snake robot with four control parameters. Indeed, we are working with a model that combines a robotic snake and Doubin car. We determine the controlling distribution and describe its properties. Consequently, we determine the signature corresponding to the mechanisms with four controlling parameters. This is essential for analysis of the underlying algebraic structure and allows us to choose suitable control model.

Keywords: Non-holonomic system · Trident robot ·
Differential geometry · Signature

1 Introduction

In this paper we study a trident mechanism, which is primarily designed for theoretical considerations about its configuration space. We discuss the properties of the configuration space understood as a smooth manifold, [1,11,13], and therefore it is referred to as the configuration manifold.

Configuration manifold depends on each mechanism construction and geometry. The following planar mechanisms will always correspond to seven-dimensional manifold. First three parameters describe its global position in a plane. The three dimensional state vector is composed of coordinates giving a position in x and y axis, respectively, and the amount of rotation w.r.t. the fixed global coordinate system. Remaining four parameters represent an input for mechanism's active elements. Trident robot is a mechanism composed of three snake-like branches, each connected to a root block in its vertex, see [3,6,8,9,12] for further details. For the description of the robotic snake's controllability we refer to [4,5,7].

The branches can be multi-link, assumed that each link has its own passive wheel, which provides footing for the robot. In this paper we examine only single link form of the mechanism. Active elements, which affect controllability, are placed on branches. As active elements it is possible to consider the rotary joints

© Springer Nature Switzerland AG 2019
J. Mazal (Ed.): MESAS 2018, LNCS 11472, pp. 82–89, 2019.
https://doi.org/10.1007/978-3-030-14984-0_7

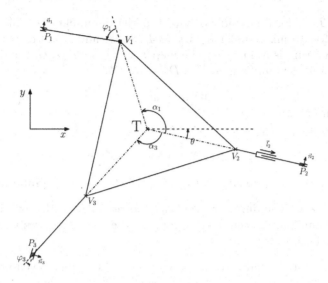

Fig. 1. Scheme of trident mechanism

of branches with the root block, rotary joint of passive wheel with a branch or a piston rod, which can change absolute length of a branch, see Fig. 1. The passive wheel of each link can move only forward or backward. The trident mechanism is a planar system with the non-holonomic constraints expressed by the following system of equations:

$$(\dot{x}_i, \dot{y}_i) \cdot \boldsymbol{n}_i = 0, \tag{1}$$

where $(\dot{x}_i, \dot{y}_i)$ stands for velocity vector of the i-th link in the (x, y) global coordinate system and $\boldsymbol{n}_i$ stands for the vector of i-th wheel's axis. The Eq. (1) is given for each passive wheel. As this paper deals with the single link trident mechanisms only, we obtain a system of three equations. Each equation is said to be in the Pfaff's form and the system is called Pfaff's system.

2   The Theory of Signatures of a Distribution (4,7)

Within this Section we recall the definition of the (4, 7) distribution signature with the notation based on [10]. The curvature of a distribution D is the linear bundle map

$$F : \bigwedge^2 D \to TM/D$$

defined by $F(X, Y) = -[X, Y] \mod D$, where $X, Y \in D$. In other words, if X, Y are local sections of D then

$$F(X \wedge Y) = -[X, Y] \mod D.$$

Write $D^\perp \subset T^*M$ for the bundle of covectors that annihilate D. This is a linear subbundle of the rank complementary to D. It is canonically dual to TM/D. Since the curvature is a linear bundle map $F : \wedge^2 D \to TM/D$, the dual of the curvature is a linear bundle map $F^* : D^\perp \to \wedge^2 D^*$.

Definition 2.1. The linear bundle map $\omega := F^* : D^\perp \to \wedge^2 D^*$ is called the dual curvature map.

Explicitly, if $\lambda_q \in D_q^\perp$ is a covector annihilating D_q, then

$$\omega(\lambda_q)(X,Y) = \lambda_q(-[X,Y]_q)$$

for $X, Y \in D$. To calculate effectively, we recall the following proposition, [10].

Proposition 2.1. The dual curvature map corresponds to the exterior derivative of forms annihilating the distribution and the consequent restriction of the resulting two-form to the distribution.

Let D be a distribution with the growth vector $(4, 7)$ on a seven-dimensional manifold M. Since the dual curvature in each point $q \in M$ is an injective map $\omega : D^\perp \to \wedge^2 D^*$, its image $\mathrm{Im}(\omega_q)$ is a three-dimensional subspace of $\wedge^2 D_q^*$. Let us define the Pfaffian as

$$Pf : D^\perp \to \overset{4}{\bigwedge} D^* : \lambda \to \omega(\lambda)^2.$$

Restricted to the fiber over a fixed $q \in M$, this is a quadratic form Pf_q on the three-dimensional vector space $D_q^\perp$ with values in the one-dimensional real vector space $\wedge^4 D_q^*$. It is the composition of the dual curvature map $\omega : D_q^\perp \to \wedge D_q^*$ with the quadratic map

$$\beta : \overset{2}{\bigwedge} D_q^* \to \overset{4}{\bigwedge} D_q^* : \alpha \to \alpha \wedge \alpha.$$

$\wedge^4 D_q^*$ is a one-dimensional vector space, so we can think of β as a real-valued quadratic form on $\wedge^2 D_q^*$ once we choose a volume form for D_q. With the choice of volume form, Pf_q becomes a real-valued quadratic form on $D^\perp$ and hence we can speak of its index. The index is simply the signature of β restricted to the image of the dual curvature form. This image is a three-dimensional subspace of the space of two-forms, and consequently the possible signatures for Pf_q are $(3, 0)$, $(2, 1)$, $(2, 0)$, $(1, 1)$, $(1, 0)$ and $(0, 0)$ provided that $\wedge^4 D^*$ is not oriented. By "signature" we mean the number of plus signs and minus signs in any diagonalization of the quadratic form w.r.t a given basis, for more details see [2,10].

3   Single Link Planar Mechanism

Let us consider an example of trident mechanism with distribution growth vector $(4, 7)$ as a single link planar mechanism with three branches, each branch has its

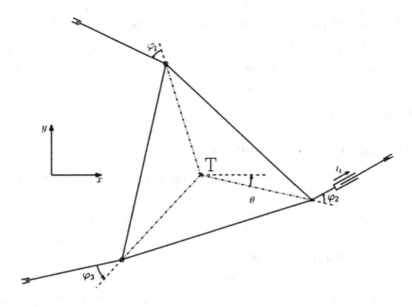

Fig. 2. Scheme of analyzed trident mechanism

own passive wheel as well as it has joint with the root block with active element
- rotary actuator. One branch has a piston rod, which can change its absolute
length, see Fig. 2. The wheels have no active joint to a link.

For such mechanism we obtain the following equations for the velocity vectors
coordinates:

$$\dot{x}_1 = \dot{x} + \sin(\theta + \alpha_1)\dot{\theta} + l_1 \sin(\theta + \alpha_1 + \varphi_1)(\dot{\theta} + \dot{\varphi}_1) + \dot{l}_1 \cos(\theta + \alpha_1 + \varphi_1),$$
$$\dot{y}_1 = \dot{y} - \cos(\theta + \alpha_1)\dot{\theta} - l_1 \cos(\theta + \alpha_1 + \varphi_1)(\dot{\theta} + \dot{\varphi}_1) + \dot{l}_1 \sin(\theta + \alpha_1 + \varphi_1),$$
$$\dot{x}_2 = \dot{x} + \sin(\theta + \alpha_2)\dot{\theta} + \sin(\theta + \alpha_2 + \varphi_2)(\dot{\theta} + \dot{\varphi}_2),$$
$$\dot{y}_2 = \dot{y} - \cos(\theta + \alpha_2)\dot{\theta} - \sin(\theta + \alpha_2 + \varphi_2)(\dot{\theta} + \dot{\varphi}_2),$$
$$\dot{x}_3 = \dot{x} + \sin(\theta + \alpha_3)\dot{\theta} + \sin(\theta + \alpha_3 + \varphi_3)(\dot{\theta} + \dot{\varphi}_3),$$
$$\dot{y}_3 = \dot{y} - \cos(\theta + \alpha_3)\dot{\theta} - \cos(\theta + \alpha_3 + \varphi_3)(\dot{\theta} + \dot{\varphi}_3),$$

where (x, y) stands for the coordinates of the trident's position on a plane and
θ is its absolute rotation. The vectors of the wheel's axis are of the form

$$n_1 = (-\sin(\theta + \alpha_1 + \varphi_1), \cos(\theta + \alpha_1 + \varphi_1)),$$
$$n_2 = (-\sin(\theta + \alpha_2 + \varphi_2), \cos(\theta + \alpha_2 + \varphi_2)),$$
$$n_3 = (-\sin(\theta + \alpha_3 + \varphi_3), \cos(\theta + \alpha_3 + \varphi_3)).$$

From Eq. (1) we derive the following three Pfaff forms.

$$p_1 = -\sin(\theta + \alpha_1 + \varphi_1)\dot{x} + \cos(\theta + \alpha_1 + \varphi_1)\dot{y} + (l_1 + \cos(\varphi_1))\dot{\theta} + l_1\dot{\varphi}_1 = 0,$$
$$p_2 = -\sin(\theta + \alpha_2 + \varphi_2)\dot{x} + \cos(\theta + \alpha_2 + \varphi_2)\dot{y} + (1 + \cos(\varphi_2))\dot{\theta} + \dot{\varphi}_2 = 0,$$
$$p_3 = -\sin(\theta + \alpha_3 + \varphi_3)\dot{x} + \cos(\theta + \alpha_3 + \varphi_3)\dot{y} + (1 + \cos(\varphi_3))\dot{\theta} + \dot{\varphi}_3 = 0.$$

For detailed description see [11,13].

4   The Signature Calculations

First, we provide some necessary calculations. Differential forms of our Pfaff system are of the form

$$
\begin{aligned}
dp_1 = & -\cos(\theta + \alpha_1 + \varphi_1)dx \wedge d\theta - \cos(\theta + \alpha_1 + \varphi_1)dx \wedge d\varphi_1 \\
& -\sin(\theta + \alpha_1 + \varphi_1)dy \wedge d\theta - \sin(\theta + \alpha_1 + \varphi_1)dy \wedge d\varphi_1 + d\theta \wedge dl_1 \\
& -\sin(\varphi_1)d\theta \wedge d\varphi_1 + d\varphi_1 \wedge dl_1,
\end{aligned}
$$
$$
\begin{aligned}
dp_2 = & -\cos(\theta + \alpha_2 + \varphi_2)dx \wedge d\theta - \cos(\theta + \alpha_2 + \varphi_2)dx \wedge d\varphi_2 \\
& -\sin(\theta + \alpha_2 + \varphi_2)dy \wedge d\theta - \sin(\theta + \alpha_2 + \varphi_2)dy \wedge d\varphi_2 - \sin(\varphi_2)d\theta \wedge d\varphi_2,
\end{aligned}
$$
$$
\begin{aligned}
dp_3 = & -\cos(\theta + \alpha_3 + \varphi_3)dx \wedge d\theta - \cos(\theta + \alpha_3 + \varphi_3)dx \wedge d\varphi_3 \\
& -\sin(\theta + \alpha_3 + \varphi_3)dy \wedge d\theta - \sin(\theta + \alpha_3 + \varphi_3)dy \wedge d\varphi_3 - \sin(\varphi_3)d\theta \wedge d\varphi_3.
\end{aligned}
$$

Second, we calculate the wedges $dp_1 \wedge dp_1, dp_2 \wedge dp_2, dp_3 \wedge dp_3$ as follows:

$$
\begin{aligned}
dp_1 \wedge dp_1 = & \cos(\theta + \alpha_1 + \varphi_1)\sin(\theta + \alpha_1 + \varphi_1)dx \wedge d\theta \wedge dy \wedge d\varphi_1 \\
& -\cos(\theta + \alpha_1 + \varphi_1)dx \wedge d\theta \wedge d\varphi_1 \wedge dl_1 \\
& +\cos(\theta + \alpha_1 + \varphi_1)\sin(\theta + \alpha_1 + \varphi_1)dx \wedge d\varphi_1 \wedge dy \wedge d\theta \\
& -\cos(\theta + \alpha_1 + \varphi_1)dx \wedge d\varphi_1 \wedge d\theta \wedge dl_1 \\
& +\sin(\theta + \alpha_1 + \varphi_1)\cos(\theta + \alpha_1 + \varphi_1)dy \wedge d\theta \wedge dx \wedge d\varphi_1 \\
& -\sin(\theta + \alpha_1 + \varphi_1)dy \wedge d\theta \wedge d\varphi_1 \wedge dl_1 \\
& +\sin(\theta + \alpha_1 + \varphi_1)\cos(\theta + \alpha_1 + \varphi_1)dy \wedge d\varphi_1 \wedge dx \wedge d\theta \\
& -\sin(\theta + \alpha_1 + \varphi_1)dy \wedge d\varphi_1 \wedge d\theta \wedge dl_1 \\
& -\cos(\theta + \alpha_1 + \varphi_1)d\varphi_1 \wedge dl_1 \wedge dx \wedge d\theta \\
& -\sin(\theta + \alpha_1 + \varphi_1)d\varphi_1 \wedge dl_1 \wedge dy \wedge d\theta \\
& -\cos(\theta + \alpha_1 + \varphi_1)d\theta \wedge dl_1 \wedge dx \wedge d\varphi_1 \\
& -\sin(\theta + \alpha_1 + \varphi_1)d\theta \wedge dl_1 \wedge dy \wedge d\varphi_1 \\
= & \ 0,
\end{aligned}
$$
$$
\begin{aligned}
dp_2 \wedge dp_2 = & \cos(\theta + \alpha_2 + \varphi_2)\sin(\theta + \alpha_2 + \varphi_2)dx \wedge d\theta \wedge dy \wedge d\varphi_2 \\
& +\cos(\theta + \alpha_2 + \varphi_2)\sin(\theta + \alpha_2 + \varphi_2)dx \wedge d\varphi_2 \wedge dy \wedge d\theta \\
& +\sin(\theta + \alpha_2 + \varphi_2)\cos(\theta + \alpha_2 + \varphi_2)dy \wedge d\theta \wedge dx \wedge d\varphi_2 \\
& +\sin(\theta + \alpha_2 + \varphi_2)\cos(\theta + \alpha_2 + \varphi_2)dy \wedge d\varphi_2 \wedge dx \wedge d\theta \\
= & \ 0,
\end{aligned}
$$

$$dp_3 \wedge dp_3 = \cos(\theta + \alpha_3 + \varphi_3)\sin(\theta + \alpha_3 + \varphi_3)dx \wedge d\theta \wedge dy \wedge d\varphi_3$$
$$+ \cos(\theta + \alpha_3 + \varphi_3)\sin(\theta + \alpha_3 + \varphi_3)dx \wedge d\varphi_3 \wedge dy \wedge d\theta$$
$$+ \sin(\theta + \alpha_3 + \varphi_3)\cos(\theta + \alpha_3 + \varphi_3)dy \wedge d\theta \wedge dx \wedge d\varphi_3$$
$$+ \sin(\theta + \alpha_3 + \varphi_3)\cos(\theta + \alpha_3 + \varphi_3)dy \wedge d\varphi_3 \wedge dx \wedge d\theta$$
$$= 0.$$

Because these wedges are equal to zero, their restrictions to $D^\perp$ are also zero. As we can see, the diagonal elements of the quadratic form's matrix are always zero. From the calculation we can see that these diagonal elements depend on the construction of the mechanism. Below we examine the non-diagonal elements of the matrix expressed as $dp_1 \wedge dp_2$, $dp_1 \wedge dp_3$, $dp_2 \wedge dp_3$. It is a direct computation of the wedge of two 2-forms and the consequent construction of a 4-form and therefore we provide the results in a truncated form only.

$$dp_1 \wedge dp_2 = \cos(\theta + \alpha_1 + \varphi_1)\sin(\theta + \alpha_2 + \varphi_2)dx \wedge d\theta \wedge dy \wedge d\varphi_2$$
$$+ \cos(\theta + \alpha_1 + \varphi_1)\sin(\theta + \alpha_2 + \varphi_2)dx \wedge d\varphi_1 \wedge dy \wedge d\theta$$
$$+ \cos(\theta + \alpha_1 + \varphi_1)\sin(\theta + \alpha_2 + \varphi_2)dx \wedge d\varphi_1 \wedge dy \wedge d\varphi_2$$
$$+ \cos(\theta + \alpha_1 + \varphi_1)\sin(\varphi_2)dx \wedge d\varphi_1 \wedge d\theta \wedge d\varphi_2$$
$$\cdots$$
$$+ \sin(\varphi_2)\sin(\theta + \alpha_2 + \varphi_2)d\theta \wedge d\varphi_1 \wedge dy \wedge d\varphi_2$$
$$- \cos(\theta + \alpha_2 + \varphi_2)d\varphi_1 \wedge dl_1 \wedge dx \wedge d\theta$$
$$- \cos(\theta + \alpha_2 + \varphi_2)d\varphi_1 \wedge dl_1 \wedge dx \wedge d\varphi_2$$
$$- \sin(\theta + \alpha_2 + \varphi_2)d\varphi_1 \wedge dl_1 \wedge dy \wedge d\theta$$
$$- \sin(\theta + \alpha_2 + \varphi_2)d\varphi_1 \wedge dl_1 \wedge dy \wedge d\varphi_2$$
$$- \sin(\varphi_2)d\varphi_1 \wedge dl_1 \wedge d\theta \wedge d\varphi_3,$$

$$dp_1 \wedge dp_3 = \cos(\theta + \alpha_1 + \varphi_1)\sin(\theta + \alpha_3 + \varphi_3)dx \wedge d\theta \wedge dy \wedge d\varphi_3$$
$$+ \cos(\theta + \alpha_1 + \varphi_1)\sin(\theta + \alpha_3 + \varphi_3)dx \wedge d\varphi_1 \wedge dy \wedge d\theta$$
$$+ \cos(\theta + \alpha_1 + \varphi_1)\sin(\theta + \alpha_3 + \varphi_3)dx \wedge d\varphi_1 \wedge dy \wedge d\varphi_3$$
$$+ \cos(\theta + \alpha_1 + \varphi_1)\sin(\varphi_3)dx \wedge d\varphi_1 \wedge d\theta \wedge d\varphi_3$$
$$\cdots$$
$$- \sin(\theta + \alpha_3 + \varphi_3)d\varphi_1 \wedge dl_1 \wedge dy \wedge d\theta$$
$$- \sin(\theta + \alpha_3 + \varphi_3)d\varphi_1 \wedge dl_1 \wedge dy \wedge d\varphi_3$$
$$- \sin(\varphi_3)d\varphi_1 \wedge dl_1 \wedge d\theta \wedge d\varphi_3,$$

$$dp_2 \wedge dp_3 = \cos(\theta + \alpha_2 + \varphi_2)\sin(\theta + \alpha_3 + \varphi_3)dx \wedge d\theta \wedge dy \wedge d\varphi_3$$
$$+ \cos(\theta + \alpha_2 + \varphi_2)\sin(\theta + \alpha_3 + \varphi_3)dx \wedge d\varphi_2 \wedge dy \wedge d\theta$$
$$+ \cos(\theta + \alpha_2 + \varphi_2)\sin(\theta + \alpha_3 + \varphi_3)dx \wedge d\varphi_2 \wedge dy \wedge d\varphi_3$$
$$+ \cos(\theta + \alpha_2\varphi_2)\sin(\varphi_3)dx \wedge d\varphi_2 \wedge d\theta \wedge d\varphi_3$$
$$\dots$$
$$+ \sin(\theta + \alpha_2 + \varphi_2)\sin(\varphi_3)dy \wedge d\varphi_2 \wedge d\theta \wedge d\varphi_3$$
$$+ \sin(\varphi_2)\cos(\theta + \alpha_3 + \varphi_3)d\theta \wedge d\varphi_2 \wedge dx \wedge d\varphi_3$$
$$+ \sin(\varphi_2)\sin(\theta + \alpha_3 + \varphi_3)d\theta \wedge d\varphi_2 \wedge dy \wedge d\varphi_3.$$

If we restrict from T_pM to the distribution D and use Maple software we obtain the following elements of the Pfaffian:

$$Pf(1,2) = -\frac{1}{l_1}\sin(\theta + \frac{\pi}{3} + \alpha_1 + \varphi_1)\sin(\theta),$$
$$Pf(1,3) = -\frac{1}{l_1}\sin(\theta + \frac{\pi}{3} + \alpha_1 + \varphi_1)\cos(\theta + \frac{\pi}{6}),$$
$$Pf(2,3) = 0.$$

Finally we observe that several non-diagonal elements of the matrix are non-zero. Should we discuss their existence let us first state that it is independent of the parameter l_1 because it will always be non-zero (which refers to non-zero length of branch (1)), so the quadratic form always exists. Eigenvalues of the form's matrix are

$$\lambda_1 = 0,$$
$$\lambda_2 = -\lambda_3.$$

With respect to the matrix's eigenvalues, the signature of the distribution D is (1, 1). In the future we will consider mechanisms with one branch passive. This means that it has no active elements, neither rotary joint with a root block nor a piston rod etc. The other two branches would have a one-and-three parameter construction or two-and-two parameter construction. The discussion of the signature of the mechanism is important for understanding the controllability of mechanisms controlled by four parameters.

Acknowledgement. This research was supported by a grant of the Czech Science Foundation no. 17-21360S, "Advances in Snake-like Robot Control" and by a Grant No. FSI-S-17-4464.

References

1. Agracev, A., Barilari, D., Boscain, U.: Introduction to Riemannian and sub-Riemannian geometry. Preprint SISSA (2016)
2. De Zanet, C.: Generic one-step bracket-generating distributions of rank four. Archivum Mathematicum **51**(5), 257–264 (2015)

3. Hrdina, J.: Local controllability of trident snake robot based on sub-Riemannian extremals. Note di Matematica **37**(suppl. 1), 93–102 (2017)
4. Návrat, A., Vašík, P.: On geometric control models of a robotic snake. Note di Matematica **37**, 119–129 (2017)
5. Hrdina, J., Návrat, A., Vašík, P., Matoušek, R.: CGA-based robotic snake control. Adv. Appl. Clifford Algebr. **27**, 621–632 (2017)
6. Hrdina, J., Návrat, A., Vašík, P., Matoušek, R.: Geometric control of the trident snake robot based on CGA. Adv. Appl. Clifford Algebr. **27**, 621–632 (2017)
7. Hrdina, J., Návrat, A., Vašík, P.: Control of 3-link robotic snake based on conformal geometric algebra. Adv. Appl. Clifford Algebr. **26**, 1069–1080 (2016). https://doi.org/10.1007/s00006-015-0621-2
8. Ishikawa, M.: Trident snake robot: locomotion analysis and control. In: NOL-COS, IFAC Symposium on Nonlinear Control Systems, vol. 6 (2004)
9. Ishikawa, M., Minami, Y., Sugie, T.: Development and control experiment of the trident snake robot. IEEE/ASME Trans. Mechatron. **15**(1), 9 (2010)
10. Montgomery, R.: A Tour of Subriemannian Geometries, Their Geodesics and Applications. Mathematical Surveys and Monographs, p. 259. AMS, Providence (2002)
11. Murray, R.M., Zexiang, L., Sastry, S.S.: A Mathematical Introduction to Robotic Manipulation. CRC Press, Boca Raton (1994)
12. Pietrowska, Z., Tchoń, K.: Dynamics and motion planning of trident snake robot. J. Intell. Robot. Syst. **75**(1), 17–28 (2014)
13. Selig, J.M.: Geometric Fundamentals of Robotics. Monographs in Computer Science. Springer, New York (2004). https://doi.org/10.1007/b138859

A Versatile Visual Navigation System for Autonomous Vehicles

Filip Majer[1] , Lucie Halodová[1] , Tomáš Vintr[1] , Martin Dlouhý[2] ,
Lukáš Merenda[3] , Jaime Pulido Fentanes[4] , David Portugal[5] ,
Micael Couceiro[5] , and Tomáš Krajník[1(✉)]

[1] Artificial Intelligence Center, Czech Technical University, Prague, Czech Republic
{majerfil,tomas.krajnik}@fel.cvut.cz
[2] Czech University of Life Sciences Prague, Prague, Czech Republic
[3] VOP CZ, Šenov u Nového Jičína, Czech Republic
[4] Lincoln Center for Autonomous Systems, University of Lincoln, Lincoln, UK
[5] Ingeniarius, Ltd., Coimbra, Portugal

Abstract. We present a universal visual navigation method which allows a vehicle to autonomously repeat paths previously taught by a human operator. The method is computationally efficient and does not require camera calibration. It can learn and autonomously traverse arbitrarily shaped paths and is robust to appearance changes induced by varying outdoor illumination and naturally-occurring environment changes. The method does not perform explicit position estimation in the 2d/3d space, but it relies on a novel mathematical theorem, which allows fusing exteroceptive and interoceptive sensory data in a way that ensures navigation accuracy and reliability. The experiments performed indicate that the proposed navigation method can accurately guide different autonomous vehicles along the desired path. The presented system, which was already deployed in patrolling scenarios, is provided as open source at www.github.com/gestom/stroll_bearnav.

1 Introduction

Considerable progress in sensor development, machine perception and scene understanding along with increasing computational power of embedded devices had a significant impact on the development of unmanned vehicles. Autonomous cars, which are able to navigate in structured environments of highways and roads and react appropriately to standard traffic situations, are already on the market. While these cars are using active sensors like radars and lidars for obstacle detection and avoidance, their ability to understand the environment is based on their capability to accurately interpret imagery from their on-board cameras.

The work has been supported by the Czech Science Foundation project 17-27006Y and by the Segurancas roboTicos coOPerativos (STOP) research project (CENTRO-01-0247-FEDER-017562), co-funded by the Agencia Nacional de Inovacao within the Portugal2020 programme.

© Springer Nature Switzerland AG 2019
J. Mazal (Ed.): MESAS 2018, LNCS 11472, pp. 90–110, 2019.
https://doi.org/10.1007/978-3-030-14984-0_8

Thus, computer vision methods constitute the core of the autonomous navigation systems of these vehicles. One of the main advantages of the cameras is their small size and consumption, which makes them ideal for deployment on small robots with limited payload. Many of these robots cannot carry expensive and bulky sensors like 3d or 2d lidars, and they have to rely solely on vision systems.

The paper [11] divides the vision systems for mobile robotics into map-less, map-based and map-building based. Map-less systems recognise traversable structures in the environment (such as roads or highway lanes) and use this information to calculate motion commands for autonomous vehicles [10]. These systems have reached a considerable maturity, and they have already demonstrated their ability to be deployed in real cars quite some time ago [36,37]. However, not all environments have a clear, easy-to-recognise structure such as roads and highways, which clearly define the directions for driving. To be able to navigate to desired locations in environments without a straightforward structure, robots use environment representations (maps) which, with the help of on-board sensors, allow them to estimate their positions. According to [11], navigation systems which require that these maps are provided apriori are called "map-based" systems. An example of such system is a method that uses a CAD-like model of a building to allow navigation in indoor environment [19]. Since an accurate model of the operational environment is often not available, Behnzadian et al. [3] propose a localisation scheme, which can deal with coarse, hand drawn maps. Another way to deal with the lack of prior maps is to endow the robots with an ability to build these maps themselves – these systems are referred to as "map-building" ones. Some of these systems employ methods commonly called visual SLAM (Simultaneous Localisation and Mapping) [12,18,29], which are able to build maps and localise the robots at the same time. The maps built by these systems represent the space in a metric way, and they can be passed to the localisation and planning submodules of the mobile robots, which then attain the ability to move across all of the mapped space. Another class of map-building methods do not represent the mapped space metrically or in a globally consistent way. These systems typically cannot move across all of the mapped space, but are constrained to the vicinity of the path they have been taught during the mapping. While somewhat limited in versatility, these techniques (called 'map-and-replay') have demonstrated their ability to support long-term operation of mobile robots [8,24,31,35]. Furthermore, they do not require skilled personnel to be set up – one has to simply guide a robot along a desired path by means of teleoperation.

Several researchers have proposed different schemes of visual teach-and-replay navigation [5,7,28,33,34]. Some of the systems build metric, local maps, which do not have to be globally consistent. For example [14] proposed to use traditional SLAM techniques to build a series of overlapping local maps and switch between them on-the fly. Similarly [33] build a map off-line from the camera feed gathered during a human-guided drive and use it later on for localisation. Other methods use even simpler schemes for teach-and-replay, which rely more on visual servoing rather than metric maps, e.g. [5,28] create a visual memory path by simply storing the images the robot has encountered during a teleoperated drive. Similarly, [34] represents the taught path as a sequence of locations, associated with a set of salient visual features. To move between these locations, a robot can use a relatively simple feature association and visual servoing. Another, even simpler method is proposed by [7], who extract salient features from the onboard camera and associate these with different segments of the taught path. When navigating a given segment autonomously, their robot moves forward and steers according to the relative positions of the currently recognised and already mapped features. A segment end is detected by means of comparing its last mapped image with the current view. The paper [35] mathematically and experimentally proves that a robot navigating along a previously recorded polygonal route does not need a globally consistent map or explicit position estimation. Later work, described in [22], extends the mathematical proof provided in [35], which allows deployment of the navigation to a wider class of mobile robots.

Many of the teach-and-repeat systems are reported to be robust enough to deal with realistic outdoor conditions and environment changes, which makes them suitable for long-term deployment of autonomous robots. The robustness of these systems to environment variations is often improved by their ability to adapt to the environment changes during the course of robot deployment, i.e. when the robot repeatedly traverses the desired route. For example, [17,21] propose to use trainable image feature descriptors and [15,31] employ the dynamic maps [4,8,9] to adapt the environment models and perception methods to the nature of the changes observed. A paper by Halodová et al. demonstrates that the employment of adaptive methods throughout the visual teach-and-repeat navigation pipeline [16] significantly improves the accuracy and robustness of mobile robot navigation.

Inspired by the robustness of the systems described in [14,31] and the simplicity of the method proposed in [35], we implemented a versatile visual navigation system with provable navigation stability. The main advantage of the system is that it does not require precisely calibrated, high-resolution cameras or accurate odometry, but it works with off-the shelf equipment. This not only makes the system low-cost, but it also significantly reduces the time required for its integration into a given robotic platform. Furthermore, given certain conditions (described in [22,35]) the system ensures that the navigated vehicles do not deviate from the taught trajectory even in low-visibility conditions. However, the system is based on a theoretical model [22], which is quite coarse and it neglects the type of the robot kinematics. The simplicity of the model and proof presented in [22] causes doubts regarding the system's ability to be deployed to various classes of autonomous vehicles in diverse environments and the soundness of the aforementioned proof. Moreover, the experiments presented in [22] were constrained in both environment size and number of platforms and the experimental evidence provided in [22] was not compelling for audience familiar with field robotics.

Thus, in this paper, we deploy the visual navigation system proposed in [22] to six different mobile vehicles with different type of kinematics and perform experiments evaluating the correctness of the mathematical proof presented in [22]. We do not focus on the theoretical model and mathematical proof and aim to provide an engineer's perspective of the system described. Apart from the experimental evaluation of the system navigation accuracy, we provide details regarding the time required to deploy the system on various platforms. We hope that the experiments, which demonstrate that the visual navigation can guide toy-like robots as well as military UGVs, will be convincing enough, so that the validity of the proof presented in [22] will be accepted with more confidence.

2 Navigation Method Description

As mentioned before, the method presented here combines two separate phases: teach and replay (or repeat). During the teach phase, a human operator manually drives the robot through the environment along the desired path. After that, the robot is able to repeat the taught path. To achieve the ability to repeat the path autonomously, the robot creates a map by processing its on-board camera image and storing the image features it saw along the path. Furthermore, the robot stores the commands issued by the operator during the teaching phase. Thus, the only inputs the system requires for both teach and replay are odometry and monocular camera image. The system can dynamically change between different feature extraction algorithms depending on the computational power available. A typical choice is the upright-SURF (Upright Speeded Up Robust Features) [2], Binary Robust Independent Features (BRIEF) [6] or Adaptive and Generic Accelerated Segment Test (AGAST) [26].

2.1 Environment Representation

The way the navigation system represents the environment determines its function, accuracy and robustness. In our case, the environment is represented from a robot-centric way as a linear topological structure, where the taught path is represented as a set of discrete, local maps that contain the images captured by the robot's on-board camera along with the extracted features. Each local map is associated with the distance from the path start and with the commands issued by the robot operator during the teaching phase. The Fig. 1 shows an example of the environment representation: let us have a robot was driven through the environment along the blue line, while using odometry to measure the distance it travelled so far. Each time the robot operator issues a new command (i.e. he changes the forward or angular velocity), the robot stores this command along with the currently travelled distance. Moreover, at the beginning of the teaching and each time the distance exceeds a certain interval (typically 0.1–0.5 m) from the last local map, the robot saves the current image and its features into a separate local map. These local maps are represented as red dots in the Fig. 1.

2.2 Image Processing

The purpose of the feature extraction is to detect image components, which will help to re-identify a location when the robot captures an image of the

Fig. 1. The taught path is represented by odometric information (blue) and a series of local maps (red dots). The maps contain captured images and their features. (Color figure online)

same location next time. Extraction of image features is a crucial component of the navigation, because it determines what kind of information about the environment is stored in the local maps and what information is then used to steer the robot. Extracting the image features from the camera works in two steps, keypoint detection and feature description. The keypoint detector locates areas (or points) in the image, which are easy to recognise even if the given scene will be viewed from another viewpoint. Typically, the keypoints are high-contrast corners or blobs, which are localised by means of Hessian matrix analysis (in the case of SURF features [2]) or by a direct comparison of a pixel's brightness value with its neighbourhood [26]. The feature descriptor characterises the detected keypoint surroundings. In the case of the SURF algorithm [2], the 64-floating number descriptive vector comprises of intensity gradients around the detected keypoint. The BRIEF descriptor [6] is a 256-bit long binary string calculated by comparing the intensities of 256 pixel pairs around the keypoint. One of the main challenges of outdoor operation is unstable illumination which strongly influences both the detection and description phases of image processing. In order to cope with the illumination changes, our image acquisition and processing modules are actively adapting their parameters (such as camera exposure and Hessian threshold) to achieve high-contrast images and to extract the desired number of image features [16].

2.3 Teaching the Path

During the teaching phase, the robot is driven by means of a joystick or other device along a path, which it is supposed to autonomously traverse later on.

When the teaching starts, the robot sets its distance counter to zero and saves the current image and its features into a local map, which is indexed by distance of zero. As soon as the operator sets the forward or angular velocity, the robot associates these with the zero distance as well, saves the command to the map and starts to move while measuring the travelled distance by odometry. Whenever the operator issues new angular or forward velocity, the robot saves these to the map along with the current distance information. The robot also measures the distance travelled since the last local map recording and whenever this distance exceeds a certain threshold, the robot creates a new local map, associates it with the distance travelled since the teaching started and fills it with the last captured image and its features. When the operator indicates that the teaching phase is over, the robot saves the stop command with the current distance, creates the last local map and provides the operator with basic statistics of the maps created. At the end of the teaching phase, the map should have a similar structure as the one shown in Fig. 1.

The resulting map is neither metrically accurate, nor globally consistent, because for self-localisation, the robot is using only odometry which is subject to drift. However, this map creation process does not aim to create metric, globally-consistent environment representation, because that is not necessary to achieve autonomous operation [14]. Rather, the mapping aims to capture only the information necessary for subsequent autonomous traversals.

2.4 Autonomous Navigation

As soon as the teaching phase is over, the operator can initiate the replay phase, where the robot navigates autonomously. The operator indicates which map to load and places the robot close to the intended path start. The method loads all the saved local maps with image features and operator commands – both commands and maps are indexed by their distance from the path start. Then, it loads the commands and local map stored at zero distance. It applies the commands, and as it moves forwards, it eventually reaches a distance where new commands were issued during teaching, and applies them further. Essentially, the robot replays the commands from the teaching phase.

However, if it would only replay these commands, its trajectory would not be the same as during teaching for several reasons. Firstly the initial angle, where the robot starts cannot be exactly identical to the one during teaching and even slightly different angle would cause the robot to diverge from the desired path after some time. Secondly, the wheel- or track-surface interaction is a source of significant uncertainty due to slippage, variable surface and tire pressure etc. Thus, the robot needs to use the local maps to correct for the aforementioned sources of position inaccuracy. This is typically performed by associating the image features from the map to the image features of the current view and using these associations to determine the robot position relatively to the local map. The relative position is then fused with the odometric information by a Kalman or particle filter, obtaining a robot position estimate.

Fig. 2. Navigation principle: a robot at a given distance from the path start (white circle) select the closest map (red circle) and established correspondences between the visible and map features. Difference between the horizontal coordinates of the feature pairs allows to determine the robot steering velocity (shown as red arrows). (Color figure online)

However, accurate and reliable position estimation usually requires that the number of established associations is relatively high, most of correspondences are correct, the features are evenly distributed across the image, they do not lie on planar surfaces or deformable objects, features are not too far away and the images are not blurred or deformed due to rolling shutter effects and the camera is well calibrated. In the real world, many of these assumptions are violated, which causes standard visual-based position estimation to be inaccurate or to fail occasionally. On the other hand, if one needs to recover only the camera heading relatively to the local map, most of the aforementioned conditions do not have to be met, because the heading estimation techniques are simpler than in the case of full 6d position estimation. The situation is even simpler for the case of ground robots, which need to estimate the heading in one direction only, because they move on (locally) planar surfaces. Moreover, the papers [22,35] provide mathematical and experimental evidence, which indicates that unlike for other navigation types, heading estimation in teach-and-repeat navigation can keep the robot position error bound.

This allowed us to implement a simpler method of using the map information to keep the robot close to the taught path. Our system uses the travelled distance information to load an appropriate local map. Then, we extract the features from the current camera view and associate them with the features from the map using the ratio match method proposed in [25]. After that, we find the most frequent horizontal displacement of the mapped and currently visible features using a histogram voting scheme. This displacement, which corresponds to the robot lateral divergence from the taught path, is then fed into the robot steering controller so that the robot turns to keep this displacement close to zero, see Fig. 2. In this way, the robot is always steered towards the taught path.

2.5   Navigation System Model

From the aforementioned description, it's clear that the robot is able to correct the lateral position error as it moves along the path due to the heading corrections. However, the correction of the longitudinal error component through the heading correction is not straightforward as it's somewhat counter-intuitive. We will show that if the robot moves along a path that is not just a straight line, the heading correction reduces both longitudinal and lateral error.

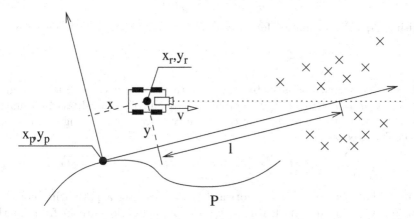

Fig. 3. Robot position error chart. The robot at a position x_r, y_r uses a local feature map of the taught path at the position x_p, y_p. Courtesy of [22].

The basic position evolution model is shown in Fig. 3. Let us assume that a robot navigated by our system is at position x_r, y_r and that it assumes that it's located on the taught path at position x_p, x_p. Thus, the robot position error x, y equals to position x_r, y_r in the coordinate system of the reference point x_p, y_p on the taught path. The evolution of the error over time $\dot{x}, \dot{y}$ is affected by the movement of the local coordinate system along the taught path and the movement of the robot itself. As the robot moves forwards with velocity v, so does the reference point (the path is indexed by travelled distance). Furthermore, the reference system rotates with a velocity given by the local path curvature κ and reference point velocity v. Thus, the position error evolution model is

$$\dot{x} = + \kappa v y - v + v \cos(\varphi) + s_x$$
$$\dot{y} = - \kappa v x \quad\quad + v \sin(\varphi) + s_y,$$
(1)

where $\kappa v y$, $\kappa v x$ and $-v$ are caused by the rotation and translation of the local coordinate system as the reference point x_p, y_p moves along the taught path,

φ is the robot heading in the local coordinate frame, the terms $+v\cos(\varphi)$ and $+v\sin(\varphi)$ reflect the robot movement, and s_x and s_y are random variables, which represent perturbations caused by odometric errors, wheel slippage etc. Since the histogram voting method turns the robot to keep the local map features at the same positions as seen from the reference point during the teaching, the robot's orientation φ in the path reference frame is determined by the average distance of these features l and the robot lateral displacement from the path y, see Fig. 3. In particular, the robot steers so that $\varphi = -\arctan(y/l)$. Considering that the distance of the landmarks l is much higher than the absolute value of the robot lateral displacement y, the robot orientation φ can be approximated as $-y\,l^{-1}$. Thus, one can approximate $v\sin(\varphi)$ by $-v\,y\,l^{-1}$ and $v\cos\varphi$ by v, and rewrite Eq. (1) as

$$\begin{aligned}\dot{x} &= +\,\kappa\,v\,y - v + v && + s_x \\ \dot{y} &= -\,\kappa\,v\,x && -\,v\,y\,l^{-1} + s_y.\end{aligned} \qquad (2)$$

Rewriting Eq. (1) in a matrix form results in

$$\begin{pmatrix}\dot{x} \\ \dot{y}\end{pmatrix} = \begin{pmatrix}0 & +\kappa\,v \\ -\kappa\,v & -v\,l^{-1}\end{pmatrix}\begin{pmatrix}x \\ y\end{pmatrix} + \begin{pmatrix}s_x \\ s_y\end{pmatrix}, \qquad (3)$$

which is a linear continuous system in the form of $\dot{\mathbf{x}} = \mathbf{A}\mathbf{x} + \mathbf{s}$. Such system is stable, i.e. x, y do not diverge, if the real component of eigenvalues of its matrix $\mathbf{A}$ are lower than 0. Factoring out the velocity v from $\mathbf{A}$, the eigenvalues $\lambda_{0,1}$ can be calculated as:

$$\begin{pmatrix}\lambda_0 \\ \lambda_1\end{pmatrix} = v\operatorname{eig}\begin{pmatrix}0 & +\kappa \\ -\kappa & -l^{-1}\end{pmatrix} = \frac{1}{2}v\left(-l^{-1} \pm \sqrt{l^{-2} - 4\kappa^2}\right). \qquad (4)$$

Assuming that the robot moves forward ($v > 0$) along a path with non-zero curvature ($\kappa \neq 0$) and the features it uses for navigation are in front of the robot at a finite distance ($0 < l < \infty$), the real parts of eigenvalues $\lambda_{0,1}$ are always smaller than 0. *Thus, if the robot moves forwards along a curved line, both longitudinal (x) and lateral (y) components of its position error do not diverge.* $\qquad\square$

The navigation model is rather crude and it reflects the real robot behaviour only in cases, where the landmark distance l is much larger than the absolute value of the lateral position error y. However, assuming $l \gg |y|$ allows to create a linear model and determine its stability through eigenanalysis. Another important assumption is that the x component of the robot position error is not large enough to cause the robot to use a wrong map for its heading correction – this corresponds to small values of the random variable s_x. Finally, we assume that the robot can control its heading so that it sees the features almost at the same spots as during the mapping phase, i.e. the absolute values of s_y are small as well.

Due to the odometry, vision and other errors, encompassed in the variables s_x, s_y, the robot position error will not be zero. Rather, it will stabilise at some value, which will be proportional to the size of perturbances s_x, s_y and inversely proportional to the landmark distance l. For a detailed proof and discussion of the model, see [22, 35].

2.6 System Implementation

To allow its easy portability and use by other research teams, we integrated our method into the Robotic Operating System (ROS). The core scheme is shown in Fig. 4. The *odometry_monitor* node reads the data from the odometry and computes travelled distance, which is transferred to the *map_preprocessor*, *navigator* and *mapper* nodes. The *mapper* node saves the current image with extracted features, which are provided by the *feature_extraction* node into the *map_storage*. It also records the speeds set by the operator. The *feature_extraction* node is grabbing the images from the on-board camera, extracts the features, and sends them to *mapper* and *navigator* nodes. The *map_preprocessor* node is used during the navigation phase, see Sect. 2.4. At the start, the *map_preprocessor* node preloads all relevant local maps and operator commands. Then, based on the information provided by the *odometry_monitor*, it forwards the corresponding commands and local map to the *navigator*. The *navigator* receives the currently visible features from the *feature_extraction* node, obtains the local map from the *map_preprocessor* and performs the histogram voting described in Sect. 2.4. The result of the histogram voting is then converted into a steering correction, which is added to the angular speed provided by the *map_preprocessor*. The resulting speeds are sent to the *robot*, which repeats the original commands from the teaching phase, while correcting its heading so that it stays on the path.

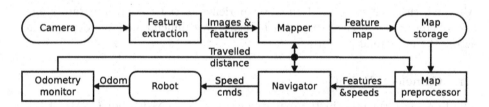

Fig. 4. Navigation system structure: ROS nodes and important topics.

All the aforementioned modules were implemented as ROS action servers with dynamically reconfigurable parameters, so that the robot operator can change their parameters during runtime or activate and deactivate the modules in case they are not necessary to run during the teaching or replay phase. Action servers also provide the operator with feedback which shows the current state of the system. Thus, the operator can see the currently used map, path profile data, number of visible image features, results of the histogram voting method etc. The described system is available as an open source code at [27].

3 Experimental Evaluation

The purpose of the experiments is to verify the ability of the navigation system to keep the autonomous vehicles on the taught path. The system is based on the model introduced in Sect. 2.5, which predicts that a robot driven by our navigation system is able to gradually reduce its position error and thus, keep it within acceptable bounds. The aim of the experimental setup for each autonomous vehicle, which we were working with, was to verify the aforementioned hypothesis. To do so, we first teach the given robot a closed path. Then, we introduce a large position error by displacing the robot from the path start. After that, we let the robot drive along the path several times, and after each path traversal, we measure (typically by hand) its distance to the path start. This distance corresponds to the robot position error after traversing one loop of the path. If the model introduced in Sect. 2.5 is correct, and the values of s_x, s_y are small, then the initial, artificially introduced position error should gradually diminish, and the robot position error should stabilise. This stabilised value indicates the overall navigation error of our system for a given platform and environment. To thoroughly evaluate the correctness of the mathematical model and robustness of the navigation method, we performed the aforementioned experiments using several mobile vehicles with a different type of kinematics. Videos of some of the experiments are provided on a special webpage reachable from the website of the navigation system [27].

The second part of our evaluation is aimed at a more practical issue: integration. We simply measured the time it took us to deploy our system onto a given platform and to perform the deployment. Since the deployment is divided into two steps: integration and debugging, we measure these times separately. Although these times depend on the experience of the system integrators and are inherently subjective, they still indicate how difficult was the deployment across the individual platforms.

3.1   Cameleon Robot

The first platform used in the experiments is the tracked robot CAMELEON ECA, which is a 0.7×0.6 m vehicle with a payload over 25 kg. The platform has two main independently controllable tracks and two auxiliary flippers for movement in difficult terrain. It is also equipped with two cameras, one in the front and one in the back of the body of the robot. As these cameras are located too low over the ground, they are not suitable for visual navigation in grassy terrains. Thus, we installed eCon's TARA stereocam, and we use the images from its left camera. This camera is attached to the robot superstructure, which also contains a stand for the control laptop and the Fenix 4000 lumen torch used for night experiments. For the fully equipped Cameleon platform, see Fig. 5.

The experiments took place at Hostibejk Hill in Kralupy nad Vltavou, Czechia, where the robot repeatedly travelled a 70 m long, complex path over concrete, grass and asphalt surfaces. Approximately one-third of the path, a small building was in the robots field of view. Otherwise, it perceived mostly trees and natural structures. This experiment was unique because of the tracks the platform consisted of, and these cause its odometry to be highly inaccurate. Despite that, the robot could quickly suppress the position error we introduced, stabilising it at values below 0.1 m, see Fig. 5.

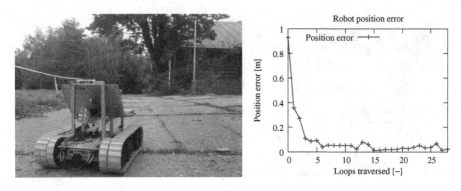

Fig. 5. Cameleon robot and its position error during the autonomous path traversal.

Integration of the system onto the Cameleon platform required to implement a ROS bridge to command the robot and to obtain odometry, which took 5 h of implementation and 1 h of testing time. The experiment itself took 1 h. The main problems encountered were caused by occasional 'freezing' of the PC, which took 1–3 s. During these periods, the robot was driving incorrectly and often deviated from the intended path. However, as soon as the system started to react, the robot started to follow the taught path and reduced the position error rapidly.

3.2   MMP-5 Robot

To demonstrate the system's ability to work indoors on a small robot, we deployed it on the MMP-5 platform, which is made by TheMachineLab. It is a small robot with dimensions 0.3×0.3 m, payload over 3 kg and four-wheel differential drive along with a control unit, which can control each wheel independently via a simple serial protocol. This platform, however, does not provide odometry, so we estimated the travelled distance by time and by the motors' PWM duty. We equipped this robot a low-resolution USB camera and a computer based on an AT3IONT-I miniATX board with Intel Atom 330 CPU.

The indoor experiments took place in the entrance building of the Czech technical university in Prague at Karlovo Náměstí campus. We taught the robot a 17 m lemniscate-shaped path, displaced it from the path start by 1.4 m and let it traverse the path 20 times. The progress of the robot was monitored by an external localisation system [20], which was tailored specifically for small platforms [1]. The use of the external localisation system, based on a high-resolution camera, allowed to monitor the evolution of the robot position error over time with centimetre accuracy, see [22]. Furthermore, the system was used to determine the robot position error after each path traversal, so we did not have to measure the displacement by hand. The experimental results, shown in Fig. 6, indicate that even a robot without wheel encoders and only a low resolution camera, which suffered from motion blur, was able to navigate through the taught path while reducing the initial error. The overall achieved position

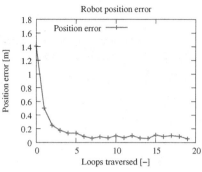

Fig. 6. MMP-5 robot and its position error during the autonomous path traversal.

error was about 0.1 m, and the integration and testing of the system took around 4 and 2 h respectively.

The main problem faced was related to the low computational power of the onboard PC, which caused issues with image processing and real-time control.

3.3   John Deere Tractor

The John Deere X300R is a small tractor, see Fig. 7, with gasoline 13.8 kW motor and automatic K46 hydrostatic transmission. It weights around 300 kg. Its wheelbase is 1.25 m and overall dimensions are $2.5 \times 1.0 \times 1.1$ m This machine, which has a car-like drive, was modified and equipped with hydraulic steering and linear motor with position feedback for pedal control. Due to the mechanics of the control system, both steering and velocity can be controlled only coarsely and with a significant time delay – this corresponds to large values of s_x and s_y in the navigation model in Eq. (3).

The tractor is also equipped with a set of sensors to support autonomy: 4 emergency stop buttons, active front bumper, rotary encoders, Hall sensor for measuring steering angle, set of indicators and switches (including manual/automatic digital input), 2D Lidar SICK LMS100 and a AV3135 Dual Sensor H.264 Day/Night camera. The low level electronic components (pedal position, encoders, steering, I/O) are controlled via CAN bus modules, connected to an APU2 computer. The navigation algorithm was running on the same PC as in the case of Cameleon ECA, i.e. Intel i3 laptop, which was connected to APU2 computer via ethernet. Apart from the laptop, which was running the navigation system, we installed our own camera to the tractor front.

This experiment took place in an experimental agricultural field in the campus of the Czech University of Life Sciences in Prague. During the experiments, the system had to deal with images blurred due to the engine vibrations, as well as rain, which affected both the environment appearance and wheel-surface interaction. Eventually, the rain became too heavy and we had to terminate the experiment prematurely. Despite of that, the robot was able to suppress the initial position error, stabilising it around 0.5 m, see Fig. 7.

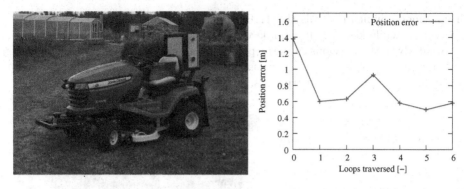

Fig. 7. John Deere platform and its position error during the autonomous drive.

Integration of the system took 2 h, initial testing 5 h and the experiment took 1 h. The main encountered problems were caused by rather slow and coarse hydromechanical control of the steering and velocity. Furthermore, we experienced a significant lag when retrieving images from the robot's IP camera, and thus, we had to install our own USB cam.

3.4   TAROS 6 × 6 Combat Support UGV

We tested the system on a large, 6 × 6 drive unmanned ground vehicle called 'TAROS', developed by the VOP company. The primary purpose of TAROS is direct support of mechanised, reconnaissance and special forces in their operations. Currently, the TAROS platform offers long-range teleoperation, GPS-based waypoint navigation and autonomous people following. However, its sensory suite, consisting of a Velodyne 3d lidar, three 2d SICK lidars and pan-tilt cameras, is sufficient for fully autonomous operation in adverse terrain.

One of the main features of TAROS is its modularity. Its core module, which contains all systems required for autonomous operation, is propelled by four electrically-driven wheels with independent suspensions. The wheels are equipped with odometric sensors and can be steered and driven individually. Each TAROS extension module adds two individually steered wheels, which allows configuring TAROS as 6 × 6 or 8 × 8 wheel drive UGV. For our experiments, we used the TAROS in 6 × 6 configuration with a hybrid power extension module.

The TAROS control system offers a convenient serial protocol which allows an ordinary PC to retrieve low-level (such as specific wheel angles) as well as high-level (such as the distance of the closest obstacle) information. Furthermore, the protocol enables low-level control of individual wheel actuators as well as high-level driving modes, which allow controlling the angular and forward speed in four different wheel configurations. The chosen protocol allowed a quick implementation of a basic ROS driver for the TAROS platform. The driver retrieves odometric information from the TAROS middle wheels and utilises a driving mode, which controls the TAROS heading by steering the forward and rear wheels while keeping the middle wheels straight.

Similarly to the previous case, the platform's IP camera images were provided with a significant delay, and thus, we used the USB camera of the i3 laptop, which was used in the previous experiments.

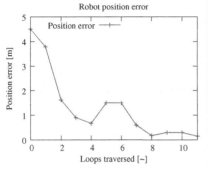

Fig. 8. TAROS platform and its position error during the autonomous drive.

The experiments took place at a small testing polygon in VOP CZ, which produces the TAROS platform. During the test, the robot was taught a 100 m long, closed path, and then it was displaced by 4.5 m and let to traverse the path autonomously 11 times. The error was gradually diminished, and despite of the platforms large size, the error was finally reduced below 0.2 m, see Fig. 8. The sudden increase of the position error in the 5th loop traversal was caused by the control computer 3 s 'freeze', leading to an overshoot at the path end. The Fig. 8 shows that as the robot continued to traverse the path, this error was gradually diminished as well.

Integration of the system took 2 h, its preliminary testing 3 h and the experiment itself took 1 h. Our software expected that the odometry is provided with at least 10 Hz update rate while the TAROS robot provides odometric information at 5 Hz. A significant part of the testing time was dedicated to identifying and correcting this issue in software.

3.5 Thorvald Agricultural Robot

Thorvald is a platform intended for various agricultural scenarios. This platform can be equipped with several different modules, which extends the variety. It has four wheels, which can be steered and actuated individually and the width between the wheels can be changed depending on the desired agricultural application. We have used the standard wheel configuration with the width of the wheelbase of 1.5 m. All the wheels are equipped with suspension modules, which allows the platform to drive in rough terrain. Low-level robot control is performed by an onboard PC, which was connected to our Intel i3 laptop over ethernet. Again, we used a standard USB camera as the main sensor.

Fig. 9. Thorvald robot and the path traversed during its test.

In the case of Thorvald, we took another approach to the testing. Instead of traversing a single loop several times, we taught it a 600 m long route around the Isaac Newton building of the Lincoln University, displaced the robot by 2 m and traversed this path once. Furthermore, during autonomous traversal, we had to manually steer the robot about 2 m from the path to avoid a large van, which stopped in the way. Despite that, the robot could reach the path end location with less than 0.3 m error.

Integration of the control system took approximately 1 h, and testing took 1 h. The experiment itself took 2 h. The main issue encountered was related to interfacing the robot control PC with the laptop running the navigation system.

3.6   STOP Robot

Finally, to demonstrate that the system can be deployed on indoor platforms as well, we tested it using the STOP robot, see Fig. 10. The STOP robot is a low-cost mobile robot platform for monitoring and surveillance of buildings and facilities, aiming at keeping intruders away, preventing the misuse of spaces, and more importantly, to act in a timely manner in case someone infiltrates the premises or violates rules of use of the spaces, e.g. by alerting security operators [32]. Several STOP robots, which are intended to cooperate, are under development by Ingeniarius, Ltd. from Coimbra, Portugal within the scope of the STOP R&D Project initiative[1]. The robot features a 360-degree RPLIDAR A2 laser range finder, an Orbbec Astra RGBD camera, a touchscreen interface, and a powerful Intel NUC mini PC. Low-level robot control is performed by a Particle Photon board, which is connected to the main NUC mini PC over serial. Being fully ROS-compliant, the robot benefits from navigation software, and a specifically tailored localisation system, which fuses laser scan matching odometry, global orientation from a digital compass, encoder odometry and an ICP correction system for global localisation. In our experiments, we kept the aforementioned localisation system on, so that we could use it as a source of ground truth position information. As in the previous experiments, the robot

---

[1] http://stop.ingeniarius.pt.

was controlled only by the navigation method proposed, which is solely based on odometric data and monocular camera image stream.

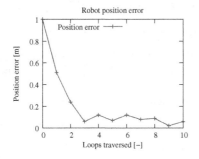 

Fig. 10. STOP robot platform and its position error during the autonomous drive.

The experimental evaluation took place in an entrance hall of the Ingeniarius Ltd., which produces the STOP robot. The robot was taught a ~10 m long path, then it was displaced by 1 m, and taught to traverse the path autonomously 10 times. The error gradually diminished and stabilised at ~0.1 m, see Fig. 10.

Before integration, we needed to perform the system update, which took more than 3 h. Since the robot supports ROS natively, integration took less than 1 h preliminary testing 1 h, and the experiment itself lasted 40 min. The main issue encountered was caused by different versions of OpenCV software libraries for image processing.

3.7   Experiments Summary

The experiments, summarised in Table 1 indicated that the 'convergence' theorem, introduced in Sect. 2.5 and papers [22,35], is valid for different types of vehicles, including tracked, car-like, with independently steered wheels, electrically or hydraulically controlled, or with combustion engines. Furthermore, the accuracy of the navigation exceeded the typical accuracy of (non-RTK) GPS service, the robots were able to correct initial position errors up to 4.5 m (see Fig. 8), and integration of the system into various platforms was relatively quick and did not require special customisation or extensive parameter tuning. Thus, the experimental results were not only in accordance of the 'convergence' theorem predictions, but they also demonstrated the maturity of the navigation system, which is provided as open-source C++ code at www.github.com/gestom/stroll_bearnav. The summary of the experiments is provided in Table 1, which shows the achieved navigation accuracy, deployment, testing and experimental times.

The main lesson learned from the experiments is that while the navigation system is mature enough to be easily integrated into various platforms, its reliability is still affected by software issues originating from the fact, that the Robotic

Table 1. Navigation accuracy and integration times on different platforms

Platform	Navigation		Time [h] required for		
	Accuracy [m]	Travelled [m]	Integration	Testing	Experiment
Cameleon	0.05	2000	5	1	1
MMP5	0.07	340	4	2	2
John Deere	0.50	180	2	5	1
TAROS	0.15	1100	2	3	1
Thorvald	0.30	600	1	1	2
STOP	0.09	100	1	1	1

Operating System is running on Linux distributions, which are not meant to provide real-time response.

3.8 Additional Simulations

Since the aforementioned experiments were aimed at proving the system accuracy in real scenarios, the number of experimental trials and scenarios was rather limited. To evaluate the method in a wider set of trajectories and environmental configurations, we simulated the system behaviour in 1000000 randomly generated scenarios with trajectory curvatures ranging from 0.001 to $1\,\mathrm{m}^{-1}$, landmark distances between 0.1 and $2\,\mathrm{m}$, initial robot position errors between 0 and $1\,\mathrm{m}$ in both x and y directions, odometric errors up to 5% and heading estimation errors up to $5°$. The simulations indicated that for every of the 1000000 configurations, the robot gradually converged to the intended trajectory.

4 Conclusion

We presented a versatile teach-and-repeat vision-based navigation system, which is easy to deploy on a variety of ground robot platforms. The system is capable to guide mobile vehicles along a path previously taught by a human operator. Instead of creating a globally-consistent metric map, our method creates a sensorimotor record of a teleoperated drive, consisting of odometric and visual information. The autonomous navigation system is based on the retrieval and replay of the recorded experience instead of estimating the global position of the robot in the 2d/3d space. We presented a mathematical model of the navigation, which allows to calculate the evolution of the robot position error over time. Then, through eigenanalysis, we show that the robot position error does not diverge, which ensures navigation accuracy and reliability.

While being computationally efficient, the system does not require camera calibration, and it can deal with realistic outdoor lighting conditions. Furthermore, the system is easy to operate, because it only requires that the robot is manually guided along a desired path, which it can repeat autonomously later on.

To verify the correctness of the 'convergence' theorem the system is based on, we tested it in simulation and on a variety of ground robot platforms with different types of kinematics and propulsion. The experiments indicated not only that the navigation accuracy exceeds the typical non-RTK GPS precision, but also that the integration of the system into a new robot is straightforward and it takes only a few hours. The presented system, which was already deployed in several scenarios including aerial inspection [13,23,30], is provided as open source at www.github.com/gestom/stroll_bearnav. Datasets and videos of the experiments are available at www.github.com/gestom/stroll_bearnav/wiki.

Acknowledgments. We thank the VOP.cz for sharing their the data and the TAROS vehicle. We would like to thank also Milan Kroulík and Jakub Lev from the Czech University of Life Sciences Prague for their positive attitude and their help to perform experiments with the John Deere tractor.

References

1. Arvin, F., Krajník, T., Turgut, A.E., Yue, S.: COSΦ: artificial pheromone system for robotic swarms research. In: IEEE/RSJ International Conference on Intelligent Robots and Systems (IROS) (2015)
2. Bay, H., Tuytelaars, T., Van Gool, L.: SURF: speeded up robust features. In: Leonardis, A., Bischof, H., Pinz, A. (eds.) ECCV 2006. LNCS, vol. 3951, pp. 404–417. Springer, Heidelberg (2006). https://doi.org/10.1007/11744023_32
3. Behzadian, B., Agarwal, P., Burgard, W., Tipaldi, G.D.: Monte carlo localization in hand-drawn maps. In: IROS, pp. 4291–4296. IEEE (2015)
4. Biber, P., Duckett, T.: Dynamic maps for long-term operation of mobile service robots. In: RSS (2005)
5. Blanc, G., Mezouar, Y., Martinet, P.: Indoor navigation of a wheeled mobile robot along visual routes. In: IEEE International Conference on Robotics and Automation (ICRA) (2005)
6. Calonder, M., Lepetit, V., Strecha, C., Fua, P.: BRIEF: binary robust independent elementary features. In: Daniilidis, K., Maragos, P., Paragios, N. (eds.) ECCV 2010. LNCS, vol. 6314, pp. 778–792. Springer, Heidelberg (2010). https://doi.org/10.1007/978-3-642-15561-1_56
7. Chen, Z., Birchfield, S.T.: Qualitative vision-based path following. IEEE Trans. Robot. Autom. **25**, 749–754 (2009). https://doi.org/10.1109/TRO.2009.2017140
8. Churchill, W.S., Newman, P.: Experience-based navigation for long-term localisation. IJRR **32**, 1645–1661 (2013). https://doi.org/10.1177/0278364913499193
9. Dayoub, F., Cielniak, G., Duckett, T.: Long-term experiments with an adaptive spherical view representation for navigation in changing environments. Robot. Auton. Syst. **59**, 285–295 (2011)
10. De Cristóforis, P., Nitsche, M., Krajník, T.: Real-time monocular image-based path detection. J. Real Time Image Process. **11**, 335–348 (2013)
11. DeSouza, G.N., Kak, A.C.: Vision for mobile robot navigation: a survey. IEEE Trans. Pattern Anal. Mach. Intell. **24**, 237–267 (2002). https://doi.org/10.1109/34.982903

12. Engel, J., Schöps, T., Cremers, D.: LSD-SLAM: large-scale direct monocular SLAM. In: Fleet, D., Pajdla, T., Schiele, B., Tuytelaars, T. (eds.) ECCV 2014. LNCS, vol. 8690, pp. 834–849. Springer, Cham (2014). https://doi.org/10.1007/978-3-319-10605-2_54

13. Faigl, J., Krajník, T., Vonásek, V., Přeučil, L.: Surveillance planning with localization uncertainty for UAVs. In: 3rd Israeli Conference on Robotics (2010)

14. Furgale, P., Barfoot, T.D.: Visual teach and repeat for long-range rover autonomy. J. Field Robot. 27(5), 534–560 (2010). https://doi.org/10.1002/rob.20342

15. Halodová, L.: Map Management for Long-term Navigation of Mobile Robots. Bachelor thesis, Czech Technical University, May 2018

16. Halodová, L., Dvořáková, E., Majer, F., Ulrich, J., Vintr, T., Krajník, T.: Adaptive image processing methods for outdoor autonomous vehicles. In: Mazal. J. (ed.) MESAS 2018. LNCS, vol. 11472, pp. 456–476 (2018)

17. Zhang, N., Warren, M., Barfoot, T.: Learning place-and-time-dependent binary descriptors for long-term visual localization. In: IEEE International Conference on Robotics and Automation (ICRA). IEEE (2016)

18. Holmes, S., Klein, G., Murray, D.W.: A square root unscented kalman filter for visual monoSLAM. In: International Conference on Robotics and Automation (ICRA), pp. 3710–3716 (2008)

19. Kosaka, A., Kak, A.C.: Fast vision-guided mobile robot navigation using model-based reasoning and prediction of uncertainties. CVGIP: Image Underst. 56(3), 271–329 (1992)

20. Krajník, T., et al.: A practical multirobot localization system. J. Intell. Robot. Syst. 76, 539–562 (2014). https://doi.org/10.1007/s10846-014-0041-x

21. Krajník, T., Cristóforis, P., Kusumam, K., Neubert, P., Duckett, T.: Image features for visual teach-and-repeat navigation in changing environments. Robot. Auton. Syst. 88, 127–141 (2017)

22. Krajník, T., Majer, F., Halodová, L., Vintr, T.: Navigation without localisation: reliable teach and repeat based on the convergence theorem. In: IEEE/RSJ International Conference on Intelligent Robots and Systems (IROS) (2018)

23. Krajník, T., Nitsche, M., Pedre, S., Přeučil, L., Mejail, M.E.: A simple visual navigation system for an UAV. In: 9th International Multi-Conference on Systems, Signals and Devices (SSD), pp. 1–6. IEEE (2012)

24. Kunze, L., Hawes, N., Duckett, T., Hanheide, M., Krajnik, T.: Artificial intelligence for long-term robot autonomy: a survey. IEEE Robot. Autom. Lett. 3, 4023–4030 (2018). https://doi.org/10.1109/LRA.2018.2860628

25. Lowe, D.G.: Distinctive image features from scale-invariant keypoints. Int. J. Comput. Vision 60(2), 91–110 (2004)

26. Mair, E., Hager, G.D., Burschka, D., Suppa, M., Hirzinger, G.: Adaptive and generic corner detection based on the accelerated segment test. In: Daniilidis, K., Maragos, P., Paragios, N. (eds.) ECCV 2010. LNCS, vol. 6312, pp. 183–196. Springer, Heidelberg (2010). https://doi.org/10.1007/978-3-642-15552-9_14

27. Majer, F., Halodová, L., Krajník, T.: Source codes: bearing-only navigation. https://github.com/gestom/stroll_bearnav

28. Matsumoto, Y., Inaba, M., Inoue, H.: Visual navigation using view-sequenced route representation. In: IEEE International Conference on Robotics and Automation (ICRA), Minneapolis, USA, pp. 83–88 (1996)

29. Mur-Artal, R., Montiel, J.M.M., Tardós, J.D.: ORB-SLAM: a versatile and accurate monocular SLAM system. IEEE Trans. Robot. 31, 1147–1163 (2015)

30. Nitsche, M., Pire, T., Krajník, T., Kulich, M., Mejail, M.: Monte Carlo localization for teach-and-repeat feature-based navigation. In: Mistry, M., Leonardis, A., Witkowski, M., Melhuish, C. (eds.) TAROS 2014. LNCS (LNAI), vol. 8717, pp. 13–24. Springer, Cham (2014). https://doi.org/10.1007/978-3-319-10401-0_2

31. Paton, M., MacTavish, K., Berczi, L.-P., van Es, S.K., Barfoot, T.D.: I can see for miles and miles: an extended field test of visual teach and repeat 2.0. In: Hutter, M., Siegwart, R. (eds.) Field and Service Robotics. SPAR, vol. 5, pp. 415–431. Springer, Cham (2018). https://doi.org/10.1007/978-3-319-67361-5_27

32. Portugal, D., Pereira, S., Couceiro, M.S.: The role of security in human-robot shared environments: a case study in ROS-based surveillance robots. In: 26th IEEE International Symposium on Robot and Human Interactive Communication (RO-MAN), pp. 981–986. IEEE (2017)

33. Royer, E., Lhuillier, M., Dhome, M., Lavest, J.M.: Monocular vision for mobile robot localization and autonomous navigation. Int. J. Comput. Vis. **74**(3), 237–260 (2007). https://doi.org/10.1007/s11263-006-0023-y

34. Segvic, S., Remazeilles, A., Diosi, A., Chaumette, F.: Large scale vision based navigation without an accurate global reconstruction. IEEE International Conference on Computer Vision and Pattern Recognition. CVPR 2007, Minneapolis, Minnesota, pp. 1–8 (2007)

35. Krajník, T., Faigl, J., Vonásek, V., et al.: Simple, yet Stable bearing-only Navigation. J. Field Robot. **27**, 511–533 (2010)

36. Thorpe, C., Hebert, M.H., Kanade, T., Shafer, S.A.: Vision and navigation for the Carnegie-Mellon Navlab. IEEE Trans. Pattern Anal. Mach. Intell. **10**(3), 362–373 (1988)

37. Wallace, R.S., Stentz, A., Thorpe, C.E., Moravec, H.P., Whittaker, W., Kanade, T.: First results in robot road-following. In: IJCAI, pp. 1089–1095. Citeseer (1985)

Visual Odometry for Vehicles' Undercarriage 3D Modelling

Tomáš Pivoňka[✉], Karel Košnar, Martin Dörfler, and Libor Přeučil

Czech Institute of Informatics, Robotics and Cybernetics,
Czech Technical University in Prague,
Jugoslávských partyzánů 1580/3, 160 00 Prague 6, Dejvice, Czech Republic
pivontom@fel.cvut.cz
http://imr.ciirc.cvut.cz/

Abstract. This work describes a part of a project developing a vehicles' undercarriage security scanner based only on cameras. The scanner is used to a security check of a vehicle's undercarriage and is typically installed at an entrance to a strategic compund. The security scanner reconstruct a 3D model of a vehicle's undercarriage from a sequence of a multi-camera stereo images. To get a complete model we need to stitch particular parts of the 3D model via transformations between particular vehicle positions in which images are captured. The method for getting these transformations is presented in this paper. The task of computing trajectory from a sequence of camera images is called visual odometry (VO). Usually, the camera is placed on a moving object and tracks its position. In our case, the camera is fixed and viewing a moving vehicle, but the task is the same. In the first part, there is a comparison of feature detectors and their parameters based on experimental data because the images properties of undercarriage are different from ordinary surroundings used by the most of VO methods. Undercarriages do not contain a lot of features, and there are many low-textured surfaces. In the second part, the proposed VO method is described. It is based on the best feature from the first part, which serves to search corresponding points between images. It uses 3D to 2D VO method to compute the transformation between consecutive frames. In this method, 3D points are triangulated from the previous pair of stereo camera images, and they are reprojected to one of the actual images. The method finds the transformation of 3D positions which minimizes reprojection error. The method was developed with respect to requirement of almost real-time computing time and low-texture environment. Finally, this method was evaluated on realistic data acquired with ground truth position.

Keywords: Visual odometry · Stereo vision · Security ·
Car inspection · 3D modelling

1 Introduction

A security of a critical infrastructure becomes more and more important nowadays. The vehicles entering the areal of the infrastructure can be utilized as a

© Springer Nature Switzerland AG 2019
J. Mazal (Ed.): MESAS 2018, LNCS 11472, pp. 111–120, 2019.
https://doi.org/10.1007/978-3-030-14984-0_9

possible and easily accessible vector for an attack. Especially, the vehicles entering regularly, e.g. vehicles of employees, can be modified without knowledge of the owner and then abused for the attack. Therefore, the vehicles entering the critical infrastructure are routinely checked for any changes. These procedure can be very time consuming and demanding on the vigilance of the operators, particularly when the security check is done manually.

1.1 Project Kassandra

This paper describes work done as a part of a joint research project Kassandra solved by VOP CZ company and CTU CIIRC and funded by the Ministry of Interior of the Czech Republic. The aim of the project is the development of a camera based 3D scanner of car undercarriages. The scanner is going to serve for a security check of cars, especially at entrances to strategical compounds like power plants or military bases, where the security check is needed. The scanner can compare a new 3D model with the previous one and detect undesirable changes, e.g. placing some explosive or tracking device on the vehicle undercarriage without knowledge of the owner.

A scheme of the scanner is in Fig. 1. The scanner is placed under a roadway level, and a car goes across the scanner. During this crossing, the scanner is taking stereo images of an undercarriage. Apart from a set of stereo cameras the scanner consists of a computing unit and a graphical terminal for a visual check.

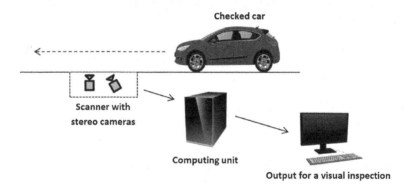

Fig. 1. Scheme of the camera based 3D scanner of car undercarriages.

1.2 Problem Definition

The scanner consists of 2 camera stereo pairs placed next to each other, each pair scanning one half of the vehicle undercarriage. The final 3D model of the undercarriage is going to be created by stitching together particular 3D models computed from each stereo images. A transformation between particular stereo images is required for stitching. The task of computing transformation between subsequent images is equal to the task of visual odometry. The transformation represents a rigid body movement which is described by a rotation matrix and a vector of translation.

As there is a request to display the resulting 3D model as soon as possible, the transformation is computed immediately from two subsequent images. Methods performing global optimization over all images are not used as they have higher demands on processing power and also they need to wait for all images, which are available after the vehicle crossing over the scanner with duration in units of seconds.

The solution of the problem is highly influenced by the properties of the undercarriages' character. Images of undercarriages contain a lot of low textured surfaces, and there are not many high contrast objects. Another specific property is that the movement of the vehicle is limited by the kinematic of the translation and the most dominant translation is in the direction of vehicle movement. Other movements like rotations or translations in remaining directions are much smaller.

1.3   Related Work

"Visual odometry (VO) is the process of estimating the egomotion of an agent (e.g., vehicle, human, and robot) using only the input of a single or multiple cameras attached to it" [1]. In this case a camera's motion is only relative to an undercarriage, and a vehicle motion is an inverse motion.

Visual odometry methods are divided into global methods and feature based methods. Global methods find correspondences between images on a pixel level. Feature-based methods detect features in images at first, and after that, they find corresponding features between images. Because global methods are less accurate and computationally more expensive [1], the proposed method is feature based. Feature methods are divided into three main groups according to corresponding points representation: 2D-to-2D, 3D-to-3D, and 3D-to-2D. These points are represented as 2D points in image coordinates or 3D points that are triangulated from the stereo images.

2D-to-2D methods usually work only with one camera and their main disadvantage is that a scale of a translation vector is ambiguous. An absolute scale can be determined by other sensors or some initial reference value.

An input of 3D-to-3D methods is a sequence of stereo images. 3D points are computed from stereo images, and they are matched between images. The resulting transformation can be computed only from 3 pairs of corresponding points.

3D-to-2D methods also use a stereo camera, but they triangulate points only in the first stereo images from subsequent images. This task is called a perspective n point problem. A transformation between spatial coordinates and camera coordinates is computed from correspondences between 3D points and their projection to a camera image plane. In our case, spatial coordinates are equal to camera coordinates of the previous position, and camera coordinates are coordinates of the camera in the actual position. A solution of this task is a transformation with a minimal reprojection error of transformed 3D points' reprojections in relation to real measured images of points. 3D-to-2D methods are generally more accurate than 3D-to-3D methods [1].

A lot of visual odometry systems serve to measure a car trajectory with onboard cameras. These methods are compared in the KITTI benchmark [4]. Compared systems work with a LIDAR or a stereo camera. The stereo-camera method with best results on KITTI benchmark is SOFT-SLAM [2]. The method consists of two threads. One thread serves to visual odometry and the second thread serves for mapping and global consistency. Its visual odometry itself was presented in an older paper [3], and it is still one of the best camera-based systems in the KITTI benchmark repository.

Aforementioned method is built on a strong filtration of features. It uses an own type of features detected by rectangular filters. Features' filtration is divided into several steps. The first step is called circular matching [2]. Features are matched between 4 images from two subsequent stereo images, and consistency of matching is checked. The method also uses a separated computation of a rotation and a translation and the RANSAC algorithm.

The Ackermann steering geometry constrains a movement of car-like vehicles. It allows to decrease number of 6 degree of freedom (DoF) of the rigid body movement in a 3D space. This simplification speeds up computations and can improve the results, but it ignores tiny movements like vibrations or drifts.

The most simplified method is presented in [5] where the vehicle movement is determined only by one DoF. By an assumption of planar motion and an Ackermann geometry, it reduces the movement to 2 DoF. Because the method uses only one camera, a scale of movement is ambiguous, the number of DoF is reduced even to one.

A smaller amount of DoFs decreases computational demands, but it can be too much restrictive. Some other less restrictive methods are decreasing the number of car-like vehicles' DoFs are described in [6]. 4 DoF motion is a connection of planar motion and vibrations which are described as a 1 DoF motion perpendicular to a plane of motion. If changes in rotation are small, these rotations can be simplified by a first order approximation, which did not decrease the number of DoFs, but it simplifies computations.

2   Automatic Test of Feature Detectors

Precision and robustness of the visual odometry method highly depend on a method for feature detection and description. It is necessary to detect enough features that can be correctly matched between images. Some methods extract a lot of features from each image, but their descriptors are not much distinctive, and it leads to many mistakes in a matching.

At first, features are detected by a feature detector which is a method recognizing distinctive areas in the image. Afterward, feature descriptor creates a description for each feature. Matching is done by comparing descriptors of features between images. In this work, following detecting methods were tested:

Harris, Shi-Tomasi, SIFT, SURF, FAST and ORB detectors. SIFT, SURF and ORB detectors have an own method for a description of features. Other methods were tested with a BRIEF descriptor. All the aforementioned methods were used in the implementation from the OpenCV library [7].

Amount of features and their quality do not depend only on a type of method for detection or description, but they also depend on the values of internal parameters. The best method and its parameters were found by an automatic test of methods for detection and description of features.

For each parameter, a small set of values were selected, and an automatic test checked all combination of parameters' values for the selected method. For each combination, features were detected and matched in 20 stereo images of an undercarriage. These images were captured at two different distances and with two different exposition times simulating a different level of lighting. Right matches are recognized by a distance of a point from a theoretically computed epipolar line computed from known parameters of the stereo camera and a position of the feature in the second image. If this distance is too big, matching is incorrect.

In the next step, a 3D position of the point is triangulated from corresponding points. For right matched points their distance from camera plane has to be in an assigned interval determined by the known distance of the undercarriage from the camera. If this rule is satisfied, the matching is considered to be correct.

Measured parameters for each combination are a number of correctly matched features, a total number of matched features, a minimal number of matched features in one stereo image, and time of the whole test. Automatic test results are presented in Sect. 4.

3   Visual Odometry Method

The proposed method is a 3D-to-2D method. An input of the method is a stereo image at time t_{i-1} and one image at time t_i. This method computes translation only between two stereo images. It uses a strong filtration of features and a separated computation of a rotation and a translation similarly to [2].

3.1   Detection and Matching of Features

ORB detector is used for the extraction of features. Features are not detected in a whole image in one go, but the image is split to rectangle areas. It is divided to a grid 4×3 rectangles. It ensures a more uniform distribution of features. Features' matching is based on searching feature with the closest descriptor. Further, the method uses *crossmatching*. It means that if a feature A in the first image is the closest feature to a feature B in the second image, the feature B has to be also the closest feature to the feature A.

In the first step only features from the stereo image are matched. Matching features between the stereo image and an image in time t_i is postponed after filtration and triangulation. The main reason is that only a filtered set of features is matched and it speeds up computations.

3.2   Filtration and Triangulation

At first, matched features from stereo images are filtered by the similar way as in the automatic test. Points are removed if their distance from an epipolar line in one of the images is higher than some threshold or if a distance of a triangulated point from camera plane is not in an assigned interval. Because a distance of an undercarriage from a camera is not known, the interval is fixed.

Afterwards, features from images at time t_{i-1} are matched to features at time t_i. The next step of filtration is similar to *circular matching* in [2], but it is done only for tree images. If corresponding features from the stereo image are saved in the same order, corresponding points to features in the third image have to have the same index for both images. If its indexes are different, the points are removed.

In the last step of filtration, close features are removed. If a distance between two features in one image is under a threshold, only feature with a stronger descriptor remains.

Fig. 2. Detected and matched features between two shifted images after filtration

3.3   Computation of Rotation

A rotation is computed from correspondences between 3D and 2D points. Because not all correspondences are correct, the RANSAC algorithm is used. At the beginning of each iteration, four points are selected randomly. Although only the rotation is an output of this step, a translation is also estimated. A perspective 3 point problem (P3P) is solved for the first 3 points by solver [8]. Generally, a P3P task has four different solutions. The 4th point serves to select the one correct solution.

This solution is evaluated on the other points. All 3D points are transformed and reprojected to the image at time t_i. Afterward, a reprojection error between reprojected and measured positions is computed. An evaluating criterion is a number of points with reprojection error which is smaller than an assigned threshold.

3.4 Computation of Translation

In the beginning, only points, which fulfill a condition for a reprojection error from the previous step, are left. Further, these points are filtered that only a half of the points closest to a median value is left and other points are removed. Final translation is computed from all remaining points. It leads to an overdetermined system of equations where only three unknown variables are components of the vector of translation. An equation for one correspondence is:

$$\lambda \cdot \begin{bmatrix} u_k \\ v_k \\ 1 \end{bmatrix} = K \cdot R \cdot \begin{bmatrix} x_k \\ y_k \\ z_k \end{bmatrix} + K \cdot \begin{bmatrix} \tau_x \\ \tau_y \\ \tau_z \end{bmatrix}. \tag{1}$$

λ is a scalar representing a scale ambiguity, u_k and v_k are 2D point coordinates. K is a camera matrix, and R is a rotation matrix which has been computed in the previous step. x_k, y_k, and z_k are coordinates of 3D points. τ_x, τ_y, and τ_z are components of a translation vector.

These equations for all correspondences are rearranged that parameter λ is removed and they are connected to an overdetermined system in a form:

$$A \cdot \begin{bmatrix} \tau_x \\ \tau_y \\ \tau_z \end{bmatrix} = B \tag{2}$$

τ is computed algebraically as the least square optimal solution.

4 Experimental Setup and Results

4.1 Testing Data and Hardware

Because a prototype of the scanner has not been available yet, data could not be captured in real conditions. A car motion was simulated by moving a construction with cameras under a car on a hoist. This manner's main advantage is that a true reference of motion could be simply measured. The construction (Fig. 3) contains four cameras creating two stereo pairs. Pairs are distant 1 m from themselves. Construction also contains lighting.

All cameras are the same model of camera. It is the Basler acA1920-155uc. The resolution of the camera is 1920 × 1200 px. Key properties of the camera are a high frame rate (max. 164 Hz) and a global shutter.

Images of an undercarriage were taken in two different distances (57 cm and 75 cm) from the vehicle undercarriage and with two different exposition times. An exposition time simulated the different level of lighting (dark and light). The majority of data simulates only a straight movement of a vehicle across the sensor. There is also a small set of dark images which simulates the straight movement combined with rotation over yaw angle.

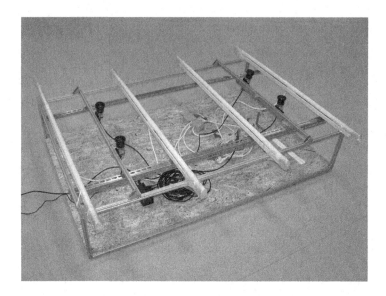

Fig. 3. Construction with two stereo pairs of cameras

4.2 Automatic Test of Feature Detectors

The automatic test tested six different methods for feature detection from the OpenCV library. A set of parameters was selected for each method, and their combinations were tested. The results are in Table 1.

The most important parameter was a minimal number of correct matches in one stereo image. Because detection in the real sensor has to be fast enough, parameters with longer time than 1 s were dismissed. No combination of SIFT parameters satisfied this condition. Harris, Shi-Tomasi, and FAST detector detected a lot of features, but they had insufficiently distinctive descriptors, and they did not have enough minimal number of correct matches.

ORB method was selected because it is fast, it has the best overall results, and it has a free license. SURF method had better results for light images.

4.3 Visual Odometry Results

The proposed visual odometry method was tested on an experimental dataset of car undercarriage. Data for testing were split into light and dark images because the final scanner is going to be equipped with lighting which ensures stable light conditions. The results are presented in Table 2.

Translations between images were 10, 20 and 30 cm with precision 0.2 cm. The average error of the proposed method was for both types of images under 1 mm. The maximal error for light images was under 3% of translation. The method was less stable for dark images, but it still returns good results. The transformation was successfully computed for all tested images, and there was no failure which could be caused by a small number of correct matches.

Table 1. Automatic test of feature detectors' parameters

	Harris	Shi-Tomasi	FAST	SIFT	SURF	ORB
Minimum of correct matches in one stereo image	4	4	0	137	49	158
Correct matches	150	161	371	4 737	2 384	4 899
Number of matches	23 787	24 892	29 950	35 701	24 310	39 773
Time of detection and matching [s]	0.28	0.29	0.77	2.13	0.96	0.39
Number of tested parameters' combinations	1500	300	144	288	384	800

A rotation over yaw angle was simulated only for dark images in a range from $0°$ to $15°$. The maximal error was only $1.5°$, which is less than the precision of measuring reference angle ($2°$).

Table 2. Visual odometry results

	Light images	Dark images
Average error [mm]	0.8	0.9
Standard deviation [mm]	1.7	1.8
Maximal error [mm]	5.2	7.8
Maximal error [%]	2.9	6.7
Maximal error of yaw angle [°]	–	1.5

5   Conclusion

A visual odometry method for a car undercarriage scanner is presented in this paper. It is based on a strong filtration of matched features, and it uses a separated computation of a rotation and a translation. The method is also suitable for low feature environments because if all matches are correct, only four matches between images are needed.

ORB detector is used for feature detection. An automatic test of features found the best inner parameters of ORB detector for images of an undercarriage. These parameters were selected from combinations of several selected values for each parameter.

The method was evaluated for testing data simulating a real car movement over the scanner. The images were split into two groups - light and dark. Because the scanner is going to be equipped with lighting, the stability of the method for both types is not needed. Nevertheless, the method has an average error for both types under 1 mm. Method's rotation error was measured on a smaller set of dark images. The error for angles from $0°$ to $15°$ was under $1.5°$.

More detailed description of the method and results from all experiments (in the Czech language) are available in diploma thesis [9].

5.1 Further Work

The main deficiency of the method is developing and testing on data which comes only from 1 vehicle. In the future, it is necessary to repeat testing for more vehicles. Because the capturing of data from a car host is too slow, the prototype of the sensor placed under a roadway level is needed. The data from the prototype is also going to be more realistic.

Another direction of improving the method is going to be optimizing computations because the scanner has to work almost in real time. Another important step is going to be connecting the method with other parts of the system. The method could be speeded up by reusing of some information from a method which creates particular 3D models. It is, for example, using the same type of features or information from triangulation.

Acknowledgements. The work described in this paper was funded by Ministry of Interior of the Czech Republic under the project number VI2VS/461. The work of Tomáš Pivoňka was also supported by the Grant Agency of the Czech Technical University in Prague, grant No. SGS18/206/OHK3/3T/37.

References

1. Scaramuzza, D., Fraundorfer, F.: Visual odometry [tutorial]. IEEE Robot. Autom. Mag. **18**, 80–92 (2011). https://doi.org/10.1109/MRA.2011.943233
2. Cvišić, I., Ćesić, J., Marković, I., Petrović, I.: SOFT-SLAM: computationally efficient stereo visual simultaneous localization and mapping for autonomous unmanned aerial vehicles. J. Field Robot. **35**, 578–595 (2018). https://doi.org/10.1002/rob.21762
3. Cvišić, I., Petrović, I.: Stereo odometry based on careful feature selection and tracking. In: 2015 European Conference on Mobile Robots (ECMR), pp. 1–6 (2015). https://doi.org/10.1109/ECMR.2015.7324219
4. Geiger, A.: Visual odometry/SLAM evaluation 2012. In: The KITTI Vision Benchmark Suite. http://www.cvlibs.net/datasets/kitti/eval_odometry.php. Accessed 25 July 2018
5. Scaramuzza, D., Fraundorfer, F., Siegwart, R.: Real-time monocular visual odometry for on-road vehicles with 1-point RANSAC. IEEE Int. Conf. Robot. Autom. **2009**, 4293–4299 (2009). https://doi.org/10.1109/ROBOT.2009.5152255
6. Choi, S., Park, J., Yu, W.: Simplified epipolar geometry for real-time monocular visual odometry on roads. Int. J. Control Autom. Syst. **13**, 1454–1464 (2015). https://doi.org/10.1007/s12555-014-0157-6
7. Bradski, G.: The OpenCV Library. Dr. Dobb's Journal of Software Tools (2000)
8. Kneip, L., Scaramuzza, D., Siegwart, R.: A novel parametrization of the perspective-three-point problem for a direct computation of absolute camera position and orientation. In: CVPR 2011, pp. 2969–2976 (2011). https://doi.org/10.1109/CVPR.2011.5995464
9. Pivoňka, T.: Visual odometry for dynamic image reconstruction. Diploma thesis (2018). https://dspace.cvut.cz/handle/10467/77042

Monocular Kinematics Based on Geometric Algebras

Marek Stodola[✉]

Faculty of Mechanical Engineering, Brno University of Technology,
Brno, Czech Republic
171519@vutbr.cz

Abstract. When reconstructing 3D scene by an autonomous system we usually use a pin hole camera. To adopt the result for a human vision, this camera must be replaced by a human eye-like device. Therefore we derive certain characteristics of this model in an appropriate mathematical formalism. In particular, we escribe the general position of a human eye and its movements using the notions of geometric algebra. The assumption is that the eye is focused on distant targets. As the main result, we describe the eye position and determine all axes of rotation available in the eye general position in terms of geometric algebra. All the expressions are based on medically traced laws of Donders' and Listing.

Keywords: Geometric algebra · Rotation · Donders' law · Listing's law

1 Introduction

This paper deals with human-like monocular vision. We assume that the eye is focused on distant targets. Theoretically, for a specific gaze direction, there are many appropriate eye positions that differ from each other by the rotation around the axis of the gaze direction. However, the medically proved Donders' Law applies which states that there is only one eye position for a specific gaze direction.

There is another medically proved law, which is called Listing's. This law says that when the eye gazes straight ahead, i.e. it is in the primary position, all movements can be obtained as rotations around an axis that lies in a plane perpendicular to the gaze direction, so-called Listing's plane. The exact forms of Donders' law and Listing's law are provided in Sect. 3, for further details see [16].

The goal of this paper is to determine the possibilities of the eye movement depending on the gaze direction, i.e. the eye is in general position and is focused on targets in infinity. Our argumentation is based on Donders' law and Listing's law. For calculations we use geometric algebra $\mathbb{G}_3$, which is suitable for point transformations such as the rotation around axes that pass through the origin of the coordinate system. Note that the properties and definitions of geometric

© Springer Nature Switzerland AG 2019
J. Mazal (Ed.): MESAS 2018, LNCS 11472, pp. 121–129, 2019.
https://doi.org/10.1007/978-3-030-14984-0_10

algebras can be found in e.g. [2,4,5,8,13–15]. Let us mentioned applications in engineering in e.g. [7,10,12] and in image processing in e.g. [6,9,11]. Geometric algebras can also be successfully used for the description of binocular kinematics, see e.g. [1].

2 Geometric Algebra $\mathbb{G}_3$

Let us formally define element G of the algebra $\mathbb{G}_3$ as

$$G = a_0 + a_1\sigma_1 + a_2\sigma_2 + a_3\sigma_3 + a_4\sigma_{23} + a_5\sigma_{31} + a_6\sigma_{12} + a_7\sigma_{123}, \qquad (1)$$

where $a_i \in \mathbb{R}$, $i \in \{0, 1, ..., 7\}$ and $\{1, \sigma_1, \sigma_2, \sigma_3, \sigma_{23}, \sigma_{31}, \sigma_{12}, \sigma_{123}\}$ is a formal basis of a vector space. As the algebra operation on $\mathbb{G}_3$ there is a multiplication, called geometric product, of two specific elements which satisfies the following identities:

$$\sigma_i^2 = 1, \quad i \in \{1, 2, 3\},$$
$$\sigma_i\sigma_j = -\sigma_j\sigma_i, \quad i, j \in \{1, 2, 3\}, \ i \neq j,$$
$$\sigma_{ij} = \sigma_i\sigma_j, \quad i, j \in \{1, 2, 3\},$$
$$\sigma_{ijk} = \sigma_i\sigma_j\sigma_k, \quad i, j, k \in \{1, 2, 3\}.$$

In $\mathbb{G}_3$, let us consider the elements in the form

$$R = a_0 + a_1\sigma_{23} + a_2\sigma_{31} + a_3\sigma_{12}, \qquad (2)$$

such that its Euclidean norm $||R||$ is equal to 1, i.e.

$$||R|| = \sqrt{a_0^2 + a_1^2 + a_2^2 + a_3^2} = 1.$$

Such elements are called *rotors*. Consequently, we define the *conjugate rotor* to rotor R as follows:

$$\overline{R} = a_0 - a_1\sigma_{23} - a_2\sigma_{31} - a_3\sigma_{12}. \qquad (3)$$

Alternatively, we can write the rotor $R = a_0 + a_1\sigma_{23} + a_2\sigma_{31} + a_3\sigma_{12}$ in the *goniometric form* as follows:

$$R = \cos\frac{\theta}{2} + N\sin\frac{\theta}{2}, \qquad (4)$$

where

$$\theta = 2\arccos a_0, \qquad (5)$$

$$N = \frac{a_1\sigma_{23} + a_2\sigma_{31} + a_3\sigma_{12}}{\sqrt{a_1^2 + a_2^2 + a_3^2}}, \quad \text{for } R \neq \pm 1, \qquad (6)$$

$$N = 0, \quad \text{for } R = \pm 1.$$

Let us note that N then plays the role of the axis of rotation. Consequently, the conjugate rotor of the rotor (4) is of the form $\overline{R} = \cos\frac{\theta}{2} - N\sin\frac{\theta}{2}$. The set of rotors together with the multiplication as in $\mathbb{G}_3$ form a group [2] with the unit element $e = 1$ and the inversion defined as $R^{-1} = \overline{R}$. If we consider the Euclidean embedding of $\mathbb{R}^3$ into $\mathbb{G}_3$, we can represent a vector $\overrightarrow{u} = (x, y, z) \in \mathbb{R}^3$ as the element of $\mathbb{G}_3$ in the form $U = x\sigma_1 + y\sigma_2 + z\sigma_3$. To rotate the vector $U = x\sigma_1 + y\sigma_2 + z\sigma_3$ around axis $N = n_1\sigma_{23} + n_2\sigma_{31} + n_3\sigma_{12}$, $\|N\| = 1$ by the angle θ we only need to provide the following calculation:

$$\widetilde{U} := RU\overline{R} = \widetilde{x}\sigma_1 + \widetilde{y}\sigma_2 + \widetilde{z}\sigma_3, \tag{7}$$

where $R = \cos\frac{\theta}{2} + N\sin\frac{\theta}{2}$. We received vector $\widetilde{U}$ which indeed is the vector U rotated around the axis N by the angle θ in the positive direction, see [14] for verification. For the composed rotation of a vector U by two successive rotors R_1 and R_2, we can use just one total rotor $R = R_2R_1$. Indeed, because rotors form a group and therefore $R^{-1} = \overline{R}$, the following assertion holds:

$$R_2(R_1U\overline{R}_1)\overline{R}_2 = (R_2R_1)U(\overline{R}_1\overline{R}_2) = (R_2R_1)U(\overline{R_2R_1}) = RU\overline{R}, \tag{8}$$

for more details see [2,5].

3   Donders' Law and Listing's Law

Let the eye look straight ahead w.r.t. the human proportions. Then we say that it's in the *primary position*, see Fig. 1. In the eye there is a point through which all axes of rotation pass. This point is called the *centre of the eye rotation*. In this paper we understand the eye as a sphere whose centre is in the centre of the eye rotation.

Let $H = (C, h_1, h_2, h_3)$ be the Cartesian coordinate system attached to the human head. Let C be both the centre of the eye rotation and the origin of the coordinate system. When the head is in the erected position, axis h_1 is forward-facing and forward-oriented, axis h_2 is horizontal and oriented to the left, axis h_3 is vertical and oriented upwards, see Fig. 1. Let $\overrightarrow{h}_1, \overrightarrow{h}_2, \overrightarrow{h}_3$ be the unit vectors fixed to the head that lie on the axes h_1, h_2, h_3, respectively, and $\overrightarrow{e}_1, \overrightarrow{e}_2, \overrightarrow{e}_3$ be the unit vectors fixed to the eye in such a way that if the eye is in the primary position, then $\overrightarrow{h}_1 = \overrightarrow{e}_1, \overrightarrow{h}_2 = \overrightarrow{e}_2, \overrightarrow{h}_3 = \overrightarrow{e}_3$, see Fig. 1.

Let the head be in the erected position and the eye focused on the targets in infinity, then the following rules are valid:

Donders' Law: The eye position is determined by the gaze direction, [3,16].

Listing's Law: When the eye is in the primary position, it can rotate around axes lying in plane that is perpendicular to the gaze direction only. This plane passes through the centre of rotation and is called the Listing's plane, see Fig. 2, [3,16].

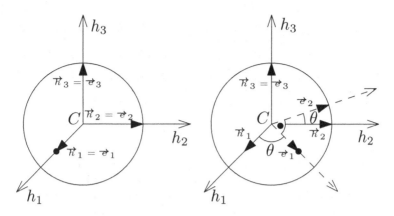

Fig. 1. Primary position and the position after rotation around axis h_3 by the angle θ.

4 Eye Position

In this Section we express the general position of the eye w.r.t. the gaze direction $\overrightarrow{e}_1 = (p_1, p_2, p_3)$ in the primary position. We describe the position by means of the unit vectors $\overrightarrow{e}_1 = (p_1, p_2, p_3)$ and $\overrightarrow{e}_3 = (x_1, x_2, x_3)$ that are fixed to the eye, see Sect. 3 and Fig. 1.

First, we will express the vectors $\overrightarrow{e}_1$ and $\overrightarrow{e}_3$ in terms of $\mathbb{G}_3$ as the elements E_1 and E_3 using the rotor $R = a_0 + a_2\sigma_{31} + a_3\sigma_{12}$ which satisfies the Listing's law, i.e. the appropriate axis of rotation is from the Listing's plane.

$$
\begin{aligned}
E_1 = RH_1\overline{R} &= (a_0 + a_2\sigma_{31} + a_3\sigma_{12})\sigma_1(a_0 - a_2\sigma_{31} - a_3\sigma_{12}) \\
&= (a_0^2 - a_2^2 - a_3^2)\sigma_1 - 2a_0a_3\sigma_2 + 2a_0a_2\sigma_3, \qquad (9)
\end{aligned}
$$

$$
\begin{aligned}
E_3 = RH_3\overline{R} &= (a_0 + a_2\sigma_{31} + a_3\sigma_{12})\sigma_3(a_0 - a_2\sigma_{31} - a_3\sigma_{12}) \\
&= -2a_0a_2\sigma_1 + 2a_2a_3\sigma_2 + (a_0^2 - a_2^2 + a_3^2)\sigma_3, \qquad (10)
\end{aligned}
$$

where H_1, H_3 are the images of the Euclidean embedding of the head-attached Cartesian coordinate system unit vectors h_1 and h_3, respectively. Clearly we obtained that $x_1 = -p_3$. Because the vectors $\overrightarrow{e}_1$ and $\overrightarrow{e}_3$ are orthogonal and unit, the following equalities for their standard scalar product hold:

$$
\overrightarrow{e}_1 \cdot \overrightarrow{e}_3 = -p_1p_3 + p_2x_2 + p_3x_3 = 0, \qquad (11)
$$

$$
p_3^2 + x_2^2 + x_3^2 = 1. \qquad (12)
$$

For $p_2 \neq 0$ we express x_2 from (11) and substitute it in (12) consequently:

$$
x_2 = \frac{p_1p_3 - p_3x_3}{p_2}, \qquad (13)
$$

$$
p_3^2 + \frac{p_1^2p_3^2 - 2p_1p_3^2x_3 + p_3^2x_3^2}{p_2^2} + x_3^2 = 1. \qquad (14)
$$

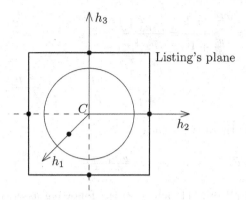

Fig. 2. Listing's plane in coordinate system H, the eye is in the primary position.

After several algebraic operations we obtain from (14) the following quadratic equation:

$$(\frac{p_3^2}{p_2^2} + 1)x_3^2 - \frac{2p_1p_3^2}{p_2^2}x_3 + \frac{p_1^2p_3^2}{p_2^2} + p_3^2 - 1 = 0 \tag{15}$$

with discriminant

$$D = \frac{4p_1^2p_3^4}{p_2^4} - 4(\frac{p_3^2}{p_2^2} + 1)(\frac{p_1^2p_3^2}{p_2^2} + p_3^2 - 1)$$

$$= -4[\frac{p_3^2}{p_2^2}(p_1^2 + p_2^2 + p_3^2 - 1) - 1] = 4,$$

which has the following solution

$$x_{3_{1,2}} = \frac{\frac{2p_1p_3^2}{p_2^2} \pm 2}{\frac{2p_3^2}{p_2^2} + 2} = \frac{\frac{p_1p_3^2}{p_2^2} \pm 1}{\frac{p_3^2}{p_2^2} + 1} = \frac{p_1p_3^2 \pm p_2^2}{p_2^2 + p_3^2}. \tag{16}$$

If we substitute (16) in (13), we get

$$x_{2_{1,2}} = \frac{p_1p_3}{p_2} - p_3\frac{\frac{p_1p_3^2}{p_2^2} \pm 1}{\frac{p_3^2}{p_2^2} + p_2} = \frac{p_1p_3}{p_2} - \frac{p_1p_3^3 \pm p_2^2p_3}{p_2p_3^2 + p_2^3}. \tag{17}$$

Now we need to determine the appropriate x_2 and x_3, respectively. If we substitute the coefficients of (9) for (p_1, p_2, p_3) and the coefficients of (10) for (x_1, x_2, x_3) into (16) and (17), respectively, we obtain the unique form of x_2 and x_3 as follows:

$$x_3 = \frac{\frac{p_1 p_3^2}{p_2^2} + 1}{\frac{p_3^2}{p_2^2} + 1} = \frac{p_1 p_3^2 + p_2^2}{p_2^2 + p_3^2}, \tag{18}$$

$$x_2 = \frac{p_1 p_3}{p_2} - p_3 \frac{\frac{p_1 p_3^2}{p_2^2} + 1}{\frac{p_3^2}{p_2^2} + p_2} = \frac{p_1 p_3}{p_2} - \frac{p_1 p_3^3 + p_2^2 p_3}{p_2 p_3^2 + p_2^3}. \tag{19}$$

Therefore for $p_2 \neq 0$ there is a solution:

$$\vec{e}_3 = \left(-p_3, \ \frac{p_1 p_3}{p_2} - \frac{p_1 p_3^3 + p_2^2 p_3}{p_2 p_3^2 + p_2^3}, \ \frac{p_1 p_3^2 + p_2^2}{p_2^2 + p_3^2} \right) \tag{20}$$

For $p_2 = 0$, we derive from (11) and (12) the following assertions:

$x_1 = -p_3, \ x_3 = p_1, \ x_2 = \sqrt{1 - p_1^2 - p_3^2} = p_2 = 0, \quad$ for $p_2 = 0, \ p_3 \neq 0,$

$x_1 = 0, \ x_2 = 0, \ x_3 = 1, \quad$ for $p_1 = 1, \ p_2 = 0, \ p_3 = 0,$

$x_1 = 0, \ x_2 = \sqrt{1 - x_3^2}, \ x_3 \in \langle -1; 1 \rangle \quad$ for $p_1 = -1, \ p_2 = 0, \ p_3 = 0.$

The solution for $p_2 = 0$ is then of the form:

$$\vec{e}_3 = \begin{pmatrix} -p_3 \\ 0 \\ p_1 \end{pmatrix}, \quad \text{for } p_1 \neq -1, \tag{21}$$

$$\vec{e}_3 = \begin{pmatrix} 0 \\ \sqrt{1 - x_3^2} \\ x_3 \end{pmatrix}, \ x_3 \in \langle -1; 1 \rangle, \quad \text{for } p_1 = -1. \tag{22}$$

As a result, we obtained the description of the vector $\vec{e}_3$ expressed by the coefficients of the gaze direction $\vec{e}_1 = (p_1, p_2, p_3)$, see (20), (21) and (22). This determines the general position completely.

5   Velocity Plane in General Eye Position

In this section we want to determine the axes available for the rotation of the eye in general position. When the eye is in the primary position, it can rotate around the axes lying in the Listing's plane only. These rotations are then expressed by the rotor $R = a_0 + a_2 \sigma_{31} + a_3 \sigma_{12}$. Indeed, let us suppose that the eye can get from primary position to the position A by rotor R and another position B by different rotor $R_1 = b_0 + b_2 \sigma_{31} + b_3 \sigma_{12}$, respectively. Now we want to express the transformation that moves the eye from the position B to some position C with the same gaze direction like in position A. Because Donders' law is valid, the assertions $A = C$ and $R = R_2 R_1$ must hold, see the calculation of the composed rotation at the end of Sect. 2. We determine the rotor R_2 from the following equation:

$$R_2 R_1 = R.$$

Indeed, by right multiplication by $\overline{R}_1$ we calculate

$$R_2 = R\overline{R}_1 = (a_0 + a_2\sigma_{31} + a_3\sigma_{12})(b_0 - b_2\sigma_{31} - b_3\sigma_{12})$$
$$= a_0b_0 - a_0b_2\sigma_{31} - a_0b_3\sigma_{12} + a_2b_0\sigma_{31} + a_2b_2 + a_2b_3\sigma_{23} + a_3b_0\sigma_{12}$$
$$- a_3b_2\sigma_{23} + a_3b_3$$
$$= a_0b_0 + a_2b_2 + a_3b_3 + (-a_3b_2 + a_2b_3)\sigma_{23} + (a_2b_0 - a_0b_2)\sigma_{31}$$
$$+ (a_3b_0 - a_0b_3)\sigma_{12}.$$

Therefore, the axes of rotation expressed by means of the rotors R and R_1 of the eye in general position are of the form

$$(-a_3b_2 + a_2b_3)\sigma_{23} + (a_2b_0 - a_0b_2)\sigma_{31} + (a_3b_0 - a_0b_3)\sigma_{12}.$$

Let us fix the coefficients b_0, b_2, b_3 and let us consider the coefficients a_0, a_2, a_3 as parameters. Then by elementary operations we derive from the initial system of linear equations the following:

$$h_1 = -a_3b_2 + a_2b_3/b_0$$
$$h_2 = a_2b_0 - a_0b_2$$
$$h_3 = a_3b_0 - a_0b_3/b_2$$

$$-\ -$$

$$b_0h_1 + b_2h_3 = -a_0b_2b_3 + a_2b_0b_3$$
$$h_2 = -a_0b_2 + a_2b_0/(-b_3)$$

$$-\ -$$

$$0 = b_0h_1 - b_3h_2 + b_2h_3 \tag{23}$$

The Eq. (23) provides the description of a plane that contains all available axes of rotation of the eye in general position. This plane is called the *velocity plane*.

Velocity plane is expressed by the coefficients of a general rotor $R_1 = b_0 + b_2\sigma_{31} + b_3\sigma_{23}$ determined by the axis from the Listing's plane, but we want to express this plane by the parameters of the gaze direction $\overrightarrow{e}_1 = (p_1, p_2, p_3)$. We know from (9) that

$$\overrightarrow{e}_1 = (p_1, p_2, p_3) = (b_0^2 - b_2^2 - b_3^2, -2b_0b_3, 2b_0b_2).$$

Therefore we derived the following equations:

$$b_0^2 - b_2^2 - b_3^2 = p_1, \quad -2b_0b_3 = p_2, \quad 2b_0b_2 = p_3.$$

Consequently, we express the coefficients of the rotor R_1 as follows:

$$b_0 = \sqrt{\frac{p_1}{2} + \frac{1}{2}},$$

$$b_2 = \frac{p_3}{2\sqrt{\frac{p_1}{2} + \frac{1}{2}}},$$

$$b_3 = -\frac{p_2}{2\sqrt{\frac{p_1}{2} + \frac{1}{2}}}, \quad \text{for } b_0 \neq 0, \ p_1 \neq -1.$$

Now we have the general equation of the velocity plane in terms of the coefficients of gaze direction $\overrightarrow{e}_1 = (p_1, p_2, p_3)$, $p_1 \neq -1$ in the form:

$$\sqrt{\frac{p_1}{2} + \frac{1}{2}}\, h_1 + \frac{p_2}{2\sqrt{\frac{p_1}{2} + \frac{1}{2}}}\, h_2 + \frac{p_3}{2\sqrt{\frac{p_1}{2} + \frac{1}{2}}}\, h_3 = 0.$$

After multiplication by the term $2\sqrt{\frac{p_1}{2} + \frac{1}{2}}$ we obtain:

$$(p_1 + 1)h_1 + p_2 h_2 + p_3 h_3 = 0. \tag{24}$$

Therefore we conclude that when the eye is focused in the general direction $\overrightarrow{e}_1 = (p_1, p_2, p_3)$, it can rotate around axes lying within the plane (24).

Acknowledgements. This research was supported by a grant of the Czech Science Foundation no. 17-21360S, "Advances in Snake-like Robot Control" and by a Grant No. FSI-S-17-4464.

References

1. Bayro-Corrochano, W.: Modeling the 3D kinematics of the eye in the geometric algebra framework. Pattern Recogn. **36**(12), 2993–3012 (2003). https://doi.org/10.1016/S0031-3203(03)00180-8
2. Dorst, L., Fontijne, D., Mann, S.: Geometric Algebra for Computer Science: An Object-oriented Approach to Geometry, 1st edn. Morgan Kaufmann Publishers Inc., Burlington (2007)
3. Haslwanter, T.: Mathematics of three-dimensional eye rotations. Vision. Res. **35**(12), 1727–1739 (1995). https://doi.org/10.1016/0042-6989(94)00257-M
4. Hestenes, D.: New Foundations for Classical Mechanics, 2nd edn. Kluwer Academic Publishers, Dordrecht (1999)
5. Hildenbrand, D.: Foundations of Geometric Algebra Computing. Springer, New York (2013). https://doi.org/10.1007/978-3-642-31794-1
6. Hrdina, J., Návrat, A.: Binocular computer vision based on conformal geometric algebra. Adv. Appl. Clifford Algebr. **27c**, 1945–1959 (2017). https://doi.org/10.1007/s00006-017-0764-4
7. Hrdina, J., Návrat, A., Vašík, P., Matoušek, R.: Geometric control of the trident snake robot based on CGA. Adv. Appl. Clifford Algebr. **27**, 621–632 (2017)
8. Hrdina, J., Vašík, P.: Notes on differential kinematics in conformal geometric algebra approach. In: Matoušek, R. (ed.) Mendel 2015. AISC, vol. 378, pp. 363–374. Springer, Cham (2015). https://doi.org/10.1007/978-3-319-19824-8_30
9. Hrdina, J., Návrat, A., Vašík, P., Matoušek, R.: Geometric algebras for uniform colour spaces. Math. Methods Appl. Sci. **41**, 4117–4130 (2018). https://doi.org/10.1002/mma.4489
10. Hrdina, J., Návrat, A., Vašík, P.: Control of 3-link robotic snake based on conformal geometric algebra. Adv. Appl. Clifford Algebr. **26**, 1069–1080 (2016). https://doi.org/10.1007/s00006-015-0621-2
11. Hrdina, J., Návrat, A., Vašík, P., Matoušek, R.: Fish eye correction by CGA nonlinear transformation. Math. Methods Appl. Sci. **41**, 4106–4116 (2018). https://doi.org/10.1002/mma.4455

12. Hrdina, J., Návrat, A., Vašík, P., Matoušek, R.: CGA-based robotic snake control. Adv. Appl. Clifford Algebr. **27**, 621–632 (2017). https://doi.org/10.1007/s00006-016-0695-5
13. Lounesto, P.: Clifford Algebra and Spinors, 2nd edn. Cambridge University Press, Cambridge (2006)
14. MacDonald, A.: Linear and Geometric Algebra. Third printing, corrected and slightly revised (2010)
15. Perwass, C.: Geometric Algebra with Applications in Engineering. Springer, New York (2009). https://doi.org/10.1007/978-3-540-89068-3
16. Wong, A.M.F.: Listing's law: clinical significance and implications for neural control. Surv. Ophthalmol. **49**(6), 563–575 (2004). https://doi.org/10.1016/j.survophthal.2004.08.002

Increased Sensitivity of Ultrasonic Radars for Robotic Use

Karel Hájek$^{(\boxtimes)}$ ⓘ

University of Defence, Brno, Czech Republic
karel.hajek@unob.cz

Abstract. At present, robotic ultrasonic radars are used only for small distances and with low resolution and great sensitivity to parasitic echo. This article shows some ways to increase the sensitivity and resolution of targets in the viewing environment. This is achieved by three basic ways. On the one hand, special noise analysis enables optimum design of input circuits for input noise reduction by 20–30 dB over standard solutions. This will allow for a considerably higher basic radar range. Another benefit is the DSP for the received signal, which allows a continuous evaluation of the amplitude and phase of this signal by Fourier analysis, coupled with additional digital filtration. This makes it possible to evaluate the reflection distance with more than 1 lambda precision. The third benefit is the use of at least pairs of sensors both horizontally and vertically. Vertical phase evaluation of the received signals will allow, in addition to higher vertical resolution, suppression of parasitic echo from the ground. Horizontal amplitude-phase evaluation along with the phase control of the pair of exciters will substantially increase the resolution of the targets in the observed environment at a broader viewing angle. It is possible to arrive at a system that produces a fairly accurate ultrasonic image of the surroundings of the radar.

Keywords: Ultrasonic radar · Radar sensitivity · DSP · Estimation of phase · Synchronous sampling

1 Introduction

It is an effort to get robotic systems to gain the most detailed and accurate information about their surroundings. For this purpose, they often use common ultrasonic radars, which usually work only for small distances and with low resolution and great sensitivity to parasitic reflections. Most commonly, a simple analogue ultrasonic signal processing is used along with a simple 8-bit processor system for evaluating measured time intervals. These systems are based on the use of standard modules such as HC-SR04, HY-SRF05, e.g. [1–5]. There are designed systems with subsequent processing to create a ultrasonic image of the robot environment, including multisensor systems (e.g. [4, 5]) or systems with rotation of the sensors [3].

Achieved sensitivity and resolution of ultrasonic radar is a complex problem that affects several factors that are in mutual relation throughout the signal chain. The basic factor is the choice of analogue or digital processing of ultrasonic signals. About fifteen or more years ago, the choice of system with DSP processing was significantly limited by available technology and pricing. Nowadays, when prices for powerful 32-bit

© Springer Nature Switzerland AG 2019
J. Mazal (Ed.): MESAS 2018, LNCS 11472, pp. 130–139, 2019.
https://doi.org/10.1007/978-3-030-14984-0_11

systems (such as ARMs) have dropped to the cost of 8-bit processors, the use of DSP systems is a priority as it provides additional significant opportunities to improve the sensitivity, resolution and overall ultrasonic radar functionality compared with previously used systems with analogue ultrasonic signal processing.

This paper discusses a system with a DSP block in the signal string. If we want to achieve maximum sensitivity and functionality, we can not underestimate any part of the whole chain. The basic sensitivity is basically limited by the sensitivity of the sensor and the input noise of the amplifier. The next block is analog signal preprocessing, providing the necessary amplification and narrowband filtering to suppress noise outside the ultrasonic signal band, including sufficient anti-aliasing filtration. Additionally, an A-D converter with sufficient dynamic range must be selected so that the quantization noise does not reduce the sensitivity attainable. Furthermore, it will be shown that it is advantageous for the sampling frequency of this A-D converter to be synchronous with the ultrasonic signal.

In terms of ultrasonic sensors, it is advisable to provide sensors with acoustic antennas to increase both their sensitivity and directionality. For these purposes and for other benefits, two (possibly even more) sensors can help. These, together with the appropriate DSP, will enable, among other things, better detection and suppression of parasitic reflections as well as increased accuracy of localization of objects around the radar. Another way to increase the quality of the ultrasonic radar system is to turn the antenna system in the horizontal axis, which, in conjunction with the DSP, will allow to obtain a better image of the radar surroundings.

An important element of the whole chain is, of course, the DSP block. This, in addition to the above-mentioned suppression of parasitic reflections, will allow both the control of the entire system and the increase of accuracy, as well as a detailed analysis of the character and properties of the signals of the individual reflections. Together with the capability to turn the antenna system, it can be significantly closer to acquiring a comprehensive ultrasonic image of the surroundings.

2 Minimisation of Preamplifier Noise

The basic task of the pair of sensor - preamplifier is to achieve maximum signal to noise ratio (SNR) with appropriate noise matching. Its importance increases with the improvement of subsequent signal processing, when we can process signals that are close to noise.

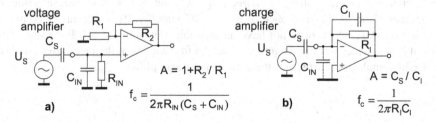

Fig. 1. Basic model of preamplifier for capacitive signal source: (a) voltage amplifier, (b) charge amplifier.

Optimization of noise adaptation for signal sources with resistive character of internal impedance are solved in commonly available literature. Somewhat more complex are cases with sensors with a general character of internal impedance. Ultrasonic sensors typically have the capacitive internal impedance. In this case, the analysis of noise ratios is considerably more complex [6]. There are three main sources of noise here. This is expressed by the following equations, where the noise voltage by frequency normalization is generally expressed as the voltage noise spectral density E_n

$$E_{neq} = \sqrt{E_{nv}^2 + (E_{nI}Z)^2 + E_{nR}^2}. \tag{1}$$

The fully accurate expression should have more members, but the resistive part of the source's internal impedance is very low, and the auxiliary resistors for amplification settings can be chosen at such low levels that their noise can be neglected. We have the three major sources of noise used in (1). The first of these is the voltage noise spectral density E_{nV} of voltage amplifier, which is essentially independent and only partly affects only the type of circuit (voltage or charge amplifier). It can also be noted that bipolar input amplifiers have somewhat lower voltage noise than amplifiers with unipolar input (ca. 10 times).

Next the voltage noise spectral density E_{nI} expresses the current noise of amplifier. This is converted to the resulting voltage noise by multiplying to any impedance, so also to the internal capacitance of the ultrasonic sensor. Of course, this is a frequency dependent one, which adds another dimension to finding the optimal solution. As well as voltage noise, the amount of current noise depends on the input circuitry of the amplifier. In this case, it is the opposite, lower current noise has unipolar input, two to three decades. The third source of noise density E_{nR} is the auxiliary resistor R for securing the DC stability of the amplifier (minimizing DC offset effect) because the sensor itself is capacitive. It has different locations in a charge or voltage amplifier (Fig. 1), but in both cases it creates a high-pass (HP) filter with a defined limit frequency from the amplifier. In addition to the thermal noise itself, according to the well-known relationship $E_n = \sqrt{4kTR}$, it also produces a multiplier effect for current noise. The voltage noise of the resistor generally rises with the square root of its resistance, and therefore it seems necessary to reduce it. But the opposite is true, because at the same time the corresponding capacity creates a low pass filter for this noise, whose attenuation grows linearly with the value of this resistance.

The whole optimization task for noise minimization has multiple parameters. It is a choice of amplifier (unipolar or bipolar with the lowest possible voltage and current noise), choice of circuit (voltage or charge amplifier) and, last but not least, selection of the optimal value of the resistance to ensure DC stability. This is a rather complex optimization task, which does not have an unambiguous solution [6]. For these variables, it is also necessary to take into account the frequencies of the permissive band of the useful signal, the sensor, and the gain of the preamplifier.

3 Sensor Antenna System

As mentioned earlier, mechanical realization of the shape of the acoustic antenna of a sensor and their integration into antenna systems (at least two sensors per direction) can significantly influence the resulting quality and parameters of the entire radar system. The basic problem is the shape of an acoustic antenna of one sensor. Typically, a simple conical shape is used, where it is necessary to find the optimal angle and length with respect to the desired properties.

It is important to use a pair of sensors in the vertical direction. In the case of the use of DSP, their use with an optimum angle is advantageous for easy resolution of parasitic echo from the ground and low objects, as opposed to distant higher objects. The result of a simple experiment with comparing two signals at the same location only when using a small vertical deviation of the direction ($+15°$) is shown in Fig. 2.

As can be seen, the fairly pronounced reflections in the near and middle distances of the original signal have completely concealed the echo of the background signal. The above-mentioned sensor angle increase by $15°$ changed the received signal y_{1h} so that all reflections from near and mid-distance declined compared to the echo of the higher background object. Additionally, it is possible to deduce from the duration of the given echo and the shape of its envelope the actual shape of the object.

The example is an illustration of how to create an orientation picture of the surrounding area in the vertical direction. Two basic paths can be used to create an image of the environment in the horizontal direction. The former has a fixed, non-rotating base with a number of sensors positioned sequentially across the broader observed horizon (e.g. [4, 5]), theoretically up to a complete circle. This path requires a relatively large number of sensors, otherwise it should have a small horizontal resolution.

The second path uses rotation of the entire antenna system (e.g. [3]). This allows a seemingly almost unlimited horizontal resolution. However, this is practically limited by the rotational speed at the end time of one measurement and is basically limited mainly by the angular resolution of the ultrasonic antenna's radiation characteristics. This is an important aspect that justifies the use of pairs of sensors not only in the vertical but also in the horizontal direction. Using pairs of sensors makes it possible to obtain additional information from the echo signals. Above all, it is information about objects that can be specified from time differences of both reflections from the same targets [7]. This measurement can be further refined in connection with the precise measurement of the phase shifts of the signals. This makes it possible to achieve a reliable echo resolution of less than one wavelength. It is possible to use a certain degree of direction and narrowing of the radiation characteristics of the considered pair of sensors at the mutual time shifts of exciting the two sensors in the pair as the simplest phase array [8]. Of course, this effect could also be used for vertical viewing. Apparently simpler and two or four times faster, the antenna system would be rotated with two or four opposite sensors, so it would be enough to rotate it only at an angle of $180°$ or $90°$.

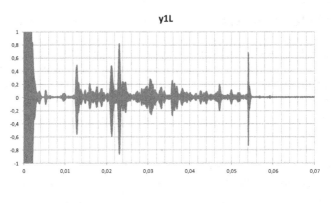

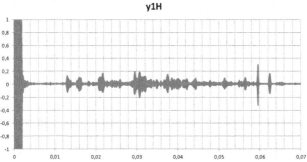

Fig. 2. Experimental comparison of two echo signals at the same location only with a small vertical deviation of direction (+15°) y_{1H} vs. y_{1L}.

4   Using DSP to Improve Ultrasonic Radar Properties

The expected use of a powerful 32-bit processor system, e.g. with some ARM processors, would allow us to take advantage of the different forms of DSP in addition to basic system control.

4.1   Synchronous Sampling and Precise Measurement of the Amplitude and Phase of the Ultrasonic Signal

The important thing is that we get complete sampled signals from the used sensors using the AD converter. From these signals, you can get more information using the DSP. To simplify and improve the overall DSP process, it is advisable to use the sampling frequency synchronization with the frequency of the transmitted ultrasonic signal. In this case, only theoretically, only three samples per signal period are sufficient to obtain complete and highly accurate amplitude and signal phase information. Since the scanned signals will not be purely harmonic but with gradual changes in amplitude, it is advisable to increase the number of samples by more than three per period, but experiments show that just about 5 samples per period are sufficient. When considering the frequency of a 40 kHz ultrasonic signal, this is the sampling frequency of 200 kHz. This also has an important practical meaning. Reducing the required sampling frequency is required to reduce the frequency and performance requirements of the processor system used, as well as the storage size. A simple example for monitoring a 100 ms time interval (about 17 m) requires a storage memory 20 kB to measure one signal.

The basic principle of calculating the amplitude and phase for several signal periods at its echo expresses these relationships

$$\mathrm{Re} = \frac{2}{N} \sum_{n=0}^{N-1} s(n) \cos\left(\frac{2\pi}{M} n\right), \tag{2}$$

$$\mathrm{Im} = \frac{2}{N} \sum_{n=0}^{N-1} s(n) \sin\left(\frac{2\pi}{M} n\right), \tag{3}$$

where M is the number of samples per period. The information about the real and imaginary component of the signal is obtained by multiplying its N samples with the corresponding cosine resp. a sinusoidal signal about the frequency of the ultrasonic signal and the corresponding summation. It is basically the calculation of the first harmonic component of the discrete Fourier series, which is in principle quite accurate for the case of synchronous sampling. Subsequent conversion to module and phase value is a trivial task. At approximately 10 periods of echo signal and five samples per period, the number of samples considered $N \cong 50$.

4.2   Accurate Measurement of Time Delay of Ultrasonic Signal

The possibility of accurate measurement and phase calculation for a short ultrasonic signal range (e.g., the interval of one of the echo) allows an easy and accurate measurement of the time delay of the ultrasonic signal with a substantially smaller error than one signal period.

The basic idea is to measure the phase shift of two signals whose mutual delay is not an integer number of periods. If we subtract from this delay an integer multiple of the period, what is usually technically feasible, the rest of the delay can be determined on the basis of the mutual phase shift of the echo sections (approximately 10–20 periods) of both signals. We find this as a difference between the phases of both signals, calculated from relations (2) and (3).

This procedure can be applied if I have a pair of sensors and we need to determine the position angle of the object to the base of both sensors. This procedure can increase the accuracy of the time delay reading of approx. 10 times. The accuracy of the angle determination relative to the base of both sensors can be increased accordingly.

If we need to increase the accuracy of reading the echo time for one signal, two-frequency measurements can be used. The basic idea lies in a pair of measurements, when we use a very small frequency deviation of signal (much smaller than the ultrasonic sensor resonance bandwidth) in the second measurement. For this purpose, a DDS generator commonly manufactured as an IC (e.g., AD9831) can be used to generate a harmonic frequency signal with a precision of up to 32 bits from a single crystal oscillator, so that the error rate of the two set frequencies is significantly lower than the crystal oscillator alone.

The precise measurement of the time delay of the ultrasonic signal consists in evaluating the phase difference of both measured signals with slightly different frequencies. If the frequency change is sufficiently small, then the phase difference $\Delta\varphi$ becomes less than one period over the given distance and the corresponding ultrasonic signal propagation time. Then, the relation for the time delay can be expressed

$$\Delta t = \Delta\varphi/(2\pi f_1) = n(\Delta f)/(2\pi f_1^2), \tag{4}$$

where f_1 is the frequency of the first ultrasonic signal, and Δf is the frequency difference of the second signal, and n is the number of whole periods at the time of the echo of the signal.

The whole problem can be illustrated by the following example. For a 40 ms time period, the 1600 signal cycles take place at a frequency of 40 kHz. To reach the 1601 period (shift by a total of 360°), we would have to increase the frequency to 40.025 kHz. If we used a new frequency with a decimal deviation of Δf (40.0025 kHz), we would measure a phase difference of 36° for a given time of 40 ms. At normal ultrasound speed and its wavelength of 7.5 mm, the measurement accuracy would be approximately 0.75 mm.

For the basic practical verification of phase difference measurement, we performed measurements of two consecutive ultrasonic pulses with the same frequency of ultrasound and we evaluated the measurable phase coincidence of unambiguous reflections from the same target at 40 kHz, sampling at 200 kHz and echo time of about 59 ms. The result is shown in Fig. 3, where very precise phase equality is evident.

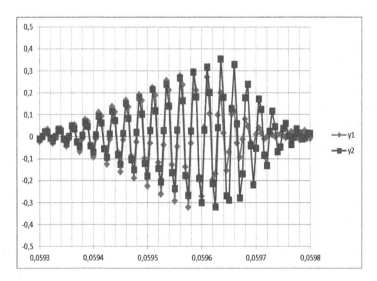

Fig. 3. Testing the phase equivalence of two independent signals of ultrasonic echo under the same conditions: f = 40 kHz, fs = 200 kHz.

Although we have not been able to provide a completely accurate synchronization of signal generation sampling and matching of the shape and length of the exciting pulse at this stage of the experiment, yet we calculated the phase difference of about 5° according to the relations (2) and (3). It can be assumed that while ensuring full synchronization, the resulting measurement accuracy can be ensured for a smaller phase difference than in the above-mentioned example of the 36° (1/10 period) considered.

4.3   Creating a Ultrasonic Image Around the Radar

As mentioned above, to obtain the most accurate picture of the surroundings of the radar, it is necessary to analyse all significant echo signals from objects around the robot (e.g. [7]). The significance assessment is necessary to correct with regard to the distance, the shape of the reflection signal, the effect of the angle of reflection on the perpendicular direction etc. It is also necessary to compare the significance of the shape changes, the magnitude and the delay of the signal from the pairs of sensors and the signals after the minimum rotation in the horizontal plane Simple example is shown in Fig. 2. It can be assumed that it will be necessary to create a minimal library of basic waveforms reflective signals so that it can be operated on the basis of artificial intelligence (AI) principles (e.g. [9–11]).

As a further aspect, it is possible to evaluate the possible movement of individual targets. In addition to tracking the change in reflection time, a Doppler effect evaluation (e.g., [12]) can be used to confirm movement of the object. Since the frequency shift can not be easily evaluated by analog processing (short duration of reflection signals), it is possible to use spectral FFT analysis where high sensitivity to frequency shift can be achieved by appropriate procedures. At the considered time of the evaluated signal (approx. 10–20 periods, i.e. $n = 50 - 100$), an error of estimation of the frequency deviation of about 10^{-6} bin can be considered, corresponding to the object's speed of movement of about 20 cm/sec.

It can be assumed that the use of DSP in connection with the use of AI elements will allow further gradual improvement of ultrasonic radar functionality and the ability to produce the most accurate images of the radar environment, as illustrated by some published articles.

The basic idea is to measure the phase shift of two signals whose mutual delay is not an integer number of periods. If we subtract from this delay an integer multiple of the period, what is usually technically feasible, the rest of the delay can be determined on the basis of the mutual phase shift of the echo sections (approximately 10–20 periods) of both signals. We find this as a difference between the phases of both signals, calculated from relations (2) and (3).

5   Conclusion

For the robotic use of ultrasonic radars, it is desirable to increase their sensitivity and improve their functionality, since currently available models do not have sufficient features for more demanding applications.

The article shows and discusses some aspects of this task. First of all, it is based on the idea that ultrasonic radar is to be seen as a complex system, where each part of the signal processing chain can play a significant role, as well as the important links between its individual parts.

In the paper there is shown a first step towards reducing the base noise and increasing the SNR of pair of sensor-preamplifier. The pre-amp noise minimization solution with respect to the capacitive impedance of the ultrasonic sensor along with its subsequent optimal analog pre-filtering can lead to a reduction of up to 30 dB.

Another way to improve functionality in terms of suppression of parasitic reflections and resolution of objects in both the vertical and horizontal plane is the use of pairs of sensors. An example is shown that the use of this pair for vertical scanning with an angle of approximately 15° allowed to significantly different parasitic reflections from nearby low objects on the ground from distant higher objects.

The article shows that the essential condition for a significant improvement in radar functionality is the use of DSP for processing the ultrasonic signal. For this, the synchronous sampling of signal from the ultrasonic sensor is shown as an important basic condition. This, together with the use of relations (2) and (3), allows simple and very accurate evaluation of the amplitude and especially the phase with minimal demands on sampling rate, computational and memory demands on hardware.

The following are principles for increasing the accuracy of time delay measurement with resolution below one wavelength. This is especially advantageous when using the pair of sensors, because it increases the angular resolution of the reflections by 10 times.

In conclusion, the assumption of the use of DSP and AI principles for a more detailed analysis of all significant reflections in one ultrasonic radar signal and obtaining of further information from comparison of signals from pairs of sensors is presented.

References

1. Paulet, M.V, Salceanu, A., Neacsu, O.M.: Ultrasonic radar international conference and exposition on electrical and power engineering (EPE 2016), Iasi, Romania, pp. 20–22 October 2016
2. Adarsh, S., Mohamed Kaleemuddin, S., Bose, D., Ramachandran, K.I.: Performance comparison of infrared and ultrasonic sensors for obstacles of different materials in vehicle/robot navigation applications. Mater. Sci. Eng. **149**, 012141 (2016). https://doi.org/10.1088/1757-899X/149/1/012141
3. Daru, R.R., Biswas, H., Barman, S., Al-Quabid, A.: Determining 2D shape of object using ultrasonic sensor. In: 2016 3rd International Conference on Electrical Engineering and Information Communication Technology (ICEEICT), INSPEC Accession Number 16726539. https://doi.org/10.1109/ceeict.2016.7873143
4. Wu, X., Abrahantes, M., Edgington, D.: MUSSE: a designed multi-ultrasonic-sensor system for echolocation on multiple robots. In: 2016 Asia-Pacific Conference on Intelligent Robot Systems. https://doi.org/10.1109/acirs.2016.7556192

5. Jansen, N.F.: Short range object detection and avoidance, Department of Mechanical Engineering Control Systems Technology, Eindhoven University of Technology, Eindhoven, November 2010. mate.tue.nl/mate/pdfs/12516.pdf
6. Hajek, K.: Improving the sensitivity of analog signal pre-processing systems (in Czech). https://docplayer.cz/22590386-Zvysovani-citlivosti-systemu-s-analogovym-predzpracovanim-signalu.html
7. Egaña, A., Seco, F., Ceres, R.: Processing of ultrasonic echo envelopes for object location with nearby receivers. IEEE Trans. Instrum. Measur. 57(12), 2751–2755 (2008)
8. Guerreroa, J.S., González, A.C., Vega, J.H., Tovarb, L.N.: Instrumentation of an array of ultrasonic sensors and data processing for unmanned aerial vehicle (UAV) for teaching the application of the Kalman filter. In: 2015 International Conference on Virtual and Augmented Reality in Education. https://doi.org/10.1016/j.procs.2015.12.260
9. Kleeman, L.: Advanced sonar sensing. In: Jarvis, R.A., Zelinsky, A. (eds.) Robotics Research. Springer Tracts in Advanced Robotics, vol. 6. Springer, Heidelberg (2003). https://doi.org/10.1007/3-540-36460-9_32
10. Heale A., Kleeman L.: A real time DSP sonar echo processor. In: Proceedings of 2000 IEEE/RSJ International Conference on Intelligent Robots and Systems (IROS 2000) (Cat. No. 00CH37113). https://doi.org/10.1109/iros.2000.893192
11. Moreno, L., Armingol, J.M., Garrido, S., et al.: A genetic algorithm for mobile robot localization using ultrasonic sensors. J. Intell. Robot. Syst. 34, 135 (2002). https://doi.org/10.1023/A:1015664517164
12. Ekimov, A., Sabatier, J.M.: Human motion analyses using footstep ultrasound and doppler ultrasound. J. Acoust. Soc. Am. 123(6), 149–154 (2008)
13. Andria, G., Attivissimo, F., Giaquinto, N.: Digital signal processing techniques for accurate ultrasonic sensor measurement. Measurement 30, 105–114 (2001). https://doi.org/10.1016/S0263-2241(00)00059-2
14. Burguera, A., González, Y., Oliver, G.: Sonar sensor models and their application to mobile robot localization. Sensors 9, 10217–10243 (2009). https://doi.org/10.3390/s91210217
15. Hirata, S., Kurosawa, M.K., Katagiri, T.: Accuracy and resolution of ultrasonic distance measurement with high-time-resolution cross-correlation function obtained by single-bit signal processing. Acoust. Sci. Technol. 30(6) (2009). The Acoustical Society of Japan

A Study on Direct Teleoperation Device Kinematics

Robert Pastor$^{(\boxtimes)}$ ⓘ, Aleš Vysocký, and Petr Novák

Faculty of Mechanical Engineering, Department of Robotics,
VŠB-Technical University Ostrava, 17. listopadu 15, 708 33 Ostrava,
Czech Republic
{robert.pastor,ales.vysocky,petr.novak}@vsb.cz

Abstract. Teleoperation using a controller with the same kinematics as the controlled robotic agent has the potential to be more effective than using a universal joystick. With direct teleoperation, the operator is restricted by the same workspace as the controlled robot, thus making the process more intuitive. In this study we compare controllers with various kinematic configurations and degrees of freedom. The performance of the controllers is tested in sample situations where users directly teleoperate a slave manipulator model matching the real master device kinematics.

Keywords: Teleoperation · HMI · Kinematics

1 Introduction

With the increased availability of high speed internet and the rise of new mobile 5G networks [1], one of the biggest problems of long range teleoperation, the time delay, is slowly fading away. The time delay was previously addressed by using a predictive model of the environment [2]. We can expect the rise of more data and delay demanding applications in our daily lives. This includes robot teleoperation.

An example of this can be found in driverless cars. Autonomous cars can drive in almost any condition these days. However there are still edge cases in which the control algorithms are uncertain what to do. This is one of the reasons preventing self-driving vehicles from increasing their numbers on the roads. Several self-driving car companies are testing ways to remotely take control of the car from afar [3, 4]. In the case of uncertainty, an operator would drive the vehicle out of an uncertain state and then return the control to the internal autonomous system. With this approach a few operators could oversee a fleet of vehicles.

Similar services could spread to different fields. One potential use is in elderly care where the task difficulty is more complex. An autonomous robot caretaker could run into a difficult situation and call an operator for assistance. The operator would take control of the robot remotely. Resolve the problem using a teleoperation device and return the control to the automated system. This could be a seamless experience for the end user. In this way one caretaker could oversee a number of patients right in their homes without the need to travel.

© Springer Nature Switzerland AG 2019
J. Mazal (Ed.): MESAS 2018, LNCS 11472, pp. 140–146, 2019.
https://doi.org/10.1007/978-3-030-14984-0_12

In the case of teleoperating a car, the operators' equipment is a standard gaming gear for racing video games. Robotic arm teleoperation interfaces tend to be considerably more expensive. From state of the art surgical teleoperation interfaces [5], experimental direct teleoperation devices [6–8] to more accessible positional and rotational input devices with force and torque feedback capabilities. The quality of experience improvements with using haptic feedback is evaluated in [9]. Direct teleoperation using a multilegged robot was demonstrated in [10], showing that not only humanoid and single arm robot can be teleoperated.

In this study we evaluate the user experience of using open-loop teleoperation devices with various kinematics to asses the role of kinematics on the intuitiveness of solving tasks.

The rest of the paper is organised as follows this paper we present four different open loop direct teleoperation controllers with various kinematics, define the conditions to test them in simulated tasks and evaluate the results.

2 Methods

Four controllers were developed. Each with different kinematic chain (Fig. 1). The arm A has three degrees of freedom (DOF) while the remaining arms B, C and D have four DOF. Table 1 shows Denavit–Hartenberg parameters for the controllers. The arms were designed to have the same maximum reach of 470 mm.

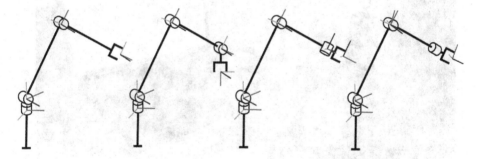

Fig. 1. Four arms with different kinematic chains. From left: A, B, C, D

Design. The arm is constructed from lightweight materials, aluminium and polycarbonate. The angle measurement in the controller joints is achieved via potentiometers. Each joint consists of two 3D printed parts connected with the potentiometers. The goal was to make the joints small and simple to allow the users easy access to all parts of the device. The joints are interconnected with square aluminium extrusion profiles. Cables are hidden inside the hollow profiles to avoid getting in the way.

In the base of each controller there is an Arduino microcontroller to which the potentiometers are connected. The microcontrollers convert potentiometer voltages into integers and send this data through a serial port into a PC. Fig. 2 shows controller type A being used to control the simulation. Figure 3 shows the setup schematic for the experiment.

Table 1. Denavit-Hartenberg parameters

Arm	Joint	θ [rad]	d [mm]	a [mm]	α [rad]
A	1	θ_1	20	0	$\pi/2$
	2	θ_2	0	230	0
	3	θ_3	0	230	0
B	1	θ_1	20	0	$\pi/2$
	2	θ_2	0	230	0
	3	θ_3	0	185	0
	4	θ_4	0	45	0
C	1	θ_1	20	0	$\pi/2$
	2	θ_2	0	230	0
	3	θ_3	0	185	$-\pi/2$
	4	θ_4	0	45	0
D	1	θ_1	20	0	$\pi/2$
	2	θ_2	0	230	0
	3	θ_3	0	0	$\pi/2$
	4	θ_4	230	0	0

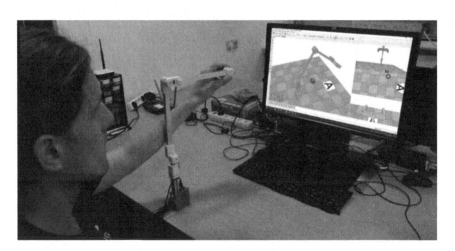

Fig. 2. Controlling the simulation

Simulation. We chose to use V-Rep for our simulation. The open source Newton dynamics engine was used in the simulations. An application in C# was developed. This application collects the measured data from the serial ports and sets the V-Rep simulation accordingly through a remote API. The simulation uses Vortex engine. The operator uses one controller at a time.

Scene. Two simulation scenes were built to perform the tests. In the first one, the operator is tasked to change the orientation of two handles by rotating them by 180° around a pivot joint. The test starts from an upright position of the arm in front of the

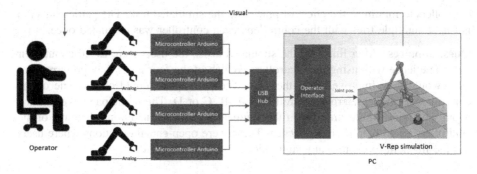

Fig. 3. Experiment setup schematic

handle panel. Time ends when both handles are turned. The second test consists of moving an object from one position to another. There are two fixed pins in the scene and three free rings inserted on one of the pins. The operator has to use the controller to grasp the rings and move them to the other pin. Time ends when all three rings are in the desired position (Figs. 4 and 5).

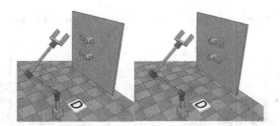

Fig. 4. Test 1 scene. Left: start. Right: end.

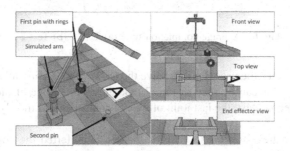

Fig. 5. Test 2 scene with the right places in the starting position.

Sample. Ten test subjects, males aged 20–27 were asked to participate in the evaluation study. None of the participants we familiar with direct teleoperation interfaces in advance. The users were performing the simulated tasks in randomized order of the controllers to minimize the effect of practicing the simulated task. The participants did not have multiple tries with the controllers, each controller was evaluated once.

Questionnaires. After finishing the simulated tasks, the users were asked about their own experience with using the controllers. Which of the four types they preferred and what would they improve about the tests. The participants could choose one of the controllers as their preferred controller as A, B, C or D. The participants were asked about what they would improve in the simulation and what additional features of the controllers could help user experience. These were open-ended questions where users has the freedom to express any suggestions they came up with.

3 Results

We looked at the average times the users took to complete the tests. Figure 6 shows the average times for each arm. The data indicated that in our test conditions, the additional DOF did not improve the user performance. In the second test the controller type A proved to be one of the faster types, outperforming types C and D, despite the fact that it had only three degrees of freedom.

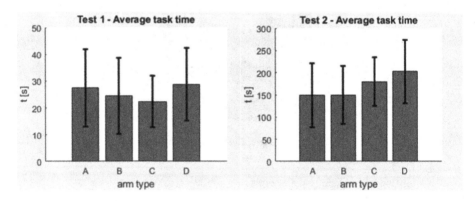

Fig. 6. Average completion time (a) Test 1 (b) Test 2

In some situations, the users tried to solve the tests without using all available joints of the controller. This can be seen in Fig. 7 that shows the usage of the 4th degree of freedom. In the second test, the last joint of type D controller was mostly unused by the users.

The questionnaires after the tests show that users preferred the controller type B (Fig. 8). When asked about possible improvements to the controller, several users reported that they were missing a confirmation of a successful gripper hold and collision warnings, indicating a need for feedback. The majority of users said they were not sure how far the end effector was from the manipulated objects.

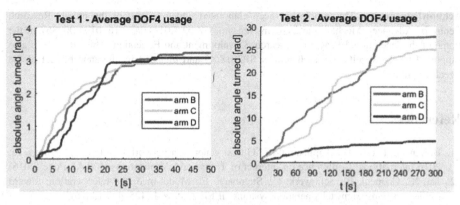

Fig. 7. Average DOF usage in (a) Test 1 and (b) Test 2

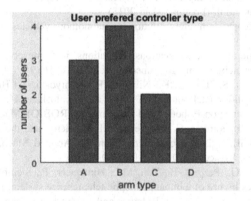

Fig. 8. Preferred controller type

4   Conclusions

Two simulated environments were designed. First one to test the controller ability to change object orientation, the second one focused on an object position change. The fact that the users were able to finish the tests with all controller types in similar time could mean that the simulated tasks we too simplified and didn't require precise enough manipulation techniques. Users often used just the first three joints of a given controller, completely avoiding the use of the additional fourth joint.

The results show that even simple robotic arm with three DOF and simple visual feedback is sufficient for a variety of tasks. However the users review indicates that they had difficulty recognising how far the object in the simulation were, this may be resolved by placing a distance measuring sensor on the arm or by using a stereo vision to give the users a sense of scale. Another problem pointed out by the users was the uncertainty of a successful end effector grip and arm collisions with the environment. Multiple feedback methods could provide this information, including visual, audio, haptic or force feedbacks. Future work will address different forms of feedback and their comparison.

Acknowledgements. This article has been elaborated under support of the project Research Centre of Advanced Mechatronic Systems, reg. no. CZ.02.1.01/0.0/0.0/16_019/0000867 in the frame of the Operational Program Research, Development and Education. This article was also supported by the specific research project SP2018/86 and financed by the state budget of the Czech Republic.

References

1. Borkar, S., Pande, H.: Application of 5G next generation network to Internet of Things. In: 2016 International Conference on Internet of Things and Applications (IOTA), Pune (2016)
2. Xu, X., Cizmeci, B., Schuwerk, C., Steinbach, E.: Model-mediated teleoperation: toward stable and transparent teleoperation systems. IEEE Access **4**, 425–449 (2016)
3. Ford. Mobility experiment: remote repositioning, Atlanta, 6 January 2015. https://media. ford.com/content/fordmedia/fna/us/en/news/2015/01/06/mobility-experiment-remote-repositioning-atlanta.html. (Přístup získán 2018)
4. Phantom Auto. Teleoperation safety solution for autonomous vehicles. https://phantom.auto/. (Přístup získán 2018)
5. Morris, B.: Robotic surgery: applications, limitations, and impact on surgical education (2005). https://www.ncbi.nlm.nih.gov/pmc/articles/PMC1681689/
6. Chen, X., Nishikawa, S., Tanaka, K., Niiyama, R., Kuniyoshi, Y.: Bilateral teleoperation system for a musculoskeletal robot arm using a musculoskeletal exoskeleton. In: EEE International Conference on Robotics and Biomimetics (ROBIO), Macau (2017)
7. Némethy, K., Gáti, J., Kártyás, G., Hegyesi, F.: Exoskeleton and the remote teleoperation projects. In: 2018 IEEE 16th World Symposium on Applied Machine Intelligence and Informatics (SAMI), Kosice, pp. 000073–000080 (2018)
8. Kron, A., Schmidt, G., Petzold, B., Zah, M.I., Hinterseer, P., Steinbach, E.: Disposal of explosive ordnances by use of a bimanual haptic telepresence system. In: 2004 Proceedings of IEEE International Conference on Robotics and Automation, ICRA 2004, New Orleans, LA, USA, vol. 2, pp. 1968–1973 (2004)
9. Hamam, A., Eid, M., El Saddik, A.: Effect of kinesthetic and tactile haptic feedback on the quality of experience of edutainment applications. Multimed. Tools Appl. **67**, 455–472 (2013)
10. Mae, Y., Inoue, T., Kamiyama, K., Kojima, M., Horade, M., Arai, T.: Direct teleoperation system of multi-limbed robot for moving on complicated environments. In: IEEE International Conference on Robotics and Biomimetics (ROBIO), Macau (2017)

Industry 4.0 Testbed at Brno University of Technology

Ludek Zalud[✉], Frantisek Burian, and Petra Kalvodova

CEITEC, Brno University of Technology, Brno, Czech Republic
{ludek.zalud,frantisek.burian,
petra.kalvodova}@ceitec.vutbr.cz

Abstract. As a part of the first-stage of Ricaip teaming project, a small demonstration testbed was built at CEITEC institute in cooperation with Faculty of Electrical Engineering and Communication, Brno University of Technology. It contains both additive and subtractive machinery technologies – 3D SLS (Selective Laser Sintering) printer, and 3DOF CNC milling machine. Material and product movement is provided by means of standard and collaborative industrial manipulators, as well as omnidirectional mobile robot with Mecanum-type wheels. The control system is designed to be easy-to-extend and to be able to cooperate with other testbeds, as was demonstrated by cooperation with Czech Technical University in Prague.

Overall description of the testbed is provided in the paper, as well as its possibilities and main technical challenges.

Keywords: Industry 4.0 · Automation · Robotics · Mobile robot

1 Introduction

In January 2018 our team decided to build small demonstration Testbed for Industry 4.0 technologies. One of the key motivation arguments was the first-stage H2020 Teaming project RICAIP, which was acquired by Czech Technical University of Technology (CVUT) in Prague together with ZEMA, DFKI and Brno University of Technology (BUT). We together decided to build BUT testbed and demonstrate its interconnection with the testbed in CVUT as a final stage of RICAIP project.

The testbed at CVUT already existed before the project started, so we decided to make the BUT testbed to be complimentary with it.

The main concept of the project is:

- combination of additive technology (3D print) and subtractive technology (NC milling cutter),
- cooperation with other testbeds – distributed manufacturing,
- conveyor belt is not necessary – transport system based on omnidirectional mobile robots,
- flexible manufacturing process – the whole manufacturing chain as well as individual machines may be reprogrammed in real-time,
- 6DOF localization of machines and workpieces.

© Springer Nature Switzerland AG 2019
J. Mazal (Ed.): MESAS 2018, LNCS 11472, pp. 147–157, 2019.
https://doi.org/10.1007/978-3-030-14984-0_13

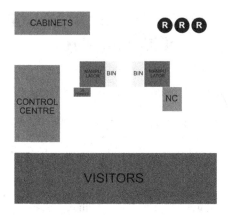

Fig. 1. Testbed disposition, room size 8 × 8 m.

The disposition of the room that we had to our disposal, is on Fig. 1. Before we started, we made full 3D model of the whole testbed (see Fig. 2), as a primary source of data for digital twin, as well as for testbed development and construction.

Fig. 2. Testbed design

The simplified scheme of the control and communication structures inside the testbed are on Fig. 3. All the wired connections are done by means of Ethernet (CAT 6) cables, the mobile robot communicates wirelessly by WiFi.

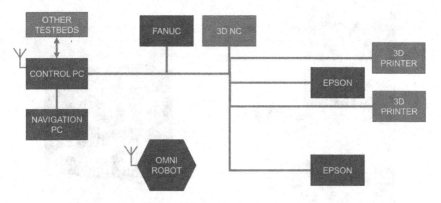

Fig. 3. Simplified communication scheme

2 Main Equipment

2.1 NC (Numeric Control) Milling Cutter

We decided to equip the testbed by SolidVision SLV EDU small milling cutter. The main parameters, important for our project are:

- 3-axis milling process,
- up to 8 automatically interchangeable tools,
- power supply 230 V AC,
- geometrical precision 0.05 mm,
- max material size 365 × 400 × 200 mm,
- small outer dimensions (passes through standard doors),
- custom control software.

During the development we faced one big challenge – the original control program has not the capability to remotely change the g-code for manufacturing, which is a key feature for us, since we typically need to personalize the artifacts being manufactures (e.g. name engraving). Even the software developer responsible for the SLV EDU control software was not able to provide as this functionality. So we were forced to modify the original program to allow the functionality of g-code alternation even during the milling process. This process is done by ethernet connection, and is secured.

2.2 3D Printers

Currently we are using two 3D printing technologies. Sintratec S1 SLS printer (see Fig. 4, right) and classical FDM (fused deposition modelling) Prusa MK3 printer.

The advantages of SLS technology is the possibility to print virtually any 3D shapes without the need of support. It also seems possible the polyamide prints might be easy-to-adjust by classical subtractive technologies, including NC milling.

The maximum print volume of the printer is 130 × 130 × 180 mm, the material is polyamide (nylon) with particle size 0.06 mm, and it is reusable.

Fig. 4. SolidVision SLV EDU milling cutter (left), and SLS 3D printer Sintratec (right)

2.3   Industrial Manipulators

Our aim is to be independent on individual producers when using 6-dof industrial manipulators. Currently we have two brands represented in the testbed – Fanuc (Fanuc CR-7iA/L) and Epson (Epson C8, and Epson N2), the other manipulator, that will be present in near future, is from Kuka company. We have developed our own communication protocol for each brand of robots, even including collaborative robots (Fig. 5).

Fig. 5. Industrial manipulators

2.4   Hermes-SQ Mobile Robot

Completely new mobile robot was developed by our team for the testbed. It is omnidirectional robot with four Mecanum-type wheels and is remotely controlled through WiFi network. All the necessary equipment (including motor controllers, batteries, control system with communication, programable colour LED panel) is in the 25 mm wide lower part.

Second robot with circular drive configuration is currently under development and will be operational by the end of this year.

Fig. 6. Hermes SQ (left), Vicon localization system (right)

2.5   Optical Localization System

The moving mobile robot has to be precisely localized in the room. The location is necessary not only in X and Y axis, but also rotation (azimuth) is necessaty for precise control of its movement, as well as for precise placement of objects that need to be transferred on the robot.

We decided to use Vicon optical localization system (see Fig. 6). Unfortunately. We were not able to get enough money to build appropriate system for the room of our size, so we only have 5 Bonita cameras with 1, 3 Mpix resolution. This is not necessary for trouble-free and highly reliable operation. Anyway we were able to make the system working for our purposes, but for the future we plan to equip the localization part with more cameras, preferably with higher resolution (Fig. 7).

2.6   Control System

Because of time and financial constraints, we were forced to develop our own control system for the whole testbed. It is made in Microsoft Visual Studio 2015, c# programming language, and we were using WindowsForms framework.

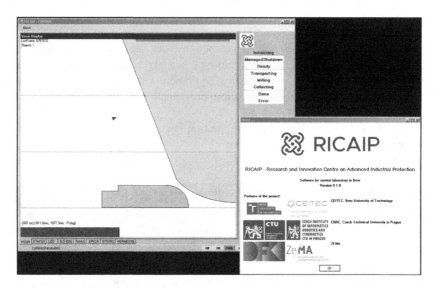

Fig. 7. Testbed control system

3 Precise Navigation

Navigation of individual parts of the machinery process is of a great importance in this project. The industrial manipulators and production machines (NC milling cutter and 3D printers) are stationery (in terms of the whole body movement), so their configuration is measured and end-effector position calculated based on known kinematics in time (forward kinematics problem). The main problem is precise localization of the mobile robot and recalculation among individual entities.

3.1 Coordinate Systems

On the testbed, there are multiple base systems and more transformations among them. All of them is depicted on the picture. All systems have Z oriented upwards, and all of them are right-handed (Fig. 8).

The main base system, in which all transformations are computed related to the Vicon system, because this system is global for all robots. In all transformations, we use only 2D transformations, because z axis of all systems are equidistant and collinear.

The transformation between system n and Vicon system is simply defined by homogenous transformation:

$$P_n = Trans(x_n, x_y)Rot_z(r_n)$$

All homogenous parameters are depicted in Table 1, so all transformations between robot systems are well defined (Fig. 9).

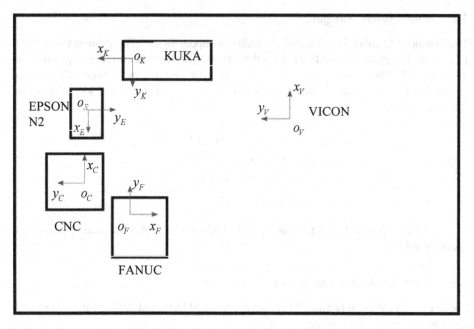

Fig. 8. Coordinate systems

Table 1. Homogenous parameters (uncalibrated)

	Homogenous parameters		
	x [mm]	y [mm]	r [deg]
VICON	0.0	0.0	0.0
KUKA	190.0	1989.0	90.0
EPSON	−263.6	2761.0	180.0
FANUC	−1613.6	1728.5	−90.0
CNC	−1750.0	1530.0	0.0

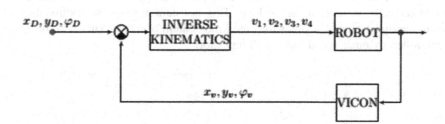

Fig. 9. Robot navigation scheme

3.2   Mobile Robot Navigation

The Hermes-SQ robot is navigated according to upper figure. The desired robot pose (x_D, y_Y, φ_D) is closed to feedback loop through feedback of measured real position of robot by VICON sensor (x_V, y_V, φ_V). The inverse kinematics converts differential vector from (x, y) space to motor speeds (v_1, v_2, v_3, v_4). The inverse kinematics has following result:

$$v_1 = +v_f - v_s + \omega$$
$$v_2 = -v_f + v_s + \omega$$
$$v_3 = -v_f - v_s + \omega$$
$$v_4 = v_f + v_s + \omega$$

Where v_f is forward speed of robot, v_s is slipping speed of robot, and ω is angular speed of robot.

3.3   Motor Controller Regulators

Speed of each motor in Hermes-SQ robot is regulated independently, by its own nested regulator for speed and current.

The motor system can be modelled by the schematic in Fig. 10.

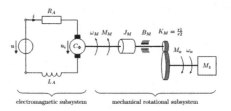

Fig. 10. Hermes-SQ controller

The electromagnetic subsystem, defined by parameters (R_A, L_A) is controlled by PI current regulator in inner regulator loop. This fast current loop can control directly current through motor, this yields to control momentum of motor. The speed of motor is controlled by outer speed loop of PI regulator.

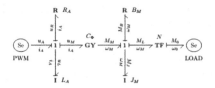

Fig. 11. Bond-graph of the motor circuit

In Fig. 11, the bond-graph simulation circuit of the motor is depicted, so the motor dynamics and regulator can be easily simulated, after identifying motor and load parameters.

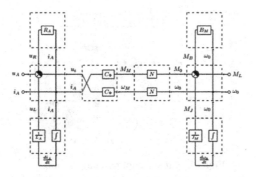

Fig. 12. Matlab simulation of the motor

On Fig. 12, there is Matlab - like simulation circuit of the motor, used for simulation and setting parameters of PI controller loops. From those picture, the motor controller equations can be written as:

$$\frac{di_A}{dt} = \frac{1}{L_A} \left(u_0 - R_A i_A - C_\phi N \omega_0 \right)$$

$$\frac{d\omega_0}{dt} = \frac{1}{JM} \left(C_\phi N i_A - B_M \omega_0 - M_L \right)$$

The parameters were identified two-step method (separated electrical and mechanical parameters) using least-squares method, using voltage pseudorandom sequences as input signals to identification.

The results of identification are on following Fig. 13.

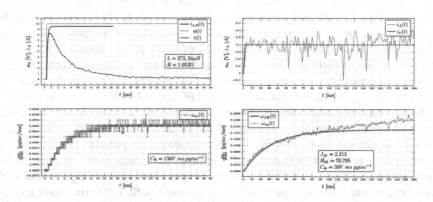

Fig. 13. Identification of motor parameters

The voltage dependency of the Current regulator is implemented due to cross decoupling schematic. This allows to have slower integration factor, helping the stability of complete system of four motors on the robot (Fig. 14).

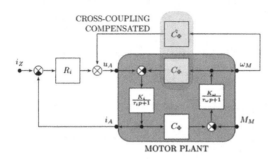

Fig. 14. Cross-coupling compensation

Fig. 15. Finalized testbed (left)

4   Conclusion and Future Work

Despite the extremely challenging time and financial constraints, the fully working testbed (see Fig. 15, left) was publicly demonstrated in 30. 8. 2018. Due this date the testbed was not only fully working, as it was demonstrated by fully autonomous manufacturing of personalized promotional object (see Fig. 15, right), but it was also connected to the testbed in Czech Technical University in Prague, where the first character of the name of a person was autonomously assembled from Lego bricks, while the rest of the object was manufactured in CEITEC, Brno University of Technology.

We are obviously continuing in the development of the testbed – as it was already mentioned, second omnidirectional mobile robot Hermes-CC is being developed, Kuka Agilus manipulator is being integrated to the manufacturing process. The testbed will also serve as demonstrator of Industry 4.0 highlights for general public (the first public action, known as Night of Science, already passed over), and for digital twin with Epson head-mounted display and to quality assessment of SLS 3D prints have already started.

Acknowledgment. This work was supported by Horizon 2020 project Number 763559, Acronym Ricaip – Research and Innovation Centre on Advanced Industrial Production, topic-WIDESPREAD-04-2017, TeamingPhase1.

This work was also supported by the Technology Agency of the Czech Republic under the project TE01020197 – 'CAK III - Centre for Applied Cybernetics'.

This work was also supported by the Technology Agency of the Czech Republic under the project TH01020862 – 'System for automatic/automated detection/monitoring radiation situation and localisation of hot spots based on a smart multifunctional detection head usable for stationary and mobile platforms including unmanned.'

References

1. Hodicky, J.: HLA as an experimental backbone for autonomous system integration into operational field. In: Hodicky, J. (ed.) Modelling and Simulation for Autonomous Systems. LNCS, vol. 8906, pp. 121–126. Springer, Heidelberg (2014). https://doi.org/10.1007/978-3-319-13823-7_11

2. Hodicky, J.: Standards to support military autonomous system life cycle. In: Březina, T., Jabłoński, R. (eds.) MECHATRONICS 2017. AISC, vol. 644, pp. 671–678. Springer, Cham (2018). https://doi.org/10.1007/978-3-319-65960-2_83

3. Li, G., Liu, Y., Dong, L., Cai, X., Zhou, D.: An algorithm for extrinsic parameters calibration of a camera and a laser range finder using line features. In: 2007 IEEE/RSJ International Conference on Intelligent Robots and Systems, San Diego, CA, pp 3854–3859 (2007). https://doi.org/10.1109/iros.2007.4399041

4. Wasielewski, S., Strauss, O.: Calibration of a multi-sensor system laser rangefinder/camera. In: Intelligent Vehicles 1995 Symposium, pp. 472–477 (1995). https://doi.org/10.1109/ivs.1995.528327

5. Yang, H., Liu, X., Patras, I.: A simple and effective extrinsic calibration method of a camera and a single line scanning lidar. In: Proceedings of the 21st International Conference on Pattern Recognition (ICPR2012), pp. 1439–1442 (2012)

6. Zalud, L.: ARGOS - system for heterogeneous mobile robot teleoperation. In: 2006 IEEE/RSJ International Conference on Intelligent Robots and Systems, Beijing, China, pp. 211–216 (2006). https://doi.org/10.1109/iros.2006.282495

7. Zalud, L., Kocmanova, P., et al.: Calibration and evaluation of parameters in a 3D proximity rotating scanner. Elektronika ir Elektrotechnika **21**, 3–12 (2015). https://doi.org/10.5755/j01.eee.21.1.7299

8. Kocmanova, P., Zalud, L., Chromy, A.: 3D proximity laser scanner calibration. In: 18th International Conference on Methods and Models in Automation and Robotics, MMAR 2013, pp. 742–747 (2013). https://doi.org/10.1109/mmar.2013.6670005

9. Zalud, L., Kocmanova, P.: Fusion of thermal imaging and CCD camera-based data for stereovision visual telepresence. In: 2013 IEEE International Symposium on Safety, Security, and Rescue Robotics, SSRR 2013 (2013). https://doi.org/10.1109/ssrr.2013.6719344

10. Chromy, A., Zalud, L.: Robotic 3D scanner as an alternative to standard modalities of medical imaging. SpringerPlus **3**(1), 13 (2014). https://doi.org/10.1186/2193-1801-3-13

Autonomous Compact Monitoring of Large Areas Using Micro Aerial Vehicles with Limited Sensory Information and Computational Resources

Petr Ješke, Štěpán Klouček, and Martin Saska[(✉)]

Department of Cybernetics, Czech Technical University in Prague,
Prague, Czech Republic
{jeskepet,kloucste,martin.saska}@fel.cvut.cz
http://mrs.felk.cvut.cz/

Abstract. In this paper, a new approach for autonomous real-time monitoring of large areas using small unmanned areal vehicles with limited sensory and computational resources is proposed. Most of the existing solutions of area monitoring require large aerial vehicles to be equipped with a list of expensive sensors and powerful computational resources. Recent progress in Micro Aerial Vehicles (MAVs) allows us to consider their utilization in new tasks, such as the considered compact monitoring, which are dedicated to large well-equipped aerial vehicles so-far only. The proposed solution enables online area monitoring using MAVs equipped with minimal sensory and computational resources and to process the obtained data only with cell phones capabilities, which considerably extends application possibilities of the drone technology. The proposed methodology was verified under various outdoor conditions of real application scenarios with a simple autonomous MAV controlled by the onboard model predictive control in a robotic operation system (ROS), while the user interface was provided on a standard smartphone with Android OS.

Keywords: Autonomous monitoring · Micro Aerial Vehicles ·
Surveillance · Sensor fusion · State estimation · Visual reconnaissance

1 Introduction

This paper addresses the problem of ground mapping of agricultural areas using MAVs with limited sensory information. The goal of the presented application is to offer a light solution of the area scanning problem suited for MAVs equipped with minimal hardware resources. We are aiming to design a system, which could be used in environmental monitoring tasks for precise agriculture (PA) and water protection being our key motivations. The core of the proposed system is an

© Springer Nature Switzerland AG 2019
J. Mazal (Ed.): MESAS 2018, LNCS 11472, pp. 158–171, 2019.
https://doi.org/10.1007/978-3-030-14984-0_14

onboard data fusion mechanism that enables to process obtained sensory information in a real time and to provide a compact view of the given perimeter immediately. Using onboard sensors and a novel precise MAV state estimation approach together with the information included in the obtained data, the user receives a compact overview of the scene composed of multiple measurements already during the flight. For example, in case of visual monitoring, a compact image of the given area can be provided in a real time without the need of any additional stitching algorithm. Using of the state-of-the-art stitching method onboard is limited by the computational performance of the MAV processors. If required, a post-processing using the computer vision stitching approaches can be realized after the mission, taking online obtained results as a proper initialization of these algorithms to improve their performance. Such solution represents a unique combination of computer vision methods, which usually require large computational resources, and robotic data fusion approaches to enable deployment of the final solution using tiny MAVs. The designed methodology is ready for use in numerous applications, such as archaeological mapping, precise agriculture, water protection, mapping of areas damaged by earthquakes or floods, and surveillance and perimeter mapping in security and defense missions. The simplicity of the required sensory equipment and the MAV platform itself enables to consider applications with a high risk of damage and loss of the vehicle, which is another interesting factor mainly in the defenses and security applications.

A lot of papers dealing with PA using Unmanned Aerial Systems (UAS) can be found. A general overview of the problem can be found in [27]. The task of collecting and processing the UAS imagery is discussed also in [12]. This study examines possibilities of using unmanned helicopters to measure chlorophyll concentration in a rice crop. Another topic, independently discussed by several research groups, is the usage of UAVs in hyperspectral imaging for the purposes of PA [3,13,14] and [11]. In [16], the research is aimed to compare unmanned and multispectral satellite imaging for estimation of basic crop parameters during the growing season and focused mainly on processing collected data and evaluating impacts of fertilizers and water supply on tested plants. In the proposed paper, significantly smaller and less equipped systems are used in comparison with platforms listed above and limitations of using MAVs in PA are studied. Attention is also paid to image processing and real-time information extraction from MAVs images, which was not presented in any of the mentioned methods. The main contribution of the proposed approach is the possibility of real-time onboard stitching of large sets of obtained images and its ability to stitch images of unified surfaces, such as large water or green areas, whose monitoring is our motivation. Due to the sensor fusion mechanism, a preview of the scanned area can be instantly provided.

An important task related to PA is complete coverage path planning (CCPP). CCPP is a problem of finding a path that passes through all the waypoints in the desired area from a starting point to a final point while avoiding obstacles. Waypoints can have many interpretations such as positions from which the robot performs measurements, places from which the MAV takes a photo or spots on the floor for a vacuum cleaning robot. The area of robotic applications cover vacuum cleaning [25], robot painting [4], harvesting [18], etc. The problem of CCPP is described in detail in Sect. 2.

As the altitude in which the MAV can fly is limited, mainly because of the resolution of the onboard camera, it is necessary to deal with the problem of combining multiple photos into a single image in order to obtain a complete image of the desired area. This problem is known as photo stitching. A general overview of photo stitching methods can be found in [23]. Currently used stitching algorithms are well documented in [29] and [28]. The problem of UAS image stitching was discussed in [8], where image stitching is established using Speed-up Robust Features (SURF) techniques. The approach used in this paper for evaluation of results obtained by the novel sensor-fusing approach is based on [10], which can be integrated into the mobile application.

As mentioned above the concept of the presented system is conceived as a "light solution", meaning that the goal to achieve the given task using minimal hardware equipment, but being able to provide onboard pre-processing of obtained sensory results. In the proposed system (Fig. 1), we rely only on a standard GPS, simple altimeter (e.g., Garmin rangefinder), IMU, down looking camera for providing the required visual data, and WiFi (or GSM) connection to be able to present a preview of captured data in a real time.

Fig. 1. MAV used in real platform experiments equipped with an essential list of components. (1) computer running ROS, (2) battery, (3) differential GPS - used for ground truth, (4) standard GPS, (5) a backup rangefinder not used in the presented experiments (6) Garmin rangefinder [17].

2   Complete Coverage Path Algorithms

Three algorithms for a complete coverage path planning were considered to show the importance of the proper flight design if using such a light system. The Wavefront algorithm (sometimes called a distance transform algorithm) [26] uses the principle of cellular decomposition to divide the mapped area into a grid of cells. Cells can be interpreted as nodes of an adjacency graph and boundaries between the cells as edges of the graph which transforms the problem of a CPPP into a problem of finding a path through the adjacency graph.

The most common approach to CCPP in ground mapping applications is a ZigZag algorithm, which simply covers the desired area using back and forth motions. This method in its simple version does not consider obstacles inside the mapping area.

Although most PA applications do not consider the possibility of a forbidden area inside the mapping region, in some cases it is a necessary assumption (mapping area can contain a no-fly zone, or contain high voltage pylons, or other obstacles which need to be avoided). For these cases, a Trapezoidal algorithm can be used to decompose a complex polygon (representing the mapping area) into triangles and trapezoids, which are easier to cover using back and forth motions [9].

3   Aerial Photography

The process of realtime onboard composition of a complete picture of the mapped area from data obtained in multiple locations is described in this section. At first, the undesirable impacts of photography from flying object are presented, together with possible solutions followed by real platform distortions and their compensations. The tilt of the MAV in space is determined by a set of three parameters, roll, pitch and yaw as shown in Fig. 2. The camera is set to the center of the coordinate system. The x and y-axes are parallel to booms of the quadcopter. The z-axis is perpendicular to both of them and points downwards. Arrows symbolize a positive direction of the coordinates.

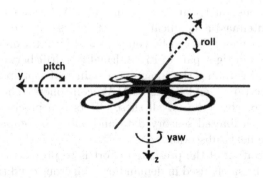

Fig. 2. MAV coordinate system.

3.1   Sensor Based Image Stitching

An onboard method for sensor-based stitching is required for a real-time initial preview of the mapped area in a system with limited computational power. The state-of-the-art image processing methods can be subsequently applied to the set of collected photos on a computer with higher computational power. In the process of the exact image stitching based on a sensor data, several distortions may occur. Their impacts are studies in this section with possible solutions suggested. As the application assumes usage of additional stitching software, the trajectories are calculated concerning an average demand on a photo overlap of the stitched photos, although for the proposed sensor based fusion stitching mechanism does not require any overlapping. Most of the stitching algorithms require at least 30% photo overlap. Trajectories in the presented experiments are calculated in a way that taken images have ideally 40% overlap. The 10% of the photo overlap is a precision which is necessary adhere to in order to obtain a full map of the desired area without blind spots.

Since we suppose using MAV without camera stabilization (a gimbal) to keep the systems as light and simple as possible, the effects caused by changing tilt of the MAV in space need to be considered in the image processing algorithm. During the flight, the roll and pitch of the MAV differ based on the flight velocity, flight direction, and current wind conditions, which causes undesirable translation of the obtained data in the scene and change of the image perspective (Fig. 3).

For the real-time initial preview, the effect of the changed perspective can be negligible due to required relatively high altitude (20 m and more) and small tilt of MAVs and due to a flat surface of the agricultural areas (i.e., surface without obstacles where the change of the perspective could appear). The translation of the photo in the scene can be simply corrected as

$$\Delta x = h \tan \alpha, \tag{1}$$

$$\Delta y = h \tan \beta, \tag{2}$$

where α is roll and β pitch of the MAV.

Although, the proper setting of the MAV yaw can be part of the flight plan and usually is set constant for all images, in real MAV systems (without multiple precise GPS antennas) a significant drift of yaw is observed due to limits of compass precision, which needs to be compensated by rotation of the photo by an angle δ, as shown in right part of Fig. 3. Finally, a drift between required and real positions of UAV taking the image occurs in real systems due to sensors and actuators uncertainty and limits of MAV controllers. The most challenging task required for correction of all these effects is a precise real-time MAV state estimation with limited sensory data and onboard computational power, as described in the next subsection.

The main motivation of the presented effort is to provide a light and cheap system, which can be safely used in demanding dangerous conditions with a high risk of vehicle lost, such as search and rescue and military missions for example.

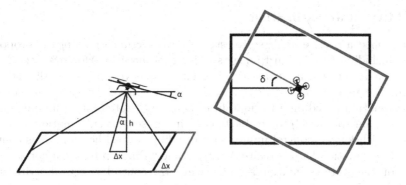

Fig. 3. Left: Translation of the obtained data in the scene relatively to the expected image (red), which depends on the roll α and the altitude h. Right: Influence of rotation distortion on the taken photo. (Color figure online)

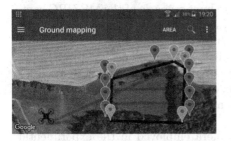

Fig. 4. Example of the mapping trajectory calculated using the Zigzag algorithm. Blue markers represents calculated trajectory, green markes defines mapped area. (Color figure online)

Fig. 5. The right dropdown list, containing a list of advanced commands for calculating the trajectory and managing markers such as "Calculate coverage" or "Fly coverage".

To achieve these requirements and evaluate limits of the system, we have purposely equipped the testing platform used for an experimental verification with simple light-weight wide angle Mobius action camera. A known disadvantage of wide-angle cameras is an image distortion known as the fisheye effect, but the wide field of view is a crucial property of an aerial system being used mainly in the above mentioned search and rescue and security applications, where a fast mapping process is required. Therefore a fast rectification process needs to be applied on each of the captured images in order to obtain planar photography. This task was implemented in real time onboard of MAVs, which enables to provide a correct planar map already in the preview of the mapped area. For details on the vision pipeline, see a summary of our work on light-weight onboard computer vision in [24].

3.2 MAV State Estimation

As mentioned a precise MAV state estimation is a cruical part of the proposed system including the onboard image stitching mechanism. Moreover, the MAV state estimation is also necessary requirement to achieve a precise path following performance in the task of images gathering. We have designed an onboard state estimation based on a Kalman filter being integrated into a unique Model Predicted Control (MPC) to be able to combine all sensory information available together with known control inputs on a receding control horizon. This control and state estimation loop is realized in real time with 8 s prediction horizon (for details on the state estimation see [22], while the control system description is provided at [5]).

The novelty of this solution lies in the possibility to achieve a precise and stable flight performance in real outdoor conditions (so without using a motion capture system) with external disturbances such as wind. The performance of the system was proven in the challenging treasure hunt scenario of the MBZIRC competition (http://mbzirc.com/), where the system achieved the best performance among 143 participants (see [15] and http://mrs.felk.cvut.cz/mbzirc for details). The presented paper, shows a deployment of the system, where the precise controller and mainly the state observation and prediction may be directly used for the onboard real-time sensor-based stitching of obtained images. Another advantage of the designed precise state estimation and motion prediction methods is the ability to compensate often a variable delay of light-weight cameras (for example 0.2–0.3 s in case of the Mobius camera employed in the experimental verification).

In the proposed system, mainly the knowledge of a precise horizontal position is required for both, the precise path following and the onboard stitching of obtained images. Position estimation in the lateral axes is based on the estimate provided by the PixHawk low-level controller. Although its precision may be satisfactory locally for short periods of time, it is prone to heavy drift in time spans of minutes. To compensate this drift, the horizontal position from PixHawk needs to be corrected using other onboard sensors, such as IMU, optic flow from down looking camera, and by model-based prediction of the MAV state. The sensor fusion of various sources of information is realized by an onboard Linear Kalman Filter (see [22] for details).

Although, the requirements on precision of horizontal MAV state estimation are higher than the required precision in the vertical axis in the tackled application, the knowledge of MAV height is also important for obtaining a scale of images estimation. Moreover, GPS precision is much lower in the vertical axis than in the horizontal one. Therefore, a sensory output from a down-facing rangefinder needs to be fused using the Kalman filter to compensate the GPS error.

4   Experiments

To highlight the easy deployability of the proposed solution and to verify the possibility of real-time displaying a stitched images without a need of external PC, a mobile application under Android OS [21] was designed to be able to communicate via Wi-Fi using RosJava [1] with the onboard computer with ROS [19]. The mobile application enables to select areas to be mapped (Fig. 4), manage autonomously generated trajectories (Fig. 5), observe the progress of the mission (Fig. 6) and display a preview of the scanned area in real-time.

Fig. 6. Left: MAV flying the mission. Right: Observing the progress of the mission in a real-time.

Numerous HW experiments were carried out in order to examine the performance of the system with a real platform in different outdoor conditions, flying in various flight modes, and using different planning algorithms (a video record of the experiments can be found at https://www.youtube.com/watch?v=wZMf-JLO3P8). In all experiments, a hexacopter with DJI F550 frame, NUC onboard PC, Garmin rangefinder, Mobius camera and PixHawk stabilization board with IMU was used (see [6,7,20] for details on communication, control, and HW design). A real-time rectification process using an OpenCV library was applied on all images obtained by the Mobius camera to compensate an image distortion in order to obtain planar photography.

For comparison of the obtained results of the proposed sensor-based technique with a state-of-the-art solution, a stitching API provided by cv::Stitcher of the OpenCV library was implemented on the onboard control PC. To be able to demonstrate the possibility of consequent offline stitching of captured images (here it is not required onboard applicability and real-time response), the Autopano software [2] was selected as an example of advanced photo stitching approaches.

The Trapezoidal algorithm was applied on a surface composed of a cultivated grassy area, a field, and uncultivated high grass (Fig. 7 left, polygon A), which offers a lot of patterns for the stitching software. The Wavefront algorithm was applied on a large planar grassy area (Fig. 7 left, polygon B) with a minimal number of patterns, which could be used by the stitching software as reference points, in order to test the capabilities of the stitching software. Another area (Fig. 7 right) with a minimal number of patterns was selected for the experiment, where the ZigZag algorithm was applied.

Fig. 7. Pictures of the mapped areas taken from high altitude in low resolution.

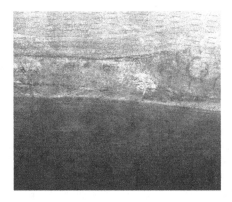

Fig. 8. A real time overview of an area mapped from 10 m altitude obtained by the sensor-based stitching using the Trapezoidal algorithm.

Fig. 9. Result of the Autopano stitching software applied on a set of photos collected using the Trapezoidal algorithm.

With the Trapezoidal algorithm, the MAV was flying 1 m/s at 10 m altitude. The state-of-the-art onboard stitching algorithm using the OpenCV library was able to stitch only several photos due to the small number of reference points, while the sensor-based stitching algorithm achieved sufficient overview for the user in real time (Fig. 8). After the mission, the ability to stitch the obtained results offline into a more detailed and precise map using the Autopano stitching software (Fig. 9) was verified in the way as proposed. Similar results were achieved with the ZigZag algorithm, which mapped the area from the 20 m altitude with a flight velocity of 1.5 m/s. A result of the sensor based stitching is shown in Fig. 12, while the map obtained offline by the Autopano stitching software is shown in Fig. 13.

The MAV was flying 2 m/s at an altitude of 15 m if the Wavefront algorithm was used. In this experiment, the system was set up to achieve a lower overlap of the taken images than the required (20%). From the tested algorithms, only the sensor based stitching (Fig. 10) was able to achieve a sufficient result. Due to the smaller overlap and a small amount of reference points even the offline Autopano stitching software was not able to stitch all the images properly (Fig. 11), which shows the importance of image overlapping for stitching images of unified ground surfaces.

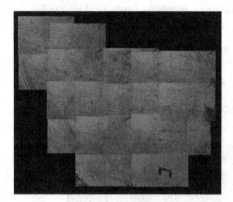

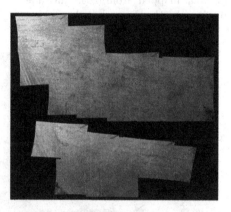

Fig. 10. A real time overview of an area mapped from 15 m altitude obtained by the sensor-based stitching using the Wavefront algorithm.

Fig. 11. Result of the Autopano stitching software applied to photos collected during mapping using the Wavefront algorithm.

In order the calculate the precision of the sensor based stitching algorithm, and to compare a precision of the three implemented algorithms, a set of various experiments was designed, both in the simulated environment and with the real platform. A set of reference objects was added to the mapped area which was subsequently mapped several times with each mapping algorithm to demonstrate the performance of the methods. Statistical results of the realized experiments were processed as follows: A distortion in image placement was measured at each result of the given mapping process by comparing the expected and real positions of the reference object in the particular result. The inaccuracy of a single image placement measured in pixels was recalculated to a percentage of inaccuracy with respect to the size of the image. The total inaccuracy of image placement in the map was calculated as an average of fractional inaccuracies.

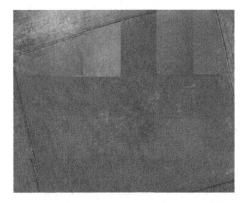

Fig. 12. A real time overview of an area mapped from 20 m altitude obtained by the sensor-based stitching using the Zigzag algorithm.

Fig. 13. Result of the Autopano stitching software applied to photos collected during mapping using the ZigZag algorithm.

Fig. 14. Compact real-time preview of the given perimeter.

The inaccuracy of each algorithm was calculated from the distortions measured in the various experiments.

Although, the precision of the algorithms calculated from positions of individual images inside a map is only approximate estimate, it was possible to confirm that the Wavefront algorithm reaches the highest accuracy with the lowest number of pictures required to be taken. The Zigzag and Trapezoidal algorithms reached a similar accuracy at an area without obstacles. The biggest distortion occurs in scenarios where the aerial robot avoids obstacles, due to a bigger amount of flight direction changes above the mapping area, which causes tilt changes resulting in increasing image placement distortions.

5   Conclusion

In the presented paper, a light solution of area scanning designed for use with small MAVs equipped with minimal hardware equipment was proposed. The proposed approach is motivated by tasks of precision agriculture, especially monitoring field and water areas, which usually offer minimal information for image processing based stitching algorithms. Therefore a sensor based stitching algorithm, which uses an estimated position of the MAV in the stitching process, was designed to provide an immediate preview (Fig. 14) of a mapped area for system users. The required functionality of the proposed solution was verified in real-world conditions with a presence of external disturbances using a very simple and light MAV system in comparison with existing UAS technologies. The sensor-based method demonstrated high robustness in situations, where obtained pictures do not include sufficient overlapping and enough correspondences required by stitching algorithms relying on image-processing. A video of the experiments can be found at https://www.youtube.com/watch?v=wZMf-JLO3P8.

Acknowledgments. This research was supported by CTU grant no. SGS17/ 187/OHK3/3T/13, by the Grant Agency of the Czech Republic under grant no. 17-16900Y and by OP VVV MEYS funded project CZ.02.1.01/0.0/0.0/16_019/0000765 "Research Center for Informatics".

References

1. Rosjava an implementation of ROS in pure java with android support. https:// github.com/rosjava/rosjava_core. Accessed 12 Dec 2017
2. The Autopano stitching software. http://www.kolor.com/autopano. Accessed 12 Dec 2017
3. Adão, T., et al.: Hyperspectral imaging: a review on uav-based sensors, data processing and applications for agriculture and forestry. Remote Sens. **9**, 1110 (2017)
4. Atkar, P.N., Greenfield, A., Conner, D.C., Choset, H., Rizzi, A.A.: Uniform coverage of automotive surface patches. Int. J. Robot. Res. **24**(11), 883–898 (2005). https://doi.org/10.1177/0278364905059058
5. Baca, T., Hert, D., Loianno, G., Saska, M., Kumar, V.: Model predictive trajectory tracking and collision avoidance for reliable outdoor deployment of unmanned aerial vehicles (2018)
6. Baca, T., Loianno, G., Saska, M.: Embedded model predictive control of unmanned micro aerial vehicles. In: 2016 21st International Conference on Methods and Models in Automation and Robotics (MMAR), pp. 992–997, August 2016. https://doi.org/10.1109/MMAR.2016.7575273

7. Baca, T., Stepan, P., Saska, M.: Autonomous landing on a moving car with unmanned aerial vehicle. In: 2017 European Conference on Mobile Robots (ECMR), pp. 1–6, September 2017. https://doi.org/10.1109/ECMR.2017.8098700
8. Cheng, C., Wang, X., Li, X.: UAV image matching based on surf feature and Harris corner algorithm. In: 4th International Conference on Smart and Sustainable City (ICSSC 2017), pp. 1–6, June 2017. https://doi.org/10.1049/cp.2017.0116
9. Galceran, E., Carreras, M.: A survey on coverage path planning for robotics. Robot. Auton. Syst. **61**(12), 1258–1276 (2013). https://doi.org/10.1016/j.robot.2013.09.004
10. Ha, Y., Kang, H.: Evaluation of feature based image stitching algorithm using OpenCV. In: 2017 10th International Conference on Human System Interactions (HSI), pp. 224–229, July 2017. https://doi.org/10.1109/HSI.2017.8005034
11. Honkavaara, E., et al.: Processing and assessment of spectrometric, stereoscopic imagery collected using a lightweight UAV spectral camera for precision agriculture. Remote Sens. **5**(10), 5006–5039 (2013). https://doi.org/10.3390/rs5105006. http://www.mdpi.com/2072-4292/5/10/5006
12. Swain, K.C., Thomson, S.J., Jayasuriya, H.P.: Adoption of an unmanned helicopter for lowaltitude remote sensing to estimate yield and total biomass of a rice crop. Trans. ASABE **13**, 21–27 (2010)
13. Katsigiannis, P., Misopolinos, L., Liakopoulos, V., Alexandridis, T.K., Zalidis, G.: An autonomous multi-sensor UAV system for reduced-input precision agriculture applications. In: 2016 24th Mediterranean Conference on Control and Automation (MED), pp. 60–64, June 2016. https://doi.org/10.1109/MED.2016.7535938
14. Piermattei, L.: EGU general assembly (2017)
15. Loianno, G., et al.: Localization, grasping, and transportation of magnetic objects by a team of mavs in challenging desert-like environments. IEEE Robot. Autom. Lett. **3**(3), 1576–1583 (2018). https://doi.org/10.1109/LRA.2018.2800121
16. Lukas, V., et al.: The combination of UAV survey and landsat imagery for monitoring of crop vigor in precision agriculture. ISPRS-Int. Arch. Photogramm. Remote. Sens. Spat. Inf. Sci. **XLI-B8**, 953–957 (2016). https://doi.org/10.5194/isprs-archives-XLI-B8-953-2016
17. MRS CTU: Micro aerial vehicle (2017). http://mrs.felk.cvut.cz/research/micro-aerial-vehicles. Accessed December 2017
18. Ollis, M., Stentz, A.: First results in vision-based crop line tracking. In: Proceedings of IEEE International Conference on Robotics and Automation, vol. 1, pp. 951–956, April 1996. https://doi.org/10.1109/ROBOT.1996.503895
19. Quigley, M., Gerkey, B., Smart, W.D.: Programming Robots with ROS: A Practical Introduction to the Robot Operating System, 1st edn. O'Reilly Media Inc., Sebastopol (2015)
20. Saska, M., et al.: System for deployment of groups of unmanned micro aerial vehicles in GPS-denied environments using onboard visual relative localization. Auton. Robots **41**(4), 919–944 (2017). https://doi.org/10.1007/s10514-016-9567-z
21. Sheusi, J.C.: Android Application Development for Java Programmers. Cengage Learning (2012)
22. Spurný, V., et al.: Cooperative autonomous search, grasping, and delivering in a treasure hunt scenario by a team of unmanned aerial vehicles. J. Field Robot. **36**(1), 125–148 (2019). https://doi.org/10.1002/rob.21816. https://onlinelibrary.wiley.com/doi/abs/10.1002/rob.21816
23. Szeliski, R.: Image alignment and stitching: a tutorial, October 2004. https://www.microsoft.com/en-us/research/publication/image-alignment-and-stitching-a-tutorial/

24. Štěpán, P., Krajník, T., Petrlk, M., Saska, M.: Vision techniques for on-board detection, following, and mapping of moving targets. J. Field Robot. **36**, 252–269 (2018). https://doi.org/10.1002/rob.21850
25. Yasutomi, F., Yamada, M., Tsukamoto, K.: Cleaning robot control. In: Proceedings of 1988 IEEE International Conference on Robotics and Automation, pp. 1839–1841, vol. 3, April 1988. https://doi.org/10.1109/ROBOT.1988.12333
26. Zelinsky, A., Jarvis, R., Byrne, J.C., Yuta, S.: Planning paths of complete coverage of an unstructured environment by a mobile robot. In: Proceedings of International Conference on Advanced Robotics, pp. 533–538 (1993)
27. Zhang, C., Kovacs, J.M.: The application of small unmanned aerial systems for precision agriculture: a review. Precision Agric. **13**(6), 693–712 (2012). https://doi.org/10.1007/s11119-012-9274-5
28. Zhou, X., Zhang, H., Wang, Y.: A multi-image stitching method and quality evaluation. In: 2017 4th International Conference on Information Science and Control Engineering (ICISCE), pp. 46–50, July 2017. https://doi.org/10.1109/ICISCE.2017.20
29. Zhu, L., Wang, Y., Zhao, B., Zhang, X.: A fast image stitching algorithm based on improved surf. In: 2014 Tenth International Conference on Computational Intelligence and Security, pp. 171–175, November 2014. https://doi.org/10.1109/CIS.2014.67

Information Gathering Planning with Hermite Spline Motion Primitives for Aerial Vehicles with Limited Time of Flight

Alexander Duben[(✉)], Robert Pěnička, and Martin Saska

Faculty of Electrical Engineering, Czech Technical University,
Technicka 2, 166 27 Prague, Czech Republic
{dubenale,penicrob,saskam1}@fel.cvut.cz

Abstract. This paper focuses on motion planning for information gathering by Unmanned Aerial Vehicle (UAV) solved as Orienteering Problem (OP). The considered OP stands to find a path over subset of given target locations, each with associated reward, such that the collected reward is maximized within a limited time of flight. To fully utilize the motion range of the UAV, Hermite splines motion primitives are used to generate smooth trajectories. The minimal time of flight estimate for a given Hermite spline is calculated using known motion model of the UAV with limited maximum velocity and acceleration. The proposed Orienteering Problem with Hermite splines is introduced as Hermite Orienteering Problem (HOP) and its solution is based on Random Variable Neighborhood Search algorithm (RVNS). The proposed RVNS for HOP combines random combinatorial state space exploration and local continuous optimization for maximizing the collected reward. This approach was compared with state of the art solutions to the OP motivated by UAV applications and showed to be superior as the resulting trajectories reached better final rewards in all testing cases. The proposed method has been also successfully verified on a real UAV in information gathering task.

Keywords: Orienteering Problem · UAV · Surveillance ·
Information gathering · Reconnaissance · Unmanned Aerial Vehicle ·
Hermite spline · 3D trajectory generation

1 Introduction

The number of real world applications of Unmanned Aerial Vehicles (UAV) grows dramatically [7]. The use of UAVs is already well anchored in many industrial and scientific fields with more potential applications being under heavy development [26]. UAVs are very suitable for topography applications such as aerial

© Springer Nature Switzerland AG 2019
J. Mazal (Ed.): MESAS 2018, LNCS 11472, pp. 172–201, 2019.
https://doi.org/10.1007/978-3-030-14984-0_15

Fig. 1. An example trajectory used for information gathering task using hexarotor UAV found by proposed solution algorithm

mapping, environment monitoring and land measurement [15]. In similar manner UAVs can be also used in agriculture for crop surveys and wild life monitoring [35]. Another important fields with significant autonomous UAV utilization potential are civilian safety services and military [19]. Applications such as forest fire detection [45], plume tracking [37], search and rescue [11] and flash flood detection [27] are currently being developed. The military use of drones is already well proven in wide range of applications such as reconnaissance or attack scenarios [31]. UAVs can also be utilized in industrial servicing for power line [10], building [24] or power plant inspections [30].

A significant limitation in majority of the mentioned applications is the need of a trained human operator for the UAV handling. A big portion of these applications is some form of information gathering and surveillance. Automation of such tasks lowers the cost and raise the availability of UAV based technology throughout the industry. Hence the motivation behind this paper is enabling the autonomous surveillance in various fields using UAVs.

The surveillance task can be defined as a need for finding a path through a given set of target locations as shown in Fig. 1. In addition it is desirable for the path to be optimal in some manner in order to safe resources or time. The need for finding an optimal path through a set of target locations is very generic problem with possible applications not only in UAV related topics and has been extensively researched in recent years [2]. An example of such path used for UAV surveillance is shown in Fig. 1. One of the most studied problems that can be directly applied to path finding is well-known Traveling Salesman Problem (TSP) [20]. The goal of the TSP is finding the shortest Hamiltonian cycle in a given graph with known distance between vertices. This means finding a path that traverses each vertex while being the shortest possible. The use of TSP in UAV path planning is suitable since every navigation problem can be transformed to a set of points that have to be visited during the UAV travel defining the graph, with shortest path visiting the targets being the solution to TSP problem.

However current UAV technology suffers of one big limiting factor and that is battery life. The typical operational time of current commercial UAV solutions is in the range of tens minutes of flight. Hence it would be suitable to include a constraint on the maximal length/time of flight for the path. The variant of the TSP generalized in this way is called Orienteering Problem (OP) and it is the main focus of this paper. The precondition of OP is having a set of points which might be either defined in plane or in full three dimensional space. Each point has a reward assigned that indicates the importance of visiting the target and both the start and end locations of vehicle is given. The OP objective is to visit a subset of these points such that the collected reward from all visited locations is maximized while the time of flight is kept smaller or at most equal to a given time budget [16].

The regular OP utilizes Euclidean paths between each target location to construct the solution trajectory. Even though the considered Vertical Take-Off and Landing (VTOL) UAV is capable of flying along Euclidean trajectories, since it is a holonomic vehicle, it requires flying with very small velocities as the turns in the Euclidean trajectories are very sharp. This leads to an introduction of parametric polynomial curve called Hermite spline as motion primitives for surveillance planning of OP. The Hermite spline enables to generate fairly complex trajectories that are smooth, which allows to follow the trajectory with considerably higher velocities. In addition, the extension of Hermite splines from 2D to 3D space is fairly simple. However, the OP is limited by a given maximal time of flight thus it is desirable to fly as fast as possible. This requires a generation of such trajectories that are in accordance with physical limitations of the used UAV. Hence a maximal velocity with which the UAV is capable of flying along the spline when evaluating the suitability of a given curve needs to be estimated - so called velocity profile.

The new variant of the Orienteering Problem utilizing Hermite splines with velocity profile calculation is introduced as Hermite Orienteering Problem (HOP). The difference from the generic Orienteering Problem is a bigger complexity because apart from the target locations that should be visited and the order of visit, also the heading angles and actual velocities need to be determined at each location in order to construct feasible Hermite spline based trajectories. The HOP is NP-hard problem as it combines two NP-hard combinatorial optimization problems, the Traveling Salesman Problem and the Knapsack problem [36]. In addition, also continuous optimization needs to be utilized to find the appropriate Hermite splines with velocity profiles. This makes finding the optimal solution very computationally demanding even for small number of target locations. This sort of problem is typically solved using heuristic methods, therefore, heuristic based on Random Variable Neighborhood Search (RVNS) [17] algorithm was used to solve the proposed HOP. The proposed approach for the HOP combines random combinatorial exploration of the state space with continuous optimization which enables to find high quality solution in reasonable time. The created HOP approach extends the RVNS algorithm, used to solve the combinatorial part of the problem, by continuous optimization which determines the optimal heading angles and velocities at each visited location.

The OP solution using RVNS algorithm is based on work done in [28], where the planar Dubins vehicle model [12] is used as a motion primitive instead of Hermite curves and the maximal length of the path is the limiting factor instead of time of flight assuming that the UAV moves with constant velocity. The solution proposed in this paper enables much faster and dynamic movement along generated smooth trajectories by calculating the maximal velocity profile which translates into OP solutions with considerably higher rewards compared to Euclidean or Dubins based solutions and thus improves the deployment of UAVs in data collection and surveillance missions. Another contribution of solution described in this paper is a fact that it can be applied in 2D as well as 3D space topology which is also enabled by the use of Hermite splines.

The remainder of this paper is organized as follows. An overview of the related work is presented in the next section. The formal definition of the Orienteering problem with Hermite motion primitives is introduced in Sect. 3. The novel VNS-based approach for HOP is proposed in Sect. 4 and the computational results with experimental verification with UAV is summarized in Sect. 5.

2   Related Work

A brief overview of current research performed on topic of information gathering planning with UAV and topics closely related is listed in this section.

The studied Orienteering Problem is very similar to one of the most classical routing problems with a long history which is the Traveling Salesman Problem [25]. There exist a large number of heuristics solving the TSP [33]. Probably the most used with is Lin-Kernighan heuristic [18, 22]. The Branch and Bound [5], and Branch and Cut [3] algorithms are examples of optimal solutions to TSP, that however, require significant computational resources even for small data sets.

The Orienteering Problem has been introduced in 1984 by Tsiligirides [41] who defined the Euclidean OP (EOP), further denoted only as OP, and proposed two heuristics for solving the problem. First method uses Monte Carlo based strategy for picking the best solution among large number of randomly generated paths. The second approach utilizes vehicle-scheduling algorithm with one depot. Tsiligirides also designed several benchmarks in a form of datasets used to compare the quality of different solutions.

Another approach to OP was introduced by Ramesh and Brown [32], who defined a four-phase heuristic which uses insertion, improvement and deletion phases to iteratively improve the OP path. A heuristic called Fast and Effective was proposed by Chao et al. [8]. The heuristic uses only the targets locations that lie inside an ellipse around the start and end with axes defined by travel budget. The initial set of generated paths is searched and the path with largest reward is modified by a set of simple operations such as two-point exchange, one-point movement and other optimization operations. Chao et al. also proposed several benchmarking datasets.

The solution of the Hermite Orienteering Problem in this paper is based on the Variable Neighborhood Search heuristic by Hansen and Mladenovi [17] which was already successfully applied to OP by Sevkli et al. in [39]. The VNS algorithm works with neighborhood structures around incumbent solution that define a sub-space in the whole state space, that can be searched for better solutions. The scope of these neighborhoods is changed with each step, typically extending the scope to find possible better solutions further away from current solution. Each newly defined neighborhood is thoroughly searched for a local optima. This process continues until a given stopping condition is met such as number of iterations. The VNS uses shake and local search operations which are used to modify the paths through the OP target locations and find the best solution. Sevkli et al. also introduced a variant of VNS algorithm which is Randomized Variable Neighborhood Search (RVNS) where the local search operators are randomized instead of systematic approach. A variant of RVNS algorithm extended by continuous optimization of heading angles and velocities was used for the HOP solution in this paper.

Euclidean trajectories used in OP are typically not suitable for real world UAVs as it creates very sharp and even unfeasible trajectories which can't be traversed in fluent motion. This led to introduction of other motion primitives in trajectory generation for autonomous vehicle movement. An extensively used approach is the trajectory generation based on Dubins maneuvers [12]. The Dubins maneuvers can be found analytically and yield the shortest path between two points with known heading angles and minimal allowable turning radius. Dubins curves were also used in previous work upon which this paper is based which combines the VNS solution of Orienteering Problem with usage of Dubins curves as a motion primitive [28]. This variant of the OP was named Dubins Orienteering Problem (DOP) and can be used to model fixed wing UAVs or VTOL UAVs with constant speeds. The DOP solution was further extended by the use of neighborhoods around the targets, since usually in information gathering, it is enough to be close enough to the target to gather required information [29]. The neighborhood is a disk with defined radius in space around the target, inside which any point can be visited in order to successfully perform the surveillance of target location in the center of the disk. A solution to DOP using self-organizing maps was proposed in [13].

This paper aims to solve the OP for VTOL UAVs that are typically not constrained by constant forward speed and the minimal turning radius as in the DOP. Herein considered motion primitives for information gathering trajectory planning are the parametric polynomial splines. Two cubic spline types are used very frequently thanks to their natural suitability, the Bézier and Hermite curves. The research of Bézier curves for path planning is done mainly for autonomous ground vehicles [9,44] in planar space. Also the Hermite curves have been used in similar tasks such as the one introduced in [6,40,43]. For the solution of the OP for UAVs we select the Hermite curves primitives due to their more suitable parametrization using two guiding points with their respective tangents that can be converted to UAVs speeds in the respective guiding points.

In order to check feasibility and estimate the time of travel when following the Hermite curve, the maximal allowable velocity along the whole curve and the motion dynamics need to be determined from known motion model of the vehicle. This approach is usually called a calculation of velocity profile. A very convenient way to generate the velocity profile is to bind the maximal velocity with the actual curvature of the curve as was done in [21]. However, this was used only in planar case for a ground vehicle. This idea can be used also in 3D if we assume decoupling of vertical and horizontal movement as was proposed in [14]. The decoupled approach is also used in this paper as it is fast and does not require a complicated motion model of the UAV. Another solution to velocity profile calculation is method used in [23,38]. The idea is again based on curvature constrained maximal velocity in each trajectory sample but then the minima are found along the curve using gradient descent search algorithm and the velocity profile is determined using numerical integration applying a bang-bang control policy on the reference dynamic model. If the feasible profile is not found the velocity minima are iteratively lowered until one is found.

3   Problem Statement

In this section, we define the regular Euclidean OP and its extension for Hermite splines motion primitives. The definition of HOP is presented together with a short introduction of cubic Hermite spline and velocity profile generation that has to be created to produce feasible trajectories for UAVs with given maximal velocity and acceleration constraints on geometric Hermite splines.

3.1   Euclidean Orienteering Problem

The goal of the Orienteering Problem is to find a subset of target locations with given rewards to be visited in order to maximize the sum of collected rewards while keeping the length of the path less or equal to a specific budget. To define the problem mathematically, we introduce a set of target locations $S = \{s_1, \ldots, s_n\}$ with each location $s_i = (t_i, r_i)$ consisting of position in plane $t_i \in R^2$ and specific positive reward $r_i \in R$. The location can be also expanded to R^3. Two targets are predefined to be start s_1 and end s_n of the solution and are assigned a reward equal to zero $r_1 = r_n = 0$ as these two would be always included in any solution. The path length is constrained by T_{max} which is the travel budget. The goal of OP is to find a subset of k target locations $S_k \subseteq S$ in order to maximize the sum of collected rewards $R = \sum r_i$ while keeping the length of the solution within the budget T_{max}. The path is defined as a sequence of visited locations in order given by permutation $\Sigma = (\sigma_1, \ldots, \sigma_k)$ of location indexes σ_i of targets s_{σ_i} in the subset $S_k = \{s_{\sigma_i}\}_{i \in (1,...,k)}$, such that $0 \leq \sigma_i \leq n$, $\sigma_i \neq \sigma_j$ for $i \neq j$ and $\sigma_1 = 1$, $\sigma_k = n$.

The combinatorial optimization of the OP then consists of finding subset S_k of k targets to visit and the order of visits Σ such that the found path has maximal collected reward within limited budget. The regular OP uses Euclidean distances as a travel cost between locations [16]. The Euclidean distance between locations t_{σ_i} and t_{σ_j} is denoted $\mathcal{L}_E(t_{\sigma_i}, t_{\sigma_j})$. The optimization problem can be then formalized as:

$$\underset{k,S_k,\Sigma}{maximize}\ R = \sum_{i=1}^{k} r_{\sigma_i}$$

$$subject\ to\ \sum_{i=2}^{k} \mathcal{L}_E(t_{\sigma_{i-1}}, t_{\sigma_i}) \leq T_{max},$$

$$\sigma_1 = 1, \sigma_k = n. \tag{1}$$

3.2 Hermite Orienteering Problem

The Hermite Orienteering Problem uses cubic Hermite splines as motion primitive to connect the visited target locations instead of straight lines as in the regular OP. The Hermite curve is defined by two endpoints and corresponding derivatives in the endpoints (tangent vectors of the curve) which determine the shape of the curve. The tangent vectors can be defined by angle relative to x axis and vector norm. For the UAV trajectory planning, these parameters can be related to heading angle and actual speed of the UAV in target locations. The heading angle θ_i at each location directly determines the orientation of the derivative vector. The derivative vector p_i' is then determined from desired velocity v at each location using multiplicative constant h_m determining the influence of the velocity on the curvature. The derivative vector in 2D space is then obtained using Eq. (2).

$$p_i' = v \cdot h_m \cdot \begin{bmatrix} cos(\theta) \\ sin(\theta) \end{bmatrix} \tag{2}$$

The considered cubic Hermite spline can be defined by four blending functions (3). The curve parameter t is used for sample generation along the curve with $t \in [0,1]$ where $t = 0$ yields the starting point of the curve and $t = 1$ the ending point. The Hermite spline (4) in arbitrary dimensional space is sum of the blending functions using the starting p_0 and ending p_1 points, and their respective derivative vector p_0' and p_1'. The example of Hermite spline in 2D is shown in Fig. 2.

$$\begin{aligned} f_1(t) &= 2t^3 - 3t^2 + 1 \\ f_2(t) &= -2t^3 + 3t^2 \\ f_3(t) &= t^3 - 2t^2 + t \\ f_4(t) &= t^3 - t^2 \end{aligned} \tag{3}$$

$$f(t) = f_1 \cdot p_0 + f_2 \cdot p_1 + f_3 \cdot p_0' + f_4 \cdot p_1' \qquad (4)$$

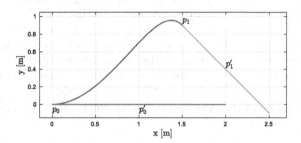

Fig. 2. An example of Hermite curve in 2D through target locations p_0 and p_1 with heading vectors (the derivative vectors) in respective locations p_0' and p_1'

Since the HOP trajectory for UAV consists of several points that need to be interpolated piecewise using separate curves, the smoothness of the trajectory have to be ensured. This is done by enforcing of the incoming and outgoing derivative vectors to be the same.

In order to estimate the time of flight needed for given Hermite curve traversal, the velocity profile needs to be calculated. A very frequent approach to the velocity profile generation along parametric curves uses the curvature of the curve to determine the maximal allowable velocity along the whole curve. This defines profile with low velocities in sharp turns and high velocities in straight segments which is desirable. An extension of this approach to be used in full three dimensional space was introduced in [14]. This method defines the possible velocity and acceleration at each sampled point along the curve using a simplifying assumption that the trajectory between two samples of the curve is Euclidean and that acceleration and velocity vectors always keep the direction towards next sample. This enables to use simple dynamical equations for constantly accelerated motion in straight line when modeling the UAV's movement along the curve utilizing triangle similarity based on geometry shown in Fig. 3.

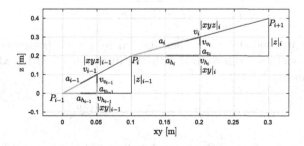

Fig. 3. Velocity, acceleration and its components between two samples of the curve. xy axis corresponds with projection of the samples into the horizontal plane

The motion constraints of the modeled VTOL UAV can be described by maximal velocity $v_{h_{max}}$ and acceleration $a_{h_{max}}$ in horizontal plane and $v_{v_{max}}$, $a_{v_{max}}$ in vertical axis separately. The horizontal components are limited based on actual curvature along with maximal values for horizontal acceleration and velocity. The vertical dynamics are only limited by maximal vertical acceleration and velocity. It is assumed that the acceleration is always positive hence samples have to be firstly iterated in forward direction, which will limit the velocity profile for acceleration, and than backwards which will limit the deceleration. On each step, the current radial acceleration in horizontal plane a_r is calculated using actual curvature of the curve, which yields the maximal horizontal tangent acceleration $a_{h_{lim}}$ using:

$$a_{h_{lim}} = \sqrt{a_{h_{max}}^2 - a_r^2}. \tag{5}$$

The maximal vertical acceleration $a_{v_{lim}}$ is determined by UAV's maximal achievable vertical acceleration $a_{v_{max}}$. Components $a_{v_{lim}}$ and $a_{h_{lim}}$ each define different acceleration in the direction of Euclidean path towards next sample (tangential acceleration) using the triangle similarity Eq. (6) for vertical component $a_{i_{vlim}}$ and horizontal component $a_{i_{hlim}}$ separately. The smaller acceleration of these two is chosen for further calculations as $a_{t_{lim}}$.

$$a_{i_{vlim}} = \frac{|xyz|_i}{|z|_i} a_{v_i} \quad , \quad a_{i_{hlim}} = \frac{|xyz|_i}{|xy|_i} a_{h_i} \tag{6}$$

The selected acceleration is used to calculate the candidate velocity v_{s_n} in the next sample using Eq. (7), where the variable s is an Euclidean distance between current sample s_c and next sample s_n, t is time of flight between these samples and v_{s_c} is a velocity in current sample.

$$s = \frac{1}{2} a_{t_{lim}} t^2 + v_{s_c} t$$
$$t = \frac{-v_{s_c} + \sqrt{v_{s_c}^2 + 2a_{t_{lim}} s}}{a_{t_{lim}}}$$
$$v_{s_n} = a_{t_{lim}} t + v_{s_c} \tag{7}$$
$$v_{s_n} = \sqrt{v_{s_c}^2 + 2a_{t_{lim}} s}$$

This candidate velocity v_{s_n} is then limited by known horizontal and vertical maximal velocities, transformed into the direction toward next sample by triangular similarity Eq. (8).

$$v_{i_{vlim}} = \frac{|xyz|_i}{|z|_i} v_{v_i} \quad , \quad v_{i_{hlim}} = \frac{|xyz|_i}{|xy|_i} v_{h_i} \tag{8}$$

The new variant of the OP combining Hermite curves with velocity profile calculation for estimation time of flight for particular Hermite spline and known UAV constrains is called Hermite Orienteering Problem. The mathematical definition of the OP (1) is extended for the HOP by a set of heading angles corresponding with each visited location $\Theta = (\theta_{\sigma_1}, \ldots, \theta_{\sigma_k})$ and variable vector $\mathcal{V} = (v_{\sigma_1}, \ldots, v_{\sigma_k})$ with actual velocities at given target locations. Both Θ and $\mathcal{V}$ then defines the derivative vectors according to (2) needed to create Hermite splines through visited target locations S_k with order Σ. The needed information about the heading angle, velocity and location is defined as a state $q_{\sigma_i} = (t_{\sigma_i}, v_{\sigma_i}, \theta_{\sigma_i})$. The HOP utilizes an operator $T_H(q_{\sigma_i}, q_{\sigma_j})$ which is the time of flight between two states q_i and q_j using Hermite curve as a motion primitive with found velocity profile. The visiting order of the states is given by permutation Σ, Θ is a set of used heading angles and $\mathcal{V}$ is a set of actual velocities at target locations. The Hermite Orienteering Problem can be described as optimization problem:

$$\underset{k,S_k,\Sigma,\Theta,\mathcal{V}}{maximize}\ R = \sum_{i=1}^{k} r_{\sigma_i}$$

$$subject\ to \sum_{i=2}^{k} T_H(q_{\sigma_{i-1}}, q_{\sigma_i}) \leq T_{max}, \tag{9}$$

$$q_{\sigma_i} = (t_{\sigma_i}, v_{\sigma_i}, \theta_{\sigma_1}), t_{\sigma_i} \in S_k,\ v_{\sigma_i} \in \mathcal{V},\ \theta_{\sigma_i} \in \Theta,$$

$$v_1 = 0, v_k = 0,$$

$$\sigma_1 = 1, \sigma_k = n.$$

4   Proposed Approach

The proposed solution to HOP is based on the Variable Neighborhood Search algorithm introduced by Mladenovi et al. in [17]. This algorithm can be used in wide range of combinatorial optimization problems and was successfully used to solve OP in [39]. The VNS algorithm operates on a given neighborhood structures N_l, where $l = \{l_1, \ldots, l_{max}\}$ is the maximal distance between solutions inside the neighborhood, which is a number of different targets visited in the OP solution. The VNS algorithm uses two types of procedures to search for new solutions, it is *shake* and *local search* procedure. The VNS algorithm starts with a given initial solution and periodically applies the *shake* procedure to escape from possible local maximum and subsequently uses the *local search* operators to find the optima in vicinity. This is repeated until given terminating condition is met, such as runtime or number of iterations without improvement. The limiting factor is that algorithm implements only the combinatorial optimization, which means working only with discretely defined states. Hence if it is needed to optimize continuous variables such as the heading angle and velocities at the targets as in HOP, the VNS algorithm needs to be extended.

The problem of continuous optimization while using the VNS algorithm was addressed in [28], where the VNS algorithm was successfully applied to the Dubins Orienteering Problem. The continuous optimization part of the DOP, lies in determination of the optimal heading angles at each of the targets in order to construct Dubins curves of the shortest length. This was tackled by discrete samples of possible angles at each target location. However, in order to find a solution of good quality, it is better to use high sampling rates which dramatically enlarges the complexity of the state space. The authors of DOP [28] therefore used randomized variant of the *local search* procedure where only predefined number of random operations are tested during the procedure. This variant of VNS is called Randomized Variable Neighborhood Search (RVNS) and it was shown that it is able to find solutions of the same rewards while being significantly faster than normal VNS [39].

The [28] introduces several operators for RVNS algorithm which can be used in the context of the OP. The operators in *shake* procedure are **Path Move** and **Path Exchange** while the *local search* consists of **Point Move** and **Point Exchange** operators. All of these operators modify the subset S_k of visited targets as well as order of visit Σ, with the appropriate heading angles Θ being selected from predefined samples always using the one that minimizes the path length as shown further in Fig. 4.

However, the proposed HOP requires to optimize not only the heading angles Θ at the target locations, but also the velocities V in order to construct trajectories from Hermite splines. This leads to the introduction of new operators **Angle Speed Shake** and **Improve Angle Speed**. These operators solve the continuous optimization part of the problem and enable to explore neighboring solutions even outside of the scope defined by initially sampled velocities and heading angles. The **Angle Speed Shake** randomly changes heading angle and velocity of current solution, while the **Improve Angle Speed** systematically improves them by implementing a simple hill climbing algorithm when introducing new samples to the state space. This enables to accept even shake solutions that slightly violate the budget as these can be further improved by continuous optimization to the point where the budget constraint is no longer violated. The biggest advantage of continuous optimization is a fact that the initial sampling rates can be fairly low as new promising samples are added during the optimization process. The whole RVNS algorithm used to solve HOP is described in detail in following sections.

4.1   RVNS Algorithm

The solution of HOP is internally represented in RVNS by a solution vector $v = (s_{\sigma_1}, \ldots, s_{\sigma_k}, s_{\sigma_{k+1}}, \ldots, s_{\sigma_n})$, which holds all the possible targets $s_i \in S$. The order of the targets in the vector corresponds with the order of visit during flight with first k being included in the solution with order given by Σ. The other $n - k$ states are not used (above budget), but since they are still in the solution vector the neighborhood operators can include these unused targets in possible new solutions.

In order to estimate the time of flight needed to traverse given Hermite curve, the velocity profile have to be calculated. This process, especially for larger amount of samples per curve, can be quite time demanding if working

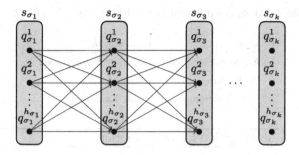

Fig. 4. Search graph used to find the best combination of samples in the current solution vector $v = (s_{\sigma_1}, s_{\sigma_2}, s_{\sigma_3}, \ldots, s_{\sigma_k})$. The sample q_i^j is defined by given value of heading angle and velocity

with hundreds of thousand possible curves, which is not uncommon even with fairly low initial sampling rates of angle and velocity. Therefore, the time of flight estimates between all possible initially sampled states are pre-calculated before the actual VNS algorithm starts in order to significantly improve the performance of neighborhood operators. Such a time of flight graph is also updated during the algorithm as new samples are added during continuous optimization operations. However, some pair of states can violate the maximal acceleration and velocity constraints of UAV and therefore are exclude from the graph.

As the neighborhood operators modify the order and number of visited target locations and their order in the solution vector, the combination of velocities $\mathcal{V}$ and heading angle Θ samples which generates the shortest trajectory needs to be found. This requires a search through the time of flight graph structure, which can be thought of as searching through a subgraph with vertices being the heading angle and velocity samples. The resulting path consists of best combinations of samples at each target location which yield the shortest trajectory. An example of such search graph is shown in Fig. 4.

Before the initial HOP solution is generated, all unreachable targets are filtered out. The target is qualified as unreachable if it's so far away from the starting target that even if the UAV would fly with maximal possible velocity the whole time, it would not be able to reach this target and return to ending target without violating the maximal budget condition. The initial solution is then created using simple greedy heuristic. The algorithm searches through the elements of the solution vector $v = (s_{\sigma_1}, \ldots, s_{\sigma_k}, s_{\sigma_{k+1}}, \ldots, s_{\sigma_n})$ which are not currently used in the solution (indexes $k + 1 \ldots n$). Each element is experimentally added to each position in current solution and the situation with the minimal added time of flight per reward is accepted and solution is extended by this target. This is repeated until the budget is hit or all targets are included.

With the initial solution found, the RVNS algorithm starts with iterative improvement using *shake* and *local search* procedures using operators summarized in Table 1. The whole RVNS method for the proposed HOP is outlined in Algorithm 1.

Table 1. All operators used in the RVNS algorithm for HOP

l	shake	local search
1	Angle Speed Shake	Improve Angle Speed
2	Path Move	Point Move
3	Path Exchange	Point Exchange

Algorithm 1. RVNS algorithm for HOP solution

S: set of target locations
T_{max}: maximal time budget
l: neighborhood scope defining the type of neighborhood operator, $l \in \langle 1, 3 \rangle$
l_{max}: maximal value of $l = 3$
P: the best solution path found
R(P): collected reward of path P
$T_H(P)$: time of flight of path P
Output: The best found path P

$S_r \leftarrow getReachableLocations(S, T_{max})$
$P \leftarrow createInitialSolution(S_r, T_{max})$
while *Stopping condition not met* **do**
  $l \leftarrow 1$
  while $l \leq l_{max}$ **do**
    $P' \leftarrow shake(P, l)$
    $P'' \leftarrow localSearch(P', l)$
    if $[T_H(P'') \leq T_{max}$ **and**$R(P'') > R(P)]$ **or** $[R(P'') == R(P)$ **and** $T_H(P'') <$
    $T_H(P)]$ **then**
      $P \leftarrow P''$
      $l \leftarrow 1$
    else
      $l \leftarrow l + 1$
    end if
  end while
end while
return P

The current best solution P is modified by applying the *shake* procedure which enables to escape from possible local maximum and then the *local search* procedure to finely explore the imminent neighborhood of the solution. If the reward of the new solution is higher than reward of the currently best solution, it is accepted as the new best solution and used in further optimization. If the reward of the new solution is the same as the current best reward, the one with smaller time of flight is selected. In the case of the time budget being large enough to visit all targets, this condition enables to perform the optimization in the sense of Traveling Salesman Problem. If the newly found solution is worse than currently best solution the process is repeated using *shake* and *localSearch* operations with larger scope of neighborhood structures that can be explored based on value l. The continuous optimization part of the problem is embodied within this algorithm as both *shake* and *localSearch* operations each contain one operator for continuous optimization. Whole process continues until one of the stopping conditions is met. This can be defined arbitrarily, however, the used stopping condition in following experiments is maximal run time along with maximal number of iterations and a number of iterations without improvement. The particular operators used by RVNS are described in following sections.

4.2   Combinatorial Operators

The combinatorial operators are used to improve the current solution by adding and removing targets in the solution and changing the order of visit. After each manipulation, the shortest trajectory through all targets is found using currently known samples of heading angle and velocity. The combinatorial operators work with previously defined solution vector v. Only first k members are considered as a part of the current solution P. This enables to introduce new targets into the current solution as the operator do not distinguish between used and omitted targets.

Shake operators are used to explore the neighborhood structures with large possible distance between solutions. This enables to escape from local maxima and can change the current solution dramatically. The two combinatorial optimization *shake* operators are **Path Move** and **Path Exchange**.

Path Move ($l = 2$) operator selects a random path inside the solution vector v and changes its position inside the vector. The operator itself performs the changes in the current solution by selecting three random indexes inside the vector $v = (s_{\sigma_1}, \ldots, s_{\sigma_n})$ such as $i_1 \in \langle 2, n \rangle, i_2 \in \langle i_1 + 1, n \rangle, i_3 \in \langle 2, i_1) \bigcup (i_2, n \rangle$, $i_{1,2,3} \neq k$. The initial start and end locations are not affected. Index i_1 and i_2 define the path that is going to be moved and index i_3 defines its new position inside the vector v. The modified solution vector for the case

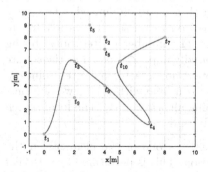 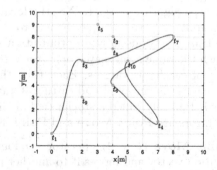

(a) Path $v = (s_1, s_3, s_6, s_4, s_{10}, s_7, s_9, s_5,$ $s_2, s_8)$ before moving $\{s_6, s_4, s_{10}\}$ after s_7

(b) Pathe after operation $v = (s_1, s_3, s_7,$ $s_6, s_4, s_{10}, s_9, s_5, s_2, s_8)$

Fig. 5. An example of **Path Move** operation

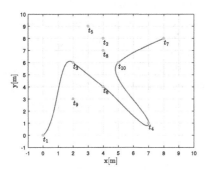 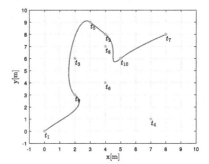

(a) Path $v = (s_1, s_3, s_6, s_4, s_{10}, s_7, s_9, s_5,$ $s_2, s_8)$ before exchange path $\{s_3, s_6, s_4\}$ with $\{s_9, s_5, s_2\}$

(b) Path after operation $v = (s_1, s_9, s_5,$ $s_2, s_4, s_{10}, s_3, s_6, s_4, s_8)$

Fig. 6. An example of **Path Exchange** operation

$i_3 > i_2$ then equals to $v = (s_{\sigma_1}, \ldots, s_{\sigma_{i_1}-1}, s_{\sigma_{i_2}+1}, \ldots, s_{\sigma_{i_3}}, s_{\sigma_{i_1}}, \ldots, s_{\sigma_{i_2}}, s_{\sigma_{i_3}+1},$ $\ldots, s_{\sigma_n})$. An example of this operation is pictured in Fig. 5.

Path Exchange ($l = 3$) operator tries to find solution even further away than **Path Move** in terms of solution distance. As the name suggest, in this case there are two paths selected in the solution vector and their positions are switched. This can be performed by randomly selecting four indexes $i_{1,\ldots,4}$, but these need to be selected in feasible manner, which means the selected paths can't be overlapping, i.e. $i_1 \in \langle 2, n \rangle, i_2 \in \langle i_1 + 1, n \rangle, i_3 \in \langle i_2 + 1, n \rangle, i_4 \in \langle i_3+1, n \rangle$, $i_{1,\ldots,4} \neq k$. The modified solution vector v than equals to $v = (s_{\sigma_1}, \ldots,$ $s_{\sigma_{i_1}-1}, s_{\sigma_{i_3}}, \ldots, s_{\sigma_{i_4}}, s_{\sigma_{i_2}+1}, \ldots, s_{\sigma_{i_3}-1}, s_{\sigma_{i_1}}, \ldots, s_{\sigma_{i_2}}, s_{\sigma_{i_4}+1}, \ldots, s_{\sigma_n})$. An example of this operation is shown in Fig. 6.

Local search operators aim to explore the imminent neighborhood of the current solution searching for local optima. This is done by modifications that change the solution in smaller scope than the shake operators. In addition, since the used algorithm uses RVNS, the local search is also randomized. The operators apply only very simple changes to the solution vector so it is possible to run multiple iterations in one optimization step. The number of iterations during which the same local search operator is repeatedly applied is equal to the square root of reachable targets. Each time the operation creates a better solution, the change is accepted and further iterations continue with this new solution - this is typically called a stochastic hill climbing algorithm [34]. The local search operators are **Point Move** and **Point Exchange**.

Point Move ($l = 2$) operator randomly selects one of the elements in the solution vector and moves it to another position. This is performed by selecting only two random indexes such as $i_1 \in \langle 2, n \rangle, i_2 \in \langle 2, i_1 \rangle \bigcup (i_1, n \rangle$, $i_{1,2} \neq k$. The newly acquired solution vector for the case where $i_1 < i_2$ is then $v = (q_{\sigma_{i_1}},$ $\ldots, q_{\sigma_{i_1}-1}, q_{\sigma_{i_1}+1}, \ldots, q_{\sigma_{i_2}-1}, q_{\sigma_{i_1}}, q_{\sigma_{i_2}+1}, \ldots, q_{\sigma_n})$. An example of this operation is shown in Fig. 7.

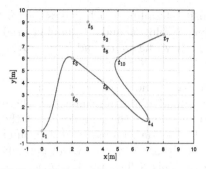 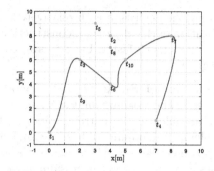

(a) Path $v = (t_1, t_3, t_6, t_4, t_{10}, t_7, t_9, t_5, t_2,$ $t_8)$ before moving point t_4 to index 7

(b) Path after operation $v = (t_1, t_3, t_6, t_{10},$ $t_7, t_4, t_9, t_5, t_2, t_8)$

Fig. 7. An example of **Point Move** operation

Point Exchange ($l = 3$) operator exchanges positions of two randomly selected elements in the solution vector. This is again implemented as random selection of two indexes but the solution vector is modified in different manner to $v = (q_{\sigma_{i_1}}, \ldots, q_{\sigma_{i_1-1}}, q_{\sigma_{i_2}}, q_{\sigma_{i_1+1}}, \ldots, q_{\sigma_{i_2-1}}, q_{\sigma_{i_1}}, q_{\sigma_{i_2+1}}, \ldots, q_{\sigma_n})$. An example of this operation is shown in Fig. 8.

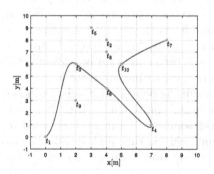 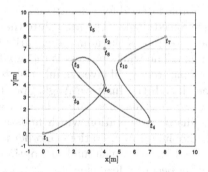

(a) Path $v = (t_1, t_3, t_6, t_4, t_{10}, t_7, t_9, t_5, t_2,$ $t_8)$ before exchange point t_3 with point t_6

(b) Path after operation $v = (t_1, t_6, t_3, t_4,$ $t_{10}, t_7, t_9, t_5, t_2, t_8)$

Fig. 8. An example of **Point Exchange** operation

4.3   Continuous Optimization Operators

The continuous optimization part of the HOP problem, which is related to the selection of suitable heading angle and velocity, is tackled by introduction of two special operators **Angle Speed Shake** and **Improve Angle Speed**. Both operators can introduce new samples for angle and velocity into the graph of all samples which are used in combinatorial optimization and effectively decrease the needed time of flight for given combination of target locations. This enables to start the algorithm with relatively low density sampling which makes the initialization of the algorithm, construction of initial solution and all combinatorial operations much faster. Used operators again follow the *shake* and *local search* design.

Angle Speed Shake ($l = 1$) operator performs the shake operation upon velocities and heading angles. That means randomly changing values for these two variables in the current solution. To further lower the complexity of the problem, the sampling of heading angle is performed only on planar disk around each target even in 3D variant of the problem, even though in 3D all possible heading angles are given by samples on sphere. This defines the heading angle interval as $\langle 0, 2\pi \rangle$. This reduces the complexity while still offering reasonably good extent of possible states. A disadvantage of this technique is a fact that the heading vector of appropriate Hermite curve determined by heading angle will always have the z component (vertical axis) equal to zero, which effectively means the vertical velocity in each target location should be zero. The velocity samples are limited by interval $\langle 0, v_{h_{max}} \rangle$, as the only contributing component is the horizontal velocity at each target location. The solution modified by this shake operation is then improved by following local search continuous optimization operator.

Improve Angle Speed ($l = 1$) operator performs the local search in terms of continuous optimization. This involves stochastic hill climbing technique. The admissible interval for velocity and heading angle stays the same as in the case of Angle Speed Shake. The procedure that is used for both variables at each target location explores the admissible variable's intervals by local iterative optimization [42]. The whole process is repeated for each target location in current solution separately until given number of iterations without improvement is reached. Even though the Improve Angle Speed procedure is defined as a neighborhood operator, it is also used after the creation of initial solution and then after each combinatorial operation which finds a better solution. It can be even used in the combinatorial operations in order to admit solution that slightly violates the maximal budget which might be optimized by Improve Angle Speed so that this solution becomes feasible.

5   Results

This section shows experimental results of the proposed RVNS-based algorithm for the HOP.

The first set of experiments aims to show the computational capabilities and performance of the HOP solution algorithm. Since the solution is generated in a stochastic way, a large number of tests is needed in order to get a meaningful statistic sample. To achieve this goal, the computational grid services provided by the organization Metacentrum with cluster Intel Xenon processors (2.2 GHz–3.3 GHz) were utilized [1].

The second experimental setup verifies the suitability and feasibility of the resulting trajectories for real hexarotor UAV. These are created with characteristics defined by the motion model of a real UAV, which is used to fly through the generated trajectory and verify the correctness of the design. Several HOP scenarios were created showing the capabilities of the solution tested on the UAV in real conditions.

5.1 Computational Results

The computational results show the performance of the proposed algorithm for solution of the HOP and effect of several key parameters. The HOP solution implementation was run for several different budgets with collected reward being the examined result. Since the results are stochastic, each test with given maximal budget was run ten times and the average and maximal collected reward was determined. This test was performed using three different data sets. The first tested scenario is defined by the MBZIRC data set, which was custom created for testing purposes with UAVs on testing range. The second used scenario was introduced by Tsiligirides in [41], where he defined three scenarios containing up to 32 target locations each, naming them Set 1, Set 2 and Set 3. The scenario named Set 1 is used in further testing and will be referred to as Tsiligirides scenario. The third used scenario was defined by Chao in [8], contains 66 target locations, and will be further referred as Set66.

The average and maximal collected rewards with respect to the maximal travel budget for each scenario are shown in Table 2. Examples of solutions for each scenario are shown in Figs. 9 and 10.

Apart from the maximal budget, the parameters influencing the algorithm are the initial sampling rates of the heading angle and velocity at target locations. The denser the sampling is, the better initial solution can be found, however,

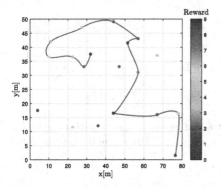
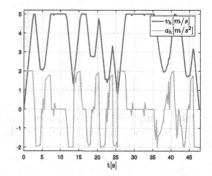

(a) MBZIRC scenario HOP solution   (b) Velocity and acceleration profile

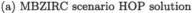

Fig. 9. An example solution of MBZIRC scenario with time budget $T_{max} = 50\,\mathrm{s}$, maximal horizontal acceleration $a_{h_{max}} = 2\,\mathrm{m/s}^2$, maximal horizontal velocity $v_{h_{max}} = 5\,\mathrm{m/s}$ along with calculated velocity and acceleration profile

Table 2. The collected average and maximal collected reward based on the maximal budget for different HOP scenarios, T_{max} - maximal traveling time budget, R_{Avg} - average collected reward in ten runs, R_{Max} - maximal collected reward in ten runs, t_{Avg} - average time of reaching the maximal achieved reward in ten runs. All tests were performed using UAV limitations $v_{h_{max}} = 5\,\text{m/s}$ and $a_{h_{max}} = 2\,\text{m/s}^2$

$T_{max}[s]$	MBZIRC			Tsiligirides			Set66		
	R_{Avg}	R_{Max}	$t_{Avg}[s]$	R_{Avg}	R_{Max}	$t_{Avg}[s]$	R_{Avg}	R_{Max}	$t_{Avg}[s]$
25	42.0	42	17.3	166.5	185	628.9	668.5	745	3210.2
30	52.0	52	156.2	211.5	215	1297.7	748.5	855	1635.0
35	58.0	58	17.9	234.0	240	3011.6	868.5	935	2936.0
40	65.0	65	2262.8	254.4	265	3205.6	1059.5	1145	2449.0
45	72.6	73	170.3	273.0	280	1734.6	1191.5	1255	3845.8
50	77.0	77	41.0	284.0	285	958.9	1350.0	1415	5916.5
55	81.0	81	21.3	285.0	285	104.0	1430.0	1485	4423.8
60	85.8	87	280.4	285.0	285	66.2	1465.5	1565	4143.8
65	87.7	88	1413.2	285.0	285	56.3	1576.0	1650	3182.2
70	88.0	88	18.3	285.0	285	68.1	1613.0	1650	2979.3
75	88.0	88	521.4	285.0	285	324.3	1669.0	1680	3032.2
80	88.0	88	147.7	285.0	285	124.7	1679.0	1680	2599.6
85	88.0	88	57.2	285.0	285	789.1	1680.0	1680	3750.9

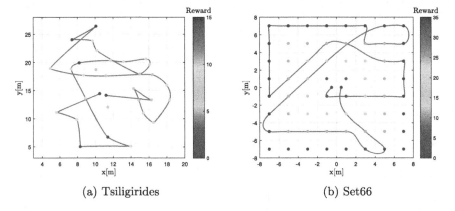

(a) Tsiligirides   (b) Set66

Fig. 10. Example HOP solutions of Tsiligirides and Set66 scenarios with time budget $T_{max} = 50\,\text{s}$, maximal horizontal acceleration $a_{h_{max}} = 2\,\text{m/s}^2$, maximal horizontal velocity $v_{h_{max}} = 5\,\text{m/s}$

the calculation time is larger. Two tests were designed to explore the influence of initial sampling rates on the performance. All rewards were again averaged over ten runs with the same settings and all test were performed using MBZIRC scenario. The results are pictured in Fig. 11.

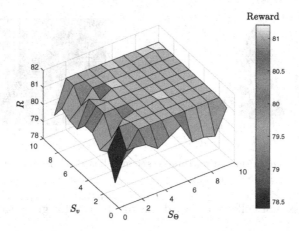

Fig. 11. Average collected reward in MBZIRC scenario based on used initial sampling rates, R - average gathered reward, S_v - number of velocity samples per target, S_Θ - number of heading angle samples per target

The graph shows that the quality of the solution is almost indifferent to the initial sampling rate that was used. Only the very low sampling rate cases (1-2 samples per location) show slightly worse collected reward. This is very positive result, as it shows that it is really sufficient to use lower density initial sampling rates such as three to four samples per target location per variable, since the quality of the final solution will be unchanged compared to the solution obtained using higher sampling rates. This is the main motivation for introduction of continuous optimization operators and this result shows that the approach is successful.

Since it was shown that due to the continuous optimization the initial sampling can be chosen arbitrarily in order to find good quality solution, the other factor that should offer sufficient insight on the suitable strategy for sampling rate determination is the time performance. The average time of reaching the

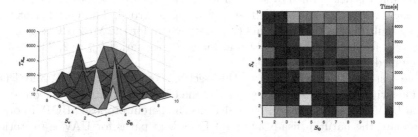

Fig. 12. Average time of finding the first solution of maximal reached reward based on chosen initial sampling rate

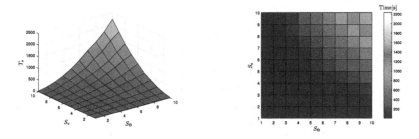

Fig. 13. Average initialization time in seconds based on used sampling rate

value of the maximal collected reward during the run for given sampling rates is shown in Fig. 12.

Even though there are some outliers throughout the graph, there is overall tendency for the calculation times to be higher in sections of either very low sampling rates or very high sampling rate. In the case of low sampling rates, the higher calculation time is clearly caused by low number of possible combinations that can be used by the algorithm. It takes significant number of iterations until new samples are introduced. This also causes the lower average collected reward. In the case of high sampling rates, the longer calculation times are caused by too many samples to iterate through.

As was noted one of the biggest performance hits is the initial calculation of the time of flight graph between all initial samples before the actual algorithm starts. For high sampling rates this initialization takes significantly more time compared to low sampling rates. Figure 13 shows this initialization time for given sampling rates.

The second important parameter that will be explored is the Hermite curve heading vector multiplier h_m. This parameter has a big impact on the resulting shape of the trajectory, but the optimal value differs for each scenario as it is dependent on the topology of the target locations. Unfortunately no analytical formula for choosing the optimal multiplier was found, hence an iterative process for finding the best multiplier for given scenario was used. The best heading vector multiplier is determined by series of tests using different values of the heading multiplier, with average collected reward being recorded. The best performing heading multiplier is then used in other tests where for the MBZIRC scenario it is $h_m = 3$, for Tsiligirides $h_m = 1$ and for Set66 scenarios $h_m = 2$. The results are summed up in the Fig. 14.

The last set of experiments in this section compares the EOP and DOP solutions with the HOP solution and show the benefits of modeling the data collection UAV scenario with limited budget as the herein proposed HOP. In order to simulate the natural disadvantage of Euclidean paths for UAV trajectories, a slightly modified version of HOP implementation was used. The continuous optimization was disabled and the only samples available are samples with zero velocity. This generates Euclidean paths and the UAV is forced to stop at each target location since the turning angles are very sharp. The velocity profile gen-

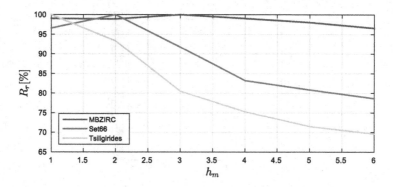

Fig. 14. Average gathered budget relative to the maximal average reward in percent R_r based on used heading vector multiplier h_m for given scenarios

eration and the combinatorial optimization stays the same. An example of such Euclidean solution along with the velocity profile is shown in Fig. 15.

The implementation of the DOP solver provided by the authors of [28] was used for comparison with the HOP approach. However, the DOP solver uses the length of the trajectory instead of time of flight as the budget constraint. The DOP approach assumes that the UAV moves with constant velocity which also determines the turning radius of the Dubins curves. In order to fairly compare the DOP solution with HOP, the average velocity with which the UAV moves along the trajectory generated in HOP solution is determined from the velocity profile. This average velocity of flight is then used in the DOP as the constant velocity. From the known constant velocity v_{DOP} and known maximal time budget T_{max} in HOP solution, it is possible to determined the length budget

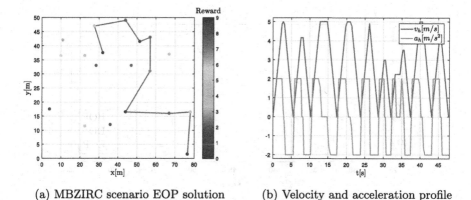

(a) MBZIRC scenario EOP solution   (b) Velocity and acceleration profile

Fig. 15. An example solution of MBZIRC scenario using EOP approach with time budget $T_{max} = 50\,\mathrm{s}$, maximal horizontal acceleration $a_{h_{max}} = 2\,\mathrm{m/s^2}$, maximal horizontal velocity $v_{h_{max}} = 5\,\mathrm{m/s}$ along with calculated velocity and acceleration profile

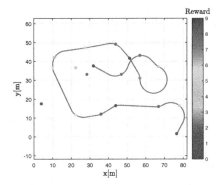

Fig. 16. An example solution of MBZIRC scenario using DOP approach with length budget $L_{max} = 247.31\,\mathrm{m}$ and constant horizontal velocity $v_h = 3.81\,\mathrm{m/s}$ which corresponds to minimal turning radius of 7.23 m. This corresponds to a time budget of 65 s

$L_{max} = v_{DOP} \cdot T_{max}$ for DOP which is equivalent to the time budget in HOP. An example of the DOP solution is shown in Fig. 16. The velocity profile is not included as it is considered constant.

The comparison of all three solutions of the EOP, DOP and HOP shows the average collected reward based on value of time budget or equivalent length budget in case of the DOP. The average reward is again calculated using results from ten runs for each budget value. The MBZIRC scenario was used for all three solutions and the results are shown in Fig. 17. It is clear that the HOP has superior results. This shows that by taking advantage of Hermite curves and the introduction of the velocity profile it is possible to collect more reward even if the actual distance might be longer.

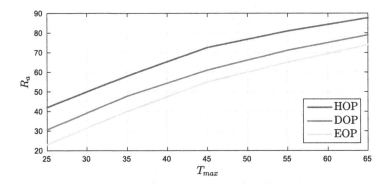

Fig. 17. Comparison of average rewards R_a for given maximal budget T_{max} for HOP, DOP and EOP solutions

5.2   Experimental Verification Using Real UAV

The experimental verification with the UAV aims to show feasibility of the generated trajectories and the correctness of the time of flight estimate. Three HOP solutions using different sets of parameters were selected in order to test the full range of capabilities of this solution. The MBZIRC scenario was used in all cases. The target locations were manually marked in open field by objects with numbers corresponding to rewards. The duration of the experiments was recorded by on-board Mobius camera with wide field of view used mainly to record the target markers on the ground. Additionally the flight of the UAV was recorded externally by commercially sold UAV with 4k camera from the height of roughly hundred meters. Selected UAV motion constraints for the HOP velocity profile generation are $v_{h_{max}} = 5\,\mathrm{m/s}$, $a_{h_{max}} = 2\,\mathrm{m/s^2}$, $v_{v_{max}} = 1\,\mathrm{m/s}$ and $a_{v_{max}} = 1\,\mathrm{m/s^2}$.

Table 3. Experiments setup

Code name	2D_HOP	3D_EOP	3D_HOP
Scenario	MBZIRC	MBZIRC (modified)	MBZIRC (modified)
Dimension	2D	3D	3D
Heading mult.	5	0 (Euclid)	5
Budget	45	55	55
Reward	70	51	62

The set of used parameters for each of three experiments is summarized in Table 3. In the 3D cases 3D_EOP and 3D_HOP, the topology of the target locations remained the same in the plane with each target being randomly assigned the theoretical height ranging from 4–20 m. Also for 3D scenarios the starting location was moved.

Recorder flight trajectories plotted over the image of the testing area are shown in Fig. 18 along with the comparison of recorded trajectories with the theoretical splines which were supplied to the UAV model predictive controller [4]. It is shown that the UAV follows the trajectory with reasonable accuracy. Minor deviations can be noticed, but these do not hold much importance for the actual goal of surveillance, since all of the target locations are visually inspected by the onboard camera. This is also supported by a video[1] from the on-board camera which shows all of the target locations in the filed of view.

It was shown that the trajectory following is reasonably accurate from geometrical point of view, but the resulting trajectories are defined with respect to the maximal theoretical velocity profile, so it is important to test the accuracy

---

[1] A link to a video footage from all three experiments: https://www.youtube.com/watch?v=gagYFLpGVC4.

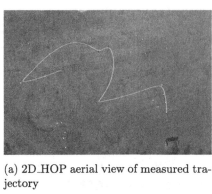

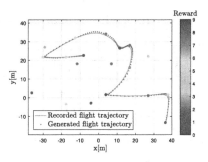

(a) 2D_HOP aerial view of measured trajectory

(b) 2D_HOP measured and theoretical
trajectory comparison

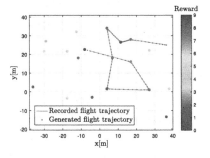

(c) 3D_EOP aerial view of measured trajectory

(d) 3D_EOP measured and theoretical
trajectory comparison

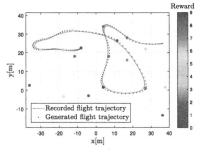

(e) 3D_HOP aerial view of measured trajectory

(f) 3D_HOP measured and theoretical
trajectory comparison

Fig. 18. Recorder flight trajectories by the UAV video platform along with comparison
of measured and theoretical generated trajectories

of this estimate. The horizontal and vertical theoretical velocity and acceleration profile along with the real measured profiles from the 3D_HOP experiment
are shown in Fig. 19. It is noticeable that the profiles differ quite significantly.

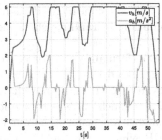

(a) Theoretical horizontal velocity and acceleration profile

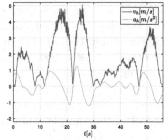

(b) Measured horizontal velocity and acceleration profile

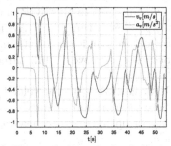

(c) Theoretical vertical velocity and acceleration profile

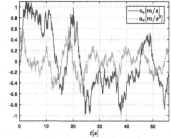

(d) Measured vertical velocity and acceleration profile)

Fig. 19. Theoretical and measured velocity and acceleration profiles comparison for the 3D_HOP scenario solution. v_h - horizontal velocity, a_h - horizontal acceleration, v_v - vertical velocity, a_v - vertical acceleration

The measured horizontal acceleration is much smoother and does not change as dramatically. This difference was caused by the fact that the measured velocities and accelerations are determined from measured positions of UAV and the proposed velocity generation does not consider jerk constraints of real UAV. This resulted in differences in predicted and measured time of flight for each scenario, which was, however, a difference of approximately 1.8%. Still, the main objective of visually inspecting target markers on the ground was in all cases successful collecting the same rewards as planned.

6   Conclusions

A novel approach to UAV information gathering planning based on Orienteering Problem is introduced in this paper. The proposed solution uses Hermite cubic splines as motion primitives to find fast and feasible solutions of the OP. This variant of the OP is introduced as Hermite Orienteering Problem.

The solution of HOP is based on the Random Variable Neighborhood Search algorithm extended by continuous optimization of heading angles and velocities required for generating Hermite splines. This enables to use small initial sampling resolution of heading angles and velocities and still find solutions of good quality, as new samples are added continually by combination of random shake and systematic exploration operations.

Two types of tests were performed to verify the performance and correctness of implemented solution. The computational tests showed that the continuous optimization works successfully as the quality of found solutions did not change with used velocity and heading angle sampling rates, which enables to use low number of initial samples that makes the whole algorithm faster. The current solution was compared to other approaches to OP which were the Euclidean OP and the Dubins OP. The HOP solution proposed in this paper showed to be superior to both, as the average collected reward was higher in all test cases which means the UAV is able to visit more targets compared to the EOP and DOP solutions using the same travel budget. It was shown that the UAV is able to successfully track the generated Hermite trajectories and can use its full kinematic potential, since it can reach high velocities and move in full three dimensional space.

For future work we plan to introduce the jerk UAV constraint to the velocity profile generation in order to minimize the error between real and planned profiles. Furthermore we plan to investigate possibility of optimizing the heading vector multiplier h_m during the RVNS optimization.

Acknowledgment. The presented work has been supported by the Czech Science Foundation (GAČR) under research project No. 17-16900Y and by OP VVV MEYS funded project CZ.02.1.01/0.0/0.0/16_019/0000765 "Research Center for Informatics". Access to computing and storage facilities of the National Grid Infrastructure Meta-Centrum, provided under the programme CESNET LM2015042, is greatly appreciated. The support of the Grant Agency of the Czech Technical University in Prague under grant No. SGS17/187/OHK3/3T/13 is also gratefully acknowledged.

References

1. Metacentrum by cesnet. https://metavo.metacentrum.cz. Accessed 1 July 2018
2. Alena, O., Niels, A., James, C., Bruce, G., Erwin, P.: Optimization approaches for civil applications of unmanned aerial vehicles (UAVs) or aerial drones: a survey. Networks **72**(4), 411–458 (2018)
3. Applegate, D.L., Bixby, R.E., Chvatal, V., Cook, W.J.: The Traveling Salesman Problem: A Computational Study. Princeton University Press, Princeton (2006)
4. Baca, T., Loianno, G., Saska, M.: Embedded model predictive control of unmanned micro aerial vehicles. In: International Conference on Methods and Models in Automation and Robotics (MMAR), pp. 992–997, August 2016
5. Balas, E., Toth, P.: Branch and bound methods for the traveling salesman problem. Technical report, Carnegie-Mellon University, January 1983
6. Boonporm, P.: Online geometric path planning algorithm of autonomous mobile robot in partially-known environment. Rom. Rev. Precis. Mech. Opt. Mechatron. 185–188 (2011)

7. Canis, B.: Unmanned Aircraft Systems (UAS): Commercial Outlook for a New Industry. Congressional Research Service Washington (2015)
8. Chao, I.M., Golden, B.L., Wasil, E.A.: A fast and effective heuristic for the orienteering problem. Eur. J. Oper. Res. **88**(3), 475–489 (1996)
9. Choi, J., Curry, R., Elkaim, G.: Path planning based on Bézier curve for autonomous ground vehicles. In: Advances in Electrical and Electronics Engineering - IAENG Special Edition of the World Congress on Engineering and Computer Science 2008, pp. 158–166 (2008)
10. Deng, C., Wang, S., Huang, Z., Tan, Z., Liu, J.: Unmanned aerial vehicles for power line inspection: a cooperative way in platforms and communications. J. Commun. **9**(9), 687–692 (2014)
11. Doherty, P., Rudol, P.: A UAV search and rescue scenario with human body detection and geolocalization. In: Orgun, M.A., Thornton, J. (eds.) AI 2007. LNCS (LNAI), vol. 4830, pp. 1–13. Springer, Heidelberg (2007). https://doi.org/10.1007/978-3-540-76928-6_1
12. Dubins, L.E.: On curves of minimal length with a constraint on average curvature, and with prescribed initial and terminal positions and tangents. Am. J. Math. **79**(3), 497–516 (1957)
13. Faigl, J., Pěnička, R., Best, G.: Self-organizing map-based solution for the orienteering problem with neighborhoods. In: IEEE International Conference on Systems, Man, and Cybernetics, pp. 1315–1321, October 2016
14. Faigl, J., Váňa, P.: Surveillance planning with Bézier curves. IEEE Robot. Autom. Lett. **3**(2), 750–757 (2018)
15. Goncalves, J.A., Henriques, R.: UAV photogrammetry for topographic monitoring of coastal areas. ISPRS J. Photogram. Remote Sens. **104**, 101–111 (2015)
16. Gunawan, A., Lau, H.C., Vansteenwegen, P.: Orienteering problem: a survey of recent variants, solution approaches and applications. Eur. J. Oper. Res. **255**(2), 315–332 (2016)
17. Hansen, P., Mladenovi, N.: Variable neighborhood search: principles and applications. Eur. J. Oper. Res. **130**(3), 449–467 (2001)
18. Helsgaun, K.: An effective implementation of the Lin-Kernighan traveling salesman heuristic. Eur. J. Oper. Res. **126**(1), 106–130 (2000)
19. Hodicky, J.: Standards to support military autonomous system life cycle. In: Březina, T., Jabłoński, R. (eds.) MECHATRONICS 2017. AISC, vol. 644, pp. 671–678. Springer, Cham (2018). https://doi.org/10.1007/978-3-319-65960-2_83
20. Lawler, E.L.: The Traveling Salesman Problem: A Guided Tour of Combinatorial Optimization. Wiley-Interscience Series in Discrete Mathematics (1985)
21. Lepetič, M., Klančar, G., Škrjanc, I., Matko, D., Potočnik, B.: Path planning and path tracking for nonholonomic robots. Mob. Robot.: New Res. 345–368 (2005)
22. Lin, S., Kernighan, B.W.: An effective heuristic algorithm for the traveling-salesman problem. Oper. Res. **21**(2), 498–516 (1973)
23. Lorenz, S., Adolf, F.M.: A decoupled approach for trajectory generation for an unmanned rotorcraft. In: Holzapfel, F., Theil, S. (eds.) Advances in Aerospace Guidance, Navigation and Control, pp. 3–14. Springer, Heidelberg (2011). https://doi.org/10.1007/978-3-642-19817-5_1
24. Morgenthal, G., Hallermann, N.: Quality assessment of unmanned aerial vehicle (UAV) based visual inspection of structures. Adv. Struct. Eng. **17**(3), 289–302 (2014)
25. Oberlin, P., Rathinam, S., Darbha, S.: Today's traveling salesman problem. Robot. Autom. Mag. **17**(4), 70–77 (2010)

26. Pajares, G.: Overview and current status of remote sensing applications based on unmanned aerial vehicles (UAVs). Photogram. Eng. Remote Sens. **81**(4), 281–329 (2015)
27. Perks, M.T., Russell, A.J., Large, A.R.: Advances in flash flood monitoring using unmanned aerial vehicles (UAVs). Hydrol. Earth Syst. Sci. **20**(10), 4005–4015 (2016)
28. Pěnička, R., Faigl, J., Vana, P., Saska, M.: Dubins orienteering problem. IEEE Robot. Autom. Lett. **2**(2), 1210–1217 (2017)
29. Pěnička, R., Faigl, J., Váňa, P., Saska, M.: Dubins orienteering problem with neighborhoods. In: 2017 International Conference on Unmanned Aircraft Systems (ICUAS), pp. 1555–1562, June 2017
30. Quater, P.B., Grimaccia, F., Leva, S., Mussetta, M., Aghaei, M.: Light unmanned aerial vehicles (UAVs) for cooperative inspection of PV plants. IEEE J. Photovoltaics **4**(4), 1107–1113 (2014)
31. Quigley, M., Goodrich, M., Griffiths, S., Eldredge, A., Beard, R.: Target acquisition, localization, and surveillance using a fixed-wing mini-UAV and gimbaled camera. Robotics and Automation, April 2005
32. Ramesh, R., Brown, K.M.: An efficient four-phase heuristic for the generalized orienteering problem. Comput. Oper. Res. **18**(2), 151–165 (1991)
33. Rego, C., Gamboa, D., Glover, F., Osterman, C.: Traveling salesman problem heuristics: leading methods, implementations and latest advances. Eur. J. Oper. Res. **211**(3), 411–427 (2011)
34. Russell, S., Norvig, P.: Artificial intelligence: a modern approach. Artif. Intell. **25**(27), 79–80 (1995)
35. Saari, H., et al.: Unmanned aerial vehicle (UAV) operated spectral camera system for forest and agriculture applications. In: Remote Sensing for Agriculture, Ecosystems, and Hydrology, vol. 13, October 2011
36. Salkin, H.M., Kluyver, C.A.D.: The knapsack problem: a survey. Naval Res. Logistics Q. **22**(1), 127–144 (1975)
37. Saska, M., Langr, J., Preucil, L.: Plume tracking by a self-stabilized group of micro aerial vehicles. In: Hodicky, J. (ed.) Modelling and Simulation for Autonomous Systems Modelling and Simulation for Autonomous Systems, vol. 8906, pp. 44–55. Springer, Cham (2014). https://doi.org/10.1007/978-3-319-13823-7_5
38. Schopferer, S., Adolf, F.M.: Rapid trajectory time reduction for unmanned rotorcraft navigating in unknown terrain. In: 2014 International Conference on Unmanned Aircraft Systems (ICUAS), pp. 305–316, May 2014
39. Sevkli, Z., Sevilgen, F.E.: Variable neighborhood search for the orienteering problem. In: Levi, A., Savaş, E., Yenigün, H., Balcısoy, S., Saygın, Y. (eds.) ISCIS 2006. LNCS, vol. 4263, pp. 134–143. Springer, Heidelberg (2006). https://doi.org/10.1007/11902140_16
40. Sprunk, C.: Planning motion trajectories for mobile robots using splines. University of Freiburg, Student project (2008)
41. Tsiligirides, T.: Heuristic methods applied to orienteering. J. Oper. Res. Soc. **35**(9), 797–809 (1984)
42. Váňa, P., Faigl, J.: On the dubins traveling salesman problem with neighborhoods. In: IEEE/RSJ International Conference on Intelligent Robots and Systems (IROS), pp. 4029–4034 (2015)
43. Wagner, P., Kotzian, J., Kordas, J., Michna, V.: Path planning and tracking for robots based on cubic Hermite splines in real-time. In: 2010 IEEE 15th Conference on Emerging Technologies Factory Automation (ETFA 2010), pp. 1–8, September 2010

44. Yang, G.J., Delgado, R., Choi, B.W.: A practical joint-space trajectory generation method based on convolution in real-time control. Int. J. Adv. Robot. Syst. **13**(2), 56 (2016)
45. Yuan, C., Zhang, Y., Liu, Z.: A survey on technologies for automatic forest re monitoring, detection, and ghting using unmanned aerial vehicles and remote sensing techniques. Can. J. For. Res. **45**(7), 783–792 (2015)

RoScan 2.0 - Multispectral Hi-Resolution Scanner

Ludek Zalud[(⊠)], Petra Kalvodova, and Frantisek Burian

CEITEC, Brno University of Technology, Brno, Czech Republic
{ludek.zalud, petra.kalvodova,
frantisek.burian}@ceitec.vutbr.cz

Abstract. The main aim of this paper is to describe novel hi-resolution multispectral scanning device for various purposes called RoScan 2.0, developed by our team at Brno University of Technology. It is a successor of RoScan 1.0 and Orpheus-X4 multispectral scanning head. The main difference comparing to RoScan 1.0 is its ability to scan in all used spectra concurrently and the ability to use more sources of physical movement for the sensory head. This means many technical and scientific challenges needed to be solved to make the device useable. Currently two positioning devices are developed a linear actuator and 2-DOF (degree of freedom) rotational manipulator, both with high precision and resolution.

Hardware of RoScan 2.0 is described, as well as calibration process necessary for fast, precise and reliable multispectral measurement of different objects. The problem of common field-of-view vs. resolution, and its impact to thermal imager and color camera depth-of-focus is mentioned.

Keywords: Laser scanning · Multispectral · Thermal imaging · CCD camera

1 Introduction

RoScan multispectral robotic scanning device was firstly introduced by our team in 2014 [1–3]. It was developed based our previous experience with multispectral scanning on field robots [4, 5]. Since this the scanning system evolved a lot. Even during the development our team was searching for practical applications in both medical research and clinical usage. Thanks to cooperation with hospitals and medical services we have found several possible applications, namely in podology, children's dermatology, and plastic surgery. Unfortunately, it proved to be highly impractical to transfer the original RoScan, consisting of Epson industrial manipulator on massive table, to the hospitals. In most cases, it also is not possible to let the patients come to our university to be measured – this is not only impractical for larger volume of patients, but also mostly prohibited by law, since concession by ethical committee is necessary and it is strictly limited to the place.

So, our team decided to develop second version of RosScan, called RoScan 2.0, which will be much more compact, while conserving maximum versatility of original RoScan. Most of the size was reduced by much more compact manipulating devices.

© Springer Nature Switzerland AG 2019
J. Mazal (Ed.): MESAS 2018, LNCS 11472, pp. 202–214, 2019.
https://doi.org/10.1007/978-3-030-14984-0_16

This obviously limits the number of manipulation degrees-of-freedom. But for many cases the compactness is more important than versatility.

It also has to be pointed out that the non-collaborative industrial manipulator on the original RoScan represents the most serious problem when passing through ethical committee approval process, because of the, at least theoretical, possibility of manipulator-to-patient crash due to reverse kinematics movement planning and accomplishment. With much simpler kinematics, this problem may be overcome.

Another problem of the original RoScan comes from the fact, the data from all the spectra are not acquired exactly at the same moment [2]. This may cause problems if the patient moves slightly during the scanning process. This issue is also addressed in RoScan 2.0.

2   RoScan 2.0 Description

As it was said, the original RoScan (see Fig. 1) consisted of a small-size 6-DOF, 3 kg payload industrial manipulator, stable table, optical sensorics head and control and measurement computer with appropriate software.

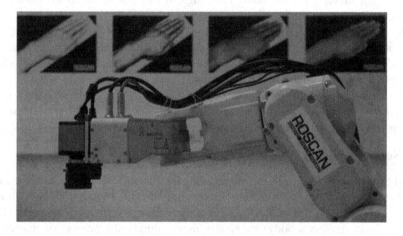

Fig. 1. Original RoScan

Both the sensorics head and manipulators needed to be considerably changed.

2.1   Manipulating Devices

So far we have developed two scanning head movement device configurations – see Fig. 2.

Fig. 2. Scanning configurations: 1DOF linear (left), 2DOF rotational (right)

The first one incorporates only one linear degree-of-freedom (see Fig. 2, left). It is represented by custom-made precise linear actuator with high-resolution sensing and custom motor driver and controller. The range of movements is up to 630 mm, the precision is better than 0.05 mm.

The second manipulator consists of two rotational DOFs (see Fig. 2, right). The mechanics is again custom built for scanning applications. AC motors together with extremely precise harmonic-drive gearboxes by Spinea company are used to achieve very high angular resolution and repeatability.

In both cases the motor drivers are developed by TGDrives company and custom-modified for our application. The real-time communication is done by means of Ethernet protocol, and we have approach to all controller parameters.

2.2   Sensorics Head

The original RoScan firstly makes line scanning by triangulation laser scanner and after this takes a series of images from color camera and thermal imager. This has two main drawbacks. The first is the latency between the parts scanned in different spectra. If the scanned object (patient's body) eventually moves during the process of scanning, it causes not only alteration of the 3D model, but also inaccuracy of alignment of individual spectral data. Other drawback is substantially lower resolution of thermal image and CCD camera data, so the corresponding multispectral layers will posses mower resolution than the 3D scanner, in general.

In RoScan 2.0 we chosen different approach to scanning – we acquire all the data of the individual line simultaneously. This overcomes the previously mentioned problems but one new technical problem. Due to large scanning depth (approx. 100 mm) relatively to the scanned distance (350–450 mm from the line-scanner chip), we need extreme wide field of focus of the thermal imager and CCD camera. While on the CCD camera this is possible to solve by big iris number set to the lens, on the thermal imagers we have not such a tool. So, we decided to equip the sensory head with two identical thermal imagers (see Fig. 3) with slightly modified chip-to-lens distance on one of them to acquire near focus.

The head is also equipped with hi-resolution USB 3.0 color CCD camera, and microEpsilon ScanControl fast triangulation laser line scanner (see Fig. 3).

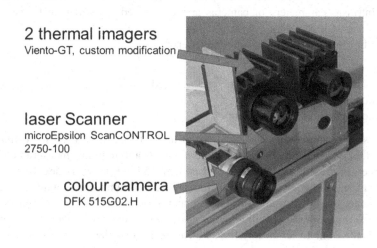

2 thermal imagers
Viento-GT, custom modification

laser Scanner
microEpsilon ScanCONTROL
2750-100

colour camera
DFK 515G02.H

Fig. 3. RoScan 2.0 – sensoric head

3   Extrinsic Calibration of Linear Scanner and Camera

Extrinsic calibration between 3D laser scanner and camera is wide topic investigated by many authors [7–9]. But our application requires direct calibration between line scanner and camera, because line scanner is rigidly mounted with camera. Here is a brief overview of approaches of other authors to this problem.

Extrinsic calibration of multiple-axis laser-stripe sensor integrated in CNC system is described in [10]. 4 Celluloid balls (white table tennis balls) are used as calibration targets. Extrinsic parameters are computed from coordinates of the ball centers extracted from laser and CCD camera measurement.

A technique for intrinsic and extrinsic calibration of laser line scanner and coordinate measuring machine is described in [11]. High precision gauge object with 240 points is used. Two images of calibration object are captured in one position. First with source of external light that serves for identification points on gauge in image. Second without external light that serves for identification of calibration object and laser plane intersection.

Extrinsic calibration of single line scanning lidar and camera is also described in [12]. These authors use calibration target consisting of two planes in V-shape from foamcore. Center line, left and right boundaries are marked with black tape to enable detection in images. Extrinsic parameters are estimated by minimizing the distance between corresponding edge and center line features projected onto the image plane. In [12] are used algorithm from [13, 14] improved by feature weighting and penalizing function for outliers.

Calibration in [15] used fact that infra-red source for commonly used lidar lies in the response range of regular cameras. This method makes visible laser scan trace (a line made of a sequence of scanned dots) in camera image by using infra-red filter. V-shape targets (e.g. intersections of two walls) are used for calibration.

3.1   Method of Proposed Calibration

This chapter describes proposed method for extrinsic calibration of linear scanner and Camera. Intrinsic parameters of cameras are determined separately in Camera Calibration Toolbox for Matlab.

Calibration method contains following steps:

- design of calibration target,
- measurement (acquisition of image and scanned profile),
- extraction of important points,
- extrinsic parameters determination.

The first very important step is to design target that can be used for calibration. Points would be clearly identifiable in camera images and in points measured by linear laser scanner. Proposed target has stairs shape with 15 holes for more details see Fig. 4. The calibration 3D target was manufactured by 3D milling cutter with geometrical precision better than 0.05 mm from Ebalta material, which is easy-to-manufacture and highly temperature-stable. Specific mutual position between laser scanner and target is used for calibration. Laser beam should measure middle line of the 5 holes on the target, see on right side Fig. 4.

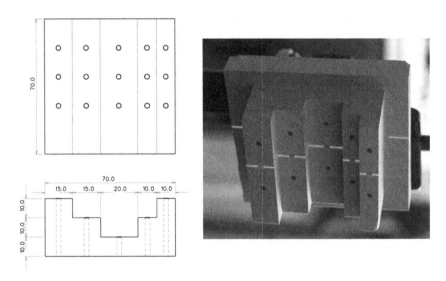

Fig. 4. Design of the calibration target – dimensions in mm

3 different types of data are needed for extrinsic calibration:

- laser profile,
- RoScan camera image,
- minimum four images from different angles acquired by Canon EOS 7D.

Data specified above is measured from different distances between target and laser scanner. Processing of measured data and determination of extrinsic parameters is comprised in following steps. Figure 6 shows schematically approach of proposed calibration for RoScan sensors.

1. **Detection of points in the edge of holes in scanner profile $XYZ_{LS\_intersections}$**
   Points not lying in calibration target and points in the holes are too far, so distance (Z coordinate) is zero. Points in the edge of the holes are identified according change Z coordinate value from zero to real values for minimum 6 consecutive measurements in the profile. Example of measured profile of calibration target is on the left side in Fig. 5.

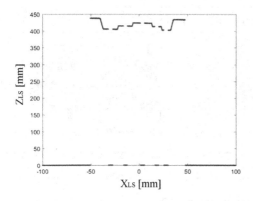 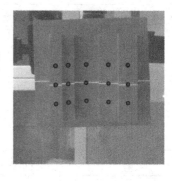

Fig. 5. Scanned profile of calibration target

2. **Detection of centers of the holes in image from camera xy_{CAM_holes}**
   Approach for detection of center holes image coordinates from camera image has the following steps:

   - Thresholding.
   - Removing small objects (noise) using morphological opening.
   - Connection of separated parts using morphological closing.
   - Filling closed objects.
   - Determining of image coordinates using Matlab function *imfindcircles*.

   Example of extracted holes in calibration target is on the right side in Fig. 5.

3. **Determination of 3D coordinates of intersection between laser beam and holes $XYZ_{PH\_intersections}$**
   3D coordinates of intersection between laser beam and holes $XYZ_{PH\_intersections}$ are determined in photogrammetric software PhotoModeler from min. 4 images of calibration target from different angle. Coordinates of the holes XYZ_{PH_holes} is used as control points in local coordinate system, these coordinates are known.

4. **Conform transformation between local coordinate system of supplementary camera and coordinate system of laser scanner**
   Transformation parameters are computed from coordinates of corresponding points $XYZ_{PH\_intersections}$ and $XYZ_{LS\_intersections}$. Coordinates of holes in local coordinate system XYZ_{PH_holes} are transformed to coordinate system of laser scanner. This proposed approach enables to continue with known task i.e. extrinsic calibration of 3D laser scanner and camera.

5. Extrinsic calibration between laser scanner and camera

Extrinsic calibration parameters are computed from coordinates of corresponding points XYZ_{PH_holes} and xy_{CAM_holes}. Mostly planar calibration target is used for extrinsic calibration of camera e.g. in [16–18], but we have 3D target. Our calibration is based on [19] camera calibration toolbox for Matlab.

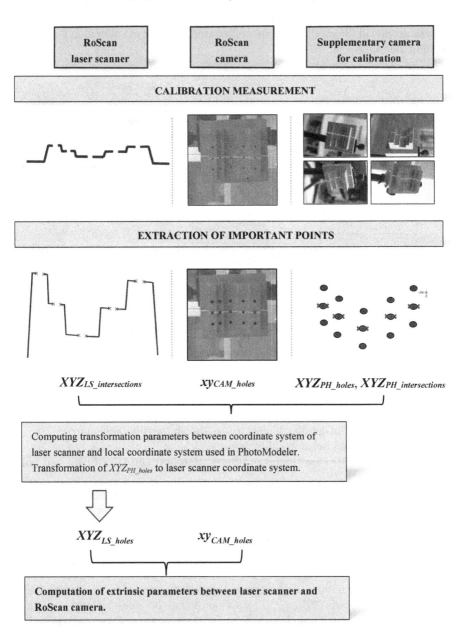

Fig. 6. Schematic approach of proposed calibration for RoScan sensors

4 Geometrical Transformations in RoScan 2.0 Scanning Process

As it was described in the Introduction, RoScan 2.0, comparing to the first RoScan version, is based on completely different manipulating devices. While original scanner consists of 6DOF industrial manipulator and scanning head, the second version is based on much more lightweight solutions. Currently two options are available – 1-linear degree of freedom manipulator, and 2 rotational degrees of freedom manipulator. It means the calculation of transformations during data acquisition has changed.

We have two configurations of sensor measurement, in the next paragraphs they are defined. At last paragraph, assumption is made, that makes software more compact, and simplifies the scanning algorithm.

The microEpsilon triangulation line-scanner measures data as cut of space on plane. The measured points $P_s = (x_S, 0, z_S)$ are referenced to base o_s. According to sensor fixed mounts and movable axes, we should define transformation to resultant global base space o_B, in which all resultant data points $P_B = (x_B, y_B, z_B)$ are stored.

4.1 Configuration I: Linear Scanning

If we want to calculate a point in the cartesian base system o_B from measured data, taken in the sensor coordinate system o_s, we need to go through homogenous transformation over system of moving frame o_v (Fig. 7):

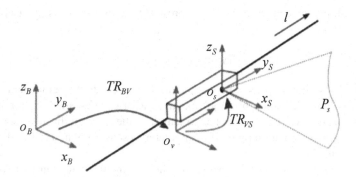

Fig. 7. Homogeneous transformations in linear scanning problem

$$P_B = TR_{BV} TR_{VS} P_S.$$

The transformation between the base and the moving frame is, in this case, very easy. Making simplifying assumption that movement of the frame is along y axis, we can write:

$$TR_{BV} = Trans(0, l, 0).$$

The transformation between frame and sensor defines actual mechanical orientation and position of sensor on moving frame:

$$TR_{VS} = Trans(dx, dy, dz)Rot_Z(\alpha)Rot_Y(\beta)Rot_X(\chi).$$

If we assume that sensor is mounted colinearly with y-axis of the manipulator, and oriented in x-direction of the base system, the transformation using all parameters is equal to the Identity matrix:

$$TR_{VS} = I.$$

This enables easy parameter callibration of the deviation.

4.2   Configuration II: 2DOF Rotational Area Scanning

Similar transformations are used in the Area scan method, but for transformation of base to the rotation-center point we can write:

$$TR_{BV} = Trans(0, 0, h)Rot_Z(\omega)Rot_Y(\xi).$$

The second, calibration transformation is:

$$TR_{VS} = Trans(dx, dy, dz)Rot_Z(\alpha)Rot_Y(\beta)Rot_X(\chi)$$

here almost all parameters are nearby zero provided the mechanics is precisely placed, so we can assume:

$$TR_{VS} = Trans(0, 0, dz)I.$$

During the calibration phase, there is the need to calculate the parameters h, and dz at first. This can be done by implementing least-squares method for equation $h + d \cos \alpha = l \sin \alpha$. Here l corresponds to length of the laser ray from sensor to floor, d is distance between rotation point of manipulator to sensor point, and h is distance of rotation point to floor.

The least squares method results in the following equations of unknown parameters:

$$h = \frac{\sum \cos^2 \alpha_n * \sum l_n \sin \alpha_n - \sum \cos \alpha_n * \sum l_n \sin \alpha_n \cos \alpha_n}{N \sum \cos^2 \alpha_n - (\sum \cos \alpha_n)^2}$$
$$d = \frac{N \sum l_n \sin \alpha_n \cos \alpha_n - \sum \cos \alpha_n \sum l_n \sin \alpha_n}{N \sum \cos^2 \alpha_n - (\sum \cos \alpha_n)^2}$$

Where α_n, l_n is measured multiple times.

5 Scanning Software

Since both the positioning devices and sensory head are considerably different comparing to the original RoScan, completely new scanning program was developed.

The software allows:

- set scanning limits and increments on both (and in the future all) scanning configurations, set the parameters,
- set all vital parameters of the triangulation line-scanner,
- view current scan in real-time,
- perform 3D scan,
- save the measured 3D scan.

The program is developed by our team in Microsoft Visual Studio, using Visual C# programming language and WindowsForms (Fig. 8).

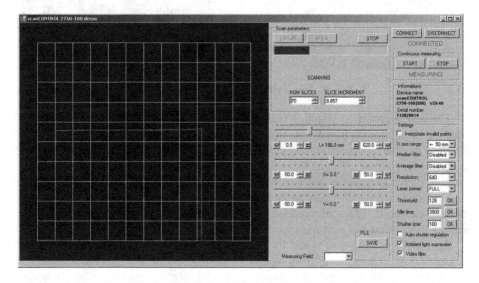

Fig. 8. The RoScan 2.0 software screenshot.

6 Conclusion and Future Work

Novel multispectral scanning device for precise scanning of parts of human body, named RoScan 2.0 was presented. The scanning is already possible (see Fig. 9), but further development is necessary to acquire fused multispectral data.

The main differences compared to the original RoScan are new options for sensory head manipulation. At the moment we have two manipulators – 1DOF linear and 2DOFs rotational, but for the future we plan to make the system extremely versatile. Among others we plan to put the scanning head on a collaborative manipulator by Fanuc, that we already have at our disposal. While this solution will definitely not be smaller than the first RoScan, it will still solve other big problem of that version – potential damage to human body, which is guaranteed to not happen with certified collaborative manipulator.

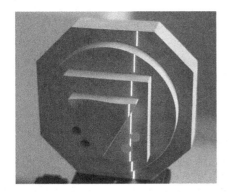 

Fig. 9. Scanning of rigid 3D object example

The new approach to scanning manipulation also means the scanning in all spectra must be done simultaneously, as described in the previous text. Especially in far-infrared (thermal) spectrum it represents a real issue solved in our case by two thermal imagers with slightly different focus plane. The development of algorithms for mixing the both thermal images is our next research goal.

For the future we also plan to include Ximea xiSpec hyperspectral line-camera to the sensorics head. This will provide more information about the tissue and we hope it will vastly increase the scanner's usability mainly in dermatology.

Recently our team was asked to participate in international research activity on children OSAS (Obstructive Sleep Apnea Syndrome). So far it seems the RoScan 2.0 with its precise 3D scanning will represent etalon for face scanning (see Fig. 10) in this activity.

Fig. 10. Human face scanning for OSAS

Acknowledgment. This work was also supported by the Technology Agency of the Czech Republic under the project TE01020197 – 'CAK III - Centre for Applied Cybernetics'.

References

1. Kocmanova, P., Zalud, L., Chromy, A.: 3D proximity laser scanner calibration. In: 18th International Conference on Methods and Models in Automation and Robotics MMAR 2013, pp. 742–747 (2013). https://doi.org/10.1109/mmar.2013.6670005
2. Chromy, A., Zalud, L.: Robotic 3D scanner as an alternative to standard modalities of medical imaging. SpringerPlus 3(1), 13 (2014). https://doi.org/10.1186/2193-1801-3-13
3. Kocmanova, P., Zalud, L.: Multispectral stereoscopic robotic head calibration and evaluation. In: Hodicky, J. (ed.) MESAS 2015. LNCS, vol. 9055, pp. 173–184. Springer, Cham (2015). https://doi.org/10.1007/978-3-319-22383-4_13
4. Zalud, L.: ARGOS - system for heterogeneous mobile robot teleoperation. In: 2006 IEEE/RSJ International Conference on Intelligent Robots and Systems, Beijing, China, pp. 211–216 (2006). https://doi.org/10.1109/iros.2006.282495
5. Zalud, L., Kocmanova, P., et al.: Calibration and evaluation of parameters in a 3D proximity rotating scanner. Elektron. Elektrotech. 21, 3–12 (2015). https://doi.org/10.5755/j01.eee.21.1.7299
6. Zalud, L., Kocmanova, P.: Fusion of thermal imaging and CCD camera-based data for stereovision visual telepresence. In: 2013 IEEE International Symposium on Safety, Security, and Rescue Robotics, SSRR 2013 (2013). https://doi.org/10.1109/ssrr.2013.6719344
7. Zhang, Q., Pless, R.: Extrinsic calibration of a camera and laser range finder (improves camera calibration). In: 2004 IEEE/RSJ International Conference on Intelligent Robots and Systems (IROS), Sendai, vol. 3, pp. 2301–2306 (2004). https://doi.org/10.1109/iros.2004.1389752
8. Herrera, D.C., Kannala, J., Heikkilä, J.: Joint depth and color camera calibration with distortion correction. IEEE Trans. Pattern Anal. Mach. Intell. 34(10), 2058–2064 (2012). https://doi.org/10.1109/TPAMI.2012.125
9. Scaramuzza, D., Harati, A., Siegwart, R.: Extrinsic self calibration of a camera and a 3D laser range finder from natural scenes. In: 2007 IEEE/RSJ International Conference on Intelligent Robots and Systems, San Diego, CA, pp. 4164–4169 (2007). https://doi.org/10.1109/iros.2007.4399276
10. Che, C., Ni, J.: A ball-target-based extrinsic calibration technique for high-accuracy 3-D metrology using off-the-shelf laser-stripe sensors. Precis. Eng. 24, 210–219 (2000). https://doi.org/10.1016/S0141-6359(00)00031-3
11. Santolaria, J., Pastor, J.J., Brosed, F.J., Aguilar, J.J.: A one-step intrinsic and extrinsic calibration method for laser line scanner operation in coordinate measuring machines. Meas. Sci. Technol. 20 (2009)
12. Kwak, K., Huber, D.F., Badino, H., Kanade, T.: Extrinsic calibration of a single line scanning lidar and a camera. In: IEEE/RSJ International Conference on Intelligent Robots and Systems, San Francisco, CA, USA, pp. 3283–3289 (2011). https://doi.org/10.1109/iros.2011.6094490
13. Li, G., Liu, Y., Dong, L., Cai, X., Zhou, D.: An algorithm for extrinsic parameters calibration of a camera and a laser range finder using line features. In: 2007 IEEE/RSJ International Conference on Intelligent Robots and Systems, pp. 3854–3859 (2007). https://doi.org/10.1109/iros.2007.4399041

14. Wasielewski, S., Strauss, O.: Calibration of a multi-sensor system laser rangefinder/camera. In: Intelligent Vehicles 1995 Symposium, pp. 472–477 (1995). https://doi.org/10.1109/ivs. 1995.528327
15. Yang, H., Liu, X., Patras, I.: A simple and effective extrinsic calibration method of a camera and a single line scanning lidar. In: Proceedings of the 21st International Conference on Pattern Recognition (ICPR 2012), pp. 1439–1442 (2012)
16. Zhang, Z.: Flexible camera calibration by viewing a plane from unknown orientations. In: Proceedings of the Seventh IEEE International Conference on Computer Vision (1999). https://doi.org/10.1109/iccv.1999.791289
17. Bouguet, J.Y.: Complete Camera Calibration Toolbox for Matlab. http://www.vision.caltech. edu/bouguetj/calib_doc/
18. Scaramuzza, D.: OCamCalib: Omnidirectional Camera Calibration Toolbox for Matlab. https://sites.google.com/site/scarabotix/ocamcalib-toolbox
19. Heikkilä, J.: Geometric camera calibration using circular control points. IEEE Trans. Pattern Anal. Mach. Intell. **22**, 1066–1077 (2000). https://doi.org/10.1109/34.879788

Modeling Proprioceptive Sensing for Locomotion Control of Hexapod Walking Robot in Robotic Simulator

Minh Thao Nguyenová, Petr Čížek$^{(\boxtimes)}$ ⬭, and Jan Faigl ⬭

Faculty of Electrical Engineering, Czech Technical University in Prague,
Technicka 2, 166 27 Prague, Czech Republic
{nguyemi5,cizekpe6,faiglj}@fel.cvut.cz,
https://comrob.fel.cvut.cz

Abstract. Proprioceptive sensing encompasses the state of the robot given by its overall posture, forces, and torques acting on its body. It is an important source of information, especially for multi-legged walking robots because it enables efficient locomotion control that adapts to morphological and environmental changes. In this work, we focus on enhancing a simplified model of the multi-legged robot employed in a realistic robotic simulator to provide high-fidelity proprioceptive sensor signals. The proposed model enhancements are based on parameter identification and static and dynamic modeling of the robot. The enhanced model enables the V-REP robotic simulator to be used in real-world deployments of multi-legged robots. The performance of the developed simulation has been verified in the parameter search of dynamic locomotion gait to optimize the locomotion speed according to the limited maximal torques and self-collision free execution.

1 Introduction

Modeling and dynamic simulations are established tools in robotics to enable verification of algorithms and control strategies before their deployment on real robots. Especially robots with complex morphology benefit from simulations to enhance their abilities to navigate the environment. It is the case of multi-legged robots [13] whose individual legs are connected through the trunk and also through the ground which all together form a complex linkage system with dynamic coupling between the individual components [10]. Simplifications and specific assumptions might be introduced in modeling of such systems, e.g., on rigidity of the construction [12], non-slippery footholds [10] or actuator dynamics [15]. However, it is important to verify that the considered simplifications and assumptions do not provide imprecise real-world execution that may result in mission failure or damage of the robot. Therefore, it is crucial to have a high-fidelity model and simulation of the real world behavior of the robot to support a seamless deployment of developed algorithms on real platforms even in complex scenarios like robotic football [16] or deployment in soft-terrains [7]. Besides,

© Springer Nature Switzerland AG 2019
J. Mazal (Ed.): MESAS 2018, LNCS 11472, pp. 215–225, 2019.
https://doi.org/10.1007/978-3-030-14984-0_17

modeling is also necessary to obtain a collision checker for motion planning techniques and advanced locomotion control strategies, e.g., as in [1,2].

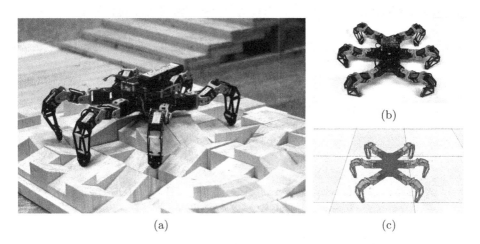

(b)

(a)   (c)

Fig. 1. (a) Hexapod walking robot in a rough terrain where the locomotion gait parametrization found using the developed high-fidelity simulation can prevent the robot from damaging the actuators. (b) The hexapod robot in the default pose and (c) its model in the V-REP robotic simulator.

The presented work is motivated by the real-world deployment of a hexapod walking robot shown in Fig. 1. We aim to have a realistic robotic simulator with a model of the robot, and its identified parameters would provide high-fidelity simulated measurements. The electrically actuated hexapod robot is built from off-the-shelf components, and each of its six legs has three joints motorized with the Dynamixel AX-12 servomotors that provide position feedback only. Thus, we are looking for a model of the robot that allows us to estimate the proprioceptive measurements, such as joint torque values that are not directly provided by the robotic platform. In particular, we need a simulation of the robot that allows finding a proper parametrization of the dynamic locomotion gait [14] prior the experimental evaluation to ensure safe and reliable operation by disallowing high joint torques, which would otherwise damage the actuators. Searching for such a parametrization during real-world experiments might have fatal consequences for the robotic platform. Hence, the simulation is a safe, fast, efficient, and inexpensive way for the development of the desired locomotion controllers.

There are multiple commercial and open-source simulation tools and libraries that support realistic robotic simulations. In our work, we utilize the Coppelia Robotics V-REP [6] realistic simulator, as it supports the modeling of static and dynamic objects and their interactions using different physics engines. We have created a simulation model of our hexapod walking robot with the emphasis on the identification of the model parameters to make it close to the real robotic platform. The achieved results in the simulation of the robot are supported by several verification experiments. Moreover, the developed simulation model of the robot has been employed for parameter search of the dynamic locomotion gait [14] to optimize the locomotion speed considering restrictions on the maximum joint torques and self-collision free execution.

The paper is organized as follows. Section 2 briefly introduces the utilized robotic simulator. The description of the modeling and parameters identification is in Sect. 3 and the results on the experimental verification of the model are reported in Sect. 4. Concluding remarks are dedicated to Sect. 5.

2   V-REP Realistic Robotic Simulator

V-REP is a powerful cross-platform 3D simulator based on distributed control architecture, i.e., control programs (or scripts) can be directly attached to individual scene objects and run simultaneously in threaded or non-threaded fashion [9]. Outside the control, the dependency of dynamic shapes can be set to simulate objects that move dependently on other objects, e.g., links between two joints of a leg. Most of the element properties can be reached and changed in the simulator GUI or via remote API client. Static properties such as proportions can be set for any shape; however, dynamic properties (e.g., mass, inertia, friction) are specific for dynamic objects only. Dynamic objects interact with the environment and other dynamic objects, and the computation of the dynamics can be done by one of four different physics engines, namely the Bullet physics library [5], Open Dynamic Engine (ODE) [11], Vortex Dynamics [3], and Newton Dynamics [8]. For the herein presented work, we use the Bullet physics library [5], which we found to be both the open source and sufficiently accurate [4].

3   Hexapod Modeling in V-REP Simulator

The addressed problem is to model the real hexapod walking robot in a realistic robotic simulation. Specifically, we aim to minimize the error between the proprioceptive data measured on a real robot and in the simulator. Thus, the goal is to estimate the proprioceptive signals that are not directly measured by the real robot, but they are acting in the robot interaction with the environment. Therefore, the requested realistic simulation has to provide the equivalent level of the interaction that allows developing locomotion control strategies using only the simulator before the experimental evaluation with the real robot. The real hexapod robot and the created visual model are shown in Fig. 1.

The modeled robot is an electrically actuated hexapod robot built from off-the-shelf components. The robot consists of the base trunk with revolute joints moving around the vertical axis and a control unit. Each leg is attached to a single joint and is formed by two more linked revolute joints moving around the horizontal axis. Parts of each leg are named coxa, femur, and tibia as it is visualized in Fig. 2a. Each joint is motorized by the Dynamixel AX-12A servomotor with the position feedback only. All 18 servomotors are connected in a daisy chain where all servomotors can be set with a new desired position within 1 ms.

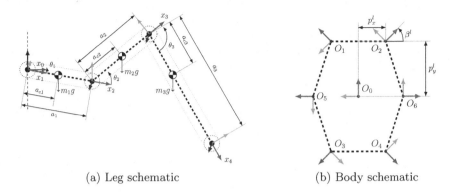

(a) Leg schematic   (b) Body schematic

Fig. 2. (a) Schematic of the robot leg. Each leg has three parts (links) – coxa, femur, and tibia connected by three joints (θ_1, θ_2, and θ_3). The coxa joint is fixed to the body with a vertical rotation axis while the two other joints are oriented with respect to the horizontal axis. (b) Schematic of the robot body. Each coxa joint corresponds to the respective origin of the coordinate frame O_l.

The current servomotor position can be obtained from the individual servomotor within 1 ms. Therefore commanding all the servomotors with new desired positions and reading back their current positions take altogether 19 ms which is a crucial property that has to be taken into account when modeling the robot dynamics. In the rest of this section, the robot modeling with the parameter identification is presented that is further followed by the description of the applied dynamic modeling of the actuators.

3.1   Kinematic Model of the Robot Body

The robot model consists of the elementary shapes modeled according to the original robot parts that correspond with the real world proportions and morphology of the real robot. The original proportions are respected with minor geometric simplifications of individual parts to speed up the simulation and avoid unnecessary computations. The modeled dimensions and masses of the individual model shapes have been set according to Tables 1 and 2 that list the properties of the robot leg and trunk, respectively, see Fig. 2. The total mass of the real robot is distributed evenly in each of the utilized elementary shapes. The inertia matrices of each link have been computed by the simulator under the assumption of evenly distributed mass. The individual shapes are connected to move dependently according to the robot morphology, i.e., the kinematic chains are formed with the hierarchical structure starting from the hexapod body and ending in the tibia link.

3.2   Dynamic Model of Servomotors

The properties of the Dynamixel AX-12A servomotor have been modeled using the V-REP simulator that supports different actuator models such as linear,

Table 1. Body parameters

l	1	2	3	4	5	6
β^l [rad]	$3\pi/4$	$\pi/4$	$5\pi/4$	$7\pi/4$	π	0.0
p^l_x [mm]	-60.5	60.5	-60.5	60.5	-100.5	100.5
p^l_y [mm]	120.6	120.6	-120.6	-120.6	0.0	0.0
Body mass (without battery)						1212 g
Mass of the battery						330 g

Table 2. Leg parameters

Link name	i	a_i [mm]	a_{ci} [mm]	m_i [g]
Coxa	1	52	26	22
Femur	2	66	20	72
Tibia	3	138	51	104

revolute, and prismatic; each with several adjustable parameters. The basic revolute actuator in the force/torque control mode is used to model the servomotor because it best fits our scenario. The simulator supports different control modes of the actuator including PID position control, spring-damper mode or custom control using user-provided Lua script attached to the actuator, which is beneficial for precise modeling of the actuator dynamics, or custom control rules. In our case, the real actuator is composed of the motor and reduction gear which dynamics can be described by the equation

$$J\ddot{q} + B\dot{q} + F(\dot{q}) + R\tau = KV, \tag{1}$$

where q is the rotor position angle before reduction, J is the rotor inertia, B is the rotor damping, F is the sum of static, dynamic and viscous friction, R is the gearbox ratio, τ is the servomotor torque, K is the back electromotive force, and finally V is the motor voltage. Although a precise model of the servomotor can be obtained by the identification of the parameters, the most influencing parts of the model are the parameters of the motion controller and frictions, which can be directly modeled in V-REP as a PID controller and link frictions, respectively. Therefore, there is no need for writing a custom actuator control script as the same behavior can be achieved by a proper parametrization of the existing solutions already provided by the simulator.

In particular, the PID controller has been parametrized according to the real servomotor as $P = 1, I = 0, D = 0$ and the inner friction of the joint has been simulated by setting the friction and angular damping in the material properties for the Bullet ver. 2.83 engine for the links attached to each joint.

The fact that the motor provides only the position feedback and uses the P-type controller allows us to estimate the static torque according to the servomotor documentation. In the static case, the position error e between the

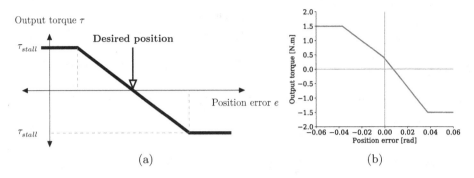

Fig. 3. (a) The relation of the torque τ and the position error e of the Dynamixel AX-12A servomotor. The value of τ is limited by a stall torque value τ_{stall}. (b) The relation of the torque and the position error for the simulated actuator.

desired and current positions of the actuator is proportional to the joint torque τ according to Fig. 3a. The stall torque $\tau_{stall} = 1.5$ Nm is reported by the actuator manufacturer, and it is used as a hard limit for the torque values in our dynamic locomotion experiment.

The real behavior under the considered simplifications has been verified by the following experiment. The femur link of a single leg has been made static in the simulation, and the tibia link has been set with a large mass. Then, the actuator has been commanded to swing between the boundary positions which does not make either of links move, but verify the studied behavior. The result is visualized in Fig. 3b that complies with Fig. 3a. A small offset of the torque values is most likely caused by the experimental setup, where it is not possible to mitigate the effects of the gravity acting on the dynamic link. Further experiments verifying the precision of the developed static and dynamic simulation are detailed together with our experimental scenario on the parametrization of the locomotion gait in Sect. 4.

3.3   Notes on the V-REP Simulation

In the synchronous operation mode, the V-REP simulator operates by advancing the simulation time with the constant time step triggered by the user. Significant attention has to be paid to the proper selection of the simulation step to cope with the real world operation and sensor communication. Specifically, the Dynamixel AX-12A servomotor allows setting a new position to all actuators once every 1 ms whereas a single read takes 1 ms as well. Therefore, we have selected 1 ms as a base simulation step with the same restrictions on the communication. Thus, bulk reading has to be done exactly in the same order in the simulation as on the real robot with the underlying calls for the simulation step.

4 Experimental Results on Verifying the Realistic Simulator

In this section, three experimental scenarios are described that have been designed to verify the realistic behavior of the developed simulation. Each report on the achieved experimental results contains both real experiments and simulated results, and their comparison. The selected verification scenarios are (i) the static torque analysis experiment that further verifies the torque-position error relation; (ii) the dynamic leg movement experiment that verifies the overall performance of the dynamic simulation; and (iii) an experimental deployment in searching for a parametrization of the locomotion controller that supports the overall deployability of the developed model in a real-world robotic task.

4.1 Torque Analysis Experiment

In this experiment, three legs of the hexapod robot are lifted to create a support polygon in the form of a triangle as it is shown in Fig. 4. Subsequently, the angle of the tibia joint on the supporting middle leg goes from $-\pi/4$ to 0, and thus gradually transfers a larger part of the robot body weight to this leg which increases the torque τ on the leg joints. The experiment has been executed and analyzed in a quasi-statically setup for which the motion is sufficiently slow to consider each robot state to be static. The experiment has been done three times with different weights of the robot trunk. Namely without the battery pack, with a single battery pack, and with two battery packs, see Table 1 for the corresponding weights. The mass of the trunk has been adjusted in the simulation accordingly. The ground truth for the torque values τ_{real} has been calculated as the torque given by the distance of the leg endpoint to the joint axis multiplied by the weight applied on the legs endpoint measured by the table scale. The simulated torque τ_{sim} has been read directly from the simulator.

The achieved results are visualized in Fig. 5a for the unloaded robot. Figure 5b and c capture torques in the same experiment with the robot loaded with one and two battery packs, respectively. In the trials with the increased

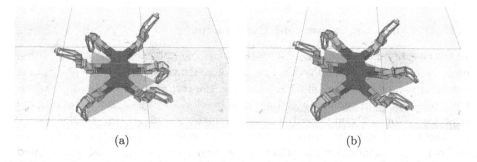

(a)   (b)

Fig. 4. The used setup in the torque analysis experiment. (a) Starting position and (b) end position.

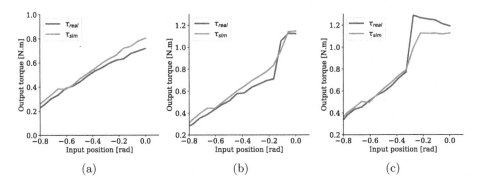

Fig. 5. Real torque values τ_{real} and the simulated torque τ_{sim} for the tibia joint and (a) unloaded robot; (b) the robot body loaded with additional 330 g; and (c) the robot body loaded with 660 g.

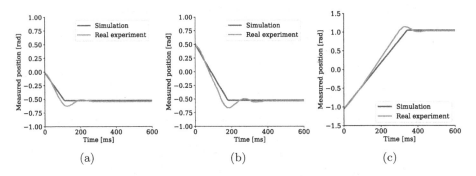

Fig. 6. The position setting on (a) coxa, (b) femur, and (c) tibia joint.

body weight, the servomotors went over the stall torque limit and turned off, which can be identified as a plateau in the plots, which is visible for both the simulation and real-world experiments. The experimental results indicate that the simulation can provide sufficiently precise estimations of the joint torque values.

4.2 Dynamic Leg Movement Experiment

In the case of quasi-static movement, the relationship between the joint position error and the torque visualized in Fig. 3a holds. However, in the case of dynamic movement, further factors start to influence the torque at the individual joints, e.g., velocities and moments of inertia. Therefore, we designed an additional experiment for analyzing the dynamic movement of a single leg as follows.

The robot has been enabled a free movement of the legs in both simulator and real experiments. A single actuator has been set with a new position followed by the immediate consecutive position readings from that particular actuator. As it takes longer than 1 ms to reach the desired position, we have obtained a swing profile of the actuator. Plots in Fig. 6 represent a comparison between the behavior of the model and real joint. The results indicate that the overall shape of the motion is correct (for the used joint speed) and the results also indicate a correct setting of the P-type controller. However, the simulation does not cover the dynamic overshoots probably due to the setting of high friction in the individual links. Unfortunately, we have not been able to overcome this issue.

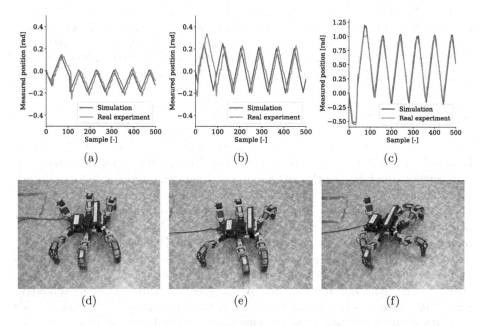

Fig. 7. The dynamic locomotion scenario. The trajectory of the coxa joint in (a) 2 times speeded-up locomotion, (b) 5 times speeded-up locomotion and (c) 6 times speeded-up locomotion for which self-collisions are already visible in the collected data. (d–f) the respective snapshots of the robot in the real experiment.

4.3   Dynamic Locomotion Experiment

Finally, the usability of the derived model has been verified in the dynamic locomotion control experiment. The model is employed in the problem of parameter search of the dynamic locomotion controller [14] which is a variant of the central pattern generator (CPG) based on an artificial neural network that produces a rhythmic pattern used as the input signal to the actuators.

A pre-learned CPG network has been utilized that is parametrized by the stride length of the gait cycle which directly influences the robot forward velocity. A greedy search has been performed to find a particular stride length value that maximizes the forward velocity of the robot but also satisfies the given restrictions on the maximum joint torques and self-collision free execution. For each examined parameter value, the locomotion has been performed for 10 s in the V-REP simulator. During the simulation, self-collisions have been monitored together with the torque values τ_{sim}. The found value provides five times speedup of the locomotion in comparison to the default setting with locomotion speed $v = 0.05\,\mathrm{ms}^{-1}$ given by [14] for which the robot parts do not collide, and the torque does not exceed the safe value for continuous robot operation of $0.75\tau_{stall}$.

The comparison of leg trajectories for the coxa joint with two times speedup, five times speedup, and six times speedup, where self-collisions are already visible in both the simulated and real data, are visualized in Fig. 7. Besides, when the speedup is higher than the estimated parameter, the individual servomotors on the robotic platform start to overheat. Therefore, the performed experiment supports the usability of the developed simulation and its deployability in planning and optimization of the locomotion control for the hexapod walking robot.

5 Conclusion

This paper reports on the development of the realistic simulation model of the hexapod walking robot in a robotic simulator. The developed model provides a high-fidelity simulation of the robot dynamics and demonstrates to be useful in the problem of optimizing parameters of the locomotion control gait to maximize the robot velocity and avoid high torques and self-collisions, which might damage the actuators and the robot platform itself. The developed simulation provides us with a safe, fast, high-fidelity, and inexpensive tool for optimization and development of locomotion controllers. In the future, we would like to automate the search for the hexapod model parametrization that can be especially beneficial in long-term autonomy missions, when the robot morphology may change. We also aim to consider transfer learning techniques to support the usability of the simulation results among different robotic platforms.

Acknowledgement. This work has been supported by the Czech Science Foundation (GAČR) under research Project No. 18-18858S.

References

1. Belter, D., Labecki, P., Skrzypczynski, P.: Adaptive motion planning for autonomous rough terrain traversal with a walking robot. J. Field Robot. **33**(3), 337–370 (2016). https://doi.org/10.1002/rob.21610
2. Čížek, P., Masri, D., Faigl, J.: Foothold placement planning with a hexapod crawling robot. In: IEEE/RSJ International Conference on Intelligent Robots and Systems (IROS), pp. 4096–4101 (2017). https://doi.org/10.1109/IROS.2017.8206267
3. Collective of authors: Vortex Simulation Software. https://www.cm-labs.com/. Accessed 30 July 2018
4. Erez, T., Tassa, Y., Todorov, E.: Simulation tools for model-based robotics: comparison of bullet, havok, mujoco, ode and physx. In: IEEE International Conference on Robotics and Automation (ICRA), pp. 4397–4404 (2015). https://doi.org/10.1109/ICRA.2015.7139807

5. Coumans, E., et al.: Bullet Physics Library. http://bulletphysics.org/wordpress/. Accessed 30 July 2018
6. Freese, M., Singh, S., Ozaki, F., Matsuhira, N.: Virtual robot experimentation platform V-REP: a versatile 3D robot simulator. In: Ando, N., Balakirsky, S., Hemker, T., Reggiani, M., von Stryk, O. (eds.) SIMPAR 2010. LNCS (LNAI), vol. 6472, pp. 51–62. Springer, Heidelberg (2010). https://doi.org/10.1007/978-3-642-17319-6_8
7. Gao, H., et al.: A real-time, high fidelity dynamic simulation platform for hexapod robots on soft terrain. Simul. Model. Pract. Theor. **68**, 125–145 (2016). https://doi.org/10.1016/j.simpat.2016.08.004
8. Jerez, J., Suero, A., et al.: Newton Dynamics. http://newtondynamics.com/forum/newton.php. Accessed 30 July 2018
9. Rohmer, E., Singh, S.P., Freese, M.: V-REP: a versatile and scalable robot simulation framework. In: IEEE/RSJ International Conference on Intelligent Robots and Systems (IROS), pp. 1321–1326 (2013). https://doi.org/10.1109/IROS.2013.6696520
10. Roy, S.S., Pratihar, D.K.: Dynamic modeling, stability and energy consumption analysis of a realistic six-legged walking robot. Robot. Comput. Integr. Manuf. **29**(2), 400–416 (2013). https://doi.org/10.1016/j.rcim.2012.09.010
11. Smith, L.R., et al.: Open Dynamics Engine. http://www.ode.org/. Accessed 30 July 2018
12. Soyguder, S., Alli, H.: Kinematic and dynamic analysis of a hexapod walking running-bounding-gaits robot and control actions. Comput. Electr. .Eng. **38**(2), 444–458 (2012). https://doi.org/10.1016/j.compeleceng.2011.10.008
13. Spis, M., Matecki, A., Maik, P., Kurzawa, A., Kopicki, M., Belter, D.: Optimized and reconfigurable environment for simulation of legged robots. In: Szewczyk, R., Zieliński, C., Kaliczyńska, M. (eds.) ICA 2017. AISC, vol. 550, pp. 290–299. Springer, Cham (2017). https://doi.org/10.1007/978-3-319-54042-9_26
14. Szadkowski, R.J., Čížek, P., Faigl, J.: Learning central pattern generator network with back-propagation algorithm. In: Proceedings Information Technologies - Applications and Theory ITAT 2018, pp. 116–123 (2018)
15. Wensing, P.M., Wang, A., Seok, S., Otten, D., Lang, J., Kim, S.: Proprioceptive actuator design in the MIT cheetah: impact mitigation and high-bandwidth physical interaction for dynamic legged robots. IEEE Trans. Robot. **33**(3), 509–522 (2017). https://doi.org/10.1109/TRO.2016.2640183
16. Zaratti, M., Fratarcangeli, M., Iocchi, L.: A 3D simulator of multiple legged robots based on USARSim. In: Lakemeyer, G., Sklar, E., Sorrenti, D.G., Takahashi, T. (eds.) RoboCup 2006. LNCS (LNAI), vol. 4434, pp. 13–24. Springer, Heidelberg (2007). https://doi.org/10.1007/978-3-540-74024-7_2

Trajectory Planning for Aerial Vehicles in the Area Coverage Problem with Nearby Obstacles

Jakub Marek[ID], Petr Váňa[(✉)][ID], and Jan Faigl[ID]

Faculty of Electrical Engineering, Czech Technical University in Prague,
Technicka 2, 166 27 Prague, Czech Republic
{marekj24,vanapet1,faiglj}@fel.cvut.cz
https://comrob.fel.cvut.cz

Abstract. In this paper, we address the coverage path planning with curvature-constrained paths for a fixed-wing aerial vehicle. The studied problem is to provide a cost-efficient solution to cover a given area by the vehicle sensor from the specified altitude to provide a sufficient level of details in the captured snapshots of the area. In particular, we focus on scenarios where the area to be covered is surrounded by nearby obstacles such as trees or buildings, and the vehicle has to avoid collisions with the obstacles but maximizes the area coverage. We propose an extension of the existing coverage planning algorithm to determine a shortest collision-free path that is accompanied by Dubins Airplane model to satisfy the motion constraints of the vehicle. The reported results support the feasibility of the proposed approach to avoid nearby obstacles by optimal adjustments of the vehicle altitude while the requested complete coverage is satisfied. If such a solution is not found because of too close obstacles, a feasible solution maximizing the coverage is provided.

1 Introduction

Finding a cost-efficient trajectory to cover a given area of interest is called the *area coverage problem* [9] and it can be found in several tasks such as lawn-mowing [4], vacuum cleaning [13], search and rescue [19], demining [8], car parts painting [3], or area mapping [20]. The herein proposed solution of the problem is motivated by information gathering missions with a fixed-wing Unmanned Aerial Vehicle (UAV) where it is necessary to satisfy the vehicle motion constraints such as the minimum turning radius and maximum climb/dive ratios. Besides, we are specifically focused on scenarios where the area to be covered is surrounded by nearby obstacles because existing coverage planning approaches address obstacles in the area by its decomposition into multiple obstacle-free cells.

In particular, cellular decomposition methods identify the obstacles first, and based on the particular method used; the area is split into cells without any

© Springer Nature Switzerland AG 2019
J. Mazal (Ed.): MESAS 2018, LNCS 11472, pp. 226–236, 2019.
https://doi.org/10.1007/978-3-030-14984-0_18

obstacle. Individual cells can have various shapes and the mostly used decomposition methods [6] are the boustrophedon decomposition [5], trapezoidal decomposition [12], Morse decomposition [1], topological decomposition [21], and polygonal decomposition [11]. Besides, grid-based methods represent the area to be covered as a discrete uniform grid of cells that can be rectangular [15], triangular [17], or hexagonal [19]. Then, each cell is examined for a possible collision with obstacles, and a path to visit all obstacle-free cells is determined. Therefore, regardless of the decomposition method, the obstacles can surround the determined obstacle-free cells.

The most efficient way for covering an obstacle-free cell is to use so-called zigzag covering pattern [22], which is also called boustrophedon pattern [5] that is visualized in Fig. 1. The pattern covers the area by repetitive movement from one side to another until the whole area is covered. The important parameter of the zigzag pattern is the coverage direction of the straight covering segments since it influences the final length significantly. The favorable direction can be determined according to the shape of the given area [14,22].

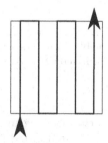

Fig. 1. An example of the zigzag pattern.

Since an efficient solution to the area coverage problem has been already developed [22], we consider the existing solution in our specific variant of the area coverage problem with nearby obstacles and fixed-wing vehicle. We propose to model the vehicle as Dubins Airplane model [18] in 3D to address the vehicle motion constraints and obstacles surrounding the coverage area by adjusting the vehicle altitude and thus avoid collision with the nearby obstacles. Moreover, if a collision-free solution for the complete coverage is not found, the complete coverage is relaxed in a benefit of a feasible solution with partial coverage.

2   Problem Statement

The studied problem is to find a feasible, shortest collision-free trajectory for a fixed-wing UAV to completely cover a given convex area. It is assumed that the area to be covered has a polygonal shape with the dominant orientation, and it is obstacle-free, e.g., the area can be an obstacle-free cell provided by one of the existing approaches to the area coverage problem. For this setup with the pre-selected dominant orientation, the straight covering maneuvers are determined as the zigzag pattern [22], and thus the covering maneuvers are collinear. Hence, the area coverage problem is transformed to the problem how to connect the straight covering maneuvers such that the motion constraints of the vehicle are satisfied and the final trajectory avoids nearby obstacles by adjusting altitude of the vehicle. An example of the covering and connecting maneuvers is in Fig. 2.

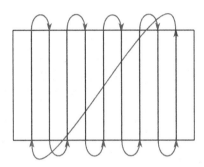

Fig. 2. A solution of the addressed area coverage problem with zigzag pattern, where the covering maneuvers (shown in the blue) are concatenated by the connecting maneuvers (in the red) into the final curvature-constrained path. (Color figure online)

The transformed problem is to determine the most suitable connecting maneuvers and order of visits to the covering maneuvers such that the length of the final collision-free trajectory is minimized and the motion constraints of Dubins Airplane model [18] are satisfied. The used model limits a curvature of the trajectory by the minimum turning radius ρ, as for Dubins vehicle [7]. In addition, the model considers the vehicle altitude, and thus the state of the vehicle is represented by its position (x, y, z), heading angle θ, and the pitch angle ψ for which the vehicle motion can be expressed as

$$\begin{bmatrix} \dot{x} \\ \dot{y} \\ \dot{z} \\ \dot{\theta} \end{bmatrix} = v \begin{bmatrix} \cos\theta\cos\psi \\ \sin\theta\cos\psi \\ \sin\psi \\ u_\theta\rho^{-1} \end{bmatrix}, \tag{1}$$

where v is the fixed forward velocity of the vehicle, $u_\theta \in [-1, 1]$ is the control input, ρ is the minimum turning radius, and the pitch angle ψ is constrained by the allowed range (2) to model the real properties of the vehicle.

$$\psi \in \langle \psi_{min}, \psi_{max} \rangle \tag{2}$$

The coverage problem can be then formulated as a variant of the *Traveling Salesman Problem* (TSP) [2] which stands to find a sequence of the covering maneuvers such that the whole area is covered, the length of the final path is minimized, and the final trajectory is feasible and collision-free. The covering maneuvers are straight line segments that satisfy the motion constraints (1) and (2) of Dubins Airplane model, and they are collision-free because the area is obstacle free. Besides, the complete coverage is satisfied by traveling along all the covering maneuvers. However, we still need to address the combinatorial part of the TSP and determine the most suitable connecting maneuvers, and thus the problem is formally defined as follows.

Let has a set of n collinear covering maneuvers located at the specific height above the terrain to provide complete coverage using a particular sensor attached to the vehicle. Each covering maneuver can be accomplished from two opposite directions that correspond to the covering trajectories $s_{\sigma_i}^1$ and $s_{\sigma_i}^2$ for the i-th covering maneuver. The covering trajectories are connected into the final trajectory consisting of $2n$ segments $\{m_1, m_2, \ldots, m_{2n}\}$ because two covering trajectories are connected by the connecting trajectory (maneuver). Every trajectory segment m_i is defined by a projection function $m_i : \mathbb{R} \to \mathbb{R}^3 \times \mathbb{S}^1$ where the initial and final states of m_i are $m_i(0)$ and $m_i(1)$, respectively. Then, the final trajectory can be represented by the permutation Σ of the covering maneuvers accompanied by the sequence of directions D defining how the individual maneuvers are traveled.

$$
\begin{aligned}
\Sigma &= (\sigma_1, \ldots, \sigma_n) & 1 \leq \sigma_i \leq \sigma_n, & \qquad (3) \\
D &= (d_{\sigma_1}, \ldots, d_{\sigma_n}) & d_{\sigma_i} \in \{1, 2\}. & \qquad (4)
\end{aligned}
$$

Based on these preliminaries, the coverage problem can be formulated as the optimization problem to minimize the total length of the covering trajectory consisting of $2n$ trajectory segments, each with the length $\mathcal{L}(m_i)$:

$$
\text{minimize}_{\Sigma, D} \ \sum_{i=1}^{2n} \mathcal{L}(m_i), \qquad (5)
$$

subject to

$$
\begin{aligned}
m_i(1) &= m_{i+1}(0) & \forall i \in 1, \ldots, 2n, & \qquad (6) \\
m_{2i-1} &= s_{\sigma_i}^{d_i} & \forall i \in 1, \ldots, 2n, & \qquad (7) \\
m_{2n}(1) &= m_1(0), & & \qquad (8) \\
\Sigma &= (\sigma_1, \ldots, \sigma_n) & 1 \leq \sigma_i \leq \sigma_n, & \qquad (9) \\
D &= (d_{\sigma_1}, \ldots, d_{\sigma_n}) & d_{\sigma_i} \in \{1, 2\}. & \qquad (10)
\end{aligned}
$$

The sum of the coverage maneuver lengths is fixed because all covering maneuvers have to be traveled and their lengths do not depend on the orientation. Therefore, the optimization is performed over the connecting maneuvers, the order of visits to the covering maneuvers Σ, and directions D. Moreover, the nearby obstacles around the coverage area may block simple connecting maneuvers, and therefore, they have to be adjusted to provide a collision-free solution, which is addressed in the proposed method described in the following section.

3   Proposed Method

The proposed solution of the addressed area coverage problem with Dubins Airplane model leverages on the existing zigzag pattern approach [22]. Hence, the collinear covering maneuvers are determined according to the zigzag pattern in the direction defined by the longest edge of the given polygonal area [14] and the determined covering maneuvers have to be connected into the final covering trajectory. Note the direction may significantly impact the solution, see Fig. 3.

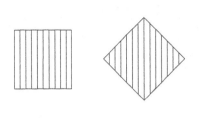

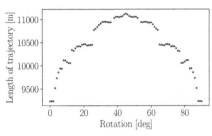

(a) Covered area for 0° and 45° angle.   (b) Final length of coverage segments

Fig. 3. An influence of various coverage directions on the final path length using the zigzag pattern on the square area of 500×500 m.

Although covering maneuvers are selected at the fixed height above the terrain, and the maximum slope is assumed to be small enough for the used fixed-wing UAV, the collision-free connecting maneuvers satisfying the motion constraints (1) have to be determined. The construction of the connecting maneuvers can be based on 2D Dubins maneuvers [7] generalized to 3D maneuvers [18]. Since it is necessary to avoid possible collisions only with the nearby obstacles, we can avoid a possible collision by adjusting the connecting maneuver. The only possible way to avoid the collision is to adjust the trajectory altitude such that the vehicle would fly over the obstacles. Therefore, the minimum necessary altitude increase is determined by sampling the terrain height beneath the trajectory and estimating the minimum non-collision altitude profile for the vehicle according to the pitch angle limitation (2). The result is always convex envelop of the altitude with the highest point at the place of the most upper part of the specific obstacle, and thus the process is computationally effective.

If the altitude change has to be significant, it may happen that the connecting maneuver cannot be directly linked to the specific covering maneuver, because an altitude discontinuity would occur. In such a case, the connection of two covering maneuvers is not possible, or we need to adjust the altitude of the covering maneuver at the cost of possible loss of the complete coverage. For the latter case, we utilize the maximal pitch angle to gain the altitude and minimize the uncovered area. An adjustment of the covering maneuver is depicted in Fig. 4.

After determination of the all connecting maneuvers and necessary updates of the covering maneuvers, the coverage problem is transformed into the

(a) Unchanged segments   (b) Changed segments

Fig. 4. Nearby obstacle (in the red) forces the vehicle to increase its altitude during the connecting maneuver. If the required altitude is too high, two neighboring covering maneuvers cannot be connected (left) and the altitude has to be gain on the covering maneuver (right), which may decrease the area coverage. (Color figure online)

Generalized Asymmetric TSP (GATSP). The transformation is provided by calculating all possible connecting maneuvers which can be bounded by $\mathcal{O}(n^2)$. The covering maneuvers are considered as the nodes in the GATSP, as their lengths change only slightly, and the connecting maneuvers define the cost of the edges between the respective nodes (covering maneuvers). If a specific connecting maneuver needs altitude change for the respective covering maneuver (i.e., shortening the covering maneuver), the length change is considered in the connecting maneuver because the change is related only to that connecting maneuver.

The altitude change of the covering maneuver may result in the coverage loss, and therefore, an additive penalization p to the length of the connecting maneuver is introduced to trade-off the solution length and possible coverage loss. The penalization p is defined as

$$p = k\left(x_n - x_o\right), \tag{11}$$

where x_o stands for the original length of the covering maneuver and x_n is the extended length because of the altitude change. The parameter k sets the power of the penalization, and for $k = 1$, only the length of the connecting maneuvers is considered. For $k > 1$, trajectories with small uncovered parts are favored.

After applying the penalization, the GATSP is transformed to the TSP using Noon-Bean transformation [16] that is solved by Lin-Kernighan heuristic algorithm [10] to find a high-quality solution in a reasonable time. An example of the final trajectories for a polygonal coverage area both with and without obstacles is depicted in Fig. 5. Evaluation results of the proposed method are reported in the next section.

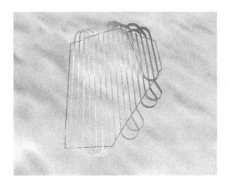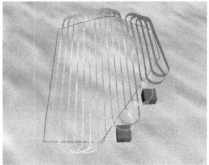

Fig. 5. Example of the area coverage with and without nearby obstacles.

4   Results

The proposed method has been evaluated in five scenarios designed to verify specific properties of the proposed approach including comparison with the reference approach [22] which, however, does not explicitly consider nearby obstacles. In all scenarios, the parameters of the used fixed-wing UAV are as follows. The coverage altitude is 20 m, the minimum turning radius is $\rho = 40$ m, the maximum climb and dive angles are $30°$, and the sensor field-of-view is $70°$.

In the first scenario, we study the influence of the ratio of the side lengths for a rectangular area 0.25 km^2 large, see Fig. 6. The length of the side a is selected from the range $[100, 500]$ in meters with the step one meter and the second side is computed to keep the same total area. We split the results with the even and odd numbers of covering segments, because for the odd number of segments, one connecting maneuver has to cross the area, and thus it significantly prolongs the total trajectory length. The resulting trajectory lengths are presented in Fig. 7.

In the second scenario, we examine the influence of the minimum turning radius of the total trajectory length. The same setup as in the previous case is used but with the equal sides of the rectangular area and no obstacles for simplicity. The value of ρ is selected from the range $\rho \in [20, 100]$ in meters with

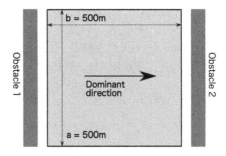

Fig. 6. First scenario with the basic setup of the equal sides 500 m long.

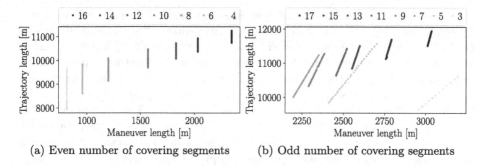

(a) Even number of covering segments   (b) Odd number of covering segments

Fig. 7. Influence of the shape of area to the trajectory length. The colors denote the particular numbers of the covering segments in the found solutions. (Color figure online)

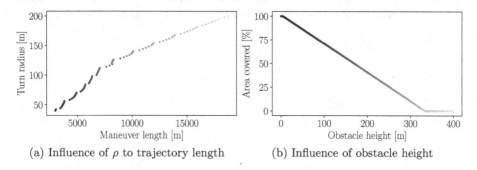

(a) Influence of ρ to trajectory length   (b) Influence of obstacle height

Fig. 8. Influence of the minimum turning radius ρ to the total trajectory length (left); and influence of the obstacle height to the area coverage (right).

the step one meter and the length of the found solutions are depicted in Fig. 8a. Discontinuities can be observed for ρ equal to an integer multiple of the distance between the covering maneuvers because the respective connecting maneuvers are selected between farther covering maneuvers.

A similar setup is also used for the third scenario, where we study the influence of the height of the surrounding obstacles that are considered to be up to 400 m tall above the terrain located around the area that is 475 × 500 m large. The surrounding obstacles with the height above 300 m require too high altitude to be avoided, and the area is not covered at all, see Fig. 8b.

In the fourth scenario, we examine the penalization for the uncovered area. The size of the area is set to 475 × 950 m, and the height of the obstacles at the shorter side of the area is 20 m. The trajectory length depends on the covering direction. If the direction is in the unblocked side, there is no collision; however, the final trajectory is significantly longer than for the blocked direction, because the final trajectory contains more connecting maneuvers. On the other hand, for the blocked direction, the trajectory length depends on the penalization weight k, see (11), and for a high value of k, the unblocked direction can be preferable, e.g., for $k > 20$ as it is shown in Fig. 9a.

In the last evaluation scenario, the proposed method is compared with the reference approach [22], which does not explicitly address the surrounding obstacles. Therefore, the reference method has been slightly modified, and the covering maneuvers are shortened such that the corresponding connecting maneuvers are collision-free. Since for the connecting maneuvers, the vehicle is tilted in the roll angle, and the sensor is not heading towards the coverage area, the coverage of the area is not considered for the connecting maneuvers. The evaluation of the methods is based on the amount of the covered area according to the ratio of the sides of the rectangular area to be covered. The results are shown in Fig. 9b and they support the feasibility of the proposed approach to explicitly consider obstacles surrounding the coverage area as the proposed method provides noticeably higher coverage. Because both methods provide complete coverage for setups without the surrounding obstacles, the main advantage of the proposed method is to address nearby obstacles while satisfying the vehicle motion constraints.

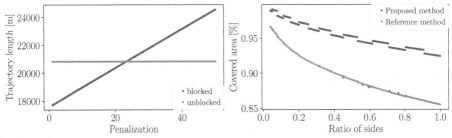

(a) Influence of the penalization weight parameter k to the trajectory length

(b) Covered area by the proposed and reference methods

Fig. 9. Results for the fourth and fifth scenarios: influence of the penalization of uncovered area and choice of the covering direction (left); and a comparison with the reference method [22] (right).

5   Conclusion

In this paper, we propose a novel method to the area coverage problem with motion constraints of a fixed-wing UAV and presence of nearby obstacles. The proposed method transforms the problem into a variant of the TSP that can be solved by existing combinatorial solvers. The final covering trajectory respects the motion constraints of Dubins Airplane model and avoids possible collisions with the obstacles. The results support the feasibility of the proposed solution with the explicit consideration of the surrounding obstacles and coverage planning for non-holonomic vehicles with the minimum turning radius and limited pitch angle.

Acknowledgement. This work has been supported by the Czech Science Foundation (GAČR) under research Project No. 16-24206S and Project No. 19-20238S.

References

1. Acar, E.U., Choset, H., Rizzi, A.A., Atkar, P.N., Hull, D.: Morse decompositions for coverage tasks. Int. J. Robot. Res. **21**(4), 331–344 (2002). https://doi.org/10. 1177/027836402320556359
2. Applegate, D.L., Bixby, R.E., Chvatal, V., Cook, W.J.: The Traveling Salesman Problem: A Computational Study. Princeton University Press, Princeton (2006)
3. Atkar, P.N., Conner, D.C., Greenfield, A., Choset, H., Rizzi, A.A.: Hierarchical segmentation of piecewise pseudoextruded surfaces for uniform coverage. IEEE Trans. Autom. Sci. Eng. **6**(1), 107–120 (2009). https://doi.org/10.1109/TASE. 2008.916768
4. Cao, Z.L., Huang, Y., Hall, E.L.: Region filling operations with random obstacle avoidance for mobile robots. J. Field Robot. **5**(2), 87–102 (1988). https://doi.org/ 10.1002/rob.4620050202
5. Choset, H.: Coverage of known spaces: the boustrophedon cellular decomposition. Auton. Robots **9**(3), 247–253 (2000). https://doi.org/10.1023/A:1008958800904
6. Choset, H.: Coverage for robotics-a survey of recent results. Ann. Math. Artif. Intell. **31**(1–4), 113–126 (2001). https://doi.org/10.1023/A:1016639210559
7. Dubins, L.E.: On curves of minimal length with a constraint on average curvature, and with prescribed initial and terminal positions and tangents. Am. J. Math. **79**(3), 497–516 (1957). https://doi.org/10.2307/2372560
8. Gage, D.W.: Randomized search strategies with imperfect sensors. In: Mobile Robots VIII, vol. 2058, pp. 270–280. International Society for Optics and Photonics (1994). https://doi.org/10.1117/12.167503
9. Galceran, E., Carreras, M.: A survey on coverage path planning for robotics. Robot. Auton. Syst. **61**(12), 1258–1276 (2013). https://doi.org/10.1016/j.robot.2013.09. 004
10. Helsgaun, K.: LKH solver 2.0.9. http://www.akira.ruc.dk/~keld/research/LKH/. Accessed 1 Aug 2018
11. Hert, S., Lumelsky, V.: Polygon area decomposition for multiple-robot workspace division. Int. J. Comput. Geom. Appl. **8**(04), 437–466 (1998). https://doi.org/10. 1142/S0218195998000230
12. Latombe, J.C.: Robot Motion Planning, vol. 124. Springer, Boston (2012). https:// doi.org/10.1007/978-1-4615-4022-9
13. Luo, C., Yang, S.X.: A real-time cooperative sweeping strategy for multiple cleaning robots. In: International Symposium on Intelligent Control, pp. 660–665. IEEE (2002). https://doi.org/10.1109/ISIC.2002.1157841
14. Maza, I., Ollero, A.: Multiple UAV cooperative searching operation using polygon area decomposition and efficient coverage algorithms. In: Alami, R., Chatila, R., Asama, H. (eds.) Distributed Autonomous Robotic Systems 6, pp. 221–230. Springer, Tokyo (2007). https://doi.org/10.1007/978-4-431-35873-2_22
15. Moravec, H., Elfes, A.: High resolution maps from wide angle sonar. In: IEEE International Conference on Robotics and Automation (ICRA), vol. 2, pp. 116–121. IEEE (1985). https://doi.org/10.1109/ROBOT.1985.1087316
16. Noon, C.E., Bean, J.C.: A lagrangian based approach for the asymmetric generalized traveling salesman problem. Oper. Res. **39**(4), 623–632 (1991). https://doi. org/10.1287/opre.39.4.623
17. Oh, J.S., Choi, Y.H., Park, J.B., Zheng, Y.F.: Complete coverage navigation of cleaning robots using triangular-cell-based map. IEEE Trans. Ind. Electron. **51**(3), 718–726 (2004). https://doi.org/10.1109/TIE.2004.825197

18. Owen, M., Beard, R.W., McLain, T.W.: Implementing dubins airplane paths on fixed-wing UAVs*. In: Valavanis, K.P., Vachtsevanos, G.J. (eds.) Handbook of Unmanned Aerial Vehicles, pp. 1677–1701. Springer, Dordrecht (2015). https://doi.org/10.1007/978-90-481-9707-1_120

19. Perez-Imaz, H.I., Rezeck, P.A., Macharet, D.G., Campos, M.F.: Multi-robot 3D coverage path planning for first responders teams. In: IEEE International Conference on Automation Science and Engineering (CASE), pp. 1374–1379 (2016). https://doi.org/10.1109/COASE.2016.7743569

20. Siebert, S., Teizer, J.: Mobile 3D mapping for surveying earthwork projects using an unmanned aerial vehicle (UAV) system. Autom. Constr. **41**, 1–14 (2014). https://doi.org/10.1016/j.autcon.2014.01.004

21. Wong, S.C., MacDonald, B.A.: A topological coverage algorithm for mobile robots. In: IEEE/RSJ International Conference on Intelligent Robots and Systems (IROS), vol. 2, pp. 1685–1690. IEEE (2003). https://doi.org/10.1109/IROS.2003.1248886

22. Xu, A., Viriyasuthee, C., Rekleitis, I.: Efficient complete coverage of a known arbitrary environment with applications to aerial operations. Auton. Robots **36**(4), 365–381 (2014). https://doi.org/10.1007/s10514-013-9364-x

Development of Foot Contact Sensors for a Crawling Platform

Gaël Écorchard$^{(\boxtimes)}$ (ID) and Libor Přeučil

Czech Insitute of Informatics, Robotics, and Cybernetics,
Czech Technical University in Prague,
Jugoslávských partyzáů 3, 160 00 Prague, Czech Republic
{gael.ecorchard,libor.preucil}@cvut.cz
https://imr.ciirc.cvut.cz/

Abstract. Walking machines are a good compromise between flying machines with a small payload and wheeled machines with limited terrain crossing capabilities for displacements on strongly uneven terrain, such as in search and rescue missions. This paper presents the hardware development of a hexapod crawling robot, in particular the development and integration of foot contact sensors. The development is based on Trossen Robotics' PhantomX hexapod robotics kit. In its original version, the robot is only meant to be remote controller without any feedback from exteroceptive sensors. We present here a new foot contact sensor providing the required feedback for the contact between the robot's foot and the ground and their electronic integration in the rest of the system.

Keywords: Walking machines · Force sensor

1 Introduction

Walking robots represent a hybrid solution between flying robot, which have limited payloads and range of operation, and wheeled or tracked robots, which have problems to evolve in very rough terrains or fragile terrain. With equal body size, a walking robot will be able to cross a much rougher terrain than a tracked robot, for example it will be able to walk on stairs even having a relatively small size compared to the step size.

Application scenarios for such robots are all scenarios where the terrain roughness does not allow the use of wheeled or tracked robots but were heavy sensors or manipulators are required. Indeed, the ability to cross difficult terrain is the main raison d'être of legged robots [3]. An example of use could be the exploration of disaster zones after an explosion or an earthquake. A typical terrain which is difficult for wheeled robots are forests. On the opposite, a walking robot can cross over branches, roots, or stones. In the agricultural domain, walking robots can be deployed in fields with valuable plants whereas the movement of a wheeled robot are limited by the fact that they cannot cross over the plants.

© Springer Nature Switzerland AG 2019
J. Mazal (Ed.): MESAS 2018, LNCS 11472, pp. 237–250, 2019.
https://doi.org/10.1007/978-3-030-14984-0_19

The platform used as testbed for walking robots is a hexapod robot PhantomX Mark III by TrossenRobotics, [10]. In order for the robot to be able to evolve in rough terrain with positive and negative obstacles (holes), the contact between the ground and the robot has to be sensed. [11] lists other use of force sensor on walking robots. The original robot contains only interoceptive sensors, i.e. sensors about its internal state. Although the motor torque could theoretically be used to detect a contact between the robot and the ground, as done in [2] and [1], the high gear ratio and high friction render the measurements too noisy to be of practical use. In [6] the authors use the difference between set joint value and actual joint value to detect the contact with the ground on the practically same platform. Due to slow communication, the actual joint values can be read only for two servomotors within one control loop of 33 ms out of the six last servomotors on the legs. Moreover, because the contact detection can be only down if the joint value difference rises a certain relatively large threshold, the movement has to be slowed down on the down-going phase of the leg swing. The robot had thus to be adapted to be able to quickly sense its environment and new contact sensors have been developed. Not only do they enable to sense the contact with the ground but they can also provide some information about its magnitude and direction.

In [5], the contact sensor are based on semiconductor strain gauges for horizontal forces and a piezoelectric load cell for the vertical forces, but these are complex systems which are not required for our application. The same holds for the BioTac sensor from SynTouch, [12], and the MD-20-SE-40N from OptoForce, [7]. Contact sensors are evoked in [9] and [8] without further technical details.

In this paper we will present the evolution of the design of the foot force sensor, the evaluation of the contact forces from the sensed forces, and the potential electronic designs with their experimental results.

2   Sensor Design

2.1   Version 1

The original aim of the foot force sensors was to detect the contact with the ground in a binary manner. The original design induced a single force-sensing resistor (FSR) glued between the rubber at the foot tip and the foot itself. FSRs are resistors which resistor value depends non-linearly on the pressure applied to their sensing surface. The chosen FSR from Interlink Electronics are shown in Fig. 1, [4]. The diameter of the sensing surface is 5.1 mm.

A foot with the mounted FSR is shown in Fig. 2. The foot design here is the one of the PhantomX Hexapod Mark II originally used as research platform.

The sensors were mounted on each of the legs of the hexapod and the foot contact detection was tested on an artificial uneven rocky terrain. The sensors could detect the contact on the surface when the force direction was close to the normal to the FSR sensing surface but the detection was not reliable in other cases. Furthermore, the shape of the foot and the small sensing surface did not

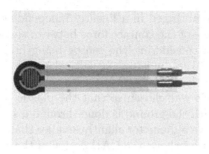

Fig. 1. Force-sensing resistor, courtesy of Interlink Electronics

Fig. 2. Version 1 of the foot force sensor

Fig. 3. Successful and failed detections with the version 1 of the foot force sensor

allow the rubber to be in contact with the ground on some obstacles such as a
hole or a step, as illustrated in Fig. 3.

To conclude, the foot tactile sensor version 1 can detect contacts when the
floor is almost perpendicular with the foot. So the contact detection was not
guaranteed for all kind of terrains and a new version of the sensor was needed.

2.2   Version 2

In order to keep the low cost for the contact sensor and improve the contact
detection compared to the first design, an array of three FSRs in a 3D-printed
housing was chosen.

The three FSRs are arranged in a linearly independent manner so that the X, Y, and Z components of the contact force between the ground and the robot can be determined. By considering the sensor being horizontal, i.e. when the leg touches an horizontal ground from a vertical position, the three FSRs are arranged with a 120° vertical polar array with an angle of 30° with the horizontal. The 30° horizontal angle was chosen so that the plastic parts can be easily 3D printed. The contact with the ground is done through a spherical shell equipped with three pads that have a diameter slightly smaller than the sensing surface of the FSR diameter. Figure 4 shows the CAD design of the contact sensor without the half sphere (in reality it is a spherical cap which is almost a half sphere) which touches the ground.

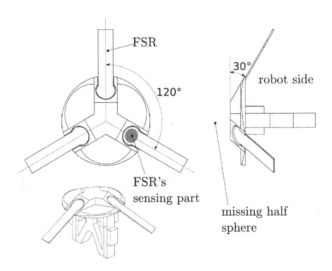

Fig. 4. CAD design of the contact sensor

The FSRs are glued to the part of the sensor that is mounted on the leg with the sticker that is integral part of the chosen FSRs. Then the half sphere is glued to this arrangement with cyanoacrylate glue. The partly assembled contact sensor and the fully assembled sensor mounted on the hexapod robot are shown on Fig. 5. The external shell is then recovered with fast cross-linking bicomponent rubber to ensure a good adhesion with the ground.

As with the previous version of the force sensors, the FSRs are glued on one side with the part touching the ground and on the other side on the foot. There was no other mechanical link between the half-sphere and the foot than through the FSRs. FSRs are designed to support a contracting effort normal to their sensing surface. Problems appeared when the contact between the foot and the ground implied tangential efforts or normal traction efforts in the opposite direction. The FSRs could not withstand these efforts and the plastic part with the resistor unstuck from the black support used as interface between the

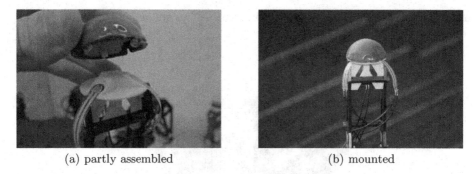

(a) partly assembled   (b) mounted

Fig. 5. Real contact sensor version 2

sensing pad and the part the FSR is glued to. A design had to be found where the efforts that the FSRs could not withstand would be taken by a mechanical link between the half-sphere and the foot.

2.3   Version 3

In the third version of the foot force sensors the pads pressing the sensing surface of the FSRs are truncated coni prolonged with a cylinder. The smaller circular face is in contact with the FSR but is not glued on it. On the part of the assembly which is on the hexapod side, the FSRs are sticked on the bottom of a cylindrical hole with a radius slightly smaller that the cylindrical part of the pads of the half-sphere. The pads take then all the tangential efforts that would otherwise destroy the FSRs. The opened foot contact sensor is shown in Fig. 6.

Flexible ring   Pressing pad

FSR

Fig. 6. CAD view of the foot force sensor version 3.

Because the half-sphere is not glued any more onto the FSRs a solution had to be found to prevent disassembly. The half-sphere is covered by a balloon such as the ones used for decoration or advertising[1]. The ballon is maintained

---

[1] The authors want to acknowledge Prague's School of Chemistry for providing such balloons in a material that does not alter with the time as quickly as some cheap balloons tested.

in position thanks to a flexible ring. Figure 7 shows the assembled foot contact sensor.

Fig. 7. Real force sensor version 3.

3   Contact Force Estimation

The force of the environment on the contact sensor is F_c, with intensity f_c. The normal forces on the FSRs have intensity N_O, N_F, and N_R, where the indices O, F, and R represent the Outer, Front, and Rear FSRs respectively. These values are the values obtained from the sensor readings. The normal forces of the environment on the FSRs are then in the direction of $-z$ in the local reference frame. Forces are represented on Fig. 8. In order to solve the otherwise hyperstastic equilibrium, some assumptions must be made. The tangential axial forces, $T_{a,O}$, $T_{a,F}$, and $T_{a,R}$, and the tangential radial forces, $T_{r,O}$, $T_{r,F}$, and $T_{r,R}$ are neglected, cf. Fig. 9. The contact in M is supposed to be frictionless so that f_c is normal to the spherical cap, $\phi = \phi_M$ and $\theta = \theta_M$.

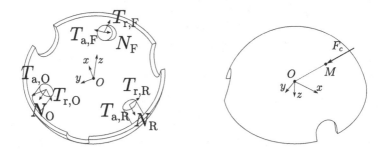

Fig. 8. Force definitions

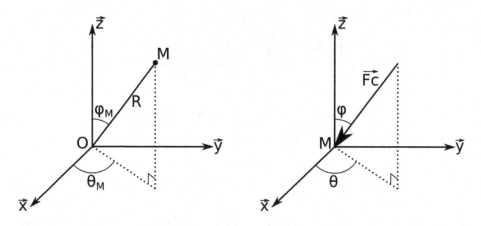

Fig. 9. Force orientation

The static equilibrium equations are then:

$$\begin{cases} \frac{\sqrt{3}}{4}N_F - \frac{\sqrt{3}}{4}N_R + \sin(\phi)\cos(\theta)f_c = 0 \\ \frac{1}{2}N_O - \frac{1}{4}N_F - \frac{1}{4}N_R + \sin(\phi)\sin(\theta)f_c = 0 \\ \frac{\sqrt{3}}{2}(N_O + N_F + N_R) - \cos(\phi)f_c = 0 \end{cases} \tag{1}$$

From the first two equations of Eq. (1), for the general case $\sin(\phi) \neq 0$, we can deduce

$$\theta = \arctan\left(\frac{2N_O - N_F - N_R}{\sqrt{3}N_F - \sqrt{3}N_R}\right) \tag{2}$$

By reintroducing this result into the last two equations of Eq. (1), we can deduce

$$\phi = \arctan\left(\frac{2N_O - N_F - N_R}{2\sqrt{3}(N_O + N_F + N_R)\sin(\theta)}\right) \tag{3}$$

Eventually, with the third equation of Eq. (1) and the preceding results, we have

$$f_c = -\frac{\sqrt{3}}{2}\frac{N_O + N_F + N_R}{\cos(\phi)} \tag{4}$$

In the case, $\phi = 0$ we obtain the same formula as Eq. (4) for f_c directly from the third equation of Eq. (1).

4   Relation Between Resistance and Force

The resistance of the FSR typically varies from a value larger than $10\,\mathrm{M\Omega}$ without applied load to $200\,\Omega$ with full load, which is around $10\,\mathrm{N}$. Its typical characteristic curve is given on Fig. 10 [4].

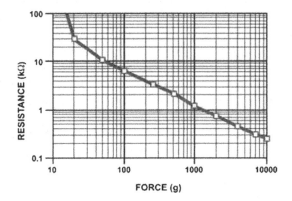

Fig. 10. Characteristic curve resistance vs. force of FSR400, courtesy of Interlink Electronics.

The characteristic curve given in Fig. 10 is approximately a line in the double logarithmic scales, between $20\,\Omega$ and $10\,k\Omega$. A linear regression is thus computed and gives:

$$10^R = -0.823 \cdot 10^M + 5.498, \tag{5}$$

where R is the resistance of the FSR in Ω and M is the corresponding mass in g. By converting to units to the international system and inverting the preceding equation, we have

$$10^F = -0.0119 \cdot 10^R + 0.0656, \tag{6}$$

where F is the force applied on the FSR in N.

5   Electronic Design and Experimental Results

Several electronic circuits were tested to read the resistance of the FSRs, with the aim to determine the less noisy circuit and the one with the highest sensitivity with a small load. For these tests, we exploited an Arduino Uno controller with its 5 V ten-bit Analogue-Digital Converters (ADC) and different electronic configurations to determine the best conversion circuit:

– voltage divider with voltage follower,
– evaluation of the time constant of a RC arrangement, and
– feedback loop.

The first two interfaces are suggested by the FSR manufacturer, while the third one has been properly designed to reach our goals. All measures have been estimated especially within the range $[0\,g,\ 500\,g]$ since the complete hexapod weight is equal to 2.5 kg and it is shared between six legs.

To test the FSRs, a small assembly was 3D-printed to apply a known force on the sensitive part of the sensors. The assembly is shown Fig. 11. The pressing pad is the same as the one from the third sensor design.

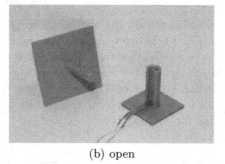

(a) closed with weight   (b) open

Fig. 11. Testbench for one FSR.

5.1 Voltage Divider

The circuit presented Fig. 12 is a voltage divider configuration with a voltage follower. The aim of the operational amplifier used as voltage follower is to augment the input impedance in order to reduce the influence of the analogue-digital converter on the output voltage of the divider.

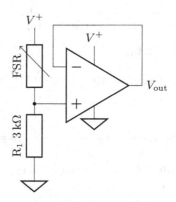

Fig. 12. Voltage divider with follower

The voltage V_{out} in the middle of the voltage divider is:

$$V_{out} = V^+ \frac{R_1}{R_1 + FSR}, \tag{7}$$

where V^+ is the microcontroller input voltage ($+5\,V$), R_1 is the fixed resistance and FSR is the force-sensing resistor. The input impedance of the microcontroller is considered to have a negligible influence on the result. This gives:

$$\mathrm{FSR} = \mathrm{R}_1 \left(\frac{V^+}{V_{\mathrm{out}}} - 1 \right). \tag{8}$$

It is possible to vary the voltage-vs.-force characteristic by tuning the resistor value, cf. Fig. 13. If one considers the main goal of the contact force sensor to detect the contact rather than measuring it precisely, a high resistance of $100\,\mathrm{k\Omega}$ would be chosen. For the reason that we were interested in evaluating the contact force, a small resistor of $3\,\mathrm{k\Omega}$ was chosen. Which ever the chosen resistor the voltage-vs.-force characteristic remains highly non linear, which is not what we are looking for.

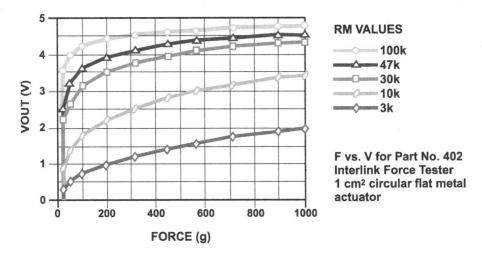

Fig. 13. Voltage vs. force characteristic of the FSR400, courtesy of Interlink Electronics.

The results of our experiments with this circuit are given Fig. 14. The values for masses smaller than $40\,\mathrm{g}$ show some aberrations and render the deduction of the contact forces difficult. The contact could be estimated by considering all voltage values greater that 600 microcontroller units as irrelevant, thus allowing the contact detection only above approximately $50\,\mathrm{g}$.

The circuit could be used to detect the contact but cannot serve the purpose of estimating the contact forces.

5.2   Evaluation of the Time Constant

With the circuit to evaluate the time constant of a resistor-capacitor arrangement the resistor of the FSR is obtained by evaluating the time constant τ of the circuit given in Fig. 15. R_min and R_max are used to obtain the upper and the lower

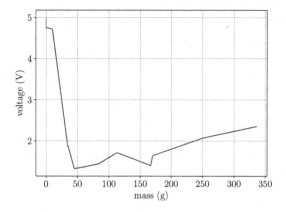

Fig. 14. Results of the voltage divider circuit.

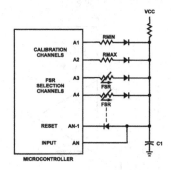

Fig. 15. Digital resistor measurement, courtesy of Interlink Electronics.

bounds of the readable values, while R_3 allows the current to flow inside the capacitor. In order to perform the tests in laboratory, we have set $\tau = 0.1$ s.

The results of our evaluation for this circuit are shown in Fig. 16. Although the circuit provides a good characteristic estimation and it is not much affected by noise, the capacitor needs a large amount of time to be charged and discharged completely. Due to the fact the a great number of digital inputs and outputs would be necessary to measure all eighteen FSRs in parallel, they have to be measured sequentially. This renders the method impractical for our application.

5.3   Feedback Loop

The feedback loop configuration, Fig. 17, is realized by inserting the FSR into an operational amplifier feedback loop, allowing the current flowing into the sensor to be theoretically constant. The operational amplifier's non-inverting input is linked to a voltage reference, obtained by a voltage divider, while the gain is tuned with the value of the resistor R_3.

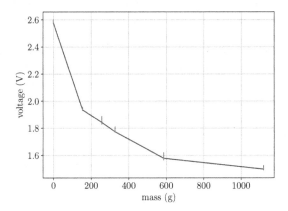

Fig. 16. Results of the resistor evaluation with the time constant estimation. The error bars show the minimum and maximum values.

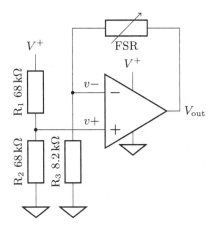

Fig. 17. Voltage divider with follower

By idealizing the operational amplifier, we can write that $v+ = v-$ and that the current flowing through FSR is the same as the one flowing through R_3, i_3. The voltage $v+$ is defined by

$$v+ = V^+ \frac{R_2}{R_1 + R_2} \tag{9}$$

and the current is given by

$$i_3 = \frac{v-}{R_3} = \frac{v+}{R_3} = \frac{V_{out} - v-}{FSR}. \tag{10}$$

By combining Eqs. 9 and 10, we can deduce

$$V_{out} = V^+ \frac{R_2}{R_1 + R_2} \frac{FSR + R_3}{R_3}. \tag{11}$$

The output voltage of the operational amplifier is then proportional to the value of the FSR.

From Eq. 11 we can deduce that the maximal theoretically measureable FSR, FSR_{max}, obtained when V_{out} saturates to V^+, is given by

$$FSR_{max} = R_3 \left(\frac{R_1 + R_2}{R_2} - 1 \right). \tag{12}$$

With the proposed resistor values, this corresponds to approximately $57\,k\Omega$. According to Fig. 10, the contact starts thus to be detected with the equivalent of less than $20\,g$. The range of the output values is approximately from $v+$ to V^+.

The results of our experiments with the feedback loop circuit are shown in Fig. 18, with error bars at the minimum and maximum values. In the small force range, the measurements are much closer to be linear than the results with the voltage divider circuit and, most importantly, they are monotonuous. This circuit has thus been chosen for the implementation on the real robot.

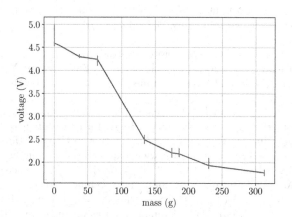

Fig. 18. Results of the feedback loop circuit.

6   Conclusion and Future Work

A new low-cost force sensor was presented for small walking machines. This sensor is based on force sensor resistors and allows the quick detection of the contact of the robot's feet with the ground. Future works will consist in the integration into the hexpod platform and further testing, particularly the possibility to estimate the contact forces and the mechanical durability.

Acknowledgments. This work was supported by the Technology Agency of the Czech Republic under project TE01020197 Center for Applied Cybernetics 3.

References

1. Bombled, Q., Verlinden, O.: Indirect foot force measurement for obstacle detection in legged locomotion. Mech. Mach. Theory **57**, 40–50 (2012). https://doi.org/10.1016/j.mechmachtheory.2012.06.003. http://linkinghub.elsevier.com/retrieve/pii/S0094114X12001310
2. Bombled, Q., Verlinden, O.: A spare method for obstacle sensing in legged locomotion. In: Field Robotics - 14th International Conference on Climbing and Walking Robots and the Support Technologies for Mobile Machines, pp. 491–498 (2012). https://doi.org/10.1142/9789814374286_0057
3. Celaya, E., Porta, J.: A control structure for the locomotion of a legged robot on difficult terrain. IEEE Robot. Autom. Mag. **5**(2), 43–51 (1998). https://doi.org/10.1109/100.692340
4. FSR400, Interlink Electronics. http://www.interlinkelectronics.com/FSR400.php
5. Klein, C.A., Olson, K.W., Pugh, D.R.: Use of force and attitude sensors for locomotion of a legged vehicle over irregular terrain. Int. J. Robot. Res. **2**(2), 3–17 (1983). https://doi.org/10.1177/027836498300200201. http://ijr.sagepub.com/content/2/2/3
6. Mrva, J., Faigl, J.: Tactile sensing with servo drives feedback only for blind hexapod walking robot. In: 10th International Workshop on Robot Motion and Control, RoMoCo 2015, pp. 240–245, July 2015. https://doi.org/10.1109/RoMoCo.2015.7219742
7. OMD, 3D force sensor. http://optoforce.com/3dsensor/
8. Palis, F., Rusin, V., Schmucker, U., Schneider, A., Zavgorodniy, Y.: Walking robot with articulated body and force controlled legs. In: Adaptive Motion in Animals and Machines (2005)
9. Palis, F., Rusin, V., Schmucker, U., Schneider, A., Zavgorodniy, Y.: Walking robot with force controlled legs and articulated body. In: Proceedings of the 22nd International Symposium on Robotics and Automation in Construction (2005)
10. Phantomx Hexapod Mark II. http://www.trossenrobotics.com/hex-mk2
11. Schmucker, U., Schneider, A., Rusin, V., Zavgorodniy, Y.: Force sensing for walking robots. In: Adaptive Motion in Animals and Machines (2005)
12. Wettels, N., Fishel, J.A., Loeb, G.E.: Multimodal tactile sensor. In: Balasubramanian, R., Santos, V.J. (eds.) The Human Hand as an Inspiration for Robot Hand Development. STAR, vol. 95, pp. 405–429. Springer, Cham (2014). https://doi.org/10.1007/978-3-319-03017-3_19

Localization Fusion for Aerial Vehicles in Partially GNSS Denied Environments

Jan Bayer[(✉)] and Jan Faigl[ⅅ]

Faculty of Electrical Engineering, Czech Technical University in Prague,
Technicka 2, 166 27 Prague, Czech Republic
{bayerja1,faiglj}@fel.cvut.cz
https://comrob.fel.cvut.cz

Abstract. In this paper, we report on early results of the experimental deployment of localization techniques for a multi-rotor Micro Aerial Vehicle (MAV). In particular, we focus on deployment scenarios where the Global Navigation Satellite System (GNSS) does not provide a reliable signal, and thus it is not desirable to rely solely on the GNSS. Therefore, we consider recent advancements in the visual localization, and we employ an onboard RGB-D camera to develop a robust and reliable solution for the MAV localization in partially GNSS denied operational environments. We consider a localization method based on Kalman filter for data fusion of the vision-based localization with the signal from the GNSS. Based on the reported experimental results, the proposed solution supports the localization of the MAV for the temporarily unavailable GNSS, but also improves the position estimation provided by the incremental vision-based localization system while it can run using onboard computational resources of the small vehicle.

1 Introduction

Accurate and reliable localization of a multi-rotor Micro Aerial Vehicle (MAV) is an essential prerequisite for its deployment not only in autonomous missions but also in semi-autonomous deployments where the vehicle is requested to follow a pre-designed path. The Global Navigation Satellite System (GNSS) is a natural choice and practical solution for outdoor missions, where it provides an estimation of the vehicle position. On the other hand, the GNSS is not always available, and it becomes unreliable or inaccurate inside and also close to tall buildings or at places where not enough satellites are in the line of sight [11]. Therefore, we examine the properties of existing vision-based localization methods using onboard sensors to provide an estimation of the MAV position. In particular, we focus on the visual odometry and vision-based methods for Simultaneous Localization and Mapping (SLAM) [23] that recently exhibited significant improvements in the pose estimation [12].

The proposed method is based on a fusion of the unreliable GNSS localization and state-of-the-art vision-based localization. The used sensor fusion approachis

© Springer Nature Switzerland AG 2019
J. Mazal (Ed.): MESAS 2018, LNCS 11472, pp. 251–262, 2019.
https://doi.org/10.1007/978-3-030-14984-0_20

Fig. 1. The MAV used during the experiments.

based on Kalman filter [23], and the main contribution of this paper is in the presented results on the experimental deployment of the developed solution on the MAV in an outdoor scenario, with the recent Intel RealSense D435 sensor [3] depicted in Fig. 1.

The developed method itself represents a general fusion framework to combine various sources of the external localization like the GNSS with the incremental localization such as visual odometry. The herein reported results support the feasibility of the proposed solution in the particular experimental deployment where a precise pose estimation from the Global Positioning System (GPS), which provides the necessary ground truth for the evaluation, is modified to decrease its reliability by adding Gaussian noise and disabling the GNSS completely for a certain deployment period.

The rest of the paper is organized as follows. Principles of the existing localization methods, the most related sensory fusion methods, and metrics of localization precision assessment are overviewed in Sect. 2. The model for the proposed localization fusion is derived in Sect. 3. The evaluation results of the proposed solution are reported in Sect. 4. The concluding remarks on the achieved results and our future work are presented in Sect. 5.

2   Related Work

Two main classes of the methods to localize a robot in GNSS denied environments can be identified in the literature. The first class of the methods are solutions that rely on beacons and transmitters [10], or other sensors like cameras [2,16] placed in the robot operational space. We do not consider these methods in this paper because these methods require an additional infrastructure in the robot operational space, which is not practical for our motivational deployment scenario.

The methods of the second class rely on exteroceptive sensors like LIDARs or cameras mounted on a robot. These sensors are used to compute a position of the robot incrementally from consecutive LIDAR scans or camera frames using some representative landmarks detected as visual feature points in the scans and camera image. The concept of the incremental localization can be further extended to improve the localization estimation by simultaneously creating a map of the environment that is used in localization of the vehicle using new sensor measurements. Such a technique can be found in the literature as SLAM [23]. A SLAM method can also be considered as an incremental localization system with loop closures that further improve the localization estimation once a robot returns to already visited places, and thus decrease the localization drift.

Methods for localization of the MAV using LIDAR scans have been successfully deployed in [4,17,27]. Although localization methods based on LIDAR sensors exhibit remarkable advancements during the last years, LIDARs are relatively expensive sensors in comparison to cameras. Besides, they are still mechanical systems, and they are more substantial than most of the conventional cameras. Therefore, we prefer vision-based localization systems that are using image processing from the onboard camera or especially new RGB-D cameras.

Several existing vision-based localization methods employ principles of the visual odometry from a single camera up to complex SLAM systems using several cameras of different types. The existing monocular camera localization methods include SVO [8], DSO [6], ORB-SLAM2 [14], and many others [18]. The main disadvantage of these methods is that they need several frames or additional information to obtain the scale of the scene, and thus the methods suffer from the map initialization problem, i.e., the bootstrapping problem [9]. Therefore, it is beneficial to use sensors that can estimate the scale of the scene from a single frame. It is the case of stereo, and RGB-D cameras and data from these sensors can be used in the existing methods such as the RGB-D SLAM [5], ORB-SLAM2 [13], and Stereo DSO [25] to name a few.

Beside exteroceptive sensors, the pose estimation can also be improved using data from the Inertial Measurement Unit (IMU). The authors of [27] and [4] report on fusion of IMU measurements with the localization based on the LIDARs data by Kalman filter. A complex navigation system for the six-legged walking robot is proposed in [21], where the authors use the information filter that is analogous to Kalman filter. Even though the authors of [21] report on lower computational requirements of the information filter in comparison to Kalman filter, it is the feature of the particular setup.

In addition to the fusion of IMU measurements with other signals, the IMU data measurements can be directly utilized inside a localization method itself, for example as in a variant of the Stereo Parallel Tracking and Mapping (S-PTAM) method [19] proposed in [7]. However, such an approach requires changes to the visual localization method and thus developing a new localization method, which is not our main intention.

Regarding the presented overview of the existing methods and our previous work [15], we choose ORB-SLAM2 in combination with the RGB-D camera as the main visual localization system to overcome GNSS denied regions of the vehicle operational environment. Besides, we chose Kalman filter approach for data fusion because it is more straightforward than an information filter.

2.1   Localization Metrics

In our work, we aim to improve the localization system, and therefore, we need a method to measure and compare the precision of different localization systems. The general framework for measuring localization precision is described in [22], where the authors measure the error of the localization from the trajectories obtained during the robotic experimental trial in a particular scenario. The compared trajectories are the ground truth, which is provided by a reference localization system, e.g., D-GPS, and a trajectory estimate provided by the examined localization system. The metrics to measure the error of the estimated trajectory are the *Absolute Trajectory Error* (ATE) and *Relative Pose Error* (RPE) [22], which are well-established and used in the literature.

The ATE is given by the equation:

$$\mathbf{F}_i = \mathbf{Q}_i^{-1}\,\mathbf{S}\,\mathbf{P}_i, \tag{1}$$

where the matrices $\mathbf{Q}_i$ and $\mathbf{P}_i$ are SE(3) positions of the ground truth and estimated trajectory, respectively. The matrix $\mathbf{S}$ is a transformation between the coordinate frames of the ground truth and trajectory estimate. The transformation is obtained from the minimization of the squared distances between the corresponding positions of the trajectory estimate and the ground truth [22].

The RPE is given by the equation:

$$\mathbf{E}_i = (\mathbf{Q}_i^{-1}\mathbf{Q}_{i+\Delta})^{-1}(\mathbf{P}_i^{-1}\mathbf{P}_{i+\Delta}), \tag{2}$$

where Δ represents a fixed distance between two positions that are used during the calculation [22] and it is $\Delta = 1$ in our case.

Once the ATE and RPE values are computed for the whole trajectory, the computed error values are used to get a statistical indicator to evaluate the precision of the localization based on the whole trajectory. In [22], the authors suggest to use the average value and the Root Mean Square (RMS) to the determined statistical indicators as

$$\overline{\mathrm{ATE}}_t = \frac{1}{n}\sum_{i=1}^{n} \| trans(\mathbf{F}_i) \|, \quad RMS(\mathrm{ATE}_t) = \left(\frac{1}{n}\sum_{i=1}^{n} \| trans(\mathbf{F}_i) \|\right)^{\frac{1}{2}},$$
$$\overline{\mathrm{RPE}}_t = \frac{1}{n}\sum_{i=1}^{n} \| trans(\mathbf{E}_i) \|, \quad RMS(\mathrm{RPE}_t) = \left(\frac{1}{n}\sum_{i=1}^{n} \| trans(\mathbf{E}_i) \|\right)^{\frac{1}{2}}, \tag{3}$$

where $trans()$ computes the size of the translation from the SE(3) matrix. In the herein reported results, we assume only the translation errors, because the used fusion method fuses only the positions of the robot. Besides, the statistical indicators that compute the average ATE and average RPE are utilized because the RMS does not provide any additional information in our particular case.

3   Fusion of the GNSS with Vision-Based Localization

We propose to use linear Kalman filter to overcome a loss of the GNSS signal that is substituted by the incremental vision-based localization in the GNSS denied environments [11]. Kalman filter is a standard technique for a relatively computationally inexpensive fusion of two or more sources of measurements, which allows developing a system to uninterruptedly provide the requested estimation of the vehicle position when a particular source is not available. Contrary to the systems with SLAM based localization [26], we select an incremental vision-based localization because we assume the GNSS signal is only temporarily unavailable. Thus, it is not expected a full loop-closure is necessary, and therefore, we can choose an existing SLAM system with low computational requirements.

The developed localization system is based on a linear Kalman filter to combine an external GNSS localization with the incremental vision-based localization. The system consists of two principal parts: Kalman filter, and a model of the vehicle, which is derived from the motion equation of the robot body. Both parts are described in the following paragraphs to make the paper self-contained.

Kalman filter for a linear and discrete-time model of the system described by the state equation

$$\mathbf{x}_{k+1} = \mathbf{A}\mathbf{x}_k + \mathbf{B}\mathbf{u}_k \tag{4}$$

can be defined by the set of the following equations [23]:

$$\overline{\boldsymbol{\mu}}_k = \mathbf{A}_k \boldsymbol{\mu}_{k-1} + \mathbf{B}_k \boldsymbol{u}_k \tag{5}$$

$$\overline{\boldsymbol{\Sigma}}_k = \mathbf{A}_k \boldsymbol{\Sigma}_{k-1} \mathbf{A}_k^T + \mathbf{R}_k \tag{6}$$

$$\mathbf{K}_k = \overline{\boldsymbol{\Sigma}}_k \mathbf{H}_k^T (\mathbf{H}_k \overline{\boldsymbol{\Sigma}}_k \mathbf{H}_k^T + \mathbf{Q}_k)^{-1} \tag{7}$$

$$\boldsymbol{\mu}_k = \overline{\boldsymbol{\mu}}_k + \mathbf{K}_k (\mathbf{z}_k - \mathbf{H}_k \overline{\boldsymbol{\mu}}_k) \tag{8}$$

$$\boldsymbol{\Sigma}_k = (\mathbf{I} - \mathbf{K}_k \mathbf{H}_k) \overline{\boldsymbol{\Sigma}}_k, \tag{9}$$

where $\overline{\boldsymbol{\mu}}_k$, $\overline{\boldsymbol{\Sigma}}_k$ are mean, and covariance of the system states predicted based on the last estimated mean $\boldsymbol{\mu}_k$ of the system states $\mathbf{x}_k$. The predictions can be used asynchronously with the correction phase represented by (7), (8), and (9) in the case we need the robot position in an arbitrary time. However, in this paper, we extract the position of the robot synchronously with the correction phase from $\boldsymbol{\mu}_k$. $\boldsymbol{\Sigma}_k$ represents the covariance of the system states $\mathbf{x}_k$ and the matrix $\mathbf{R}_k$ represents the uncertainty of the model of the system. The vector $\mathbf{z}_k$ is the measurement obtained according to

$$\mathbf{z}_k = \mathbf{H}_k \mathbf{x}_k + \boldsymbol{\delta}_k, \tag{10}$$

where $\boldsymbol{\delta}_k$ is the Gaussian noise with the zero mean and covariance $\mathbf{Q}_k$. The more detailed description of the vehicle model and the related matrices is as follows.

The model of the vehicle is based on the body motion model that can be described as

$$
\begin{bmatrix} \mathbf{p}_{k+1} \\ \mathbf{v}_{k+1} \end{bmatrix} = \begin{bmatrix} \mathbf{I}_{3\times3} & t_s \cdot \mathbf{I}_{3\times3} \\ \mathbf{0}_{3\times3} & \mathbf{I}_{3\times3} \end{bmatrix} \begin{bmatrix} \mathbf{p}_k \\ \mathbf{v}_k \end{bmatrix} + \begin{bmatrix} \frac{t_s^2}{2m} \cdot \mathbf{I}_{3\times3} \\ \frac{t_s}{m} \cdot \mathbf{I}_{3\times3} \end{bmatrix} \mathbf{f}_k, \tag{11}
$$

where $\mathbf{f}_k$ is the force that moves the vehicle, $\mathbf{p}_k$ and $\mathbf{v}_k$ are the position and velocity of the vehicle, respectively, t_s is the time, and m is the vehicle weight.

Let suppose that we use the position from the incremental localization as the position $\mathbf{p}_k$ during the filtration. The problem is that the incremental localization drifts, and thus its uncertainty grows to infinity. Moreover, if the uncertainty of the incremental localization increases to a certain level, the contribution of the incremental localization would vanish. On the other hand, the incremental localization provides relatively precise estimates of the transformations between the consecutive robot positions, and a precision of these transformations is not affected by the position drift. Therefore, we can incorporate differences of the consecutive positions to the vehicle model. The straightforward way to incorporate these differences is to substitute the vehicle velocity using the equation

$$
\mathbf{v}_k = \frac{\mathbf{q}_k - \mathbf{q}_{k-1}}{t_s} = \frac{\Delta \mathbf{q_k}}{t_s}. \tag{12}
$$

Then, the model of the system (11) changes to

$$
\begin{bmatrix} \mathbf{p}_{k+1} \\ \Delta \mathbf{q}_{k+1} \end{bmatrix} = \begin{bmatrix} \mathbf{I}_{3\times3} & \mathbf{I}_{3\times3} \\ \mathbf{0}_{3\times3} & \mathbf{I}_{3\times3} \end{bmatrix} \begin{bmatrix} \mathbf{p}_k \\ \Delta \mathbf{q}_k \end{bmatrix} + \begin{bmatrix} \frac{t_s^2}{2m} \cdot \mathbf{I}_{3\times3} \\ \frac{t_s}{m} \cdot \mathbf{I}_{3\times3} \end{bmatrix} \mathbf{f}_k. \tag{13}
$$

In the case of the used MAV, the force $\mathbf{f}_k$, which moves the vehicle, is usually not hard to obtain. In other cases, when the force is laborious to obtain, it is possible to follow the tracking scenario [24], which is considered in the rest of this section. The model of the vehicle can be expressed without the force as

$$
\begin{bmatrix} \mathbf{p}_{k+1} \\ \Delta \mathbf{q}_{k+1} \end{bmatrix} = \begin{bmatrix} \mathbf{I}_{3\times3} & \mathbf{I}_{3\times3} \\ \mathbf{0}_{3\times3} & \mathbf{I}_{3\times3} \end{bmatrix} \begin{bmatrix} \mathbf{p}_k \\ \Delta \mathbf{q}_k \end{bmatrix}, \tag{14}
$$

where the matrices of the state Eq. (4) used in Kalman filter are

$$
\mathbf{x}_k = \begin{bmatrix} \mathbf{p}_k \\ \Delta \mathbf{q}_k \end{bmatrix}, \ \mathbf{A} = \begin{bmatrix} \mathbf{I}_{3\times3} & \mathbf{I}_{3\times3} \\ \mathbf{0}_{3\times3} & \mathbf{I}_{3\times3} \end{bmatrix}, \ \mathbf{B} = \mathbf{0}_{6\times1}.
$$

There is not an input to the model (14) because the model only updates the position of the robot by the last value of $\Delta \mathbf{q}_k$. Thus, the position of the robot provided by the model is uncertain and has to be corrected by new measurements. The model uncertainty is described by the covariance matrix $\mathbf{R}_k$, see (6).

The measurements used to correct the position of the robot estimated by the model (14) are: (i) the position data provided by the external localization that corrects the state $\mathbf{p}_k$; and (ii) differences of the consecutive measurements provided by the incremental localization correct the states $\Delta\mathbf{q}_k$. The precision of the measurements is described by the covariance matrix $\mathbf{Q}_k$, which represents Gaussian noise $\boldsymbol{\delta}_k$ in the measurement Eq. (10). The matrices $\mathbf{H}_k$ and $\mathbf{Q}_k$ of the measurement equation are composed as follows.

$$\mathbf{H} = \mathbf{H}_{k=1,2,\dots} = \mathbf{I}_{6\times6}, \quad \mathbf{Q}_k = \begin{bmatrix} \mathbf{Q}_{p,k} & \mathbf{0}_{3\times3} \\ \mathbf{0}_{3\times3} & \mathbf{Q}_{\Delta q,k} \end{bmatrix},$$

where $\mathbf{Q}_{p,k}$ and $\mathbf{Q}_{\Delta q,k}$ represent the uncertainty of the external localization and the uncertainty of the incremental localization, respectively.

4   Experimental Results

The developed localization system has been experimentally verified with the real MAV shown in Fig. 1 in a practical field experimental deployment. The experiments took place in the environment where the GNSS was available because we used a precise GPS as the reference localization system to evaluate the performance of the developed solution with the available ground truth. Thus, the ground truth and RGB-D data from the visual localization have been collected using ROS [20] in four flights with the vehicle operating at the height between 5 and 10 m above the terrain. The reference localization has been provided by the localization system that combines a precise GPS with a laser altimeter, and the IMU onboard of the MAV.

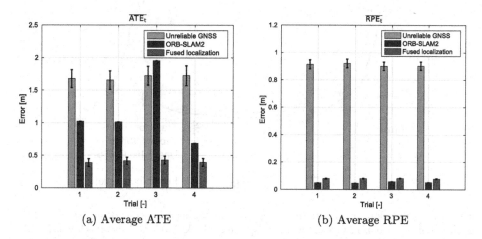

(a) Average ATE   (b) Average RPE

Fig. 2. Accuracy of the provided trajectory localization for the unreliable GNSS, ORB-SLAM2, and the localization fusion.

The RGB-D data provided by the Intel RealSense D435 [3] have been processed by the ORB-SLAM2 [13] to get trajectory estimation solely based on the visual localization. The captured data have been processed at the same frequency as they have been captured by the onboard RGB-D camera using ROS. In particular, the ORB-SLAM2 was running on the computer with the dual-core Intel i5-5257U CPU running at 2.7 GHz with 4 GB of memory. The used computational environment has similar computational power as the onboard computer at the used MAV to reflect requirements on the onboard processing. The conditions under we used the visual localization are almost identical to the on-line deployment on the vehicle, and the ORB-SLAM2 provides localization at the average frequency of 6.67 Hz. Notice that even though the ORB-SLAM2 can perform a large loop closing, which may improve the localization precision significantly, this feature has been disabled to demonstrate the ability of the proposed filter to address the localization drift.

An unreliable GNSS-based localization has been substituted by adding the Gaussian noise with the standard deviation of 0.5 m in all three axes to the ground truth trajectory. Moreover, we selected approximately 20% of the trajectory, where the localization signal was disabled entirely to simulate temporal unavailability of the GNSS. The developed Kalman filter was used for fusing the localization provided by the ORB-SLAM2 and the simulated unreliable GNSS.

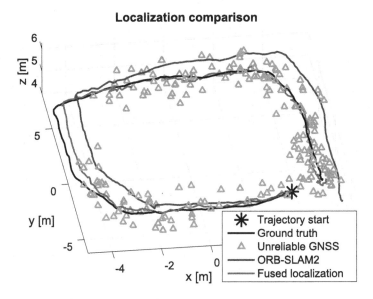

Fig. 3. Localization trajectory provided by the evaluated localization systems: the ground truth, unreliable GNSS localization, the ORB-SLAM2, and the developed fusion of the vision-based localization. Notice the gap in the GNSS poses in the left part of the plot where the green triangles are missing. For this part of the trajectory, the GNSS is temporarily unavailable. (Color figure online)

Before the fusion, we estimate the matrices $\mathbf{Q}_{p,k}$, $\mathbf{Q}_{\Delta q,k}$, and $\mathbf{R}_k$ that describe the uncertainty of the localization systems and the uncertainty of the MAV

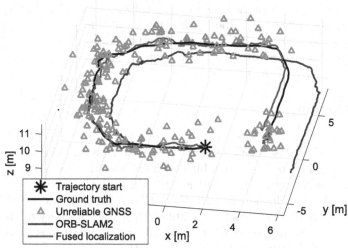

Fig. 4. Detail of the localization drift observed for the ORB-SLAM2.

model. The matrix $\mathbf{Q}_{p,k}$ has been estimated from the parameters of the unreliable GNSS localization with the added noise, and it reflects the GNSS availability as well. If the unreliable GNSS is disabled, $\mathbf{Q}_{p,k}$ is increased 40 times. The matrix $\mathbf{Q}_{\Delta q,k}$ has been estimated so that it models the uncertainty of the position error between two consecutive positions. Finally, the matrix $\mathbf{R}_k$ has been tuned as the diagonal matrix, which forces the filter to not rely on the velocity estimates provided by the model but on the robot position.

Four flight trials have been performed by the vehicle which follows approximately square shaped trajectory using the model predictive trajectory tracking [1]. For each trial, we obtained one trajectory by the ORB-SLAM2, and 500 different trajectories from the artificially created unreliable GNSS signal. Thus, the localization fusion has been performed 500 times per each experimental trial. All fused trajectories have been evaluated using the average of the errors computed by (3). In particular, we suppose the orientation of the vehicle is known at the beginning of each trial, which is utilized to synchronize the coordinate frame of the localization provided by the ORB-SLAM2 with the other localization systems. Since the fusion method fuses only the position of the robot and not its orientation, we add the orientation to the resulting fused trajectory from the trajectory provided by the ORB-SLAM2; so, we can use the evaluation metrics (1) and (2). The results are summarized in Fig. 2 and the ground truth, unreliable GNSS, ORB-SLAM2, and the achieved trajectory estimations provided by the fusion method are depicted in Fig. 3.

Figure 2a indicates that the method performed as expected because the error of the localization method that employs fusion is lower than the error for both input sources of the localization, which in fact is the expected behavior. However, the ATE of the third trial shows that the error of the resulting localization is affected by the huge drift of the incremental localization, which is further detailed in Fig. 4. On the other hand, the localization obtained by the proposed fusion method has almost 50% higher RPE than the trajectory provided by the ORB-SLAM2. Nevertheless, the RPE of the developed localization is still significantly lower than the RPE of the unreliable GNSS.

5   Conclusion

The presented results indicate the developed solution of the localization fusion provides an improved estimation of the robot pose with the ATE lower than the localization provided by the unreliable GNSS and also the pose estimation solely based on the visual localization. On the other hand, the localization obtained by the fusion method has higher the RPE than the trajectory provided by the visual localization method. The reported results support the feasibility of the method that seems to be a vital approach to deal with a temporary absent GNSS and also with the drift of the visual localization method.

In the presented experimental scenario, the drift of the visual localization is high mainly in the position estimation. However, we expect that for a very high drift in the robot orientation, the fusion method would probably fail. Therefore, we plan to extend the solution by an additional sensor for measuring the vehicle orientation to handle the orientation drift of the visual localization.

Acknowledgement. This work has been supported by the Technology Agency of the Czech Republic (TAČR) under research Project No. TH03010362.

References

1. Báča, T., Heřt, D., Loianno, G., Saska, M., Kumar, V.: Model predictive trajectory tracking and collision avoidance for reliable outdoor deployment of unmanned aerial vehicles. In: IEEE International Conference on Intelligent Robots and Systems (IROS), pp. 6753–6760 (2018)
2. Collective of authors: Vicon Motion Systems Inc. https://www.vicon.com. Accessed 05 Aug 2018
3. Collective of authors: Intel RealSense Depth Camera D435. https://click.intel.com/intelr-realsensetm-depth-camera-d435.html. Accessed 04 Aug 2018
4. Cui, J.Q., Lai, S., Dong, X., Liu, P., Chen, B.M., Lee, T.H.: Autonomous navigation of UAV in forest. In: International Conference on Unmanned Aircraft Systems (ICUAS), pp. 726–733 (2014)
5. Endres, F., Hess, J., Engelhard, N., Sturm, J., Cremers, D., Burgard, W.: An evaluation of the RGB-D SLAM system. In: IEEE International Conference on Robotics and Automation (ICRA), pp. 1691–1696 (2012)
6. Engel, J., Koltun, V., Cremers, D.: Direct sparse odometry. IEEE Trans. Pattern Anal. Mach. Intell. **40**(3), 611–625 (2018)
7. Fischer, T., Pire, T., Čížek, P., De Cristóforis, P., Faigl, J.: Stereo vision-based localization for hexapod walking robots operating in rough terrains. In: IEEE International Conference on Intelligent Robots and Systems (IROS), pp. 2492–2497 (2016)

8. Forster, C., Pizzoli, M., Scaramuzza, D.: SVO: fast semi-direct monocular visual odometry. In: IEEE International Conference on Robotics and Automation (ICRA), pp. 15–22 (2014)

9. Gauglitz, S., Sweeney, C., Ventura, J., Turk, M., Höllerer, T.: Live tracking and mapping from both general and rotation-only camera motion. In: IEEE International Symposium on Mixed and Augmented Reality (ISMAR), pp. 13–22 (2012)

10. Khoury, H.M., Kamat, V.R.: Evaluation of position tracking technologies for user localization in indoor construction environments. Autom. Constr. **18**(4), 649–656 (2009)

11. Lu, Y., Xue, Z., Xia, G.S., Zhang, L.: A survey on vision-based UAV navigation. Geo-Spatial Inf. Sci. **21**(1), 21–32 (2018). https://doi.org/10.1080/10095020.2017.1420509

12. Menze, M., Geiger, A.: Object scene flow for autonomous vehicles. In: IEEE Conference on Computer Vision and Pattern Recognition (CVPR), pp. 3061–3070 (2015)

13. Mur-Artal, R., Tardós, J.D.: ORB-SLAM2: an open-source SLAM system for monocular, stereo, and RGB-D cameras. IEEE Trans. Robot. **33**(5), 1255–1262 (2017)

14. Mur-Artal, R., Montiel, J.M.M., Tardós, J.D.: ORB-SLAM: a versatile and accurate monocular SLAM system. IEEE Trans. Robot. **31**(5), 1147–1163 (2015)

15. Nowicki, M., Belter, D., Kostusiak, A., Čížek, P., Faigl, J., Skrzypczynski, P.: An experimental study on feature-based SLAM for multi-legged robots with RGB-D sensors. Ind. Robot: Int. J. **44**(4), 320–328 (2017)

16. Olson, E.: AprilTag: a robust and flexible visual fiducial system. In: IEEE International Conference on Robotics and Automation (ICRA), pp. 3400–3407 (2011)

17. Opromolla, R., Fasano, G., Rufino, G., Grassi, M., Savvaris, A.: LIDAR-inertial integration for UAV localization and mapping in complex environments. In: International Conference on Unmanned Aircraft Systems (ICUAS), pp. 444–457 (2016)

18. Piasco, N., Sidibé, D., Demonceaux, C., Gouet-Brunet, V.: A survey on visual-based localization: on the benefit of heterogeneous data. Pattern Recognit. **74**, 90–109 (2018)

19. Pire, T., Fischer, T., Civera, J., De Cristóforis, P., Jacobo Berlles, J.: Stereo parallel tracking and mapping for robot localization. In: IEEE International Conference on Intelligent Robots and Systems (IROS), pp. 1373–1378 (2015)

20. Quigley, M., et al.: ROS: an open-source robot operating system. In: IEEE International Conference on Robotics and Automation (ICRA): Workshop on Open Source Software (2009)

21. Stelzer, A., Hirschmüller, H., Görner, M.: Stereo-vision-based navigation of a six-legged walking robot in unknown rough terrain. Int. J. Robot. Res. **31**(4), 381–402 (2012)

22. Sturm, J., Engelhard, N., Endres, F., Burgard, W., Cremers, D.: A benchmark for the evaluation of RGB-D SLAM systems. In: IEEE International Conference on Intelligent Robots and Systems (IROS), pp. 573–580 (2012)

23. Thrun, S., Burgard, W., Fox, D.: Probabilistic Robotics. MIT Press, Cambridge (2005)

24. Toloei, A., Niazi, S.: State estimation for target tracking problems with nonlinear Kalman filter algorithms. Int. J. Comput. Appl. **98**(17), 30–36 (2014)

25. Usenko, V., Engel, J., Stückler, J., Cremers, D.: Direct visual-inertial odometry with stereo cameras. In: IEEE International Conference on Robotics and Automation (ICRA), pp. 1885–1892 (2016)

26. Wang, C., Wang, T., Liang, J., Chen, Y., Zhang, Y., Wang, C.: Monocular visual slam for small UAVS in GPS-denied environments. In: IEEE International Conference on Robotics and Biomimetics (ROBIO), pp. 896–901 (2012). https://doi.org/10.1109/ROBIO.2012.6491082
27. Wang, F., Cui, J.Q., Chen, B.M., Lee, T.H.: A comprehensive UAV indoor navigation system based on vision optical flow and laser FastSLAM. Acta Autom. Sinica **39**(11), 1889–1899 (2013)

Swarming - R&D and Application

Multi-UAV-Based Reconnaissance and Assessment of Helicopter Landing Points in Manned-Unmanned-Teaming Missions

Human-in-the-Loop Evaluation and Results

Marc Schmitt[(✉)] and Peter Stuetz

Institute of Flight Systems (IFS), University of the Bundeswehr Munich (UBM),
Neubiberg, Germany
{marc.schmitt,peter.stuetz}@unibw.de

Abstract. This article presents implementation aspects and experimental results for a cooperative multi-UAV team sensor and perception system. The presented on-board system is integrated in a R&D full-mission helicopter simulator used for manned-unmanned teaming (MUM-T) research in complex military search & rescue scenarios requiring field landings in uncontrolled and unsafe areas. Thus, to reduce the risk, the presented multi-UAV system is capable of performing highly-automated landing zone reconnaissance and landing point evaluation. Thereby, the presented sensor and perception management system (SPMS) incorporates knowledge on information demands regarding safe landing points and probabilistic reliability estimations of applied perceptive capabilities to derive its own course of actions for the landing zone reconnaissance. Expert knowledge is used to weight the single criteria in the multi-dimensional landing point assessment process while a perceptive trustworthiness estimation is used to handle the uncertainty of the measuring processes when fusing the reconnaissance results to assess the single landing points and to create a landing point recommendation for the helicopter crew. The presented system was evaluated in an extensive human-in-the-loop campaign with German Army Aviators. This paper presents questionnaire-based results gathered during the experimental campaign, focusing on human factors, user acceptance, and system design aspects.

Keywords: MUM-T · Manned-Unmanned-Teaming · Multi-UAV ·
Perception management · Human factors · Human-in-the-Loop

1 Introduction

Recent research aims at the teaming and cooperation of (multiple) Unmanned Vehicles (UxVs) with manned forces to perform complex missions and scenarios. Within this manned-unmanned-teaming (MUM-T) domain, the R&D project CASIMUS (Cognitive Automated Sensor Integrated Unmanned Mission System) investigated means of cooperation of a manned two-seated transport helicopter (H/C) with several reconnaissance UAVs (Unmanned Aerial Vehicles) in military search & rescue oriented

© Springer Nature Switzerland AG 2019
J. Mazal (Ed.): MESAS 2018, LNCS 11472, pp. 265–284, 2019.
https://doi.org/10.1007/978-3-030-14984-0_21

scenarios, e.g. CASEVAC (casualty evacuation) or CSAR (combat search and rescue) [1], whereas classical reconnaissance procedures may not be applicable due to the mission-inherent time-criticality. This is especially challenging in highly vulnerable flight phases such as landing or take-off. Thus, one of the investigated aspects is the usage of UAVs to gather up-to-date recce information on desired landing zones shortly prior the H/Cs arrival. In addition, to increase overall operational flexibility and to avoid typical command & control (C2) latencies, the UAVs shall be guided and controlled by the pilot-in-command (PiC) on LOI 4/5 (levels of interoperability [2]) in a MUM-T fashion from on-board the H/C cockpit, both in terms of mission management and sensor assessment. Figure 1 depicts the C2-principle for a MUM-T setting, showing the H/C cockpit and three UAVs conducting a CASEVAC mission. The UAVs are commanded directly by the PiC (yellow arrows) and the gathered recce results are immediately retrieved and displayed on the multi-function-displays (MFDs) in the H/C cockpit (blue arrows).

Fig. 1. MUM-T principle in our H/C mission simulator at the IFS. The PiC (left) is commanding multiple UAVs to reconnoiter mission critical areas, e.g. the H/Cs flight path (white) and the desired H/C landing zone in the background (red). (Color figure online)

Yet, shifting the C2-loop of the UAVs into the cockpit increases the task spectrum of the pilot-in-command. In contrast to legacy unmanned aerial systems (UAS), there exist no dedicated ground control station during mission. Instead, the PiC is now responsible for the UAVs too and must handle all UAV-related tasks in addition to his conventional range of tasks. Thus, a naïve MUM-T approach without explicitly developed and adopted automation mechanisms managing the unmanned systems, bears the risk to greatly increase and potentially exceed the crews workload which needs to be monitored and addressed in some way [3–5], either by additional crew members or by an associate system on-board of the manned platform [6–11]. A promising way for workload mitigation is the adaption of the task sharing between human and machine depending on the actual load of the crew and the current tactical situation by applying varying levels of automation (LOA, [12, 13]) in the C2 control and feedback loop. Moreover, complex time-critical and mainly perception-driven tasks like *Landing Zone Reconnaissance* (LZR) require the coordinated sequencing and combination of various perceptive capabilities to satisfy the crews information demands on a safe and suitable landing point, e.g. military threats, obstacle situation, or terrain conditions. As the simultaneous guidance & control of multiple UAVs, while at the same time assessing, evaluating, and interpreting their respective sensors and reconnaissance results, certainly exceeds the cognitive and manual handling resources of the crew by far [6, 7, 9], the UAVs must be capable of performing certain tasks in a (semi-)autonomous manner and thereby acquiring and evaluating the respective reconnaissance results by themselves, thus enabling higher LOA.

The remainder of this article is structured as follows: Section 2 explains the general system concept of the multi-UAV perception system POCA (Perception-Oriented Cooperation Agent [14]) for the highly-automated reconnaissance of H/C landing points and its operational principle. The system was deployed and integrated in a full-mission H/C simulator at IFS (cf. Fig. 1) and evaluated in a human-in-the-loop (HITL) experimental campaign with German Army Aviators in selected MUM-T scenarios. Consequently, Sect. 3 states integration aspects and provides an overview of the experimental setup and design. Subsequently, the findings and questionnaire-gathered results of the campaign are presented in Sect. 4, focusing on human factors, user acceptance, and system design aspects. Finally, Sect. 5 concludes the article.

2   Multi-UAV-Based Landing Zone Reconnaissance Concept

2.1   Problem Scope

The task of reconnoitering a landing zone (LZ) is regarded one of the most important mission tasks to perform in typical H/C rescue missions. Thereby, the terms "Landing Zone Reconnaissance" (LZR) or "Landing Site Detection" are used synonymously and interchangeably in aviation to denote the task of reconnoitering a designated area to examine its suitability for take down, incorporating possible threats and physical characteristics of the landing zone. In this regard, performing such a LZR in a full-fledged military search & rescue mission like the aforementioned CSAR or CASEVAC scenarios [1] differs from its civilian counterpart as additional tactical and other mission-critical aspects must be considered and incorporated in the landing point assessment. In the following the main selection criteria for safe and suitable H/C landing points in military applications are outlined based on the requirements stated in [1, 15] and gathered in consultative talks with German Army Aviators. In general, various heterogenous information needs must be considered when reconnoitering a H/C landing zone:

- **Tactical Considerations.** The dominant concern in military scenarios; contains the tactical situation and thus the safety assessment of the LZ for take-down, i.e. if there are any (potential) threats or if the LZ could be easily reached by foes. In addition, mission-critical criteria like mission-achievable or the distance to the mission objective are regarded as part of the tactical assessment.
- **Aeronautical Considerations.** Contains all flight safety related aspects in the LZ; This includes available free space, the obstacle situation at and around the LP for approach and departure, the ground slope, and the surface conditions.
- **Meteorological Considerations.** Contains requirements on the meteorological conditions in the LZ which directly influence the H/Cs capability or risk for landing and take-off; In particular, wind and visibility conditions are a critical concern as well as the prevailing density altitude at the LP.

The specific requirements resulting from each of the points above must be met for a thorough LZR, whereas some of them come with an inherent sequential ordering, i.e. some needs like threat assessment must be satisfied first [14, 16].

Thereby, the data gathered to satisfy these requirements must be regarded with care, as automated perceptive functions are imperfect by design. Thus, their reliability or "trustworthiness" value [17, 18] must be incorporated in the assessment process and in the final rating [16]. Furthermore, the result presentation should reflect these uncertainties to avoid false impressions of landing point safety by the human operators.

2.2 System Concept

In this scope, we developed the on-board team perception system POCA (Perception-Oriented Cooperation Agent [14]) and its integrated signal and data processing algorithms for the highly-automated multi-UAV-based reconnaissance of H/C landing points, thereby incorporating probabilistic reliability estimations of applied perceptive subfunctions [18, 19] in a multi-criteria fusion process for landing point quality assessment [16]. Thereby, the human operator, usually the pilot-in-command (PiC), issues a LZR task to the system in a task-based supervisory control manner [13, 20], along with some external constraints as tactical boundaries and available resources. This task is then analyzed by POCA to derive necessary perceptual and navigational actions to accomplish the task. In addition, the aforementioned external constraints, operational task requirements and interdependencies as well as available UAV team capabilities (extracted from a Perception Resource and Capability Ontology [21–23]) are taken into account for task planning and scheduling. By combing these, the planning agent creates a Perception Task Agenda (PTA) comprised of interleaved perceptual and navigational subtasks. These subtasks are then used to subsequently control and coordinate the underlying automation functions on-board the single UAVs, i.e. their flight management systems (FMS) and sensor- and perception management systems (SPMS) [24]. The gathered reconnaissance results are transmitted to the team leading UAV, which fuses and evaluates them to derive a recommendation on the best suited landing point. Figure 2 depicts this operational principle.

Consequently, a Bayesian Inference based fusion mechanism is used in POCA to probabilistically incorporate the reliability and trustworthiness of the automated perception functions in the landing point assessment and recommendation process [16]. The derived results are then transmitted to the H/C, thus supporting the PiC in selecting the appropriate landing point.

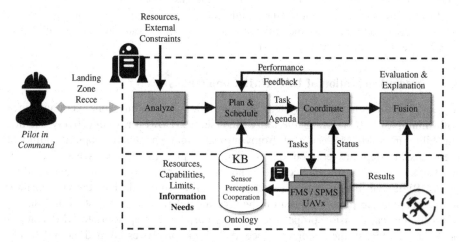

Fig. 2. Basic operational principle of POCA [14]. The notation used in the image is based on the work system principle in [25]. Thus, the supervisory control arrow (green) denotes both control and information feedback flow. (Color figure online)

To allow operation in non-compliant states or in degraded environments, POCA and its integrated signal processing mechanisms are solely on-board the UAVs. Therefore, a connection to the manned H/C is only necessary to assign the LZR task and designate the UAV team composition as well as to feed back the perceptive results to the H/C crew. Thereby, by applying the concepts of task-based guidance [26] and sensor & perception management [24], the need for continuous monitoring and supervision of the UAVs as required by legacy UAV systems is eliminated. Nevertheless, a projective estimation of the UAVs current state is necessary to maintain the pilot's situational awareness and trust in the UAVs, which must be integrated in the aforementioned associate system onboard the manned H/C [27].

Furthermore, POCA tries to mitigate system failures or UAV mishaps by either using on the adaptive capabilities of the underlying SPM system [23, 28] to compensate technical failures as a defective sensor or by re-scheduling and re-tasking the UAV team, thereby trying to complete the LZR task even if a UAV was destroyed. As this could lead to unsatisfactory (or incomplete) reconnaissance results, the data fusion mechanism [16] as well as the result presentation must express this incompleteness or uncertainty and explain the situation to the operator (cf. Sect. 3.1).

3 System Integration and Experimental Evaluation

As stated before, POCA was integrated in the institute's H/C flight and mission simulator (Fig. 1) and evaluated in a vast human-in-the-loop (HITL) experimental campaign with military-trained H/C pilots. Thereby, the focus of the experimental campaign was on the observation of the crew's usage and interaction with the LZR HMI and the crew's evaluation of the single HMI components.

In the following integration aspects and realization of this HMI components are stated. Afterwards, the experimental setup (Subsect. 3.2) is outlined along with some data on the participants (Subsect. 3.3).

3.1 System Integration of LZR HMI Components

In Sect. 2.2, the concept of a system for the on-demand landing zone reconnaissance with (multiple) UAVs during mission was presented. However, to effectively use this capability in a MUM-T context from on-board the manned helicopter, additional preprocessing steps are required to prepare and visualize the reconnaissance results (cf. the landing zone criteria in Sect. 2.1) and the assessment of the individual landing points. By taking into account the uncertainties of the (merged) data, the visualization formats enable the pilots to question the recommendations of the system and understand the critical decision properties. For this purpose, the appropriate HMI components must present the reconnaissance data in a preprocessed and automatically prepared, symbolic form. In addition, the HMI components must be accessible on-demand, allowing the pilots to display and interpret the preprocessed recce data as required during the different mission phases. Furthermore, the HMI should be easy to operate and intuitively understandable for trained H/C pilots in order to keep the additional mental and cognitive load as low as possible.

Therefore, to integrate the LZR functionality in the H/C cockpit and thus allowing the H/C's crew to access and interpret the landing point recommendation and its decision rational, a three-stage HMI was developed and integrated in the tactical map display of the helicopter's Multi-Function-Displays (MFDs). Thus, different representations with varying degrees of detail are available, allowing the crew members appropriately choose the desired level depending on the current flight and mission state.

The first display mode "Tactical Landing Point Symbology" enables the pilots to quickly and easily locate a landing point as well as its reconnaissance status and overall quality, using a slightly modified version of the tactical symbol for a (friendly) helicopter landing site according to MIL-STD-2525C [29] on the tactical map display of the MFD, whereas the inner part of the symbol is color coded according to the quality assessment. Figure 3 illustrates the presentation mode for an exemplary landing zone with three landing points, where one landing point was rated "unsuitable" (red), one landing point "conditionally suitable" (yellow) and one landing point "suitable" (green).

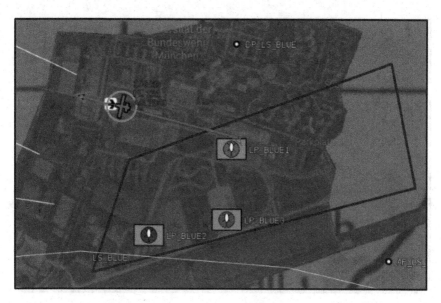

Fig. 3. Display mode "Tactical Landing Point Symbology" for an exemplary H/C landing zone. The landing points are color coded according to their quality assessment. (Color figure online)

The second presentation mode "Landing Point Evaluation Table" shown in Fig. 4 allows the pilots to gain a more detailed insight into the conditions of a landing zone and the landing points therein by providing a tabular comparison. Thereby, the individual reconnaissance results of the assessment criteria (cf. Sect. 2.1) can be compared semantically and textually. Furthermore, to reduce the number of criteria presented and thus to increase clarity and decrease clutter, the interim results of the Bayesian fusion network from [16] are interpreted in the semantic processing step. Pictograms on the left facilitate identification of the individual quality criteria for the currently selected landing point by the operators.

Finally, the third display format in Fig. 5 is an automatically generated landing chart, providing a visual representation of certain features in the vicinity of the selected landing point, like obstacles or possible obstructions in approach or departure. Beside this, wind strength and direction as well as the orientation of the underlying reconnaissance image are displayed. In addition, the assessment rational of the individual evaluation criteria can be queried by selecting the above-mentioned pictograms. A semantic-textual interpretation of the criteria is then displayed on the right-hand side, as illustrated in Fig. 5.

In addition to this dedicated HMI components for presenting LZR results, the tactical map display in the H/C cockpit provides additional means for UAV tasking [30] and for UAV sensor deployment and reconnaissance visualization [10, 11], thereby enabling the crew to access already preprocess and tactically annotated recce results as well as viewing a live video stream of the UAVs cameras.

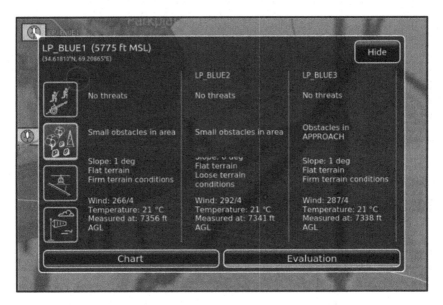

Fig. 4. Presentation mode "Landing Point Evaluation Table" based on the sample landing zone "LS_BLUE". It has been generated for the landing point "LP_BLUE1" (top left).

Fig. 5. Display option "Landing Chart" for landing point "LP_BLUE1". The assessment criteria "Obstacle situation" is currently selected (underlined in green). As a consequence, the semantically and textually processed explanation of the reconnaissance result is shown on the right. (Color figure online)

3.2   Experimental Setup

As stated above, we conducted an extensive human-in-the-loop experimental campaign with four military-trained transport helicopter crews in our R&D full mission H/C simulator at the institute. In our experimental setup, the manned H/C was deployed along with three tactical UAVs (cf. Fig. 1) to perform time-critical missions [1] in an manned-unmanned-teaming (MUM-T) manner as described in Sect. 1.

Therefore, the crew and especially the pilot-in-command (PiC) had to handle all UAV related tasks in addition to his conventional task spectrum. Therefore, each crew conducted a step-by-step tutorial and training phase of about two days at the beginning of every experimental week, in which the different aspects of the examined systems like UAV tasking, UAV sensor and reconnaissance result interpretation, and interaction with the cognitive cockpit automation were trained. Thereby, detailed handling instructions and background information were provided for each of the introduced functionalities and the handling of these were first trained separately in isolated training scenarios. One after another, the single aspects were combined, thereby repeating the already trained functionalities. After finishing the individual parts of the tutorial, two full training missions were carried out in which further handling instructions and explanations were given by the test personnel if required.

The remaining days of the week, the crews performed five complete MUM-T missions on their own, each with a duration about 35–60 min. A military style briefing was held at the beginning of each mission. The missions were then independently carried out by the crews. Thereby, different kinds of missions were executed (e.g. CASEVAC, MEDEVAC, or air transport, see [1]), all of them requiring the crew to perform field-landings in uncontrolled and potentially dangerous areas. Thus, the UAVs-based LZR (cf. Sect. 2.2) was necessary to increase the crew's situational awareness and decrease the risk of the landing zone operations. Three UAVs were available in the missions and had to be deployed by the commander in a task-based guidance style [26, 31] in order to accomplish the various mission tasks as LZR or route reconnaissance [11] using the interactive tactical map interface on the MFD.

In this regard, a LZR command was issued to the UAVs using a context based radial menu [30]. After assigning the task to an UAV, pre-selected LP candidates were extracted from a database, which formed the foundation for the UAV-based LZR. Subsequently, a combined flight and task plan was automatically created by POCA, which was then carried out, whereby the individual evaluation features were acquired, classified and fused (cf. Sect. 2.2).

In order to make the highly-automated landing zone reconnaissance process more understandable for the crew to increasing the transparency of the agent system [32, 33], the UAVs transferred already processed and reconnoitered results during mission execution, thus making the progress traceable on the tactical map of the MFD.

3.3   Participants

Seven helicopter pilots with a German Armed Forces background and military experience took part in the experimental campaign, mixed in four experimental crews. Thereby, one of the pilots took part in two experimental weeks, in which he was PiC in the first week and active pilot (Pilot Flying, PF) in the second. To avoid overly influencing the commander's tactical decisions and therefore the course of the experiment, he was obliged to remain silent in the second week.

The pilots' age span was in the range of 31 to 59 years. In terms of flight and operational experience, the overall flying hours across all pilots (PF + PiC) varied between 535 and 6850 whilst the operational experience was between 0 and 2000 h. Only pilots with more than 100 h of operational experience served as mission commander.

4 Results

In the following, results of the HITL experimental campaign are presented, regarding the highly automated LZR system. Thereby, the HITL experiment focused on evaluating the general system concept (cf. Sect. 2.2) and the UI elements built into the MFDs of the H/C (cf. Sect. 3.1).

For this purpose, each crew member was asked to assess and evaluate the system subjectively by completing an overall questionnaire at the end of the experimental week, covering the suitability of the implemented HMI elements, an assessment of the applicability of the general LZR system and an estimation of the sense and usefulness of the selected point-based reconnaissance approach. Thereby, the crews were asked to state their subjective opinion on the general concept and the behavior of the LZR agent system. An additional questionnaire was used to evaluate the individual interaction modalities to ensure that the assessment of the automation systems core components was not influenced by visualization errors or inadequacies in user interface handling.

The questionnaires were largely implemented with Likert-style scales ranging from 1/negative to 7/positive, with a neutral value of 4 [34]. For some questions that used a negative formulation, both the question and the collected data were inverted to present the results in a uniform way.

In this regard, the crews were first asked to validate the constraints and peculiarities of the individual missions as well as the suitability of the simulation environment to verify the perceived realism and immersion during the experiments (Fig. 6). The scenarios as well as the missions were consistently considered very good and realistic, only the virtual environment was rated somewhat unrealistic but sufficient enough for the purpose of the study.

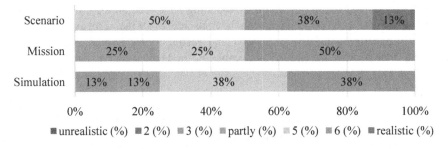

Fig. 6. Degree of realism and immersion of experimental environment.

It can be concluded that the experimental environment was considered realistic and the pilots experienced a high degree of immersion during the experiments. Therefore, it is assumed that the results in the following subsections were not or only to a limited extent influenced by lack of immersion and simulation shortcomings.

4.1   Operational Benefits and Necessity

To determine the possible necessity of highly-automated LZR systems in the examined MUM-T missions, the participants were asked to estimate whether they could have conducted the experimental missions without the UAV-based LZR system (Fig. 7). In addition, a polarity profile of the overall rating was created to assess the operational benefits (Fig. 8).

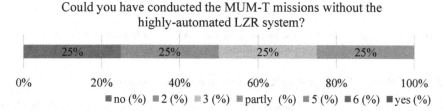

Fig. 7. Need for an automated LZR system in MUM-T operations.

Basically, one can conclude that in a MUM-T context an automated system for landing zone reconnaissance is regarded as very useful or even necessary. Furthermore, it becomes clear that the chosen highly automated system approach was considered to be a straightforward and efficient solution and greatly alleviated the mission's workload from both the active pilots and the commander's perspective.

It became clear during the accompanying interviews that the pilots saw the system as very valuable and a major time-saving advantage, allowing them a direct approach for field landings. In addition, an increase in the pilot's sensation of safety in the landing zone operation was stated.

The participating pilots regarded the automated landing zone reconnaissance as a critical capability, albeit some assumed that they possibly could conduct the operations without an explicit LZR system on the UAVs. When asked again, the high temporal advantage of the system was emphasized, which would be lost both in current off-the-shelf systems and in a hypothetical UAV system without an explicit LZR component similar to the components examined in [11]. As a result, the participants concluded that they would in any case require significantly more time in the operating area and that this would result in an increased workload.

The UAV-based landing zone reconnaissance was...

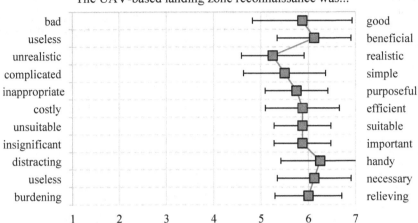

Fig. 8. Polarity profile for the system rating of the UAV-based landing zone reconnaissance. The values ranged from 1/negative to 7/positive.

4.2   Usefulness of the LP-Centered Reconnaissance Approach

As explained in Sect. 3.2, the proposed LZR system always focused on single landing points (LPs) inside the landing zones. Therefore, in addition to evaluating the overall system and the general idea of a UAV-based landing zone reconnaissance, the crews' opinion on the sense of this strongly landing point centered approach was asked as it was realized here and used by the crews in the HITL campaign. Figure 9 shows the resulting rating by the crews as a polarity profile.

Clearly, reducing the reconnaissance problem down to single, dedicated landing points was seen as very reasonable and goal-oriented. The pilots considered it to be a highly intuitive and logical solution, as landing point assessment in current systems basically does the same and evaluates only an area of a few rotor diameters.

The landing point centered approach was...

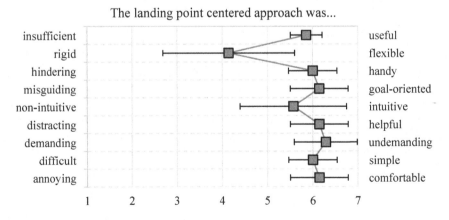

Fig. 9. Rating of the LP-centered reconnaissance approach by the participants. The values ranged from 1/negative to 7/positive.

However, the solely automated generation and pre-selection of the landing point candidates was perceived as somewhat too rigid and the participants expressed the desire to be able to add additional candidates on the MFD's map display on their own.

4.3   Benefits of LZR Criteria Fusion

As stated in Sect. 2.2, the single assessment criteria for H/C LZR are fused probabilistically and the result is expressed as a singular quality value [16]. Therefore, to complete the technical and conceptual evaluation of the UAV LZR system, the probands were asked to assess this fusioning and data merging. Figure 10 shows the rating of the fusion concept and mechanism by the participants.

The fusion in a singular quality value has helped me…

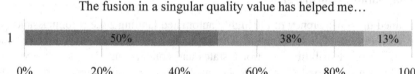

■ completely (%) ■ 2 (%) ■ 3 (%) ■ partly (%) ■ 5 (%) ■ 6 (%) ■ not at all (%)

Fig. 10. Rating of the LZR data fusion by the test pilots.

Most of the test pilots welcomed the reduction of the multiscalar reconnaissance problem. However, one pilot considered the fusion scheme as inadequate. Upon request, it turned out that the poor rating at this point was due to the chosen colors on the map display of the MFD (cf. Sect. 4.6).

4.4   Trust-in-Automation

The term *Trust-in-Automation* [35–37] denotes the human operator's expectations in the reliability and relevance of the actions performed by a technical system. Or, as in

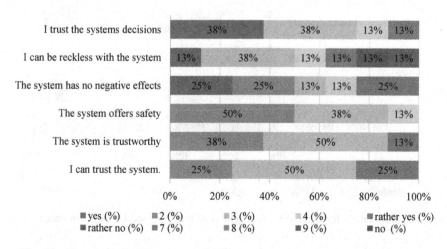

Fig. 11. Trust in the highly-automated LZR system as expressed by the participants.

the case of our LZR system, in the assessments made by an automated system. Figure 11 shows the trust assessment as expressed by the participants for the LZR system.

Generally, it can be observed that the majority of the participating test pilots trusted the decisions and recommendations of the UAV-based LZR and tended to regard the system as trustworthy. However, as the assessment and recommendation of a landing point, particularly in insecure areas, is a critical intervention in familiar mission procedures, there was a certain degree of caution when dealing with the system. Thus, about half of the participants considered that caution was necessary when handling with the systems outcome. However, despite the skepticism shown, the pilots trusted the system in general and considered it to be a real security improvement, thereby coinciding with the findings in the previous sections.

4.5 Automation Transparency

We examined the transparency of the highly automated landing zone reconnaissance as another very important human factor. Thereby, the term transparency refers to the perceivability and plausibility of actions, states and conclusions of an automated system by a human operator [32, 33]. Hence, it also directly influences the previously discussed automation trust. In transparent behavior and systems are generally perceived as less reliable and trustworthy than understandable ones. Figure 12 shows the corresponding assessment results as expressed by the participants in the questionnaires.

The agent system was predominantly considered to be highly transparent. From the pilots' point of view, the system worked reliable and the offered reconnaissance proposals were plausible and understandable to a great extent. We can therefore conclude that the trust assessments considered previously (see Fig. 11) resulted partly from the high degree of system transparency.

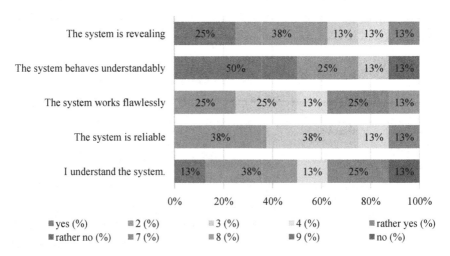

Fig. 12. Transparency of the LZR agent system as assessed by the participants.

Only the soil conditions considered during information fusion were found to be opaque according to one participant. Thereby, the final quality assessment of the fusion mechanism was found to be lacking transparency for the operators. Under this aspect it can thus be stated that the presentation modes must be improved at some points in order to increase the LZR agents self-explanatory capabilities [16] and that a more intense training of pilots is necessary.

4.6 HMI Ergonomics

In order to complete the subjective overall system assessment, it was necessary to assess not only the UAV systems but also the display components on the MFD in the cockpit, which are specifically part of the UAV-based landing zone reconnaissance (cf. Sect. 3.1). A separate questionnaire was used to operationalize the suitability of the proposed man-machine interfaces. Separated questionnaires were used to ensure that potential display errors or handling difficulties with the user interface would not influence the assessment of the actual automation system. This separation of the system components was communicated to the respective crews at the beginning of each experimental week. However, it was not always possible for the test persons to clearly distinguish between the different aspects. Therefore, the evaluation results were partly influenced by each other as stated in Sect. 4.3.

Figure 13 shows the general evaluation of the HMI of the LZR system. As can be seen, the offered presentation formats were generally received positively. The participants rated the interacting modalities of the system both as concise and visually appealing.

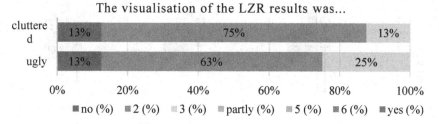

Fig. 13. Evaluation of the human-machine interface of the LZR system.

Additionally, to get a more detailed breakdown of the suitability and to be able to analyze potential shortcomings, another query was conducted using a modified Cooper Harper rating, specifically designed for UAV displays [38, 39]. As the evaluated UI components (Sect. 3.1) consist solely of displays designed for information reception and decision support, not providing any means of controlling interaction, the Human Factors variant "HF MCH-UVD" was deliberately selected [40], which was designed specifically for pure information retrieval displays [39].

Figure 14 shows the evaluation results for the modified Cooper-Harper rating for the different presentation modes. It depicts the mean in blue, the minimum rating value in green (best value), and the maximum rating value in red (worst value). The resulting standard deviations are plotted against the mean values with error bars (black). Additionally, the median is depicted in orange, as the mean value is disproportionally influenced by single opinions due to the small sample size. Furthermore, one participant hadn't rated the "Landing Point Evaluation Table" since he hadn't used it in the conducted experimental missions.

Hereby the previous ratings were confirmed as the test persons mostly rated the proposed display formats as good or very good. Especially the symbolic mode "Tactical Landing Point Symbology" on the tactical map was regarded as very useful by the majority of the participants. In the applied Cooper-Harper rating scheme according [40] *"the display format facilitates effective decision making"* when considering the median value. The evaluation was thus exactly in the range of "Rating 2: *Good with Negligible Deficiencies*". Removing the extrema from the equation results in a mean of 2.33. This lowers the rating slightly towards "Rating 3: *Minor but Tolerable Deficiencies*", which is equal to the rating based on the mean value taking into account all participants and thus including the two extrema.

Similar results were obtained for the second interface proposal, the tabular comparison of the landing points. Most of the pilots considered the presentation mode as good, whereas the rating in the modified Cooper-Harper is similar to the previously analyzed, i.e. the majority of the participants rated the tabular comparison as very suitable.

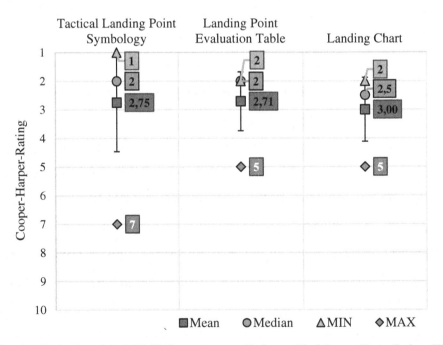

Fig. 14. Evaluation of the LZR HMI components with the modified Cooper-Harter-Ratings HF MCH-UVD [40]. (Color figure online)

The third implemented UI "Landing Chart" had a slightly lower evaluation score. Considering the mean value, *"the display does not sufficiently facilitates efficient decision making"*. Thus, the scoring in the HF MCH-UVD [40] results in "Scoring 3: *Minor but Tolerable Deficiencies*". According to the interviews with the individual pilots, this rating was mainly due to orientation difficulties in the automatically generated approach or landing chart.

5 Conclusion

We implemented a Multi-UAV-based perception system for the highly-automated execution of complex and mainly sensor driven mission tasks, such as the automatic reconnaissance of H/C landing points during mission. Thereby, safe landing zone reconnaissance requires the acquisition, interpretation and assessment of a multitude of heterogeneous decision criteria to determine the suitability of a landing point.

The proposed automation system explicitly addresses these criteria when determining the necessary perceptive tasks to be accomplished and fuses the results in a multi-criteria decision-making process. In addition, the criteria are regarded as pilot's information needs which must be available in the cockpit to support them in their own landing decision. Thereby, the system must be aware of the automation reliability and the trustworthiness of the automatically perceived results, incorporating them for fusion [16] and result representation (Sect. 3.1).

We conducted an extensive human-in-the-loop experimental campaign with military trained H/C pilots to evaluate user acceptance, operational benefits and human factors of the proposed system.

The participants almost exclusively regarded the UAV-based landing zone reconnaissance as a critical and essential component in the examined MUM-T scenarios with very short mission preparation times. The proposed highly automated system approach was generally viewed highly favorable, with a particular emphasis on the time savings due to the high degree of automation and reconnaissance result preparation.

In terms of human factors, the proposed LZR agent system was rated both as having a high transparency and be trustworthy. Thereby, we assumed that the observed high trust and general believe in the decisions and recommendations of the system can be attributed to the high transparency and thus to the self-explanatory capabilities of the reconnaissance results.

For the three proposed HMI components and presentation modes, the overall assessment by the participants was positive. Most of them regarded the proposed user interfaces as suitable for displaying reconnaissance data for landing points and found it very advantageous to be able to choose from different display formats depending on their actual mission state and situational workload. Strong workstation-dependent preferences were found for the detailed display formats. While the in-flight commanders primarily made use of the tabular comparison, the active pilots preferred the automatically generated landing chart.

A need for improvement was identified in the weighting of the evaluation criteria of the fusion system and in the color design on the MFDs in the H/C cockpit.

References

1. NATO Standardization Agency: STANAG 2999 - Use of Helicopters in Land Operations Doctrine ATP-49 (F) (2012)
2. NATO Standardization Agency: STANAG 4586 - Standard Interface of UAV Control System (UCS) for NATO UAV Interoperability (2012)
3. Schulte, A., Donath, D., Honecker, F.: Human-system interaction analysis for military multi-RPA pilot activity and mental workload determination. In: IEEE International Conference on Systems, Man, and Cybernetics (SMC 2015), pp. 1375–1380 (2015)
4. Honecker, F., Schulte, A.: Automated online determination of pilot activity under uncertainty by using evidential reasoning. In: Harris, D. (ed.) EPCE 2017. LNCS (LNAI), vol. 10276, pp. 231–250. Springer, Cham (2017). https://doi.org/10.1007/978-3-319-58475-1_18
5. Brand, Y., Schulte, A.: Model-based prediction of workload for adaptive associate systems. In: 2017 IEEE International Conference on Systems, Man, and Cybernetics (SMC 2017), pp. 1722–1727. IEEE (2017)
6. Strenzke, R., Uhrmann, J., Benzler, A., Maiwald, F., Rauschert, A., Schulte, A.: Managing cockpit crew excess task load in military manned-unmanned teaming missions by dual-mode cognitive automation approaches. In: AIAA Guidance, Navigation, and Control Conference (2011)
7. Honecker, F., Brand, Y., Schulte, A.: A task-centered approach for workload-adaptive pilot associate systems. In: The 32nd EAAP Conference, Cascais, Portugal (2016)
8. Brand, Y., Schulte, A.: Design and evaluation of a workload-adaptive associate system for cockpit crews. In: Harris, D. (ed.) EPCE 2018. LNCS (LNAI), vol. 10906, pp. 3–18. Springer, Cham (2018). https://doi.org/10.1007/978-3-319-91122-9_1
9. Ruf, C., Stütz, P.: Model-driven sensor operation assistance for a transport helicopter crew in manned-unmanned teaming missions: selecting the automation level by machine decision-making. In: Savage-Knepshield, P., Chen, J. (eds.) Advances in Human Factors in Robots and Unmanned Systems. AISC, vol. 499, pp. 253–265. Springer, Cham (2017). https://doi.org/10.1007/978-3-319-41959-6_21
10. Ruf, C., Stütz, P.: Model-driven payload sensor operation assistance for a transport helicopter crew in manned-unmanned teaming missions: assistance realization, modelling experimental evaluation of mental workload. In: Harris, D. (ed.) EPCE 2017. LNCS (LNAI), vol. 10275, pp. 51–63. Springer, Cham (2017). https://doi.org/10.1007/978-3-319-58472-0_5
11. Ruf, C., Stütz, P.: Establishing a variable automation paradigm for UAV-based reconnaissance in manned-unmanned teaming missions. In: Chen, J. (ed.) AHFE 2018. AISC, vol. 784, pp. 24–35. Springer, Cham (2019). https://doi.org/10.1007/978-3-319-94346-6_3
12. Parasuraman, R., Sheridan, T.B., Wickens, C.D.: A model for types and levels of human interaction with automation. IEEE Trans. Syst. Man Cybern.-Part A: Syst. Hum. 30, 286–297 (2000)
13. Sheridan, T.B.: Adaptive automation, level of automation, allocation authority, supervisory control, and adaptive control: distinctions and modes of adaptation. IEEE Trans. Syst. Man Cybern.-Part A: Syst. Hum. 41, 662–667 (2011)
14. Schmitt, M., Stuetz, P.: Perception-oriented cooperation for multiple UAVs in a perception management framework: system concept and first results. In: 2016 IEEE/AIAA 35th Digital Avionics Systems Conference (DASC). IEEE, Sacramento (2016)
15. U.S. Department of the Army: Attack Reconnaissance Helicopter Operations (FM 3-04.126) (2007). http://usacac.army.mil/sites/default/files/misc/doctrine/CDG/cdg_resources/manuals/fm/fm3_04x126.pdf

16. Schmitt, M., Stütz, P.: Multi-UAV based helicopter landing zone reconnaissance. In: Harris, D. (ed.) EPCE 2017. LNCS (LNAI), vol. 10276, pp. 266–283. Springer, Cham (2017). https://doi.org/10.1007/978-3-319-58475-1_20

17. Russ, M., Stuetz, P.: Application of a probabilistic market-based approach in UAV sensor & perception management. In: 2013 16th International Conference on Information Fusion (FUSION 2013), Istanbul, pp. 676–683 (2013)

18. Hellert, C., Stuetz, P.: Performance prediction and selection of aerial perception functions during UAV missions. In: 2017 IEEE Aerospace Conference. IEEE, Big Sky (2017)

19. Russ, M., Schmitt, M., Hellert, C., Stuetz, P.: Airborne sensor and perception management: experiments and results for surveillance UAS. In: AIAA Infotech@Aerospace (I@A) Conference, Guidance, Navigation, and Control and Co-located Conferences, (AIAA 2013-5144), pp. 1–16. AIAA, Boston (2013)

20. Sheridan, T.B.: Telerobotics, Automation, and Human Supervisory Control. MIT Press, Cambridge (1992)

21. Hellert, C., Smirnov, D., Stuetz, P.: Ontologiedesign für Sensor- und Perzeptionsfähigkeiten von UAVs. In: Deutscher Luft- und Raumfahrtkongress 2014. Deutsche Gesellschaft für Luft- und Raumfahrt - Lilienthal-Oberth e.V., Augsburg (2014)

22. Smirnov, D., Stuetz, P.: Knowledge elicitation and representation for module based perceptual capabilities onboard UAVs. In: AIAA SciTech 2014. American Institute of Aeronautics and Astronautics, National Harbour (2014)

23. Smirnov, D., Stuetz, P.: Ontology-based modelling of sensor and data processing resources using OWL: a proof of concept. In: The First International Conference on Applications and Systems of Visual Paradigms. IARIA, Barcelona (2016)

24. Russ, M., Stuetz, P.: Airborne sensor and perception management: a conceptual approach for surveillance UAS. In: Proceedings of the 15th International Conference on Information Fusion (FUSION 2012), pp. 2444–2451. IEEE, Singapore (2012)

25. Schulte, A., Donath, D., Lange, D.S.: Design patterns for human-cognitive agent teaming. In: Harris, D. (ed.) EPCE 2016. LNCS (LNAI), vol. 9736, pp. 231–243. Springer, Cham (2016). https://doi.org/10.1007/978-3-319-40030-3_24

26. Uhrmann, J., Schulte, A.: Concept, design and evaluation of cognitive task-based UAV guidance. Int. J. Adv. Intell. Syst. 5, 145–158 (2012)

27. Meyer, C., Schulte, A.: Concept for satisfying pilot information demands in MUM-T under selective datalink availability. In: Proceedings of the 33rd European Association for Aviation Psychology Conference (EEAP33). European Association for Aviation Psychology, Dubrovnik (2018)

28. Hellert, C., Stuetz, P.: A concept for adaptation of perceptual capabilities for UAV platforms using case-based reasoning. In: AIAA SciTech 2014, 52nd Aerospace Sciences Meeting, pp. 1–8. American Institute of Aeronautics and Astronautics, National Harbour (2014)

29. U.S. Department of Defense: MIL-STD-2525C - Common Warfighting Symbology (2008). http://mapsymbs.com/ms2525c.pdf

30. Rudnick, G., Schulte, A.: Implementation of a responsive human automation interaction concept for task-based-guidance systems. In: Harris, D. (ed.) EPCE 2017. LNCS (LNAI), vol. 10275, pp. 394–405. Springer, Cham (2017). https://doi.org/10.1007/978-3-319-58472-0_30

31. Uhrmann, J., Strenzke, R., Schulte, A.: Task-based guidance of multiple detached unmanned sensor platforms in military helicopter operations. In: COGIS (COGnitive systems with Interactive Sensors), Crawley, UK (2010)

32. Billings, C.E.: Human-Centered Aviation Automation: Principles and Guidelines. NASA, Moffett Field (1996)

33. Seong, Y., Bisantz, A.M.: The impact of cognitive feedback on judgment performance and trust with decision aids. Int. J. Ind. Ergon. **38**, 608–625 (2008)
34. Likert, R.: A technique for the measurement of attittudes. Arch. Psychol. **140**, 1–55 (1932)
35. Lee, J.D., Moray, N.: Trust, self-confidence, and operators' adaptation to automation. Int. J. Hum Comput. Stud. **40**, 153–184 (1994)
36. Lee, J.D., See, K.A.: Trust in automation: designing for appropriate reliance. Hum. Factors J. Hum. Factors Ergon. Soc. **46**, 50–80 (2004)
37. Sharit, J.: Human error and human reliability analysis. In: Salvendy, G. (ed.) Handbook of Human Factors and Ergonomics, pp. 734–800. Wiley, Hoboken (2012)
38. Cummings, M.L., Meyers, K., Scott, S.D.: Modified cooper harper evaluation tool for unmanned vehicle displays. In: Proceedings of UVS Canada: Conference on Unmanned Vehicle Systems Canada, Montebello, PQ, Canada (2006)
39. Donmez, B., Cummings, M.L., Graham, H.D., Brzezinski, A.S.: Modified cooper harper scales for assessing unmanned vehicle displays. In: Proceedings of the 10th Performance Metrics for Intelligent Systems Workshop on - PerMIS 2010, pp. 235–242. ACM Press, Baltimore (2010)
40. Donmez, B., Brzezinski, A.S., Graham, H.D., Cummings, M.L.: Modified Cooper Harper Scales for Assessing Unmanned Vehicle Displays. MIT Humans and Automation Laboratory, Cambridge (2008)

M&S-Based Robot Swarms Prototype

Marco Biagini and Fabio Corona[⊠]

NATO Modelling and Simulation Centre of Excellence, 00143 Rome, Italy
{mscoe.cde01,mscoe.cde04}@smd.difesa.it

Abstract. The future operational environment will be characterized by the employment of Unmanned Autonomous Systems (UAxS) on the battlefield. Contemporary technology is still not mature enough to support this vision and additional development and experiments are necessary. In particular, in the area of the interaction between the human beings and robots, taking into consideration the role of the military Command and Control (C2) Operational Function.

The NATO Modelling and Simulation (M&S) Centre of Excellence is going further in this area and in collaboration with Industry and Academia has developed a prototype of a M&S-based platform built on OPEN standard architecture simulating operational environments to support experiments regarding robot swarms operating mainly in the air and ground domains.

The platform, codename R2CD2 project, is built on selected constructive simulators and it includes the C2SIM interoperability language extension which enables interoperability between C2 systems and simulated robotic entities.

This paper presents the results of the first implementation of a platform prototype suitable to support the standardization and operationalization of the C2SIM language and its extensions to UAxS, through proof of concepts and M&S-based experimentation activities.

In conclusion, this paper contains a detailed technical overview of the prototype architecture and its implementation. Further prototype development is concentrated on several operational scenarios, several robot swarms behaviors, missions and tasks according to the robot level of autonomy within the operational vignettes.

Keywords: Unmanned Autonomous Systems · C2SIM · R2CD2 · Modelling and Simulation

1 Introduction

The increasing employment of Unmanned Autonomous Systems (UAxS) in different operational domains, either as conventional forces or as improvised weapons exploited by insurgents or terrorists, is a rising concern for North Atlantic Treaty Organization (NATO). Aim of the NATO Modelling and Simulation Centre of Excellence (M&S COE) is to support Allied Command for Transformation (ACT) facing such threat, leveraging Modelling and Simulation (M&S) technology in order to perform Concept Development and Experimentation (CD&E) on UAxS employment in the modern urban battlefield [1]. This activity is critical in the development process of new future military capability able to both employ and counter robotic systems [2]. M&S is a great enabler to do experimentation on new systems and weapons, doctrine, training, military

© Springer Nature Switzerland AG 2019
J. Mazal (Ed.): MESAS 2018, LNCS 11472, pp. 285–301, 2019.
https://doi.org/10.1007/978-3-030-14984-0_22

organization and logistics, in a cost-efficient way [3, 4]. In particular, M&S COE concentrates on four main areas of investigation:

- the interaction between human troops and robots;
- the military Command and Control (C2) of robotic units;
- technical tactical procedures for UAxS;
- development of functional requirements of new robotic platforms;
- countering UAxS;
- in an urban environment in the near or mid-term future.

A first M&S solution to address the needs for experimentation of concepts regarding UAxS operational deployment was already conceptualized back in 2016 [5]. Afterwards, a methodology for the development of operational scenario in an urban environment was developed and employed [6], in order to find requirements for further design of the necessary simulation environment.

Therefore, in the framework of the "Research on Robotics Concept and Capability Development (R2CD2)" project, an M&S-based, scalable and modular platform was built on open standard architecture, by the NATO M&S COE in collaboration with the Industry and Academia. This platform was based on selected constructive simulators to execute scenarios on the military employment of robots in order to be an innovative tool for proof-of-concept activity on robotic capability for NATO, like Concept Development Assessment Game (CDAG) [7] and Disruptive Technology Assessment Game (DTAG) [8]. M&S technology allowed to reuse models, proof-of-concept prototypes, systems, studies, developed for different projects, saving precious resources.

In particular, the M&S platform illustrated in this paper made use of the Level of Autonomy (LoA) concept developed during the Autonomous Systems Countermeasures (C-UAxS) project [9] of the ACT for the behaviour and the kind of human-robot interaction. The urban model of a mega city of the future, built during the ACT Urbanization Project (UP) [10] by the NATO M&S COE was employed for the terrain generation. The New C2SIM interoperability language standard [11] was included in the platform architecture for interoperability between C2 systems and simulated robotic entities. Within this project the first implementation of C2SIM extension to UAxS [12] was developed.

So, in the second section the scenario developed is illustrated after having recalled the process followed for requirements of scenario and extension of C2SIM to UAxS. In the third section the C2SIM interoperability language is introduced, while in the fourth section the architecture and components of the M&S platform are described.

Finally, the results of this study and experimental activity are presented, together with the way ahead already delineated.

2   R2CD2 Scenario

Following the Simulation Interoperability Standards Organization (SISO) Guideline on Scenario Development for Simulation Environments (GSD) [13], NATO M&S COE has defined a scenario development process which was already well outlined in [6] and applied into the R2CD2 project as detailed in [12].

Here only the main elements of this process are recalled in order to illustrate the R2CD2 project scenario and introduce the results on the M&S platform, in terms of conceptual and executed scenarios and messages implemented for the interoperation of systems.

2.1 Scenario Objectives

The main simulation objectives for the R2CD2 project scenario are:

- interaction between simulated UAxS and real C2 systems;
- study of UAxS employment in a megacity of the future;
- consideration of two operational domains for UAxS (Air and Land);
- use of C2SIM Interoperability Language for messages (Orders & Reports) and study of requirements for C2SIM extension to UAxS, either live or simulated.

Features peculiar of UAxS, such as the level of interaction with humans and decision making capacity, should be identified and included as new data elements in the C2SIM messages. The terrain should be an example of future urban environment.

2.2 Operational Scenario

An unknown vehicle loaded with explosive, moves around in a mega city of the future. After the detection of the danger by a police patrol, small units of robotic autonomous systems are alerted to face the threat. In particular, in the air domain, a swarm of Unmanned Aerial Vehicles (UAVs) performs a reconnaissance mission to find the threat and report about its position. In the land domain, a team of Counter Improvised Explosive Devices (C-IED) Unmanned Ground Vehicles (UGVs) is ordered to neutralize the threat. Each team has its own command post which assigns the mission to be performed autonomously by the robots according to a level of autonomy appropriate for the mission. The autonomous systems report back to their command posts according to the level of autonomy assigned.

2.3 Problem Space

A first structured description of all scenarios to be executed in the simulation environment is called "problem space". It sets the common features of all possible scenarios which satisfy the simulation objectives and the operational scenario. The following elements make the problem space:

1. Terrain

Since the terrain should be a mega city of the future (2035), a piece of the ARCHARIA model developed for the ACT Urbanization Project (UP) [10] is used. The model is characterized by a lot of envisioned problems for the future (e.g., overpopulation, high density buildings, exposure to natural disasters, like a volcano or coastal tsunami).

2. Order of Battle

The composition of enemy and friendly forces has to be set, in terms of number, level and type of units, with weapons and equipment as well. The type of Unmanned Autonomous Vehicles has to be defined, i.e. their size category and application area, as well as their payloads, like weapons and/or sensors. For this case, UAVs are Small platforms (from 3 to 10 kg), equipped only with electro-optical sensor. The UGVs are more robust and built for C-IED missions, like light armored and equipped with explosive.

The enemy unit is a blue truck loaded with explosive.

3. Enemy Course of Action (ECOA)

In this scenario, the hostile truck moves around the city, trying to not be detected, only to stop near a sensible spot and explode.

4. Level of Autonomy (LoA) of the UAxS

The Level of Autonomy (LoA) for unmanned autonomous platform fixes the behavior of the UAxS and their interaction with their own command posts, like: the rate of reports; if they need coordination and/or confirmation on the tasks to be performed; if they can make their own decisions based on the feedbacks from the environment. In order to define the LoAs, NATO M&S COE reuses the results of the Autonomous Systems Countermeasures (C-UAxS) project [9] of the ACT. Seven levels of autonomy for UAxS have been defined, numbered from 0 to 6. These levels consider systems starting from "human controlled" to "fully autonomous", based on the degree of human interaction while performing their tasks and not only on the level of technology. A LoA set the ability of the autonomous system to tackle the problems connected with the mission complexity or the difficulty of environment.

3 C2SIM Interoperability Language

The "Command and Control Systems to Simulation Systems Interoperation (C2SIM) is a family of standards for expressing and exchanging Command and Control (C2) information between C2 systems, simulation systems, and robotic and autonomous (RAS) systems in a coalition context" [14]. The C2SIM replaces the Coalition Battle Management Language (CBML) [15–17], for describing task and report messages in operational or simulation systems [18], and the Mission Scenario Definition Language (MSDL) [19], for initializing the operational environment (terrain, units, weather conditions, COAs, simulation checkpoints, etc.) in a wide variety of simulation and connected systems. C2SIM is developed beginning from a core Logical Data Model (LDM), which provides at logical level a set of data elements common to most C2 and simulation systems, combined with a standard way to adding to that core a collection of additional elements specific to a particular domain.

Therefore C2SIM was considered the best choice for a language for Command and Control of UAxS. The R2CD2 project provided the right platform to generate requirements for extending the C2SIM LDM to autonomous systems.

The scenario development process recalled in Sect. 2 was used to design a scenario for a first requirements' definition for extending the C2SIM interoperability language to the UAxS domain in collaboration with the Fraunhofer Institute (GER) [20] in 2017. This work was performed in the framework of the "Operationalization of Standardized C2-Simulation Interoperability (C2SIM)" tasking activity (MSG-145) [11] of the NATO M&S Group of NATO Science and Technology Organization (STO). Thereafter, this consolidated development process was applied again for extract a much more extensive range of requirements for C2SIM extension to UAxS during the R2CD2 project.

For the generation of the messages needed for the interoperation between command posts and simulators in the executable scenario, CBML eXtensible Markup Language (XML) schemas were extended, since at that time C2SIM core schemas were not yet available.

In the results section, this paper goes into details of the data elements that extend the CBML XML schemas to autonomous systems and that NATO M&S COE proposed to include in the C2SIM UAxS extension ontology under development. These data elements were extracted from all required information to be inserted in the exchanged messages, as defined in the conceptual scenario.

4 M&S Platform Prototype

In this section, the logical architecture of the simulation environment of the R2CD2 project is illustrated. All the simulators are Commercial-of-the-Shelf (COTS) with Artificial intelligence (AI) add-ons for generating the UAxS behavior. In this first prototype two simulators for UAxS were included: a simulator for Unmanned Ground Vehicles (UGV); a simulator for Unmanned Aerial Vehicles (UAV). All the messages were in CBML format with data element extension developed for UAxS. They were exchanged between simulators through a shared file system and translated by the developed ad-hoc interfaces for each simulator. A real C2 system is included in the architecture with the role is to display the Common Operational Picture (COP). This C2 system was not equipped with an interface for CBML and the consequences were twofold. Firstly, a system to generate orders for the UAV simulator was necessary, so an UAV Air Task Order (ATO) Graphical User Interface (GUI) was developed for this purpose. Secondly, the C2 system was fed using NFFI [21] by a DIS/HLA-C2 gateway, the LVC gateway. It translated DIS/HLA [22, 23] information to a NFFI feed and shared entities and events between the two simulators through both DIS and HLA, since the same HLA implementation could not be used. Anyway, the HLA RTI was included also with the further developments in mind, to enlarge the architecture in a modular way to include other specific simulators, like a communications, networks and cyber effects simulator or others for weapon systems and weather effects.

In details, the components of the platform are (see Fig. 1):

– SitaWare HQ by SYSTEMATIC as real C2 system;
– LVC Gateway by VITROCISET, as DIS/HLA-C2 Gateway;

- Sword by MASA as simulator for UGVs with Artificial Intelligence (AI) add-on modules for autonomous behaviour and its BML connector;
- a TranslatorBML, which is a piece of software for generating orders for UGVs from information read into reports from UAVs.
- VR-Forces by VT MÄK as simulator for UAVs with AI add-on modules for flight behaviour and sensor feed management;
- CBML Parser for the UAV simulator, as translator of CBML orders/reports to/from the UAV simulator;
- UAV ATO GUI for generation of the ATO (waypoints, parameters for flight formation, LoA, sensor, etc.)
- An HLA Run-Time Infrastructure (RTI) [23, 24].

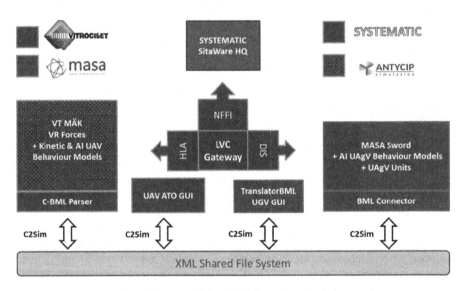

Fig. 1. Logical architecture of the R2CD2 project simulation environment

5   R2CD2 Project Results

The R2CD2 project results are represented by the simulated scenario and the implemented messages to allow the interoperation between systems of the M&S prototype. Firstly, the conceptual scenario which descends from the requirements defined in Sect. 2 is illustrated. Then its execution on the R2CD2 platform is shown. Finally, the new data elements of the messages exchanged between the R2CD2 platform systems are presented, building the first implementation of the C2SIM extension to UAxS.

5.1 Conceptual Scenario

The conceptual scenario is an implementation-independent representation of a single scenario which satisfies all the requirements defined in the "problem space".

Initialization
The initialization of the conceptual scenario can be described by the following items.

1. Order of Battle

   Red Forces
   An hostile truck, blue, with a known plate, loaded with explosive, moves around a quarter of a future mega-city.
   Blue Forces

   – a swarm of three small UAVs, equipped with electro-optical sensor for reconnaissance missions;
   – a team of five light armoured mechanized UGVs, for counter-IED missions.

2. LoA

The Level of Autonomy of UAVs is equal to 3 (ref CUAxS): they are tasked with a mission and they can elaborate intermediate tasks, but always ask the humans for confirmation. They always report back or display to humans the feedbacks of their sensors.

The Level of Autonomy of UGVs is equal to 5 (ref CUAxS): they are tasked with a mission, they perform it according to tasks which they elaborate autonomously and don't ask to humans for confirmation. They report to humans at the end of the mission.

3. Terrain

The terrain is a quarter of ARCHARIA.

4. ECOA

The hostile truck moves around the quarter of the city, trying to not be detected, only to stop near a sensible spot in order to explode.

Vignette
In the following the flow of the actions and information exchanged during the execution of the scenario are described.

1. The UAV command post (CP) receives an police alert about a truck bomb moving in an area of the city, thus it orders to the UAV swarm to search and follow the truck in the area (with an ATO) and send information on the target.
2. UAVs search autonomously the target in linear formation, initially following waypoints inserted into the ATO, then searching the target in circular concentric trajectories according to implemented technical tactical procedures.
3. While moving, the UAVs report about their own status in a General Status Report.
4. When the truck is found one UAV follows it, sends back a video streaming of target and reports target's position and status (moving) in a Target Report.

5. The UGV CP orders to counter-IED UGV team to Be Prepared To take action (BPT order).
6. When the target stops the UAV hovers over it and reports back target's position and status (holding) in Target Report.
7. The UGV CP orders to counter-IED UGV team to reach the location of the truck and to disposal the target.
8. The UGV team performs autonomously the mission (clear area and disposal of the target) and reports back the outcome.
9. The UGV team return to the base autonomously.
10. UAV swarm receives the order to return to the base by its UAV CP and it does it.

Using the NATO Architectural Framework (NAF) v.3 methodology, the "operational activity model" or OV 5 (in Fig. 2) is used to represents the model of the activities performed by the operative nodes with the flow of the information. The sequence of the activities and tasks is identified with the message exchange and the events and messages which trigger different actions.

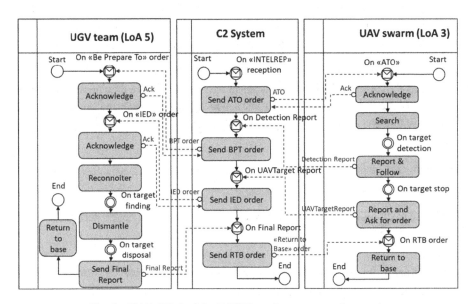

Fig. 2. NAF OV 5 of the R2CD2 project conceptual scenario

Passing from the conceptual scenario to the executable scenario, in the following the different phases of the scenario as executed on the systems are presented.

5.2   Executed Scenario

Hereafter, in several pictures, the outcomes of the different phases of the scenario on the simulators and systems are shown.

In Fig. 3 the police alert is generated in VR-Forces, so in the ATO C2 GUI system the ATO order for the UAV swarm to search and follow the truck in the area is created.

Fig. 3. ATO order creation in UAV C2 GUI

The UAVs search autonomously the target in linear formation following waypoints inserted into the ATO. While moving, the UAVs report about their own status in a General Status Report (Fig. 4).

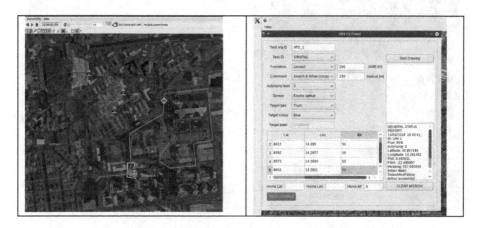

Fig. 4. UAV swarm flying in linear formation and generating General Status Report

In Fig. 5 UAV2 follows the truck, when the target is found, and updates the Target Report with the truck position and status (moving).

Fig. 5. One UAV follows the suspicious truck and updates the Target Report

When the target stops the UAV2 hovers over it and reports back target's position and status (holding) in Target Report (Fig. 6).

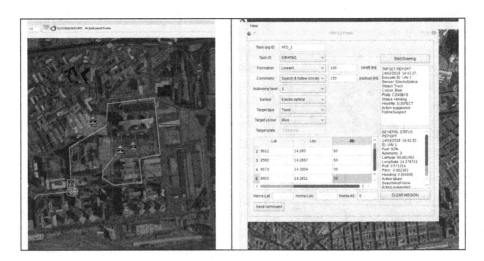

Fig. 6. UAV n.2 hovers over target and reports its final position

The UGV team performs autonomously the mission to reach the location of the truck and to disposal the target when it receives the order in MASA Sword simulator (Fig. 7).

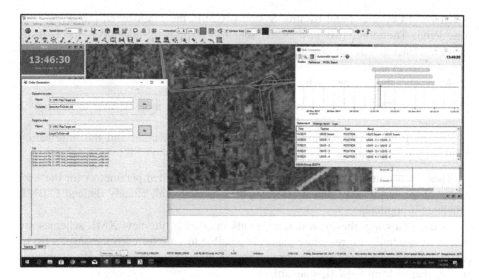

Fig. 7. UGV team performs fully autonomously the mission

UAV swarm receives the order to return to the base from UAV C2 GUI and it does it (Fig. 8).

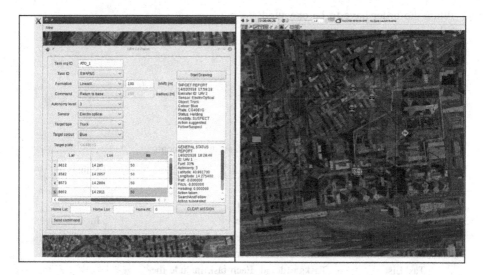

Fig. 8. UAV swarm returns to home base at order

5.3   Messages in R2CD2 Project

As already said, starting from the requirements for information exchange in the R2CD2 project scenario, new data elements were defined. All the messages among autonomous systems implemented in the VT MÄK VR Forces simulator and between simulation systems and Command and Control Systems (C2) in the M&S-based prototype were considered. Therefore, two new XML schemas have been developed for C-BML messages, since it was not possible to extend the new C2SIM standard schemas which were not yet available at the time the project started.

As results, the two new schemas were for:

– ATO (Air Task Order) messages to instruct one or more UAVs to perform a mission;
– Reports from UAVs sending back their status and position and the information about a possible threat detected.

These two schemas were built considering all the needed parameters for the execution of the UAV tasks and reusing, as data elements, all already defined C-BML structures if applicable.

In the following, the main data elements inserted in the new XML schemas are described in details. In (ref white paper) is possible to find, for each message, a list of parameters together with their description and an indication if the single data element is new or reused from C-BML standard.

Air Task Order (ATO) Message

The ATO message is used to instruct an UAV or a swarm of UAV to perform certain tasks. The message format is described for its main parts in the following Table 1.

An ATO contains the following information:

– When the order have to be executed;
– Who have to perform the task;
– What has to be done;
– With what the task has to be performed;
– What target has to be looked for.

Table 1. ATO message data elements

Type	Elements	Description	Values
ATO	AtoIssuedWhen	ATO date	
	AtoID	ATO Identification	
	AutonomyLevel	UAaS autonomy level	Int 0–6
	MinFuelLevel	Min acceptable fuel level	0–100%
	MinLinkQuality	Min acceptable communication link level	0–100%
	TaskList	Tasks ordered. Each task include the information below	
	TaskeeWhoRef	Who that is to carry the task	

(*continued*)

Table 1. (*continued*)

Type	Elements	Description	Values
		What is to be done	– Attack
			– move
			– Loiter
			– Hold
			– Search
			– Search and Follow
			– Follow
			– Home
	UAVFormation	Which type the formation the UAaS s are in	– Delta
			– Line
			– Column
	UAVBehaviour	Which type of behaviour the UAaS s are keeping	– Careless
			– Stealth
			– Aware
	TaskHow	Which type of sensor the UAaS s use to perform the task	– EO
			– IR
			– LIDAR
			– SAR
	WhatTarget	What target the UAaS s are looking for	
	Route	The route to follow to perform the task	List of waypoints

Report Message

The Report message is sent by a single UAV or a swarm of UAVs with information about what they are doing, their position, what they suggests to do, enemy data. The message format is described for its main parts in the following Table 2.

A Report contains the following information:

- Who is sending the report
- When the report is sent
- What the report is about
- With what the report info have been collected
- What target has been found.

Table 2. Report messages data elements

Report types	Elements	Description
General status report	TypeOfReport	The type of the report
	ReporterWho	Who is sending the report
	ReporterWhen	When the report is sent
	FuelLevel	Fuel level (0–100%)
	AutonomyLevel	Autonomy level (0–6)
	LinkQuality	Communication Link level (0–100%)
	StatusWord	Status information
	Position	The current position of the UAV
	Attitude	The current attitude of the UAV
	ActionTaken	What the UAV is going to do
	ActionSuggested	What the UAV suggests to do
	UAVFormation	The formation the UAVs are in
Target report	TypeOfReport	The type of the report
	ReporterWho	Who is sending the report
	ReporterWhen	When the report is sent
	SensorType	Which kind of sensor has been used to discover the target
	TargetStatus	Status of the target (Moving or Holding)
	TargetInfo	Information about the discovered target
	Position	The target position
	Type	Target Type (app6c code)
	Additional Info	Additional Info about the target
	Reliability	Reliability of the report (0–100%)
	Hostility	Hostility of the target (0–100%)
	ActionTaken	What the UAV is going to do
	ActionSuggested	What the UAV suggests to do
UnexpectedEventReport	TypeOfReport	The type of the report
	ReporterWho	Who is sending the report
	ReporterWhen	When the report is sent
	Position	The target position
	Type	Event Type
	Additional Info	Additional Info about the target
	Reliability	Reliability of the report (0–100%)
	ActionTaken	What the UAV is going to do
	ActionSuggested	What the UAV suggests to do
CBRN report	TypeOfReport	The type of the report
	ReporterWho	Who is sending the report
	ReporterWhen	When the report is sent
	Position	The target position
	Agent	Radioactive material type
	Concentration	Agent concentration (0–100%)
	Reliability	Reliability of the report (0–100%)

6 Way Ahead

The future developments of the M&S-based prototype for robot swarm will include:

- use a real C2 system to generate order and read reports, thanks to a C2SIM interface.
- distribution of the messages with a server/client architecture in order to have a real distributed architecture over the network [25]. For this the use of the C2SIM Reference implementation Server developed by the George Mason University (GMU) [26] is already planned, but the UAxS extension needs to be implemented in it.
- moving from the CBML XML schemas to the C2SIM XML schemas.

The ambition is to obtain an M&S-based platform which potentially addresses and solves issues regarding:

- Detection and Identification of enemy robotic units utilizing UAxS and sensors within a Communication, Networking and Cyber (CN&C) environment;
- Situational Awareness augmented with Artificial Intelligence (AI) Decision Support;
- Defense against UAxS applying new capabilities and related TTP's.

In particular the idea for the evolution of the R2CD2 project is to experiment about countering UAxS with a defensive system based on a safety bubble were, according to the proximity of the threat, the countermeasures increase from non-kinetic (jamming or capturing) to kinetic (shooting). A Data Base of models for as many as possible UAxS platforms and their countermeasures would allow for an as much comprehensive as possible experimentation Implemented models of countermeasures to UAxS could be jammers, energy weapons, UAxS with capturing or firing capabilities, conventional fire shooting weapon systems. Also cyber countermeasures could be available, since the cyber domain could not be excluded from the studies. An additional communications and network simulator will be included in the architecture for traffic generation and experimentation on communication protocols, procedures, cyber effects and counter-measures on the Electromagnetic spectrum.

Another desired simulation capability of the future M&S-based platform would be the collection and assessment of data from various networked sensors to detect, identify and locate hostile UAxS with timely reports to real C2 systems.

The use of Artificial Intelligence (AI) is key to support the decision making in an operational scenario dominated by robots, like speeding up the choice of the proper countermeasure or providing alternative secure exfiltration or attack routes both for humans and UAxS, according to the near real time provided situational awareness. For this reason a decision making module is planned to be paired with the C2 system in order to generate orders to simulated UAxS.

Finally, further use of the obtained results under the M&S as a Service paradigm [27] is desirable, as well as the capability to create interactions between live and simulated UAxS.

7 Conclusions

In conclusion, this paper presents a detailed technical overview of the architecture and implementation of an M&S prototype for CD&E activity on UAxS employed in urban operational scenarios. This prototype allows experimentation on new robotic platforms, their tactical procedures, their sensor payload, behaviours, missions and tasks according to their level of autonomy within the operational vignettes.

Moreover, the basis for extending the C2SIM interoperability language Logical Data Model to autonomous systems were built, defining new data elements which NATO M&S COE proposes to include in the C2SIM extension to UAxS ontology.

Acknowledgments. The authors thank all the Industry partners for the fruitful collaboration in the "Research on Robotics Concept and Capability Development (R2CD2)" project. In particular, the VITROCISET company which is the main author of the first demo implementation of the CBML XML schemas extension to autonomous systems and developer of the air domain part of the R2CD2 simulation environment. Moreover, many thanks go to MASA Group company, which developed all the land domain part of the same platform, making available also their simulator for the R2CD2 project. It cannot be forgotten that the results of this project would not be achieve without the availability of companies as Antycip and Systematic, which provided their products in the framework of technical agreements, according to the Open SimLab paradigm of the NATO M&S COE, a business model which foster collaboration and resource pooling among Government, Industry and Academia. Special thanks also to the C4I&Cyber center of George Mason University (Fairfax, VA, USA) always very supportive during further experimentation of the extension.

References

1. Hodicky, J.: Modelling and simulation in the autonomous systems' domain – current status and way ahead. In: Hodicky, J. (ed.) MESAS 2015. LNCS, vol. 9055, pp. 17–23. Springer, Cham (2015). https://doi.org/10.1007/978-3-319-22383-4_2

2. Hodicky, J.: Autonomous systems operationalization gaps overcome by modelling and simulation. In: Hodicky, J. (ed.) MESAS 2016. LNCS, vol. 9991, pp. 40–47. Springer, Cham (2016). https://doi.org/10.1007/978-3-319-47605-6_4

3. Hodicky, J., Prochazka, D., Prochazka, J.: Training with and of autonomous system – modelling and simulation approach. In: Mazal, J. (ed.) MESAS 2017. LNCS, vol. 10756, pp. 383–391. Springer, Cham (2018). https://doi.org/10.1007/978-3-319-76072-8_27

4. Hodicky, J.: Standards to support military autonomous system life cycle. In: Březina, T., Jabłoński, R. (eds.) MECHATRONICS 2017. AISC, vol. 644, pp. 671–678. Springer, Cham (2018). https://doi.org/10.1007/978-3-319-65960-2_83

5. Biagini, M., Corona, F.: Modelling and simulation architecture supporting NATO counter unmanned autonomous system concept development. In: Hodicky, J. (ed.) MESAS 2016. LNCS, vol. 9991, pp. 118–127. Springer, Cham (2016). https://doi.org/10.1007/978-3-319-47605-6_9

6. Biagini, M., Corona, F., Casar, J.: Operational scenario modelling supporting unmanned autonomous systems concept development. In: Mazal, J. (ed.) MESAS 2017. LNCS, vol. 10756, pp. 253–267. Springer, Cham (2018). https://doi.org/10.1007/978-3-319-76072-8_18

7. NATO STO SAS 086: Maritime Situational Awareness: Concept Development Assessment Game (CDAG). STO CSO - STO activities. http://www.cso.nato.int/activities.aspx?pg=3&RestrictPanel=6&FMMod=0&OrderBy=0&OrderWay=2. Accessed 2017

8. NATO STO SAS 082: Disruptive Technology Assessment Game - Evaluation and Validation. NATO STO CSO - STO activities. http://www.cso.nato.int/activities.aspx?pg= 2&RestrictPanel=6&FMMod=0&OrderBy=0&OrderWay=2. Accessed 2017

9. NATO ACT CEI CAPDEV: Autonomous Systems Countermeasures. Innovation Hub. http://innovationhub-act.org/AxSCountermeasures. Accessed 2016

10. NATO ACT: NATO Urbanization Project. NATO Allied Command Transformation. http://www.act.nato.int/urbanisation. Accessed 2016

11. NATO STO NMSG 145: Operationalization of Standardized C2-Simulation Interoperability. Science and Technology Organization - Collaborative Support Office. STO CSO-STO activities. http://www.cso.nato.int/activities.aspx?RestrictPanel=5. Accessed 2018

12. Biagini, M., Corona, F., Innocenti, F., Marcovaldi, S.: C2SIM Extension to Unmanned Autonomous Systems (UAXS) - Process for Requirements and Implementation, 1st edn. NATO Modelling and Simulation Centre of Excellence, Roma (2018)

13. SISO GSD PDG: Guideline on Scenario Development for Simulation Environments. Simulation Interoperability Standards Organization, Orlando (2016)

14. SISO C2SIM PDG/PSG: Command and Control Systems - Simulation Systems Interoperation. SISO C2SIM PDG/PSG. https://www.sisostds.org/StandardsActivities/Development Groups/C2SIMPDGPSG-CommandandControlSystems.aspx. Accessed 2018

15. SISO-STD-011: Standard for Coalition Battle Management Language Phase 1. Simulation Interoperability Standards Organization, Orlando (2014)

16. Pullen, J.M., et al.: Joint Battle Management Language (JBML) - US Contribution to the C-BML PDG and NATO MSG-048 TA. The NPS Institutional Archive. Naval Posgraduate School, Monterey, CA, USA (2014)

17. Perme, D., et al.: Integrating air and ground operations within a common battle management language. In: Spring Simulation Interoperability Workshop, San Diego, CA, USA (2005)

18. Remmersmann, T., Tiderko, A., Schade, U.: Interacting with multi-robot systems using BML. In: 18th International Command and Control Research and Technology Symposium (ICCRTS), Alexandria, VA, USA (2013)

19. SISO-STD-007: Standard for Military Scenario Definition Language. Simulation Interoperability Standards Organization, Orlando (2008)

20. Biagini, M., Corona, F., Wolski, M., Schade, U.: Conceptual scenario supporting extension of C2SIM to autonomous systems. In: 22nd International Command and Control Research and Technology Symposium (ICCRTS), Los Angeles, CA, USA (2017)

21. NSA STANAG 5527: NATO Friendly Force Information (NFFI). NATO Standardization Agency (2008)

22. IEEE 1278-2012 Standard for Distributed Interactive Simulation (DIS) - Application Protocols. IEEE Standard Association (2012)

23. IEEE 1516-2010 Standard for Modeling and Simulation (M&S) High Level Architecture. IEEE Standard Association (2010)

24. Hodicky, J.: HLA as an experimental backbone for autonomous system integration into operational field. In: Hodicky, J. (ed.) MESAS 2014. LNCS, vol. 8906, pp. 121–126. Springer, Heidelberg (2014). https://doi.org/10.1007/978-3-319-13823-7_11

25. Tolk, A., et al.: Composable M&S web services for net-centric applications. J. Def. Model. Simul. 3, 27–44 (2006)

26. George Mason University-C4I Center. OpenBML. https://netlab.gmu.edu/trac/OpenBML. Accessed 2018

27. NATO STO MSG 136: STO activities. Science and Technology Organization - Collaboration Support Office. http://www.cso.nato.int/activities.aspx?RestrictPanel=5. Accessed 2016

Battle Management Language
for Robotic Systems
Experiences from Applications on an UGV and an USV

Rikke Amilde Seehuus$^{(\boxtimes)}$ (ID), Kim Mathiassen (ID), Else-Line Malene Ruud (ID),
Aleksander Skjerlie Simonsen, and Fredrik Hermansen (ID)

The Norwegian Defence Research Establishment (FFI), Kjeller, Norway
`rikke-amilde.seehuus@ffi.no`

Abstract. Unmanned systems are gradually becoming more autonomous, meaning they can perform more tasks with less micromanagement. These systems will have increasing importance in military operations over the next years. For communication with these systems, a language that is unambiguous and machine interpretable is needed. Battle management language (BML) refers to such a language, which can be used to exchange digitized forms of orders, reports and requests. BML has mainly been developed to support inter-operation among command and control (C2) systems and C2 and simulation systems, but BML has also been used for robotic systems. The only standardized BML so far is coalition battle management language (C-BML), which specifically addresses the needs associated with coalition operation.

In this paper we describe our effort in applying BML to an unmanned surface vehicle and an unmanned ground vehicle. The unmanned systems are able to autonomously perform tasks like move from A to B, move along a route, patrol, and survey an area. However, transition from teleoperated systems to more autonomous systems is progressing gradually, meaning we still want at least the possibility to provide relatively detailed instructions to the robotic system. In addition, the ability to make small adjustments of the behavior is sometimes useful. We find that there is a gap between the standards for teleoperation and BML in its current form and would like a discussion on how to best fill this gap. Our preliminary solutions are presented in this paper.

Keywords: Robotic systems · Autonomous systems · UGV · USV ·
Battle management langauge

In the context of applying autonomy to unmanned vehicles, it is often useful to talk about different levels of human independence. This can range from non autonomous or remotely piloted through teleoperated and remotely monitored to fully autonomous. The transition from teleoperated to fully autonomous can be done gradually as the technology matures, but in any case the appropriate level of human independence will vary depending on the complexity of the task

© Springer Nature Switzerland AG 2019
J. Mazal (Ed.): MESAS 2018, LNCS 11472, pp. 302–320, 2019.
https://doi.org/10.1007/978-3-030-14984-0_23

and/or the environment [10]. This means we want the possibility to provide relatively detailed instructions to the unmanned system as well as higher level missions.

As long as an autonomous vehicle is controlled or monitored by a human operator, the operator will need sufficient information from the vehicle about what it is doing and why, e.g which task it is doing, which obstacles it sees, at what range it will start treating obstacles as something to avoid, etc.

To communicate with an autonomous vehicle, including both tasks and reports, a language that is unambiguous and machine interpretable is needed. Battle management language (BML) referrers to such a language, and is defined as "the unambiguous language used to command and control forces and equipment conducting military operations and to provide for situational awareness and a shared, common operational picture" [5,6]. BML is designed to assist in system-to-system interchange of tasks, reports and requests concerning military operations. A system can be a command and control information system (C2IS), a simulation system or a robotic system. Using BML, it should be possible for any of these systems to communicate unambiguously with any other of these systems. Coalition BML (C-BML) is a standardized BML for coalition operations [18]. C-BML has been developed and tested through a series of technical working groups in NATO [13,14]. It has mainly been used to support interoperation among command and control (C2) systems and C2 and simulation systems [3,4,7,11], but BML has also been applied to robotic systems [16,17].

This paper documents how we use BML to communicate with an unmanned ground vehicle (UGV) and an unmanned surface vehicle (USV). The UxVs are controlled trough a custom graphical user interface, called FFI ground control station (GCS), which sends tasks to and recives reports from the vehicles. The vehicles and their autonomy system are introduced in Sect. 1.

We use BML to represent tasks and reports for communication between the GCS and the vehicles. Our BML is a subset of C-BML, but with some custom extensions to make it possible to send more detailed commands to the vehicle. Also, we have made several extensions for the reports, as it is very important for our operators to get as much information as possible back to the GCS about what the vehicle has detected and how it will proceed. We express the BML directly in Google Protocol Buffers [1], and the implementation is explained in Sect. 2.

Our main focus has been on being able to express what we need to communicate in an efficient way that can easily be extended as more is needed. Our main motivation for using C-BML was that we wanted to avoid making a proprietary protocol and would rather use a standard that has been developed over several years and tested in various settings, including for autonomous systems. We also recognize the benefit of using a standard that can facilitate exchanging information between various command and control systems, simulation systems and robotic systems in the future [8,9]. We discuss our experiences in Sect. 3.

1   The Unmanned, Autonomous Vehicles Olav and Odin

We have applied BML to the UGV Olav and the USV Odin. Figures 1a and b shows pictures of the two vehicles. Both these vehicles are controlled trough a custom graphical user interface, the FFI GCS. The GCS communicates with a decision making module called Hybrid autonomy layer (HAL) [20]. HAL executes the tasks by breaking the task down into commands to the vehicle's low level control system. In executing the tasks, HAL depends on input from a perception module about the state of the world around the vehicle, and utilizes services for computationally heavy calculations like route planning. During execution, HAL reports back to the GCS about what it "sees" and "thinks". An illustration of the autonomy architecture on the vehicles is shown in Fig. 2.

(a) The UGV Olav

(b) The USV Odin

Fig. 1. The photos show the autonomous vehicles addressed in this paper.

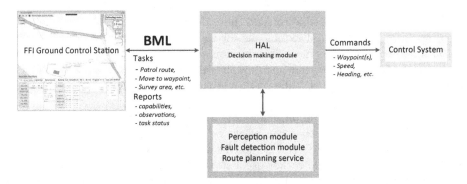

Fig. 2. The GCS communicates with the decision making module HAL, which commands the lower level control system based on information from a perception module, among others.

1.1   The UGV Olav

At FFI we are currently developing an UGV named Olav[1] for autonomous driving. Self-driving cars have become a major research area for both the automotive and the IT industry. Several automotive companies have ambitious plans for launching self-driving cars to the public. We believe that the military may use self-driving platforms to reduce personnel risk, and may exploit much of the commercial development in this area. However, the military domain has some specific challenges, for instance autonomous driving in terrain and operations under very harsh conditions, and this is the main reason for our UGV research.

Olav has two primary missions. The first is to navigate autonomously to a specific location in order to observe, either by its own sensors or by deploying an unattended ground sensor (UGS) system. It should be able to observe the scene of interest for a prolonged period of time. The second mission is aiding a base protection team, where Olav can be sent out to check the perimeter of the base. Here, Olav will primarily drive autonomously, but sometimes remote operation is needed to do fine navigation of the vehicle.

Olav is based on a Polaris Ranger XP 900 EPS vehicle, and has been modified to be controlled autonomously [12]. It has been equipped with cameras and a lidar as perception sensors and an inertial navigation system for localization. A perception module has been developed that produces a map over the surrounding area of the car. This map is used by a motion planning component, which plan a driveable trajectory that avoids obstacles and at the same time follows the desired route. A route can be given by a user or planned using a on-board route planning service.

1.2   The USV Odin

USVs may be a potent addition to naval powers in the future as well as a cost-effective solution for commercial application. First of all, USVs have the advantage that they drastically, if not completely, remove the risk to human life in maritime operations. Also, since unmanned vessels do not need large areas like a galley, cabins, or other installations designed with the sole purpose of supporting human comfort, it is possible to build marine crafts where the entire volume of the ship is dedicated to its operational purpose.

Mine counter measures (MCM) vessels are one of the most important naval capabilities of the Royal Norwegian Navy, consisting today of the *Oksøy* class mine hunters and the *Alta* class mine sweepers. These vessels have been in service since the mid 90s and are approximately 55 m long, and each is crewed by a complement of 32 sailors and officers. Unmanned vessels have been suggested as the future replacement for the Royal Norwegian Navy's MCM capability and therefore, FFI has acquired a vessel named Odin, which serves as a test and development platform for USV operations.

Odin is an 11 m long and 3.5 m wide surface vessel with a mass of 5800 kg and is powered by a twin water jet system yielding 450 hp. This gives a payload

---

[1] OLAV = Off-road Light Autonomous Vehicle.

capacity of approximately 3000 kg, thus making it suitable for a wide range of maritime applications. For example, launching and recovering autonomous under water vehicles (AUVs) is typically performed with manned surface vehicles, but could be done with a vessel like Odin in an area covered with mines that the AUV is trying to locate.

In order to complete its tasks autonomously, Odin is equipped with a variety of sensors, including lidar, radar, and cameras, both optical and infrared, which when combined provide the basis for a sophisticated perception system. In addition, an autopilot and a waypoint guidance system is implemented in order to give both human operators and the autonomy module onboard the ability to commander the vessel to a desired heading, on a desired course, or along a specific line. The autonomous mission planning and decision making module (HAL) will then, based on the input from the perception module and the internal state of the vehicle, execute the mission by breaking it down into various tasks. Examples of such tasks are moving to a desired waypoint, follow a desired route, surveying an area, waiting, avoiding collision or following another vessel. Some of these tasks are given to the vessel as an order, like for instance surveying an area, while others, e.g., avoiding collisions are spontaneously being invoked by Odin itself. Only the former set is expressed in the BML language that is used to command the vehicle.

2   BML on Olav and Odin

The BML used to communicate with Olav and Odin is based on several sources. For tasking, we base our BML on work conducted by the modeling and simulation group at FFI. They have used a simplified version of the C-BML light scheme [18] to send orders from a C2IS and a custom web based graphical user interface (GUI) to a simulation system [2,4]. In addition, we have studied the work done at the Fraunhofer Institute in developing the command and control lexical grammar (C2LG) for commanding teams of robots [15–17]. Especially for reports, C2LG, which uses C-BML light scheme, is our main sources of inspiration. Whenever we needed to express something that we did not find in our sources, we made extensions as we saw fit.

In this section we describe the current status of our BML for both tasking and reporting. Section 2.1 describes an extension we have made to make the vehicle tell the GCS what it can do, including tasks and task parameters. This is used to populate the GCS with the appropriate fields for commanding the vehicle. The BML for sending tasks is explained in Sect. 2.2. Section 2.3 describes additional types of reports that are used to keep the operator informed about what the vehicle is doing. We have chosen to express the BML directly in Google Protocol Buffers, and parts of our proto-files are included as a part of the descriptions. The extensions and our experiences with using BML for autonomous vehicles is discussed further in Sect. 3.

2.1   BML for Initialization

In order for the control station and the user of the control station to know which tasks an autonomous unit can perform, the unit reports its capabilities. We have added a new type of BML report, called *capability report* to cover this functionality. The schema is found in Listing 1.1, and the types used are found in Listing 1.2.

```
 1  message CapabilityWhere {
 2      WhereType whereType = 1;
 3  }
 4
 5  message CapabilityHow {
 6      HowType howType = 1;
 7      Constraint constraint = 2;
 8  }
 9
10  message CapabilityWhat {
11      What what = 1;
12
13      repeated CapabilityWhere requiredWhere = 2;
14      repeated CapabilityHow    requiredHow = 3;
15
16      repeated CapabilityWhere optionalWhere = 4;
17      repeated CapabilityHow    optionalHow = 5;
18  }
```

Listing 1.1. Capability report schema

```
 1  message What {
 2      int32 id = 1;
 3      string name = 2;
 4  }
 5
 6  enum WhereType
 7  {
 8      NONE = 0;
 9      LOCATION = 1;
10      ROUTE = 2;
11      AREA = 3;
12      LINE = 4;
13      HEADING = 5;
14  }
15
16  enum HowType
17  {
18      CONSTRAINT = 0;
19  }
20  message Constraint {
21
22      enum Unit {
23          UNDEFINED = 0;
24          SECONDS = 1;
25          METERS = 2;
26          M_PER_S = 3;
27          DEGREES = 4;
28      }
29
30      string name = 1;
31      double value = 2;
32      Unit unit = 3;
33  }
```

Listing 1.2. Types used by capability report

A capability report contains a number of `CapabilityWhat` messages, each field reports one of the capabilities of the unit. The first field in the `CapabilityWhat` message is the `what` field, which contains the id and name of the task which the unit can perform. Each task has a number of required or optional parameters, which are defined using the remaining fields. When commanding the unit, the user will know which parameters are required and which are optional. The fields `requiredWhere` and `optionalWhere` define if the task needs a `where` to be executed and lists the possible `Where` types that can be used with a task. The fields `requiredHow` and `optionalHow` defines required and optional information on how the task should be performed. The `CapabilityHow` contains two fields. The first field (`howType`) defines the type. Currently, only constraints are implemented, but it is possible to extend with other fields. The second field (`constraint`) defines a numeric constraint to the task. This can for instance be maximum allowed speed when driving.

In order to better understand the capability report, we will give two examples from the capabilities of Odin and Olav. In Listing 1.3 a survey task is defined. The `what` field defines the id (2) and the name (Survey) of the task. The only valid `Where` type allowed is `Area`, and this is required when sending this task to Odin. There are four optional constraints that the user can specify when giving the task. The `value` field specifies the default value of the constraints. The `unit` field defines which units that are used on the constraints, and they are selected from an enumerate list. In this specific task Odin will survey an area with a sonar, mapping the sea bottom. The parameter `MaxDuration` defines a maximum allowed duration before the task is aborted, `MaxSpeed` defines the maximum allowed speed the vehicle can move. The area is surveyed in a lawn mower pattern, going back and forth in straight lines until the entire area is covered. The parameter `LineDistance` defines the distance between the lines. The last constraint is the `LengthOutsideArea` parameter, which defines how far outside the survey area Odin should be before it turns, in order to ensure good quality of the data inside the survey area.

The second example is given in Listing 1.4. In this task Olav is ordered to move somewhere. There are two `requiredWhere` fields. This means that the user must provide one of the where types, either `Location` or `Route` (not both). When Olav is ordered to do this task, it will plan according to the provided `Where` field. If the user does not want Olav to do additional planning, the user can set the a field in the `Route` message. Generally, this tells the unit that it should not alter the route or do additional route planning. In Olav's case, the internal route planning service will not be invoked and the route will be given directly to the motion planning component. There are two constraints associated with this task. The `WaypointRange` defines the radius around a waypoint that the vehicle must be within to mark a waypoint as reached. The `Speed` constraint defines the desired speed the vehicle should use while driving.

```
 1  capabilityWhat {                        1  capabilityWhat {
 2    what {                                2    what {
 3      id: 2                               3      id: 1
 4      name: "Survey"                      4      name: "Move"
 5    }                                     5    }
 6    requiredWhere {                       6    requiredWhere {
 7      whereType: AREA                     7      whereType: LOCATION
 8    }                                     8    }
 9    optionalHow {                         9    requiredWhere {
10      constraint {                       10      whereType: ROUTE
11        name: "MaxDuration"              11    }
12        value: 3600.0                    12    requiredHow {
13        unit: SECONDS                    13      constraint {
14      }                                  14        name: "WaypointRange"
15    }                                    15        value: 1.0
16    optionalHow {                        16        unit: METERS
17      constraint {                       17      }
18        name: "MaxSpeed"                 18    }
19        value: 2.0                       19    requiredHow {
20        unit: M_PER_S                    20      constraint {
21      }                                  21        name: "Speed"
22    }                                    22        value: 3.0
23    optionalHow {                        23        unit: M_PER_S
24      constraint {                       24      }
25        name: "LineDistance"             25    }
26        value: 40.0                      26  }
27        unit: METERS
28      }
29    }
30    optionalHow {
31      constraint {
32        name: "LengthOutsideArea"
33        value: 50.0
34        unit: METERS
35      }
36    }
37  }
```

Listing 1.3. Example capability of Odin   **Listing 1.4.** Example capability of Olav

These two examples show how one can define different tasks the unit can perform by defining capabilities on what the unit can do, and then specify the required and optional parameters needed for the unit to perform the task. The GCS is automatically populated with the possible options, as illustrated in Fig. 3.

2.2 BML for Tasking

The BML for tasking was indirectly explained in the previous section through the capability report, but here we will relate what we do to the standard C-BML and provide some more examples.

A C-BML order is based on the 5Ws; *Who*, *What*, *When*, *Where*, and *Why*. *Who* can be used to represent the taskee, the tasker or an other unit that is a part of the task, an affected who. *What* is the actual task. *When* is used to state when the task should start and/or end, in absolute time or relative time according to other tasks. *Where* is a route, an area or a position relating to where the task should be conducted. An order will also typically include control

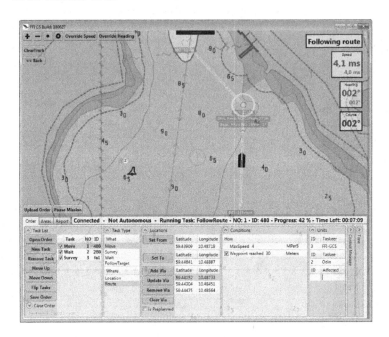

Fig. 3. The figure is a screenshot from the GCS that shows the tasking GUI for the USV Odin. The different capabilities are shown in the "What" menu located below the "Task Type" panel in the lower left part of the GUI. (Color figure online)

measures informing and restricting the execution of tasks, like boundary lines, phase lines, etc. *Why* is used to represent the commanders intent.

We more or less follow the C-BML standard when we build up an order consisting of several tasks that are expressed in the aforementioned five-word syntax, complemented by control measures. So far, we make extensive use of *What*, which we simply assign to one of the vehicle's capabilities, and *Where*. As we see it, the lack of maturity when it comes to operating unmanned systems autonomously necessitates additional information to be sent in an order. We have added a sixth word to the order, *How*, which contains details about how the task should be executed. Almost all our tasks make extensive use of *How*, the most common example is to assign a speed to the vehicle, which we do not find a proper place for in the 5 Ws.

An example illustrating our BML is shown in Fig. 4 where the USV Odin is tasked to follow an AUV. The pre-planned route of the AUV is, if available, assigned as an optional *Where*, and *What* is set to "Follow Target". The AUVs id is set to the "affected who" field of the order, and the desired distance and relative bearing from the AUV are described in *How*. Lastly, a control measure area is used to restrict the area of operation, ensuring that Odin do not follow the AUV outside this area. Another example is illustrated in Fig. 5, in which the UGV Olav is sent on a patrol mission. Here, the patrol route is described in *Where* and the speed and the waypoint acceptance radius constitute *How*.

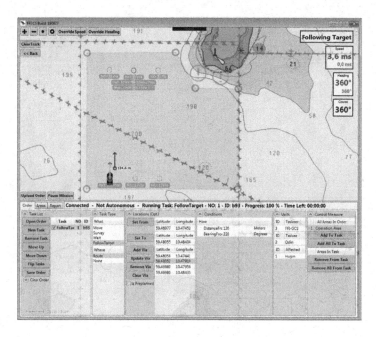

Fig. 4. The GCS shows the USV Odin during a following mission. The target to be followed is located in front and slightly to the right of the USV and is marked by a blue dot in the map. In this case, the USV follows an AUV that is preplanned to move along a known path which is provided to the USV and shown with a yellow curve. In addition, an operation area is provided as a control measure to ensure that Odin does not follow the AUV outside of this area. (Color figure online)

In both the examples, *When* is set to ASAP and *Why* is not used. As we make more complex tasks and want to coordinate more UxVs, we expect to start using *When* more extensively, but at this point, the tasks are simply executed in the same order as they are specified in the plan sent to the vehicle.

2.3 BML for Reports

Autonomous vehicles will react to a variety of different circumstances and environmental factors, which are unknown during the mission planning phase. As autonomous systems are not yet fully matured in neither the land- nor the surface domains, most systems will most likely be remotely monitored for quite some time. Then, in order to build trust between the human remote observer and the unmanned system, unambiguous and detailed information about anything that can make the vehicle behave differently must be available to the aforementioned observer. For example, when a USV is encountering another vessel in its operation area, it will engage in evasive maneuvers in order to safely overtake or pass by the other vessel. In order for the human operators to verify that the altered speed and course is both safe and meaningful, an observation report regarding

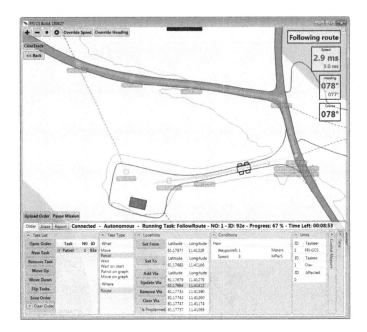

Fig. 5. The figure shows a screenshot of the GCS controlling the vehicle Olav. Here, Olav is sent on a patrol mission in which a waypoint route is assigned to the order (*Where*). Details, like the speed of the vehicle and the distance from each Waypoint Olav has to achieve in order to progress its mission are described in the *How* specifier.

the other vessel should be sent to the GCS. In addition, a task status report explaining what the vessel is doing, makes the behavior really transparent for the operator.

In this section we explain the two report types observations report and task status report. Listing 1.5 shows our report schema, which also includes the capability report explained earlier and measurement reports, to which we did not do any extensions.

```
1  message Report {
2      Who         reporterWho = 1;
3      TimeDate reporterWhen = 2;
4
5      repeated CapabilityWhat capabilityWhat = 3;
6      repeated Measurement measurement = 4;
7      repeated Observation observation = 5;
8      repeated StatusTask statusTask = 6;
9      bml.Order ackOrder = 7;
10
11 }
```

Listing 1.5. Report schema

Observations Report. Observations constitute an important type of information that continuously must be communicated from the unmanned vessel to its human remote observers or operators. The schema is shown in Listing 1.6. Our observations include a `Location`, defining its global position, and a `Velocity`, which may be used to indicated the estimated motion of the observation. In addition, an `Observation` includes a `Who`, which can used to describe if the observation matches a known unit in the area or simply to uniquely identify different observations made by the perception module of the system. The next entries are both enums. The first, `ObservationType`, indicates the type on the observation. In this example we have included some that are relevant for maritime operations, i.e., motor boats, sail boats and sea marks which all dictate a different response from the vessel. The second enum, `ObservationAction`, indicates if the autonomous system actively interacts with the observation, for instance avoiding it or moving closer to identify it if its type would be unknown. Lastly, an `Image` of the observation is added in the report. Figures 6 and 7 illustrates how the content of observational reports are displayed in the GCS.

```
 1  enum ObservationType {
 2      UNKNOWN_TYPE = 0;
 3      PERSON = 1;
 4      MOTOR_BOAT = 2;
 5      SAILING_BOAT = 3;
 6      SEA_MARK = 4;
 7  }
 8
 9  enum ObservationAction {
10      UNKNOWN_ACTION = 0;
11      AVOIDING = 1;
12      IDENTIFY = 2;
13  }
14
15  message Image {
16      ImageFormat imageFormat = 1;
17
18      bytes data = 2;
19  }
20
21  message Measurement {
22      // Meta data
23      TimeDate when = 1;
24      Location where = 2;
25      string sensorIdentifier = 3;
26
27      // Measurement
28      Image image = 4;
29  }
30
31  message Observation {
32      Location where = 1;
33      Velocity velocity = 2;
34      Who who = 3;
35      ObservationType type = 4;
36      ObservationAction action = 5;
37      Image image = 6;
38  }
```

Listing 1.6. Observation report schema

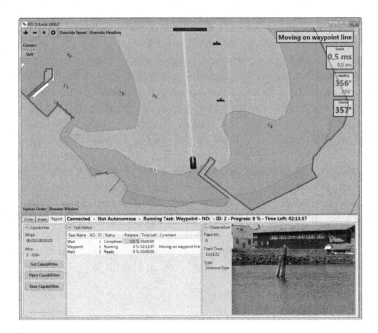

Fig. 6. The GCS shows the USV Odin close to an observation. The observation is marked in the map as an orange dot (to the left of the vessel) and its corresponding image, as provided by our perception system, is sent to the GCS with the observation report and displayed in the lower part of the GUI as an "unknown type".

Task Status Report. Our tasks can be fairly high level. The UxVs interpret and break them down into smaller tasks. All details about how the task will be solved is therefore not known to the operator in the mission planning phase. It can thus be useful for the operator to receive a detailed report about how each task is being solved at the time of execution. In our initial version of the task status report, we relied on the fields in the C2LG description [15] with minor adjustments, such as a simplification of When to a TimeLeft field, and a generalization of Label to a Comment field that can contain a more general, short description on how the autonomous unit is solving the task.

However, additional task status information was required on the operator side, and three new fields were added: Where, TaskName, and TaskMetaData. Listing 1.7 shows a schema of our expanded version of the task status report. The field TaskName is quite self-explanatory, and is simply a name corresponding, but not necessarily equal, to the task name in the order. Where is used to indicate where the autonomous unit is planning on being located for the foreseeable future. It can differ from the Where in the order task by being a more detailed plan of movement, such as a path calculated by the collision avoidance module.

```
1  message StatusTask {
2
3      enum TaskStatus {
4          UNINITIALIZED = 0;
5          READY = 1;
6          RUNNING = 2;
7          INTERRUPTED = 3;
8          COMPLETED = 4;
9      }
10
11     string taskId = 1;
12     Who who = 2;
13     Where where = 3;
14     TaskStatus taskStatus = 4;
15     string comment = 5;
16     Time timeLeft = 6;
17     double progression = 7;
18     string taskName = 8;
19     repeated TaskMetaData taskMetaData = 9;
20 }
```

Listing 1.7. StatusTask schema

The field `TaskMetaData` does not fit in with the standard. It is specifically designed to meet our needs, and can be expanded as more task information is required. It is meant to contain different types of information about the task execution that can be useful for the operator. Listing 1.8 shows our current version this new type.

```
1  message TaskMetaData {
2      oneof metadata {
3          Sector colavSector = 1;
4          Circle taskAcceptanceRadius = 2;
5      }
6  }
```

Listing 1.8. TaskMetaData schema

An example of the control station visualisation of a task status report can be seen in Fig. 7. Our USV is executing a move task, and gray sectors representing the decision space for different evasive maneuvers that abide to certain collision avoidance regulations at sea are visualized. These sectors are added in the `Sector colavSector` field in `TaskMetaData`, and are used by tasks that will allow collision avoidance maneuvers while being executed. So far, there are only two fields in the `TaskMetaData` type, the collision avoidance sectors just mentioned, and a `Circle taskAcceptanceRadius` which is used to visualize the radius around a waypoint within which the waypoint will be accepted as reached. In Fig. 3 this is visualized as a yellow circle. Both types of information have proved very useful for operators when monitoring our autonomous USV. We have as of yet not used the `TaskMetaData` field while operating our UGV. The remaining parts of the task status report can be seen in the left bottom part of Fig. 7.

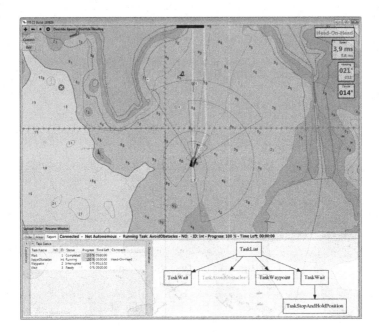

Fig. 7. The GCS shows the USV Odin in the middle of an evasive maneuver in order to avoid collision with an approaching vessel. The observation reports are illustrated as orange (static obstacles) and red (moving vessels) dots in the map and with a red line indicating the velocity of the latter. In addition, information from the task status report is visualized, including the predicted route (shown as a yellow curve), the comfort sectors around the vessel which indicate when it should avoid obstacles and finally, the text string in the upper right corner indicating the current "Head-on-Head" situation. (Color figure online)

3   Discussion and Conclusion

This paper has described examples of utilizing BML for autonomous, unmanned vehicles, including examples of initialization, tasking and reporting. Our BML is based on the C-BML light schema [18], the BML used in the modeling and simulation group at FFI [4] and examples from Fraunhofer on BML for autonomous systems [15–17]. Whenever we did not find what we needed, we made an extension.

The most important extension we made is the *capability report*, which describes the capabilities of an autonomous system. In our opinion the capability report adds an important feature to a BML language, since it defines what a unit is capable of doing. When tasking a unit with a certain task, it is important to know that the unit is capable of doing that specific task. By adding the capability report, the user is able to know all the tasks a unit is capable of performing. Another advantage is that the user is able to get a list of task specific constraints that can be altered according to the need of the user. In the future, we might add a description field, in addition to the name field, so the user can have better knowledge of what the constraint actually does.

Using capability reports also separates the development of autonomous platform capabilities and control station capabilities. When a new capability is added to an autonomous platform, it simply adds another `CapabilityWhat` message to its report, and no additional changes are required in the control station. This implies that the development of the platform and the control station can in a large degree be done separately, which decreases coordination efforts between teams.

Another benefit with this approach is that it makes it easier to develop interoperable platforms and control stations, since a control station only has to implement a few messages to be able to control a platform. The details on what a platform can do is stored in the platform and retrieved by the control station when someone wants to control that unit. The control station only needs to implement basic map primitives, as defined in the `Where` message, such as area, route, location. It must also be able to send a complete `Task` message to control a platform. This is, in our opinion, a very limited set of capabilities on a control station to be able to control a large set of platform with a large variety of tasks.

The BML for tasking the vehicle follows more or less a subset from the standard described in C-BML light schema, mainly using *what* and *where*. For our purposes we saw that we needed extensions to provide the vehicle with more details on *how* it should execute some task. The reason for this is that the operator wants the flexibility to choose the level of autonomy he or she wants to give the vehicle, depending on the task and the complexity of the environment. We did this by adding a *How* to the order. There might be a way to convey the same low-level details to the vehicle via *What* trough the full C-BML standard, but we found the current *How*-extension very convenient as it clearly separates the type of task/capability from lower-level details. For complex situations in which the autonomy is not fully capable of handling the environment, a large amount of *How* specifiers and additional control measures are typically used in order to "restrict" the vehicles realm of possible behaviors. As the autonomy becomes more and more sophisticated, or the environment comparably simpler, the use of such specific instructions or micro-management is expected to diminish.

Most of the work describing examples of the use of BML seem to focus on tasking. For our purposes reports are equally, if not more, important. It is much easier for the operator to build trust in the autonomous system if it is transparent what the system is "seeing" and "thinking". Our extensions are mainly to the report types observational report and task status report.

As illustrated in Sect. 2.3, observation reports are used for two purposes. The most obvious intent is to send back information relevant for the mission to the GCS. In our case, we add an image in the observation report. It can be added if the vehicle is supposed to photograph some object or location. However, as unmanned vehicles become more and more autonomous, their behavior will to a larger and larger degree adapt spontaneously to environmental factors and to new situations. In order to aid the operator with the aforementioned transparency in regards to what the system "sees" and "thinks", we rely heavily on observation reports. The C2LG documentation describes an observation report which is used whenever a robot detect an unknown, suspect or hostile person, organization or object [15]. They include a position for the observation and also a field for either `PersonType`, `OraganisationType` or `ObjectType`. Our implementation is very similar to this, but we rather use a more generic `ObservationType` which describes what is being observed. In addition, we have added the `ObservationAction` field, in which we indicate how the autonomous vehicle behaves in response to the given observation. Two observations with identical type may demand completely different actions based on the overall scenario and environment, and it is here that we utilize the latter report entry. This is perhaps best illustrated in our usage of observation reports with the aim of aiding the remote operators to understand collision avoidance maneuvers that occur when an USV is encountering sea marks, motor boats, unknown obstacles etc. Although implementing a general standard in this regards is a nontrivial issue, we do, based on our experience, suggest that an `ObservationAction`, which indicates if the nature of the observation demands a radical change in the autonomous vehicle's behavior, would greatly help in assuring proper human control over autonomous vehicles.

When it comes to the task status report, the C-BML standard covers our needs only partly. Our tasks can be high level, meaning that the operator often will not know the details of how the tasks are going to be solved by the autonomous unit. Much information is needed for an operator to keep a sufficient overview of how tasks are being solved, and we therefore require sending more detailed information about the execution of the task than what the BML standard facilitates. Our solution using the new, specialized field *TaskMetaData* covers our needs, but is not easily standardized and scales poorly. We welcome any input on how this can be solved in a more general manner, preferably one which can be used with different types of autonomous platforms.

Note that all the extensions that we have made have been fairly pragmatic. In this report we point out features that we need and see are missing in the standard[2]. We do not suggest how these additions should look like in a standard BML, but we provide examples of how we have solved it for now. This approach of adding to the language as needed might leed to an incoherent language, and care should be made to include not only the vocabulary but also the morphology of the underlying command and control information exchange data model [19].

We follow the current effort that is going on in developing the new standard C2SIM as a joint development of the C-BML and the military scenario definition language (MSDL). It will be interesting to see how our work will comply with the guidelines and extensions for autonomous systems that will be made for this new standard. Future work on commanding autonomous systems might aslo include looking into the work made on grammar an ontology. Our intention is to apply the standards that are beeing developed, in our systems. However, our research focus is on increasing the systems autonomy capabilities.

---

[2] Note that we did not consider the C-BML full scheme.

References

1. Google Protocol Buffers. https://developers.google.com/protocol-buffers/. Accessed 3 Aug 2018
2. Alstad, A., Løvlid, R.A., Bruvoll, S., Nielsen, M.N.: Autonomous battalion simulation for training and planning integrated with a command and control information system. FFI-rapport 2013/01547, Forsvarets forskningsinstitutt (2013)
3. Brook, A.: UK experiences of using coalition battle management language. In: Proceedings of the 15th IEEE/ACM International Symposium on Distributed Simulation and Real Time Applications (2011). https://doi.org/10.1109/ds-rt.2011.25
4. Bruvoll, S., et al.: Simulation-suported wargaming for analysis of plans. In: Proceedings of the NATO Modelling and Simulation Group Symposium M&S Support to Operational Tasks Including War Gaming, Logistics, Cyber Defence (MSG-133) (2015)
5. Carey, S., Kleiner, M., Hieb, M., Brown, R.: Standardizing battle management language - a vital move towards the army transformation. In: Proceedings of the 2001 Fall Simulation Interoperability Workshop, 01F-SIW-067 (2001)
6. Carey, S., Kleiner, M., Hieb, M., Brown, R.: Standardizing battle management language - facilitating coalition interoperability. In: Proceedings of the 2002 European Simulation Interoperability Workshop, 02E-SIW-OO5 (2002)
7. Garmendia-Doval, A., García-Juliá, I.: C-BML orders in a course of action analysis simulator. In: Proceedings of the 2008 Euro Simulation Interoperability Workshop, 08E-SIW-037 (2008)
8. Hodicky, J.: Standards to support military autonomous system life cycle. In: Březina, T., Jabłoński, R. (eds.) MECHATRONICS 2017. AISC, vol. 644, pp. 671–678. Springer, Cham (2018). https://doi.org/10.1007/978-3-319-65960-2_83
9. Hodicky, J., Prochazka, D.: Challenges in the implementation of autonomous systems into the battlefield. In: Proceedings of the 2017 International conference on Military Technologies (2017). https://doi.org/10.1109/miltechs.2017.7988855
10. Huang, H.M. (ed.): Autonomy levels for unmanned systems (ALFUS) framework volume i: terminology. Special Publication 1011-I-2.0, National Institute of Standards and Technology (2008)
11. Kruger, K., Frey, M., Schade, U.: Battle management language: military communication with simulated forces. In: NATO RTO MSG-056 Symposium on Improving M&S Interoperability, Re-Use and Efficiency in Support of Current and Future Forces (2007)
12. Mathiassen, K., Baksaas, M., Olsen, L.E., Thoresen, M., Tveit, B.: Development of an autonomous off-road vehicle for surveillance missions. In: Proceedings of IST-127/RSM-003 Specialists' Meeting in Intelligence & Autonomy in Robotics. NATO Science and Technology Organization, Bonn, Germany, October 2016
13. NATO Modeling & Simulation Group 048: Coalition battle management language (C-BML) - NMSG-048 final report. RTO-TR-MSG-048, NATO Research and Technology Organization (2012)
14. NATO Modeling & Simulation Group 085: Standardisation for C2-simulation interoperability - final report of NMSG-085. STO-TR-MSG-085, NATO Science and Technology Organization (2015)
15. Rein, K., Remmersmann, T., Schade, U., Trautwein, I.: Command and control lexical grammar for commanding teams of robots (Robot-C2LG). Fraunhofer FKIE (2015)

16. Remmersmann, T., Schade, U., Rein, K., Tiderko, A.: BML for communicating with multi-robot systems. In: Proceedings of the 2015 Fall Simulation Interoperability Workshop (2015)
17. Remmersmann, T., Schade, U., Schlick, C.M.: Interactive multi-robot command and control with quasi-natural command language. In: Proceedings of the 2014 IEEE International Conference on Systems, Man, and Cybernetics (2014). https://doi.org/10.1109/smc.2014.6973952
18. Simulation Interoperability Standards Organization (SISO): Standard for Coalition Battle Management Language (C-BML) Phase 1, SISO-STD-011-2014 (2014)
19. Tolk, A., Dially, S.: A system view of C-BML. In: Proceedings of the 2007 Fall Simulation Interoperability Workshop, 07F-SIW-054 (2007)
20. Wiig, M.S., Løvlid, R.A., Mathiassen, K., Krogstad, T.R.: Decision autonomy for unmanned vehicles. In: Proceedings of IST-127/RSM-003 Specialists' Meeting in Intelligence & Autonomy in Robotics. NATO Science and Technology Organization, Bonn, Germany, October 2016

ROS/Gazebo Based Simulation of Co-operative UAVs

Cinzia Bernardeschi[1]([☒]), Adriano Fagiolini[2], Maurizio Palmieri[3],
Giulio Scrima[1], and Fabio Sofia[1]

[1] Department of Information Engineering, University of Pisa, Pisa, Italy
cinzia.bernardeschi@unipi.it, giulio.scrima@gmail.com,
fabiosofia@hotmail.it
[2] Department of Energy, Information Engineering and Mathematical Models,
University of Palermo, Palermo, Italy
adriano.fagiolini@unipa.it
[3] Department of Information Engineering, University of Florence, Florence, Italy
maurizio.palmieri@ing.unipi.it

Abstract. UAVs can be assigned different tasks such as e.g., rendez-vous and space coverage, which require processing and communication capabilities. This work extends the architecture ROS/Gazebo with the possibility of simulation of co-operative UAVs. We assume UAV with the underlying attitude controller based on the open-source Ardupilot software. The integration of the co-ordination algorithm in Gazebo is implemented with software modules extending Ardupilot with the capability of sending/receiving messages to/from drones, and executing the co-ordination protocol. As far as it concerns the simulation environment, we have extended the world in Gazebo to hold more than one drone and to open a specific communication port per drone. In the paper, results on the simulation of a representative co-ordination algorithm are shown and discussed, in a scenario where a small number of Iris Quadcopters are deployed.

Keywords: ROS/Gazebo · Co-operative UAVs · Simulation

1 Introduction

Simulation applied at the early stages of system design allows developers to gain confidence that the system behaves as expected and it is an important tool to visualize and validate systems before industrial production or deployment. Many different languages and environments have been introduced to support modeling and simulation.

During the last decades, much research has been carried out showing the potentialities of multi-robot systems in many real applications, ranging from precision farming, surveillance, patrolling, etc. [2,7,21,29].

ROS (Robot Operating System) [28] is a standard de facto for robot software development. Analogously, for simulation of UAV aircraft, Gazebo [12] has

© Springer Nature Switzerland AG 2019
J. Mazal (Ed.): MESAS 2018, LNCS 11472, pp. 321–334, 2019.
https://doi.org/10.1007/978-3-030-14984-0_24

been widely used in the scientific community. In Gazebo, Ardupilot [3] is the open-source software that allows to carry out the control of different unmanned vehicles mainly through its four different components: the Antenna Tracker, the APM Rover, the ArduPlane and the Arducopter. In particular, Arducopter implements the actual drone control. So far, the current ROS/Gazebo architecture only allows for the simulation/emulation of a single aircraft.

In this work we present a possible extension of the architecture enabling the simulation of possibly multiple heterogeneous vehicles, adhering to their own individual dynamics, as well as interacting with each other according to shared co-operation strategies. In particular, we consider UAVs with an underlying attitude controller based on Ardupilot, which uses the MAVLink protocol (Micro Air Vehicle Link) [6] for the communication. The integration of the co-ordination algorithm in Gazebo is implemented with software modules extending Ardupilot with the capability of (i) sending/receiving MAVLink messages to/from drones, and (ii) executing the co-ordination protocol. An abstraction of the communication channel by which drones exchange information is implemented with a co-ordination script, which is executed locally by each drone instance. Every fixed time interval a drone sends information (e.g. actual position) to other drones. Each drone uses the data received from the other drones via the co-ordination script to compute a new target point, based on the task the drones have to perform. A case study has been developed, where a small number of Quadcopters are deployed and perform space-coverage operations by applying a slightly modified version of the Olfati-Saber et al. co-ordination algorithm [13].

The paper is organized as follows. Section 2 reports related work. Section 3 provides a short overview of the state of the art of the ROS/Gazebo simulation environment. Section 4 describes the modifications made to the simulation environment to allow multi-UAVs simulation. Section 5 shows the application of the developed framework to a case study. Finally, Sect. 6 contains a discussion on the presented framework and the conclusions.

2 Related Work

The problem of co-ordination of UAVs has been addressed in many works. Among others, the works [25] and [26] report on co-ordination systems by focusing on the problem of optimizing the path and/or time of flights to cover an area of interest. The works [19,23] and [27] report on the problem of tracking a target moving on the ground. In the context of civil security for disaster management, [11] introduces a decisional architecture for co-ordination of multiple UAVs (such system was developed in the framework of the AWARE Project [14], which considers scenarios of surveillance with multiple UAVs, sensor deployment and fire threat confirmation).

The work [24] reports on a further application of co-ordination and surveillance of UAVs; in this case co-ordination algorithms are applied inside a urban environment, with particular attention to surveillance approaches of a territory.

In [5] the complexity in the co-operation and co-ordination of independent UAVs is studied: clustering techniques are applied for the division of aircrafts in sub-teams according to the objective which they have to achieve.

The ROS/Gazebo simulation environment is a complex framework that involves many different elements and allows realistic simulations of robotic systems. Other simulation environments are available, like for example [20], where a control and co-operation strategy exploiting the CATA, Control Automation and Task Allocation, is used; or [10], where Matlab/Simulink is used for the real time control of UAVs. These environments are easier than ROS/Gazebo but they do not allow a graphical representation of the UAVs.

3   ROS/Gazebo Simulation Environment

The ROS/Gazebo development simulation environment involves three main elements:

- the open-source autopilot software Ardupilot [3];
- the collection of software frameworks for robot programming ROS (Robot Operating System) [16];
- the 3D simulation environment Gazebo [15].

In particular, the UAVs attitude controller is based on the Ardupilot software. ROS is instead exploited like a middleware to help programmers in developing robot applications. Gazebo allows a visual, tridimensional simulation of a scenario consisting of cyber-physical systems, e.g. ground-rovers, UAVs, and different other objects, like, e.g., simple obstacles, or surrounding environmental elements, together composing what it is called a Gazebo world.

3.1   Ardupilot

The base element that comes into play to start a software simulation of a single vehicle is the open-source software Ardupilot (http://ardupilot.org/). Developed by the community DIY Drones, it allows to carry out the control of different unmanned vehicles, including drones.

The Ardupilot software offers different flight modes, which can be distinguished in manual flight modes and automatic ones, with a combination of customizable parameters. For instance, it is possible to control the UAV flight through a "guided" mode, by which Ardupilot will adjust automatically the values of yaw, roll, and pitch, according to the position given as input. Moreover, through other flight modes, it is possible for the UAV to keep the desired altitude, a given position, to move the UAV with a circle trajectory (fixing the radius of the trajectory), to make the UAV come back to the launch point or simply land, and many others.

For the communication with the unmanned vehicle, the MAVLink protocol [6] is exploited. It allows vehicles communications through the exchange of packets, which are represented at low level by strings in C language, in which each bit has a specific function in the communication. These packets are provided with header, payload, and checksum, and are transmitted through some serial communication channels. Through the MAVLink protocol it is possible to exchange some predefined commands: the fundamental "heartbeat" message (characterized by the ID 0), which is used to keep under control the communication state with the UAV, or the "set_mode" message (characterized by the ID 11), which is used to set a given flight mode, together with many other messages which are useful to control the vehicle. Moreover, it is possible to create custom messages.

3.2 ROS

ROS is not actually an operative system, but rather it represents an open source collection of frameworks/libraries [18] for the development of software for the programming of robots. ROS was developed with the objective of facilitating and expediting the prototyping of a robotic software. In particular, the principle of software reuse is exploited, enabling interoperability among all the tools involved with the ROS environment and solving problems such as real-time collection of data from cyber-physical systems sensors, implementation of the publish/subscribe model in a network of ROS nodes for the communication from/to robots, management of commands received by a user and related actuation actions.

ROS provides hardware abstraction, device drivers, libraries, visualizers, message-passing, package management, and more. ROS is based on FreeBSD, the open source operating system developed by Berkeley Software Distribution. In ROS, the components of the robots can be represented like nodes in a network, which communicate one another, via the anonymous and asynchronous publish/subscribe mechanism. This is a powerful design pattern that can significantly reduce the development effort and promote flexibility and modularity in a system.

Another important element in ROS is the definition of the physics of the robot, and a part of tools was developed to this aim. In particular, it is possible to use the URDF (Unified Robot Description Format) files to describe the robot physical parameters. This improves the integration with 3D simulators, like e.g. Gazebo [15].

3.3 Gazebo

Exploiting a visual simulator is useful in case one wants to test the functioning of a given algorithm. In Gazebo [15], realistic scenarios for cyber-physical systems, including the surrounding environment [8], can be created. This simulator is complete with dynamic and kinematic physics, and a pluggable physics engine.

Integration between ROS and Gazebo is provided by a set of Gazebo plugins that support many existing robots and sensors. Since the plugins present the same message interface as the rest of the ROS environment, a Gazebo user can write ROS nodes that are compatible with simulation, logged data, and hardware. A relevant aspect is that a user can develop an application directly in the simulation environment and then deploy the physical robot with little or no changes at all in the code. Simulation of UAV aircraft through Gazebo has been widely used in the scientific community.

4   Multi-UAV Simulation

This section shows the approach we have followed to extend the base environment of Gazebo with the simulation of multiple robotic systems. The architecture of the simulation environment has been modified since the connection between Ardupilot and ROS/Gazebo is provided through a unique port, while, to have a multi-vehicle simulation, it is indispensable the creation of a number of connection ports equal to the number of UAV instances.

Moreover, the Ardupilot software was enhanced with some modules allowing drones exchanging MAVLink messages to each other directly.

Finally, Gazebo was extended to allow the graphical visualization of different UAVs, which cooperate communicating with each other, by means of the MAVLink protocol using the previously input/output inserted ports, in order to execute the chosen co-ordination protocol for the specific application (e.g., a space coverage application).

More precisely, our extension uses:

- the "guided" mode for controlling the UAV flight, by which Ardupilot adjusts automatically the values of yaw, roll, and pitch, according to the target position given as input to the drone;
- the Python script `mavproxy.py`, which is used for configuring the communications among the unmanned vehicles using the MAVLink protocol;
- the Python script `simvehicle.py`, which is used to create instances of vehicles with their base parameters and their MAVLink connections, and to control such instances. The `simvehicle.py` script contains also the algorithm to be executed locally for the implementation of the co-ordination protocol.
- the SITL (Software In The Loop) simulator which allows us to run Ardupilot without any physical hardware, emulating the behavior of the drones;
- the command `roslaunch` to run the Gazebo simulator, together with a launch file holding the Gazebo environment, i.e. all the elements within the so-called "Gazebo world", and one or more simulated drones in such an environment.

Figure 1 shows the architecture of our simulation for the case study in Sect. 5: there is the Gazebo world, which is activated by the `roslaunch` command, there are N drones, each of which is activated by a `simvehicle.py` script that has been enhanced with a local co-ordination script to enable the data exchange among drones and the execution of a co-ordination algorithm.

To carry out the simulation of cooperative UAVs we use a fixed, defined apriori, number of Iris Quadcopters. These quadcopters, developed by 3D Robotics, represent commercial, state-of-art drones which can be exploited in the professional sector for different base applications. In particular, Iris UAVs hold the typical equipment of a drone, including the remote controller, radio for communications, Wi-Fi card for control using Android devices, high-resolution cameras,

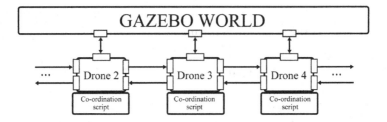

Fig. 1. Multi-UAVs architecture for the case study in Sect. 5.

and other useful sensors. The Ardupilot software runs on top of the Pixhawk flight controller board [17]. Even though Iris drones have the limitation of a time of flight of about 20 min, which is, however, a typical feature of commercial drones, the rationale behind these quadcopters was their simplicity of programming and usage.

4.1   Base Environment Modifications

To allow the creation of the extension described in the previous subsection, some modifications to the base development/simulation environment Ardupilot-Gazebo were needed. In particular, for the multi-UAV extension, the following versions of the environment tools were exploited:

– version 16.04 of Ubuntu;
– the most recent version of Ardupilot;
– ROS Kinetic;
– version 8 of the Gazebo simulator, together with the `ardupilot_sitl_gazebo` plugin for the integration of the two environments (Ardupilot and Gazebo), in a context in which the emulation of drones by means of SITL is exploited, i.e. we can work even in the absence of hardware.

First of all, it was necessary to make some modifications to the Iris drone configuration files, so as to allow the presence of different, uniquely identified drones. To this objective, we modified, in the base configuration files, the model name exploited for the insertion of the particular vehicle in the Gazebo world, and the input/output communication ports for each Iris drone, so as to have them different for each UAV. This enhancement allows a direct information transfer among drones. This will be described in a detailed way in the following subsections.

After having created different Iris UAV instances, we created a Gazebo multi-UAV world, i.e. a "world" that could contain all the created instances together. To this aim, all the single models, identified in a unique way through the model name, were inserted, each one in a given position, which was chosen in the Gazebo world configuration according to the co-ordination algorithm of the application object of the simulation.

For the case study reported in Sect. 5, the Gazebo world with multi-UAVs we created is shown in the code below. For each UAV instance, identified through the `model name` field, information about the initial position (`pose` field) inside the Gazebo world and the physical model of the drone (`uri` field) are provided. For instance, the first drone, identified by the name `iris_uav_instance1`, is placed at position (0,0,0), and it is a Iris UAV (`model://iris_with_standoffs_demo`).

```
<model name="iris_uav_instance1">
  <pose>0 0 0 0 0 0</pose>
  <include>
    <uri>model://iris_with_standoffs_demo</uri>
  </include>
</model>
<model name="iris_uav_instance2">
  <pose>0 10 0 0 0 0</pose>
  <include>
    <uri>model://iris_with_standoffs_demo</uri>
  </include>
</model>
<model name="iris_uav_instance3">
  <pose>0 20 0 0 0 0</pose>
  <include>
    <uri>model://iris_with_standoffs_demo</uri>
  </include>
</model>
<model name="iris_uav_instance4">
  <pose>0 50 0 0 0 0</pose>
  <include>
    <uri>model://iris_with_standoffs_demo</uri>
  </include>
</model>
<model name="iris_uav_instance5">
  <pose>0 100 0 0 0 0</pose>
  <include>
    <uri>model://iris_with_standoffs_demo</uri>
  </include>
</model>
```

Then, for the purposes of applying the co-ordination algorithm to the drones swarm, whose number is apriori defined, some small changes have been made to the Ardupilot Python script `sim_vehicle.py`, for the control of simulated vehicles. Finally, we extended the script `mavproxy.py`, which manages the commands given to drones, using dedicated options.

4.2   Co-ordination Algorithm

For the execution of the co-ordination algorithm:

- we implemented a Python script (named co-ordination script in the follow), executed locally by each drone. Each UAV performs the operations envisaged by the algorithm and communicates with the other drones. The Python script is implemented using the Python Dronekit [1] library, through which it is indeed possible to connect and communicate to a drone. The co-ordination script will be described in detail in Subsect. 5.1.
- we added a port in `sim_vehicle.py` to communicate with the co-ordination script, to get real-time information about the position of drones.
- we defined a new command in `mavproxy.py` in order to start the co-ordination script in an automatic way. In particular, a new function was written in the code, to start a new non-blocking process, executing the co-ordination script.

For the computation of the distance between two drones, the Haversine formula [22] is used, which is the one shown below. Considering two points, each one at a given latitude and longitude, their distance will be given by:

$$d = 2r \arcsin \left(\sqrt{\sin^2 \left(\frac{\phi_2 - \phi_1}{2} \right) + \cos(\phi_1)\cos(\phi_2)\sin^2 \left(\frac{\lambda_2 - \lambda_1}{2} \right)} \right) \quad (1)$$

where ϕ_1 and ϕ_2 are, respectively, latitude of point 1 and latitude of point 2, and λ_1, λ_2 are, respectively, longitude of point 1 and longitude of point 2. Moreover, r is the radius of the earth.

Formula (1), which is commonly exploited in navigation, allows computing the distance between two objects in the Earth, known the positions in the form of geographic coordinates, i.e. the couple (latitude, longitude).

5   An Application Scenario

Among the many possible interaction policies, we focus on the problem of co-ordination of a team of drones, and we present a variation of the classical formation control scheme, based on the well-known consensus protocol described e.g. in [13]. The algorithm in [13] is distributed and allows drones to asymptotically converge to a target point. The co-ordination algorithm we simulate, instead, is obtained by the original one, simply assuming that two drones are fixed at the extreme of a line segment. This variant allows the uniform placement of the drones along the interval and it has not been studied in that work.

As an application scenario, we considered the case of 5 drones that are supposed to coordinate on the interval [0, 100]. The first and fifth vehicles are supposed to be stationary at the outer position of the interval, while the other three must recursively adjust their positions according to the shared co-ordination policy.

5.1 Co-ordination Script

The co-ordination script involves the following operations:

- a connection to the actual drone instance is created, through which it is possible to acquire information by the UAV about its position. In our case, we use only the longitude. The same information will be sent by this drone instance to the closest drone on the left and to the closest drone on the right.
- a control about the type of drone instance, fixed drone (stubborn) or mobile drone, is added to the code. Indeed, in case the drone is a stubborn one, the script will terminate, since fixed drones do not have to perform any movement, according to our modified version of Olfati-Saber et al. algorithm, but they have rather to be fixed in the position assigned to them. Mobile drones have to change their position according to the formula in the previous paragraph. The control on the drone instance is made exploiting the port in the UAV-address which is passed to the script, which identifies uniquely such instance. According to the architecture, the ports related to the single aircraft, and thus their ordering in space, are already known apriori, whereas their relative (geographical) position, at the beginning of the execution of the co-ordination script and in any subsequent instant, are not known.

In the first phase of the script execution, some functions are used to arm drone motors and to make them take off until they reach a height of 10 m, which represents an arbitrarily-chosen height. Then the script contains a loop where each iteration computes a step of the co-ordination algorithm, using the positions of adjacent drones. In particular, a reference longitude is used, the one of the first stubborn UAV, which is considered for the application of the algorithm, and the distance of the other UAVs is computed in relation to this stubborn. Each drone executes the co-ordination algorithm using the actual position of its adjacent drones and its own desired position.

5.2 Simulation

A typical simulation scenario is reported in Fig. 2. Assuming as a reference the leftmost UAV in position 0, in the initial deployment the second, third and fourth drones, are, respectively, placed 10, 20, and 50 m after the first one. The last UAV is placed at the last extreme of the line segment, i.e. 100 m after the first drone. For each of the five drones, the sim_vehicle.py script is launched, in order to create the corresponding MAVLink connection for the control of the vehicle, and the launch file containing the Gazebo world is executed through the roslaunch command. At this point, Gazebo will start, and the 3D simulation environment will show the five drones in their initial placement. The co-ordination script can now be launched. Aircraft will arm their motor and will take off at the predetermined height of 10 m, and each mobile drone will start to communicate

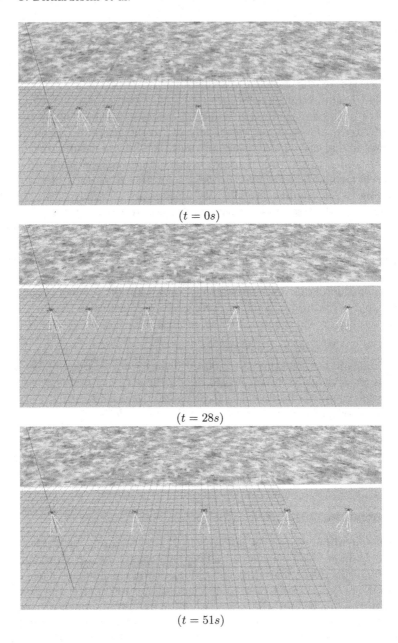

$(t = 0s)$

$(t = 28s)$

$(t = 51s)$

Fig. 2. Dynamic behavior of a team of cooperative UAVs: (a) initial deployment at $t = 0$ s, (b) intermediate displacement at $t = 28$ s, and (c) final deployment at $t = 51$ s.

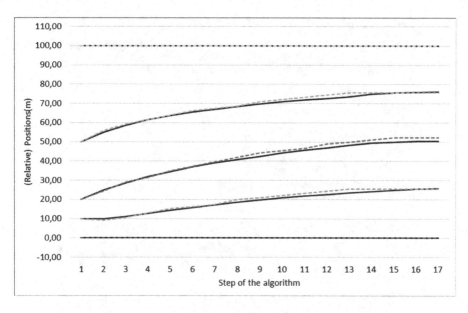

Fig. 3. Actual (dashed lines) and desired (continuous lines) positions of drones during the co-ordination script execution.

with the adjacent UAV and to make some position change, according to the formula shown above. Figure 2 shows the position of drones at the beginning (time $t = 0$ s), in the middle (time $t = 28$ s), and at the end (time $t = 51$ s) of the simulation.

Figure 3 reports the desired positions of the drones, i.e. the positions after the computation of the modified version of the Olfati-Saber et al. algorithm, together with the actual positions of UAVs, during the execution of the simulation. Step #1, Step #9 and Step #17 of the co-ordination algorithm refer to simulation at $t = 0$, $t = 28$ and $t = 51$, respectively.

Another simulation scenario is reported in Fig. 4, where the initial position of mobile drones is at the boundaries of the interval. In the figure, we show again the dynamic behavior of drones and we consider three different phases during the execution of the coordination algorithm.

The results of the simulations show the convergence of the co-ordination algorithm, as expected.

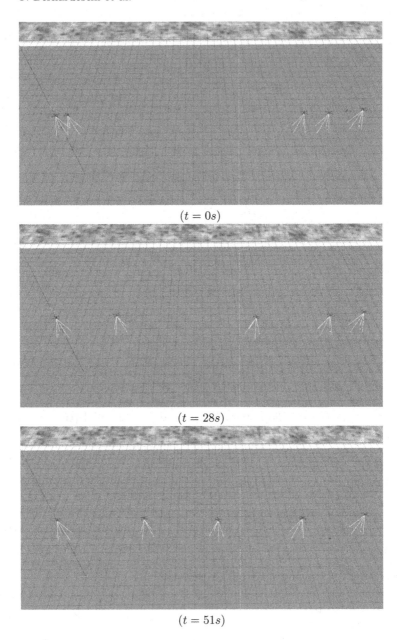

$(t = 0s)$

$(t = 28s)$

$(t = 51s)$

Fig. 4. Dynamic behavior of a team of cooperative UAVs: (a) initial deployment at $t = 0$ s, (b) intermediate displacement at $t = 28$ s, and (c) final deployment at $t = 51$ s.

6 Discussion and Conclusions

This work extends the ROS/Gazebo architecture with the possibility of simulation of co-operative UAVs, adhering to their own individual dynamics as well as interacting with each other according to shared cooperation strategies. Two important features of the framework are its flexibility and modularity. Indeed, heterogeneous vehicles and new co-ordination algorithms can be easily plugged into the simulation, by simply modifying the corresponding Python script. In the paper, preliminary results on the proposed formation control scheme that combine actual positions and desired positions of the vehicles are presented. A limitation of the framework is that launching and configuring a multi-UAVs simulation requires some manual operations. Scalability poses another major issue mostly attributable to the physical-level simulation performed within Gazebo. Finally, although the use of trigonometric functions for distance measurements (Haversine formula) involves small approximation errors, simulations show that the computations have maintained near the theoretical expectations. The automated creation of the UAV instances and the application of the framework to more complex scenarios are objects of further work. Moreover, co-simulation techniques [4,9], in which different sub-systems, possibly modelled and simulated with different tools, are co-ordinated by a co-simulation engine will be investigated.

References

1. 3DRobotics: Dronekit-python's documentation (2016). http://python.dronekit.io/
2. Adams, S.M., Friedland, C.J.: A survey of unmanned aerial vehicle (UAV) usage for imagery collection in disaster research and management. In: 9th International Workshop on Remote Sensing for Disaster Response, p. 8 (2011)
3. ArduPilot-DevTeam: ArduPilot documentation (2016). http://ardupilot.org/ardupilot/
4. Bernardeschi, C., Domenici, A., Masci, P.: A PVS-simulink integrated environment for model-based analysis of cyber-physical systems. IEEE Trans. Softw. Eng. **44**(6), 512–533 (2018)
5. Chandler, P.R., et al.: Complexity in UAV cooperative control. In: Proceedings of the 2002 American Control Conference (IEEE Cat. No. CH37301), vol. 3, pp. 1831–1836. IEEE (2002)
6. Dronecode-Project: MAVlink developer guide (2018). https://mavlink.io/en/
7. Ham, Y., Han, K.K., Lin, J.J., Golparvar-Fard, M.: Visual monitoring of civil infrastructure systems via camera-equipped unmanned aerial vehicles (UAVs): a review of related works. Vis. Eng. **4**(1) (2016)
8. Koenig, N.P., Howard, A.: Design and use paradigms for Gazebo, an open-source multi-robot simulator. In: IROS, vol. 4, pp. 2149–2154. Citeseer (2004)
9. Larsen, P.G., et al.: Integrated tool chain for model-based design of cyber-physical systems: the INTO-CPS project. In: 2016 2nd International Workshop on Modelling, Analysis, and Control of Complex CPS (CPS Data), pp. 1–6, April 2016
10. Lu, P., Geng, Q.: Real-time simulation system for UAV based on Matlab/Simulink. In: 2011 IEEE 2nd International Conference on Computing, Control and Industrial Engineering (CCIE), vol. 1, pp. 399–404. IEEE (2011)

11. Maza, I., Caballero, F., Capitán, J., Martínez-de Dios, J.R., Ollero, A.: Experimental results in multi-UAV coordination for disaster management and civil security applications. J. Intell. Robot. Syst. **61**(1–4), 563–585 (2011)

12. Meyer, Johannes, Sendobry, Alexander, Kohlbrecher, Stefan, Klingauf, Uwe, von Stryk, Oskar: Comprehensive Simulation of Quadrotor UAVs Using ROS and Gazebo. In: Noda, Itsuki, Ando, Noriaki, Brugali, Davide, Kuffner, James J. (eds.) SIMPAR 2012. LNCS, vol. 7628, pp. 400–411. Springer, Heidelberg (2012). https://doi.org/10.1007/978-3-642-34327-8_36

13. Olfati-Saber, R., Fax, J.A., Murray, R.M.: Consensus and cooperation in networked multi-agent systems. Proc. IEEE **95**(1), 215–233 (2007)

14. Ollero, A., et al.: AWARE: platform for autonomous self-deploying and operation of wireless sensor-actuator networks cooperating with unmanned aerial vehicles. In: 2007 IEEE International Workshop on Safety, Security and Rescue Robotics, SSRR 2007, pp. 1–6. IEEE (2007)

15. OSRF: Gazebo API reference (2017). http://osrf-distributions.s3.amazonaws.com/gazebo/api/8.2.0/index.html

16. OSRF: ROS Wiki: documentation (2018). http://wiki.ros.org/

17. PX4-DevTeam: Pixhawk series (2018). https://docs.px4.io/en/flight_controller/

18. Quigley, M., et al.: ROS: an open-source robot operating system. In: ICRA Workshop on Open Source Software, Kobe, Japan, vol. 3, p. 5 (2009)

19. Quintero, S.A., Papi, F., Klein, D.J., Chisci, L., Hespanha, J.P.: Optimal UAV coordination for target tracking using dynamic programming. In: 2010 49th IEEE Conference on Decision and Control (CDC), pp. 4541–4546. IEEE (2010)

20. Rasmussen, S.J., Chandler, P.R.: MultiUAV: a multiple UAV simulation for investigation of cooperative control. In: 2002 Proceedings of the Winter Simulation Conference, vol. 1, pp. 869–877. IEEE (2002)

21. Remondino, F., Barazzetti, L., Nex, F., Scaioni, M., Sarazzi, D.: UAV photogrammetry for mapping and 3D modeling-current status and future perspectives. Int. Arch. Photogramm. Remote Sens. Spat. Inf. Sci. **38**(1), C22 (2011)

22. Robusto, C.C.: The cosine-haversine formula. Am. Math. Mon. **64**(1), 38–40 (1957)

23. Rysdyk, R.: Unmanned aerial vehicle path following for target observation in wind. J. Guid. Control Dyn. **29**(5), 1092–1100 (2006)

24. Semsch, E., Jakob, M., Pavlicek, D., Pechoucek, M.: Autonomous UAV surveillance in complex urban environments. In: Proceedings of the 2009 IEEE/WIC/ACM International Joint Conference on Web Intelligence and Intelligent Agent Technology, vol. 02, pp. 82–85. IEEE Computer Society (2009)

25. Techy, L., Woolsey, C.A., Schmale, D.G.: Path planning for efficient UAV coordination in aerobiological sampling missions. In: 2008 47th IEEE Conference on Decision and Control, CDC 2008, pp. 2814–2819. IEEE (2008)

26. Tortonesi, M., Stefanelli, C., Benvegnu, E., Ford, K., Suri, N., Linderman, M.: Multiple-UAV coordination and communications in tactical edge networks. IEEE Commun. Mag. **50**(10), 48–55 (2012)

27. Wise, R., Rysdyk, R.: UAV coordination for autonomous target tracking. In: AIAA Guidance, Navigation, and Control Conference and Exhibit, 6453 (2006)

28. Pyo, Y., Cho, H., Jung, L., Lim, D.: ROS Robot Programming (English). ROBOTIS, December 2017

29. Zhang, C., Kovacs, J.M.: The application of small unmanned aerial systems for precision agriculture: a review. Precis. Agric. **13**(6), 693–712 (2012)

Real-Time Localization of Transmission Sources by a Formation of Helicopters Equipped with a Rotating Directional Antenna

Václav Pritzl[1], Lukáš Vojtěch[2], Marek Neruda[2], and Martin Saska[1(✉)]

[1] Department of Cybernetics, Czech Technical University in Prague,
Prague, Czech Republic
{pritzvac,martin.saska}@fel.cvut.cz
[2] Department of Telecommunication Engineering,
Czech Technical University in Prague, Prague, Czech Republic
{vojtecl,marek.neruda}@fel.cvut.cz
http://mrs.felk.cvut.cz/

Abstract. This paper proposes a novel technique for radio frequency transmission sources (RFTS) localization in outdoor environments using a formation of autonomous Micro Aerial Vehicles (MAVs) equipped with a rotating directional antenna. The technique uses a fusion of received signal strength indication (RSSI) and angle of arrival (AoA) data gained from dependencies of RSSI on angle measured by each directional antenna. An Unscented Kalman Filter (UKF) based approach is used for sensor data fusion and for estimation of RFTS positions during each localization step. The proposed method has been verified in simulations using noisy and inaccurate measurements and in several successful real-world outdoor deployments.

Keywords: RFID localization · Micro Aerial Vehicles ·
Unscented Kalman Filter · Directional antenna ·
Radio frequency transmission sources localization

1 Introduction

Fast and precise radio frequency (RF) transmission sources localization is a challenging task required in numerous application scenarios. Active radio frequency identification (RFID) tags are commonly used in many industrial applications, such as finding working tools or machinery in construction sites and localization and identification of stock items in warehouses. RFIDs are conveniently used in agriculture for livestock tracking in order to monitor cattle health, prevent cattle rustling or localize lost animals. Tracking of endangered species is another widespread use of RFID tags. Furthermore, RF localization brings an undeniable benefit to searching for people during natural disasters or search

© Springer Nature Switzerland AG 2019
J. Mazal (Ed.): MESAS 2018, LNCS 11472, pp. 335–350, 2019.
https://doi.org/10.1007/978-3-030-14984-0_25

Fig. 1. MAV with directional antennas

and rescue operations (such as localizing people in avalanches using special RF devices or looking for missing people by tracking their mobile phones). Military applications include localizing wounded soldiers on the battlefield or using RF localization to substitute the GPS system in case of GPS jamming or operating in indoor spaces. In all of these cases, the speed, precision, and reliability of the localization are extremely important.

The use of Micro Aerial Vehicles (MAVs) has recently experienced a great surge in popularity and new applications for MAV use emerge every day. In these scenarios, mainly the ability to quickly reach distant and possibly dangerous places makes them especially appealing. In a structured environment, MAVs flying above the obstacles also reduce the influence of signal reflection and interference. Moreover, cooperatively acting groups of MAVs are often the only tool that can be used for localization of moving RFIDs with short detection range in a hardly accessible environment and their usage can significantly extend the application domain of RF localization (Fig. 1).

Several different approaches to RF localization with variable requirements in terms of cost, required infrastructure, localization speed and precision can be used. For the utilization of MAVs, we propose a localization method based on combined Received Signal Strength Indication (RSSI) and Angle of Arrival (AoA) data obtained from directional antennas mounted on board the MAVs. The proposed approach enables to fully exploit the ability of an MAV formation to quickly change its position and shape in order to quickly and precisely determine the position of the localized object. The advantage of a cooperating group of autonomously flying micro-size aerial vehicles is twofold. They enable to gain the sensory information in multiple locations simultaneously, which is crucial for the localization of moving objects and which speeds up the localization process of stationary RFIDs. The fast mobility of onboard sensory equipment brings the possibility of quick change of mutual distances between the measurement locations and of the distance between the formation and the estimated RFID position. Different mutual distances correspond to different baseline used for source position estimation. Usually, larger formations are required for localization of distant RFTS and the MAVs should fly closer for final position estimation. Similarly, the advantage of the possibility to move towards the estimated RFTS position increases the precision and reliability of the localization as signal strength decreases with the power of distance.

1.1 State of the Art

Numerous approaches to RF localization by aerial as well as ground robotic systems have already been deployed in numerous application scenarios. The simplest systems are using RSSI values measured by an omnidirectional antenna. This method does not require complicated infrastructure and only cheap RF devices can be used. However, it requires knowledge of parameters of signal propagation in the given environment and it suffers from radio disturbance, multipath effects, shadowing and other effects influencing radio transmission propagation.

This method is explored in [1] which deals with the use of MAVs in RFID localization for environmental monitoring. Specifically, it uses an RSSI-based multilateration approach, which estimates the position of the localized chip using a least-squares method. In [2], the use of a particle filter on RSSI data measured by a UAV sweeping a large outdoor area is tested. Similarly, [3] proposes a method for Wi-Fi devices localization in a large region using UAVs to collect RSSI data. A Bayesian optimization based on Gaussian process regression is used for the localization in this approach. In [4], tracking of an intermittent RF source using a UAV swarm measuring RSSI data is researched. Two algorithms for localization are compared - EKF and a recursive Bayesian estimator. Furthermore, the paper compares two trajectory planning algorithms for RF localization - steepest descent posterior Cramer-Rao lower bound path planning and a bio-inspired heuristic path planning.

Another possible approach is estimation of AoA information from the dependency of RSSI on angle measured by a directional antenna. This approach is used in [5], where a directional antenna is mounted on top of an MAV, which rotates around its vertical axis, and a particle filter is used for RF source localization. In [6], autonomous navigation of a mobile ground robot towards an RF source using AoA information and a particle filter is explored. Similarly, directional RSSI-based localization using a mobile robot carrying a corner reflector antenna and an online statistical filter is researched in [6].

In order to decrease the influence of multipath effects on successful localization, the aforementioned approaches have to be combined. In [7,8], methods for localization in Non Line of Sight condition from coupled RSSI and AoA measurements using a particle filter and a multi-step Gaussian filtering approach, respectively, are proposed.

Another option is the use of Time Difference of Arrival (TDoA), which is based on the measurement of the difference in time between the arrival of the transmission to multiple receivers. RF localization using TDOA measurements from 2 UAVs and a comparison of an EKF and UKF approach are described in [9]. Similarly, [10] proposes a dual-EKF algorithm used to localize an RF emitter from TDoA data measured by two UAVs. [11] proposes a TDOA-based method utilizing a least squares approach for localization using UAVs in battlefield environments as a substitute for Global Navigation Satellite Systems (GNSS). Although the TDoA approach is resistant to multipath effects, it requires complicated and expensive infrastructure in order to achieve a precise time synchronization of the RF receivers.

This paper proposes a combined RSSI and AoA approach as a proper technique for RF localization by cooperative teams of MAVs. The proposed approach explores the use of a dedicated device for rotating the antenna and increases the localization precision by dynamically estimating the uncertainty of AoA measurements which is not utilized in any of the aforementioned works. Furthermore, it combines the hybrid RSSI and AoA technique with the use of a multi-MAV formation, while the above-mentioned works, which propose the use of both RSSI and AoA, use only a single UAV and verify the proposed techniques in simulations only.

1.2   Problem Statement

The proposed localization system consists of multiple autonomous cooperatively working MAVs flying in a formation. A team of three MAVs was employed in the presented experiments to show minimal system requirements, but the system is easily scalable for larger teams in case that a larger field needs to be explored. Numerical analyses have shown that (obviously) a larger team of MAVs enables to find the RFTS faster and with higher precision due to the added information from another measurement location provided simultaneously. The proposed system is also flexible in terms of the number of beacons to be localized. In the problem solution, we expect that positions of the beacons within a given area are completely unknown beforehand. The beacons are active RFID tags capable of communication with directional antennas mounted on each MAV.

The antennas are able to send a message to each beacon and measure RSSI of the beacon's response. The antenna is able to rotate itself to a specific position around its vertical axis and using this feature it is possible to measure the current dependency of RSSI on angle in 360°. Using this measured dependency it is possible to simply and reliably estimate distance and angle (bearing) between each MAV and the beacon being localized. The environment where the localization is performed is assumed to be without obstacles for MAVs to simplify motion planning algorithm and to be able to focus this paper on the localization itself.

For purpose of the RF localization, it is further assumed that the beacon is placed on the ground, at zero altitude, which correlates with all of the above-mentioned motivation scenarios, where the objects to be localized are expected to be placed in positions with almost zero z-coordinate in comparison with MAVs altitude and the estimation of beacon's altitude is not required. Such a practical constraint reduces the localization task to two dimensions.

Moreover, it is assumed that the positions of the MAVs are accurately known, e.g. using GPS or any other localization system suited for multi MAV teams. See [12–15], for examples of methods designed for precise onboard mutual localization of MAVs in a team that we have designed to reliably solve the given task in real-world conditions, where precision of GPS may not be sufficient.

2   System Model

2.1   State-Space Representation of System

A discrete state-space model of the localization system for m MAVs and one beacon is used as

$$\boldsymbol{x}_{k+1} = \mathbf{A}\boldsymbol{x}_k + \boldsymbol{v}_k, \tag{1}$$

$$\boldsymbol{z}_k = \boldsymbol{h}(\boldsymbol{x}_k, \boldsymbol{u}_k) + \boldsymbol{w}_k, \tag{2}$$

where $\boldsymbol{x}_k$ is the 2-dimensional state of the system at timestep k containing Cartesian coordinates of the localized beacon (the localization task has been reduced to two dimensions as described in Sect. 1.2), $\boldsymbol{v}_k$ is the state noise process vector, $\boldsymbol{z}_k$ is the observation vector, $\boldsymbol{h}$ is the measurement function, $\boldsymbol{u}_k$ is the input vector containing x, y and z coordinates of the currently used MAV and $\boldsymbol{w}_k$ is the measurement noise. Matrix $\mathbf{A}$ is a 2-dimensional identity matrix which represents that position of the localized beacon is static.

$$\mathbf{A} = \begin{bmatrix} 1 & 0 \\ 0 & 1 \end{bmatrix} \tag{3}$$

The state vector, input vector and measurement vector are defined as

$$\boldsymbol{x}_k = [x_B, y_B]^T,$$

$$\boldsymbol{u}_k = [x_M, y_M, z_M]^T,$$

$$\boldsymbol{z}_k = [RSS, \theta]^T,$$

where x_B and y_B denote the Cartezian coordinates of the localized beacon. x_M, y_M, z_M are the coordinates of the MAV whose sensor data are currently used for localization. RSS is the average RSSI of the dependency of RSSI on angle measured by the MAV and θ is the estimated AoA obtained from the MAV. The measurement function is defined as

$$\boldsymbol{h} = \begin{bmatrix} P_0 - 10\gamma\log_{10}(\sqrt{(x_M - x_B)^2 + (y_M - y_B)^2 + (z_M - z_B)^2}) \\ \mathrm{atan2}(y_B - y_M, x_B - x_M) \end{bmatrix}.$$

where atan2 is the four-quadrant inverse tangent defined as

$$\text{atan2}(x,y) = \begin{cases} \arctan(\frac{y}{x}) & \text{if } x > 0 \\ \arctan(\frac{y}{x}) + \pi & \text{if } x < 0 \text{ and } y \geq 0 \\ \arctan(\frac{y}{x}) - \pi & \text{if } x < 0 \text{ and } y < 0 \\ +\frac{\pi}{2} & \text{if } x = 0 \text{ and } y > 0 \\ -\frac{\pi}{2} & \text{if } x = 0 \text{ and } y < 0 \\ \text{undefined} & \text{if } x = 0 \text{ and } y = 0. \end{cases}$$

The first row of the measurement function vector contains RSSI calculation from 3D distance between the beacon and the MAV according to Eq. (6) and the second row of the function contains calculation of estimated MAV-beacon angle from position of the beacon and the MAV. It can be seen that the measurement function is highly nonlinear which highlights the necessity of using a type of Kalman filter designed for nonlinear systems. The process and measurement noise vectors are defined as

$$\boldsymbol{v}_k \sim \mathcal{N}(0, \mathbf{Q}_k),$$
$$\boldsymbol{w}_k \sim \mathcal{N}(0, \mathbf{R}_k),$$

where $\mathbf{Q}_k$ is the process noise covariance matrix defined as

$$\mathbf{Q}_k = \begin{bmatrix} q^2 & 0 \\ 0 & q^2 \end{bmatrix},$$

where q is a constant parameter determined from the UKF performance on experimental data and $\mathbf{R}_k$ is the measurement noise covariance matrix defined as

$$\mathbf{R}_k = \begin{bmatrix} \sigma_{RSS}^2 & 0 \\ 0 & \sigma_\theta^2 \end{bmatrix}.$$

The average RSSI standard deviation σ_{RSS} is a constant value identified from the performance of the filter on real experimental data while estimated AoA uncertainty σ_θ is changed in every step.

The localization system contains m MAVs, each passing measured data to the filter in every position of measurement. The measured data (vectors $\boldsymbol{u}_k$ and $\boldsymbol{z}_k$) are passed to the filter sequentially resulting in m filter steps for every formation measurement position.

2.2   Signal Strength Model

The dependency of received signal strength on the distance between transmitter and receiver is influenced by multipath effects caused by signal reflection from the ground and obstacles, shadowing and other propagation effects occurring under real-world conditions. A brief summary of effects influencing the transmission propagation can be found at [16]. To model this dependency and account for these influences the Log-distance path loss model

$$PL(d) = \overline{PL}(d_0) + 10\gamma \log_{10}(\frac{d}{d_0}) + \chi \tag{4}$$

is used. The received signal strength then equals

$$P_r = P_t - PL \tag{5}$$

where P_r is the power received by the receiver antenna in dBm, and P_t is the power delivered to the transmitting antenna. $PL(d)$ is the path loss in dB at distance d, $\overline{PL}(d_0)$ is the mean path loss at reference distance d_0 and γ is the path loss exponent. $\chi \in (0, \sigma^2)$ is normally (Gaussian) distributed random variable with zero mean and standard deviation σ that represents the effects of multipath, shadowing and radio disturbance on the transmission. By combining Eqs. (5) and (4) and substituting $d_0 = 1$ a dependency

$$P_r = P_0 - 10\gamma\log_{10}(d) + \chi \tag{6}$$

of RSSI P_r on distance d between transmitter and receiver which contains two parameters P_0 and γ is obtained. These parameters depend on transmitter and receiver properties and on the environment where the signal spreads and can be experimentally identified by measuring the dependency of RSSI on transmitter-receiver distance and fitting the function (6) to the data using a least squares method.

2.3   Antenna Radiation Pattern and MAV-Beacon Bearing

The radiation properties of an antenna in a particular direction are characterized by its radiation pattern. Portions of the radiation pattern bounded by regions of relatively weak radiation intensity are called radiation lobes. Usually, they are subclassified into main, side and back lobes, the main lobe containing the direction of maximum radiation [17].

The directivity of the antenna is characterized by its gain in dBi (decibels isotropic). It represents the ratio of the radiation intensity in a given direction to the radiation intensity of a lossless isotropic antenna (radiating equally in all directions). Specifically, the antenna used in this paper has a gain of 8 dBi in the direction of the main lobe.

The shape of the dependency of RSSI on angle, obtained by rotating the antenna around a fixed axis and measuring RSSI in particular directions, corresponds to the shape of the radiation pattern of the antenna but is deformed due to various propagation effects influencing the transmission under real-world conditions.

The angle θ representing the bearing from the particular antenna to the beacon being localized can be estimated from the location of the main lobe in the measured dependency of RSSI on angle. The angle θ is defined as the horizontal angle in the x-y plane, $\theta = 0$ corresponds to positive half of the x axis, and θ is positive in the direction of the positive half of the y axis.

3   Localization Algorithm with Uncertainty Estimation

3.1   Dependency of RSSI on Angle Measurement

The dependency of RSSI on angle, described in Sect. 2.3, is measured by using the step motor to rotate the antenna in 360° sampled into 128 positions around its z axis. In every position, a constant number of RSSI samples is measured.

3.2   Preprocessing of Measurements

The measured dependency is first preprocessed in order to filter out the noise caused by RF propagation effects under real-world conditions. First, an average RSSI is calculated in each step motor position, where a measurement was made. An average RSSI of the whole dependency is then calculated for estimation of the transmitter-receiver distance. Then, the whole dependency is passed through moving mean in order to further smoothen the data. The transmission AoA is then determined from this preprocessed dependency as the angle with maximal RSSI or an average angle in case of multiple positions sharing the same RSSI. Figure 2 depicts an example of a measured dependency containing the individual samples, the dependency after averaging the RSSI in each position, the dependency after applying moving mean and the detected AoA.

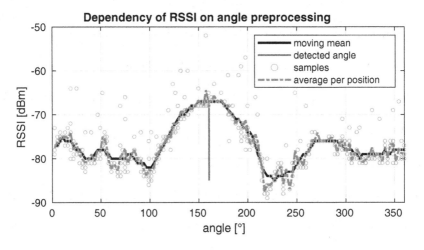

Fig. 2. Preprocessing of measured dependency of RSSI on angle

3.3   Uncertainty of Angle Measurement Estimation

After preprocessing of the measured data and estimating the angle of arrival, an uncertainty of the current AoA measurement is estimated from the angle between the detected AoA and the angle, where the measured RSSI drops 2 dBm below the maximal RSSI. It serves as a weight of the measurement in localization and is further used for determining whether the measurement should be used in localization and then for construction of the measurement noise matrix $\mathbf{R}$. The 2 dBm value was chosen because of high correlation of the corresponding angle with the error of AoA estimation in real experimental data.

3.4 Unscented Kalman Filter

Unscented Kalman filter (UKF), described in [18], is used to fuse the measured data from all MAVs and estimate the position of the beacon being localized. The UKF implementation is in detail described in [19].

The UKF contains a validation gate which rejects the measurements which are too distant from the current state of the filter. This way the robustness of the UKF is greatly improved. The topic of validation gating is described in [20] and its implementation using the calculation of the Normalized estimation error squared is described in [19].

If the algorithm localizes more beacons at the same time, a separate instance of UKF is created for every beacon.

The UKF initial state estimate x_0 is set to be in the center of the MAV formation.

4 Experiments

MAV platforms built by the CTU Multi-robot System group were used in the experiments. Each used MAV is a hexacopter controlled by a Pixhawk unit running Robot Operating System (ROS). Real-time kinematic (RTK) positioning in combination with Global Navigation Satellite System (GNSS) is used in order to get high-precision position data for each MAV. Every MAV offers approximately 15 min of flight time depending on flight conditions. The real world experiments were performed either with the use of a single drone or a formation of 3 MAVs. More information about the used MAV hardware can be found in [21,22]. Information about the Model Predictive Control (MPC) used to control the MAVs can be found in [23,24]. The MAVs use a decentralized collision avoidance system [25] in order to enable a safe execution of real experiments.

XBee S2C radio module with an integrated wire antenna is used as the localized beacon. Technical specifications of the XBee S2C module can be found at[1].

The rotating antenna device was designed by Matouš Vrba from Multi-robot Systems group at CTU within his work on RFID tags localization. The rotating antenna device consists of XBee S2C module with a CW8DPA patch directional antenna connected to XBee via RP-SMA connector. The antenna gain is 8 dBi and its operating frequency band is 2.4–2.5 GHz.

---

[1] https://www.digi.com/resources/documentation/digidocs/pdfs/90002002.pdf (last accessed on 31/10/2018).

The antenna and RF module are mounted on a 28BYJ-48 step motor that is used for rotating the antenna. An Arduino Nano is used to control the step motor. The step motor is capable of rotating the antenna to 128 different positions which equals a step of approximately 2.8°. For localization purposes, the antenna is mounted below the MAV.

4.1   Dependency of Average RSSI on Distance

For the purpose of localization algorithm design, the parameters P_0 and γ from Eq. (6) needed to be identified. In order to achieve this, a measurement using a beacon and one MAV was performed. This experiment was carried out on an empty field in order to eliminate the effect of shadowing and minimize multipath effects interfering with the RSSI values. The beacon was placed on the ground at zero altitude and the MAV gradually flew away from the beacon in a straight line while measuring the current dependency of RSSI on angle in multiple points along its trajectory.

This measurement was performed at 3 different altitudes. First, the MAV flew in the altitude of 5 m and in 2D distance from 0 up to 50 m away from the beacon. 18 different dependencies were measured along this trajectory. During the second measurement, the MAV moved from 2 to 20 m away in the altitude of 2.5 m while doing a measurement in 8 points and next the MAV repeated the same 8 point measurement in the altitude of 7.5 m. A short video from this experiment can be seen on youtube.[2]

An average RSSI was calculated for every dependency measured as described in Sect. 3.2. The dependency was comparable for all 3 altitudes although it contained more noise in shorter distances due to multipath effects.

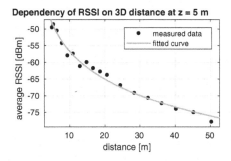

Fig. 3. Average RSSI values measured at $z = 5$ m with fitted curve

The Eq. (6) was fitted to the data measured from the 5 m altitude using a least squares method. This way the parameters were identified as

$$P_0 = -29.06, \ \gamma = 2.765.$$

Figure 3 shows the measured data along with the fitted curve.

---

[2] https://youtu.be/lpT_dYN07Gg (last accessed on 31/10/2018).

Furthermore, by calculating the deviation of the angle with maximal RSSI from the real AoA it was discovered that the deviation is larger in a closer distance and in higher altitudes. This is probably caused by a bigger vertical angle between the antenna and the beacon which increases as the distance gets smaller and as the MAV altitude grows. This highlights the advantage of using a larger formation moving at a lower altitude during the localization.

4.2   One MAV Following a Rectangular Trajectory

This experiment was performed on a large empty outdoor urban area. It was discovered that the RSSI path loss parameters differ from the parameters identified in Sect. 4.1. This can be caused by bigger radio interference in the urban environment. To account for this, new parameters were identified from data measured in this experiment as

$$P_0 = -33.96, \ \gamma = 3.$$

One MAV was following a rectangular trajectory around the beacon and during this time it made 14 different measurements. The first MAV position was chosen as the initial estimate of beacon position. The positions where the MAV made its measurements along with the detected AoA can be seen in Fig. 4b. The beacon was placed at x: 0, y: −7. The MAV started at the lower left corner of the rectangle and then followed the trajectory in a counter-clockwise direction.

The RMSE of average RSSI values from the theoretical fitted curve is 2.6 dBm. The AoA uncertainty σ_{theta} and AoA estimation error were tested for correlation by calculating the Pearson correlation coefficient and using Student's t-test [26] to calculate statistical significance of the discovered correlation. The correlation coefficient is 0.80 at statistical significance 0.0005.

Figure 4a shows the progression of UKF localization error over the number of performed filter steps. The error is calculated as the Euclidean distance between the current position estimate and the real beacon position. It can be seen that the error dropped from 13.39 m to 4.22 m in just 5 filter steps but then stayed roughly the same for the rest of the localization. The final localization error is 3.88 m. Figure 4b contains the progression of estimated beacon position in xy plane along with covariance ellipse representing the 95% confidence area of the final estimate. It can be seen that the true beacon position lies inside this area.

4.3   MAV Formation Sweeping a Large Area

This experiment was performed in the same place as experiment Sect. 4.1, therefore the UKF uses the path loss parameters identified during that experiment. A formation of 3 MAVs and one beacon were used during this experiment. The formation swept a 65 by 50 m large area by traveling to 16 predefined positions resulting in 48 different measurements. A video of the experiment can be seen on youtube.[3] Figure 5a contains a picture from the video depicting measurement

---

[3] https://youtu.be/oLTGGI9qxHQ (last accessed on 31/10/2018).

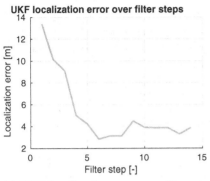
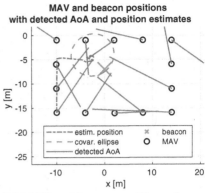

(a) UKF localization error over filter steps (b) MAV and beacon positions along with detected AoA and beacon position estimates

Fig. 4. MAV following a rectangular trajectory - localization error, detected AoA and position estimate development

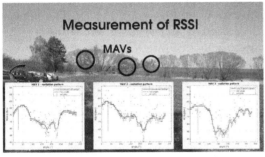
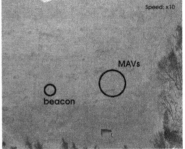

(a) Measurement of dependencies of RSSI on angle   (b) Localization of the beacon

Fig. 5. Formation of 3 MAVs following predefined trajectories - screenshots from the video of the experiment

of the individual dependencies of RSSI on angle. Figure 5b contains another picture from the video which shows the MAV formation following the predefined trajectory and localizing the beacon.

The positions where the measurements were performed along with the formation trajectory and the detected AoAs can be seen in Fig. 6. The correlation of σ_θ and AoA estimation error is 0.29 at 0.043 statistical significance. The average RSSI values along with the theoretical curve are shown in Fig. 7b. The RMSE of the values from the theoretical curve is 3.45 dBm.

During this experiment, the largest number of correct measurements was collected therefore the data from this experiment were used for tuning the parameters of the UKF with the goal of minimizing the final localization error and the

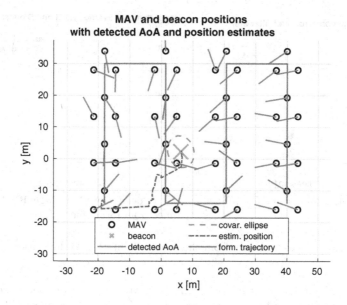

Fig. 6. MAV formation following a predefined trajectory - MAV and beacon positions, detected AoAs and beacon position estimates

performance of the filter with these parameters was then verified on data gained from the other experiment.

Figure 7a shows the progression of UKF localization error over the number of performed filter steps for 3 different settings of the filter. In the $RSSI + AoA$ setting the **R** matrix was tuned so that the UKF relied both on the RSSI and AoA information. It can be seen that its localization error drops to 1.66 m in 25 filter steps and then keeps slowly converging to the real beacon position. The final localization error is 0.04 m.

The AoA setting uses just the AoA data for localization which was done by setting the σ_{RSS} parameter to a very large value (1000 dBm) and tuning the **Q** matrix for the best possible localization results. In this case, the final localization error is 5.92 m. The $RSSI$ setting uses only the RSSI data which is set by multiplying the σ_θ parameters by a large scalar value (1000) and tuning the **Q** matrix and σ_{RSS} parameter to produce a good and stable filter performance. The performance is again worse than the performance of the hybrid $RSSI + AoA$ filter and the final localization error is 3.19 m.

Figure 6 also depicts the progression of the $RSSI + AoA$ UKF estimated position in the xy plane. It can be seen that the estimate starts with its initial value in the first formation position and then gradually converges to the real beacon position. A covariance ellipse of the final estimate representing the 95% confidence area is plotted as well.

This experiment lasted approximately 10 min and the localization error dropped under 1.66 m in approximately 5 min. However, the localization time

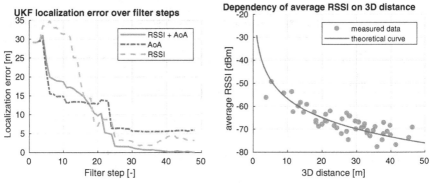

(a) UKF localization error for different fil-(b) Dependency of average RSSI on dis-
ters   tance from formation experiment

Fig. 7. Measured data from the second flight of MAV formation

could be significantly reduced by taking a lower number of measurements in
each position and recording less samples during each measurement. This experiment was designed to record as much data as possible for further analysis.

5   Conclusion

A method for localization of sources of RF transmission using a formation of
relatively localized MAVs equipped with a rotating directional antenna was proposed in this paper. A dependency of RSSI on angle is measured by rotating the
antenna using a step motor. The proposed algorithm calculates the average RSSI
value of each measured dependency of RSSI on angle and estimates AoA of the
transmission along with its uncertainty. A UKF-based approach is used for data
fusion and estimation of the RF beacon position. Using the UKF, insufficiently
precise measurements are detected and rejected in order to improve localization
performance.

The proposed algorithm was implemented in ROS and its functionalities were
verified in simulations. Multiple different real-world experiments were performed
to verify the system functionalities with real MAV platforms and test the performance of the proposed approach.

It has been shown that the proposed algorithm performs well in real-world
conditions. The algorithm benefits from the use of coupled RSSI and AoA measurements in sense of increased reliability and precision. The utilization of an
MAV formation enables to quickly obtain a large number of measurements from
various positions which further improves the localization performance.

Acknowledgments. This research was supported the Grant Agency of the Czech
Republic under grant no. 17-16900Y, by student CTU grant no. SGS17/187/OHK3/
3T/13, and by OP VVV MEYS funded project CZ.02.1.01/0.0/0.0/16_019/0000765
"Research Center for Informatics".

References

1. Greco, G., Lucianaz, C., Bertoldo, S., Allegretti, M.: Localization of RFID tags for environmental monitoring using UAV, pp. 480–483. IEEE, September 2015
2. Wagle, N., Frew, E.: A particle filter approach to WiFi target localization. American Institute of Aeronautics and Astronautics, August 2010
3. Carpin, M., Rosati, S., Khan, M.E., Rimoldi, B.: UAVs using Bayesian optimization to locate WiFi devices, p. 5 (2015)
4. Koohifar, F., Guvenc, I., Sichitiu, M.L.: Autonomous tracking of intermittent RF source using a UAV swarm. arXiv:1801.02478 [cs, eess], January 2018
5. Isaacs, J.T., Quitin, F., Garcia Carrillo, L.R., Madhow, U., Hespanha, J.P.: Quadrotor control for RF source localization and tracking, pp. 244–252. IEEE, May 2014
6. Jeong, D., Lee, K.: Directional RSS-based localization for multi-robot applications. In: Communications and Signal Processing, p. 7 (2013)
7. Dehghan, S.M.M., Farmani, M., Moradi, H.: Aerial localization of an RF source in NLOS condition. In: IEEE International Conference on Robotics and Biomimetics, pp. 1146–1151. IEEE, December 2012
8. Dehghan, S.M.M., Moradi, H.: A multi-step Gaussian filtering approach to reduce the effect of non-Gaussian distribution in aerial localization of an RF source in NLOS condition. In: RSIIISM International Conference on Robotics and Mechatronics, p. 6 (2013)
9. Fletcher, F., Ristic, B., Musicki, D.: Recursive estimation of emitter location using TDOA measurements from two UAVs, pp. 1–8. IEEE (2007)
10. Lee, S.C., Lee, W.R., You, K.H.: TDoA based UAV localization using dual-EKF algorithm. Int. J. Control Autom. $2(4)$, 8 (2009)
11. Kim, D.H., Lee, K., Park, M.Y., Lim, J.: UAV-based localization scheme for battlefield environments. In: MILCOM 2013–2013 IEEE Military Communications Conference, pp. 562–567, November 2013
12. Krajník, T., et al.: A practical multirobot localization system. J. Intell. Robot. Syst. **76**, 539–562 (2014)
13. Faigl, J., Krajnik, T., Chudoba, J., Preucil, L., Saska, M.: Low-cost embedded system for relative localization in robotic swarms. In: IEEE International Conference on Robotics and Automation, pp. 993–998. IEEE, May 2013
14. Walter, V., Staub, N., Saska, M., Franchi, A.: Mutual localization of UAVs based on blinking ultraviolet markers and 3D time-position Hough transform. In: Accepted to the Fourteenth Annual IEEE International Conference on Automation Science and Engineering (IEEE CASE 2018) (2018)
15. Walter, V., Saska, M., Franchi, A.: Fast mutual relative localization of UAVs using ultraviolet LED markers. In: 2018 International Conference of Unmanned Aircraft System (ICUAS 2018) (2018)
16. Jain, R.: Channel models - a tutorial. In: WiMAX Forum AATG, p. 21, February 2007
17. Balanis, C.A.: Antenna Theory: Analysis and Design, 3rd edn. Wiley, Hoboken (2005)
18. Julier, S.J., Uhlmann, J.K.: A new extension of the Kalman filter to nonlinear systems, pp. 182–193 (1997)
19. Labbe Jr., R.R.: Kalman and Bayesian filters in python, p. 491 (2015)
20. Bailey, T., Upcroft, B., Durrant-Whyte, H.: Validation gating for non-linear non-Gaussian target tracking. In: 2006 9th International Conference on Information Fusion, pp. 1–6, July 2006

21. Spurný, V., et al.: Cooperative autonomous search, grasping and delivering in a treasure hunt scenario by a team of UAVs. J. Field Robot. **36**, 35 (2018)
22. Loianno, G., et al.: Localization, grasping, and transportation of magnetic objects by a team of MAVs in challenging desert like environments. IEEE Robot. Autom. Lett. **3**, 8 (2018)
23. Baca, T., Loianno, G., Saska, M.: Embedded model predictive control of unmanned micro aerial vehicles. In: 21st International Conference on Methods and Models in Automation and Robotics (MMAR), pp. 992–997. IEEE, August 2016
24. Spurny, V., Baca, T., Saska, M.: Complex manoeuvres of heterogeneous MAV-UGV formations using a model predictive control. In: 21st International Conference on Methods and Models in Automation and Robotics (MMAR), pp. 998–1003. IEEE, August 2016
25. Baca, T., Hert, D., Loianno, G., Saska, M., Kumar, V.: Model predictive trajectory tracking and collision avoidance for reliable outdoor deployment of unmanned aerial vehicles. In: IEEE IROS, p. 8 (2018)
26. Press, W.H., Teukolsky, S.H., Vetterling, W.T., Flannery, B.P.: Numerical Recipes - The Art of Scientific Computing, 3rd edn. Cambridge University Press (2007)

PΦSS: An Open-Source Experimental Setup for Real-World Implementation of Swarm Robotic Systems in Long-Term Scenarios

Farshad Arvin[1]([✉])(iD), Tomáš Krajník[2](iD), and Ali Emre Turgut[3](iD)

[1] School of Electrical and Electronic Engineering, The University of Manchester, Manchester M13 9PL, UK
farshad.arvin@manchester.ac.uk
[2] Artificial Intelligence Centre, Faculty of Electrical Engineering, Czech Technical University, Prague, Czechia
[3] Mechanical Engineering Department, Middle East Technical University, 06800 Ankara, Turkey

Abstract. Swarm robotics is a relatively new research field that employs multiple robots (tens, hundreds or even thousands) that collaborate on complex tasks. There are several issues which limit the real-world application of swarm robotic scenarios, e.g. autonomy time, communication methods, and cost of commercialised robots. We present a platform, which aims to overcome the aforementioned limitations while using off-the-shelf components and freely-available software. The platform combines (i) a versatile open-hardware micro-robot capable of local and global communication, (ii) commercially-available wireless charging modules which provide virtually unlimited robot operation time, (iii) open-source marker-based robot tracking system for automated experiment evaluation, (iv) and a LCD display or a light projector to simulate environmental cues and pheromone communication. To demonstrate the versatility of the system, we present several scenarios, where our system was used.

Keywords: Open-source · Swarm robotics · Artificial pheromone · Perpetual robot swarm · Tracking system

1 Introduction

Swarm robotics [29], which by definition must deploy large number robots to control them by algorithms inspired by eusocial animals, was so far not very successful in terms of real-robot scenarios and applications. This is mainly because maintaining a large swarm over long time requires a significant effort

The work has been supported by UK EPSRC (Project No. EP/P01366X/1), EU H2020 STEP2DYNA (691154), and Czech Science Foundation project 17-27006Y.

© Springer Nature Switzerland AG 2019
J. Mazal (Ed.): MESAS 2018, LNCS 11472, pp. 351–364, 2019.
https://doi.org/10.1007/978-3-030-14984-0_26

which comes mostly from the fact that small robots' batteries need to be exchanged frequently. For large swarms, one must replace the batteries almost all the time, which is not only cumbersome, but it also disrupts the swarm operation. Furthermore, many swarm experiments require that the robots know their position or positions of other robots in their vicinity, or strength of a given environmental cue, such as a pheromone, at their current location. This requires the employment of a tracking system which is able to detect and localise a large number of robots in real time.

We present a system which overcomes the aforementioned problems by combining an affordable robotic platform, continuous charging module, real-time large-scale localisation method and artificial pheromone system into one experimental test-bed. This test-bed, PΦSS, allows long-term large-scale experiments, where complex swarm behaviours emerge from interaction of a large number of simple physical agents.

1.1 Long Term Autonomy

To tackle the problems with limited battery capacity, several different approaches have been employed to date. The simplest, tedious way is to connect robots with depleted batteries to their chargers manually [6], or to manually replace their batteries [67]. These approaches do not scale well with the number of robots and timespan of experiments. Alternatively, robots can seek charging stations themselves [49] or schedule their charging times in accordance with anticipated users' demands [42,60]. These approaches require very reliable automated docking and they cause robots to spend a significant fraction of their operation time at the charging location. While this can be partially resolved by automated battery swapping systems [13], the robot has to still interrupt its current activity, which might affect the swarm operation.

To eliminate the battery problem, one can supply the energy in a continuous manner, e.g. by a tether [22]. While this is doable for a single robot, a tethered multi-robot system would have to take into account the cable positions, otherwise these will get entangled. Another researchers proposed solutions like a powered ground [38,43,66,73,75], where the robots collect energy continuously from the ground via direct contacts. Still, the mechanical connectors get worn out and dirty over time, which might affect the energy flow to the individual robots and impact the behaviour of the entire swarm. Therefore, some researches [19,36,78] proposed to use wireless power transfer which does not suffer from the gradual deterioration of contact-based systems.

1.2 Tracking System

To evaluate the experiments performed or to provide the robots with their positions in real-time, researchers typically use an external localisation system capable of real-time tracking of multiple robots. These systems can be based on active emitters, such as visible-light [16] or ultraviolet LEDs [72], ceiling projection [76], radio or ultrasound beacons. A costly, high-end solution is the commercial motion capture system from ViCon [53,70], which is based on high-speed and high-resolution IR (infra-red) cameras, IR emitters and reflective targets. Alternatively, the systems can use passive markers, such as ARTag [20], ARToolKit+ [71] or AprilTag [51], which not only allow to determine the robot pose, but also encode some additional information, such as robot ID. These target detectors were used in several works in order to obtain pose information of mobile robots [14,21,55,64]. Alternative target shapes are also proposed in recent literature, which are specifically designed for vision-based localisation systems with a higher precision and reduced computational costs. Due to several positive aspects, circular shaped patterns appear to be best suited as fiducial markers in external localisation systems and can be found in several works [1,34,44,52,77].

Since the methods presented in [20,51,71] are reported to be computationally expensive, and [53,70] are prohibitively expensive, we chose to use a fast open-source system which uses a single camera to track the positions of high number of robots in real time [39,41]. Since the original system does not distinguish between individual markers, we use its extensions, which use slightly modified markers with encoded ID [5,45,46].

The computational efficiency of the WhyCon system was examined in [39,45]. These comparisons show that for our application, the WhyCon system [39,46] outperforms both the ArUco [57] and AprilTag [51] systems in terms of computational efficiency, while providing comparable accuracy. Additionally, [39,46] report that compared to AprilTag and Aruco, the method can use much smaller marker size, which is also beneficial in case of robotic swarms.

1.3   Pheromones and Environmental Cues

Pheromone communication was used in swarm robotics both in simulated [23] and real-robot scenarios for almost a decade [15]. In real-robot scenarios, substances such as alcohol are used to emulate pheromones [25,48,54,58,59]. However, these chemicals typically have different characteristics (e.g. diffusion and evaporation) than the actual pheromones and one cannot alter these parameters to study their influence on swarm behaviours. Furthermore, sensors detecting these chemicals are expensive and have slow responses. To solve that, the pheromones were simulated by means of RFID tags [31,37], audio [3,32] and light [8,26,62,65]. While these methods are more flexible compared to the use of volatile chemicals, they also have certain drawbacks. For example finite size of RFID tags limits the simulated pheromone resolution, adjustment of the evaporation and diffusion rates of audio-based artificial pheromone is almost impossible etc. Fortunately, light-based systems exhibit flexibility, because light can be emitted with different intensities and colours using off-the-shelf components such as conventional projectors [65]. In our case, we also employ light-based artificial pheromones, which are either realised by a projector, or an LCD screen, over which the robots move [5,11].

2   PΦSS System Description

PΦSS is a combination of four systems which have been previously developed for swarm robotic applications. These systems are: (i) open-source miniature robot, called Mona [2], (ii) perpetual swarm support system [10], (iii) efficient and accurate robot tracking module [39], and (iv) artificial pheromone system [5]. This section describes the proposed experimental test-bed in detail.

2.1   Mona Robot

Mona [2] is an open-source miniature mobile robot[1] that has been developed for use in swarm robotic applications. The main goal in developing the robot was to provide a low power robot to study feasibility of *Perpetual Robot Swarm* [10]. Moreover, the application of the robot was expanded and recently it is also used as an education purpose platform of mobile robotic lab for postgraduates in control engineering. Figure 1 shows a Mona robot's main platform without extension modules.

Fig. 1. Mona robot platform.

Mona has been developed based on Arduino architecture[2] which is a successful open-source project mainly developed for educational purpose. Mona adopted ATmega 328 Mini/Pro architecture hence it has the benefit of having access to all available libraries for the Arduino module. A micro USB cable is the only requirement to start work with the robot and connection link for programming Mona using Arduino IDE. It is a small size robot with diameter of 8 cm. It is actuated using two DC motors with directly attached

---

[1] https://github.com/MonaRobot.
[2] https://www.arduino.cc.

wheels with diameter of 32 mm. The motors rotational speeds are controlled independently using pulse-width modulation. Each motor has a magnetic encoder that generates 1500 pulses per wheel revolution. It generates enough resolution for a precise trajectory and implementing closed-loop controllers.

Mona uses a 3.7 V, 350 mAh battery which provides 3 h of autonomy. However, it is extendible to days and months using the perpetual robot swarm system [10] which will be presented in the following section. There were several autonomous power transfer methods have been proposed [7,13,17,30,35,50,60, 69] which can be used for extending autonomy time of Mona robots deployed in a swarm robotic scenario.

Since Mona is a low-cost robot, the platform includes only basic IR proximity sensors to detect obstacles, walls, and other robots. Therefore, it must be a modular platform allowing users to attach their required modules e.g. sensors and indicators. Several modules are currently available for Mona such as ROS (Robot Operating System) module [74] and vision board [33] for bio-inspired image processing [24]. It is easy to develop external modules for Mona that supports several communication approaches including I^2C, RS232, SPI, and direct general purpose digital pins.

2.2   Perpetual Swarm Support System

Perpetual robot swarm [10] is an active system which provides continuous power to individual mobile robots without an interruption. Therefore, robots follow their own task without considering their battery limitations. It overcomes the limitation of robots' battery capacity that limits the swarm scenarios to a very short period of time.

The proposed perpetual robot swarm consists of two main parts: i) inductive power transmitter which is connected to a main DC power source and ii) receiver board which is connected to charging circuit of the mobile robots. At the beginning of experiments, robots' batteries are fully charged (4.2 V). Since the arena is covered by many independent charging transmitters, robots harvest small amount of power by crossing over each transmitter. Hence, they manage to keep their battery level at the fully charge level.

A mathematical model was also proposed for the perpetual robot swarm system which shows limitation of the system in terms of maximum speed which robot can have to manage a perpetual system. It is directly related to forward speed of a robot and also maximum power transmitted by the transmitters.

Several probabilistic models were proposed for modelling of swarm robotic systems [9,12,18,47,63]. Swarm systems were also modelled with various models e.g. a Langevin equation [28], *Stock & Flow* model [61], and power-law equation model [4].

For the model that was proposed for perpetual swarm system, it was assumed that: (i) the robot is a circular with diameter of d_r, (ii) rectangular individual charging cells with dimensions of x_c, y_c, and (iii) rectangular arena with dimensions of x_a, y_a. Therefore, assuming that a robot moves in a way that the probabilistic distribution of its position inside of the arena is uniform, the probability that it is charging is:

$$p'_c = n_c \frac{a_c}{a_a} \frac{v_o}{v_c} = \frac{n_c v_o (x_c - d_c - v_o \frac{t_c}{2})(y_c - d_c - v_o \frac{t_c}{2})}{v_c (x_a - d_r)(y_a - d_r)}, \tag{1}$$

where n_c is the number of chargers, v_c is the robot speed when detecting the charging signal, v_o is the robot normal speed, a_c and a_a are the effective areas of the arena and the charging cells respectively, t_c is the coupling time of the robot to the charging cell takes a finite time, and d_c corresponds to the minimal distance of the charging coil centre from the charging cell border. Therefore, the only condition for robot to operate perpetually is a non-negative energy balance, then:

$$p'_c w_c - w_o \geq 0, \tag{2}$$

where w_c is the charging power and w_o is the robot's power consumption during routine operation.

2.3   Accurate Robot Tracking

Most of the experiments performed require tracking of the individual robots, typically for the purpose of experiment evaluation. Our tracking system is based on an overhead camera, which provides a complete overview of the experimental area, and a set of black-and-white markers which are attached on the robots. The system's core is a freely available software package capable of accurate and efficient localisation of a large number of white-and-black circular patterns [39]. The accuracy of the system is in the order of millimetres and it is able to track hundreds of robots in real-time. Since the experiments typically require to distinguish the robots from each other, we had to use either the [46] variant, which uses a Manchester coding scheme to embed an ID into the tag or the COSΦ variant, which encodes the IDs into elliptical patterns with different semiaxes lengths, see Fig. 2. The COSΦ tag variant, has an elliptical shape with dimensions of 32×26 mm, and the centre of the inner, white circle is shifted by 1 mm. The the ellipse major axis allows to establish the orientation of the robot within $[-90°, 90°]$ and the offset of the centres is used to resolve the orientation ambiguity. The dimensions of the inner ellipse encode the pattern ID, which allows to distinguish individual robots. However, the COSΦ elliptical tags can be used only in 2d with a camera directly overhead the arena. To overcome this limitation, we implemented a more advanced variant of the tag [45,46] which can embed more IDs and provides a full 6 DoF pose.

While processing the camera images typically takes less than 50 μs per robot, image capture and its transfer over USB typically takes 100 ms, which introduces an undesirable delay. In case one needs to run experiments with rapidly-moving robots, a specialised camera might be needed.

Fig. 2. Original (left), COSΦ (middle) and WhyCode (right) localisation tags.

2.4 Artificial Pheromone System

COSΦ [5] is a high precision, flexible and low-cost experimental setup which provides a reliable and user friendly platform to study bio-inspired artificial pheromone mechanisms. The proposed system consists of: (i) a visual localisation system to detect robots [39] described in the previous section and (ii) a pheromone trail system which generates artificial pheromone and projects to the arena. The pheromone trail display system is realised either as an LCD screen, on which the robots move, see Fig. 3, or a projector, see Fig. 4, that illuminates the arena from above.

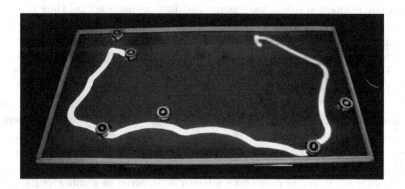

Fig. 3. Artificial pheromone system realized by an LCD screen and robots with CosΦ markers.

The system can simulate several pheromones and their interactions simultaneously, where characteristics of each pheromone are determined by four parameters:

- injection, ι, defines how fast a pheromone is released by a given robot,
- evaporation, e_ϕ, determines how quickly the pheromone fades over time,
- diffusion, κ, defines pheromone spreading rate,
- influence, c, characterises how much the pheromone affects the output image.

The output image which is a combination of multiple pheromones is represented by a matrix $\mathbf{I}$, hence brightness of a pixel at a position (x, y) is presented as $\mathbf{I}(x,y)$, an ith pheromone is modelled as a matrix $\mathbf{\Phi_i}$, and the brightness of each pixel that is projected to the arena is given by

$$\mathbf{I}(x,y) = \sum_{i=1}^{n} c_i \mathbf{\Phi_i}(x,y), \qquad (3)$$

where $\mathbf{\Phi_i}(x,y)$ is a 2D array that represents i^{th} pheromone intensity at position (x,y) and c_i defines the pheromone's influence on the displayed image.

Intensity of each pheromone are continuously updated by:

$$\dot{\mathbf{\Phi}}_{\mathbf{i}}(x,y) = e_{i\phi}\mathbf{\Phi_i}(x,y) + \kappa_i \triangle\mathbf{\Phi_i}(x,y) + \iota_i(x,y), \qquad (4)$$

where $\dot{\mathbf{\Phi}}_{\mathbf{i}}(x,y)$ corresponds to the rate of the pheromone change caused by its evaporation $e_{i\phi}$, diffusion κ_i and injection ι_i. To see how the interplay of the aforementioned parameters affects the swarm behaviour, see a video of the operational system at https://www.youtube.com/watch?v=bC_hdlxvGkA and a paper [5].

2.5   Arena Configuration

The aforementioned systems were combined into a single arena, thus producing an interactive system which can be used for various purposes e.g. group programming, multi-system collective scenarios, evolutionary robotics etc. The system is controlled centrally by a PC which monitors the system's action and applies the output signals for controlling the individual charging transmitters as well as pheromone system.

The arena itself is a rectangle made out of 4×4 cm aluminium strut profile with size of 120 cm $\times$ 100 cm. The arena's floor covered by 480 independent inductive charging transmitters. Figure 4 shows architecture of the arena. The arena is an active system with a continuous feedback. It means that we are able to read the status of the charging cells whether they are in *charging* or *idle* mode. The charging mode refers to a state which the receiver of a robot is harvesting power and the idle mode refers to the state that transmitter is not connected to any robot. There are two different monitoring approaches for the PΦSS system that are: (i) an overhead camera tracks robots in the arena and (ii) current sensing from individual charging cells in the perpetual swarm system. There are two outputs from the main controller which are: (i) digital images projected by the overhead projector which are generated by the pheromone system and (ii) control signals connected to all individual transmitters which are managed by the perpetual swarm system.

To combine COSΦ [5] which uses floor to illustrate pheromone trails on the flat screen and perpetual swarm system [10] which also uses floor to transmit power to the robot, a video projector was utilised for the pheromone system. Therefore, the perpetual swarm system uses floor and the pheromone system uses overhead video projector.

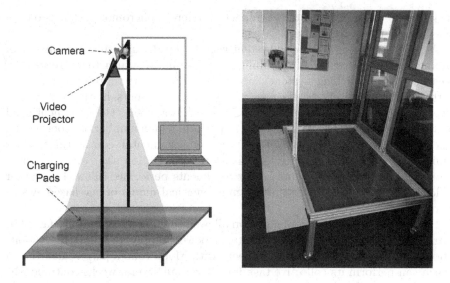

Fig. 4. (Left) Architecture of the arena and (right) PΦSS arena made.

In terms of power management, transmitters consume very low power (about 50 mW at 12 V) when they are in the idle mode, however it reaches up to 2 W when a robot is in power harvesting mode. Therefore, the total power consumption of the arena directly relies on the number of robots, n_r, which are deployed in an experiment: $(n_c - n_r)$ 50 mW $-n_r \times 2$ W, where n_c is 480 chargers.

For the heat generated by the transmitters that, is not significant, see [10] for more detail.

There are several interactive experimental environments which have been developed for swarm robotic researches [26,27,56,68]. However, those systems were bespoke development for a specific application or work only with a specific robot platform.

3   System Use Cases

The aforementioned system and its early versions, were used in a number of scenarios investigating long-term behaviour of bio-inspired robotic swarms.

For example, in artificial pheromone communication, COSΦ [5], a flexible communication medium for swarm and multi robotic systems using light and its intensity has been proposed. In this communication method, some of robots can use arena to release their messages (pheromones) and the rest of group can detect and read those messages using their light sensors. The system allows implementing multi-layer pheromone trails which can be used for multiple messages e.g. alarm or food message. Thanks to the versatility of the system, the paper [5] could investigate the impact of different pheromone settings on the behaviour of the swarm robots.

Another scenario, where the system was deployed, concerned the impact of dynamic and static cues on the ability of the robotic swarm to aggregate [9]. Because of the tracking system's ability (see Sect. 2.3) to recover positions of the robots relatively to the cue and number of robots in the aggregates, experiments could be evaluated in an automatic way using the aforementioned metrics. This allowed the authors of [9] to perform a high number of experiments, which investigate the impact of different environmental cue settings, their configuration, size and texture as well as the population size, robot velocity, cue waiting time etc. The number of experiments performed allowed to gather sufficient data required to establish a mathematical model of cue-based swarm aggregation.

Finally, the full system configuration allowed to implement a perpetual robot swarm [10], which provides a continuous power supply to miniature mobile robots. Results of performed experiments with Mona robots [2] revealed that a swarm can perform its collective task for more than several weeks continuously. It allows researchers to investigate real-world very long-term scenarios on evolutionary robotics. Currently, we are working on effects of environmental changes on long-term collective scenario, which robots are able to learn and update their sensing ability to adapt a dynamic environment.

4   Conclusion

In this paper we proposed a multi-purpose experimental environment for use in swarm robotic researches The platform is completely build from off-the-shelf components and open-source hardware and software: Mona robots [2], wireless charging modules that allow unlimited robot operation time [10], robot localisation module that can positions and velocities of individual robots on the arena [40], and a LCD display or a light projector, which simulate environmental cues and pheromone communication [5]. Using this platform, one can perform long-term large-scale swarm experiments with real robots. This system opens an opportunity to test swarm behaviours, which could be tested only in simulation. We present three example scenarios, where the system presented demonstrated its usefulness.

References

1. Ahn, S.J., Rauh, W., Recknagel, M.: Circular coded landmark for optical 3D-measurement and robot vision. In: IROS, pp. 1128–1133 (1999)
2. Arvin, F., Espinosa, J., Bird, B., West, A., Watson, S., Lennox, B.: Mona: an affordable open-source mobile robot for education and research. J. Intell. Robotic Syst. **92**, 1–15 (2018)
3. Arvin, F., Turgut, A., Bazyari, F., Arikan, K., Bellotto, N., Yue, S.: Cue-based aggregation with a mobile robot swarm: a novel fuzzy-based method. Adapt. Behav. **22**, 189–206 (2014)

4. Arvin, F., Attar, A., Turgut, A.E., Yue, S.: Power-law distribution of long-term experimental data in swarm robotics. In: Tan, Y., Shi, Y., Buarque, F., Gelbukh, A., Das, S., Engelbrecht, A. (eds.) ICSI 2015. LNCS, vol. 9140, pp. 551–559. Springer, Cham (2015). https://doi.org/10.1007/978-3-319-20466-6_58

5. Arvin, F., Krajník, T., Turgut, A.E., Yue, S.: COSΦ: artificial pheromone system for robotic swarms research. In: IEEE/RSJ International Conference on Intelligent Robots and Systems (IROS), pp. 407–412 (2015)

6. Arvin, F., Murray, J., Zhang, C., Yue, S.: Colias: an autonomous micro robot for swarm robotic applications. Int. J. Adv. Rob. Syst. 11(7), 113 (2014)

7. Arvin, F., Samsudin, K., Ramli, A.R.: Swarm robots long term autonomy using moveable charger. In: International Conference on Future Computer and Communication (2009)

8. Arvin, F., Samsudin, K., Ramli, A.R., Bekravi, M.: Imitation of honeybee aggregation with collective behavior of swarm robots. Int. J. Comput. Intell. Syst. 4(4), 739–748 (2011)

9. Arvin, F., Turgut, A.E., Krajník, T., Yue, S.: Investigation of cue-based aggregation in static and dynamic environments with a mobile robot swarm. Adapt. Behav. 24(2), 102–118 (2016)

10. Arvin, F., Watson, S., Turgut, A.E., Espinosa, J., Krajník, T., Lennox, B.: Perpetual robot swarm: long-term autonomy of mobile robots using on-the-fly inductive charging. J. Intell. Rob. Syst. 92(3–4), 1–18 (2017)

11. Arvin, F., et al.: ΦClust: pheromone-based aggregation for robotic swarms. In: IEEE/RSJ International Conference on Intelligent Robots and Systems (IROS) (2018)

12. Bayindir, L., Şahin, E.: Modeling self-organized aggregation in swarm robotic systems. In: Swarm Intelligence Symposium, pp. 88–95. IEEE (2009)

13. Bonani, M., et al.: The marxbot, a miniature mobile robot opening new perspectives for the collective-robotic research. In: IROS (2010)

14. Bošnak, M., Matko, D., Blažič, S.: Quadrocopter hovering using position-estimation information from inertial sensors and a high-delay video system. J. Intell. Rob. Syst. 67(1), 43–60 (2012)

15. Brambilla, M., Ferrante, E., Birattari, M., Dorigo, M.: Swarm robotics: a review from the swarm engineering perspective. Swarm Intell. 7(1), 1–41 (2013)

16. Breitenmoser, A., Kneip, L., Siegwart, R.: A monocular vision-based system for 6D relative robot localization. In: IROS, pp. 79–85 (2011)

17. Carrillo, M., et al.: A bio-inspired approach for collaborative exploration with mobile battery recharging in swarm robotics. In: Korošec, P., Melab, N., Talbi, E.-G. (eds.) BIOMA 2018. LNCS, vol. 10835, pp. 75–87. Springer, Cham (2018). https://doi.org/10.1007/978-3-319-91641-5_7

18. Correll, N., Martinoli, A.: Modeling self-organized aggregation in a swarm of miniature robots. In: ICRA Workshop on Collective Behaviors inspired by Biological and Biochemical Systems (2007)

19. Deyle, T., Reynolds, M.: Surface based wireless power transmission and bidirectional communication for autonomous robot swarms. In: ICRA. IEEE (2008)

20. Fiala, M.: ARTag, An Improved Marker System Based on ARToolkit (2004)

21. Fiala, M.: Vision guided control of multiple robots. In: First Canadian Conference on Computer and Robot Vision, pp. 241–246 (2004)

22. Floreano, D., Mondada, F.: Automatic creation of an autonomous agent: genetic evolution of a neural network driven robot. In: 3rd International Conference on Simulation of Adaptive Behavior: From Animals to Animats 3 (1994)

23. Fossum, F., Montanier, J.M., Haddow, P.C.: Repellent pheromones for effective swarm robot search in unknown environments. In: IEEE Symposium on Swarm Intelligence (SIS), pp. 1–8 (2014)

24. Fu, Q., Hu, C., Peng, J., Yue, S.: Shaping the collision selectivity in a looming sensitive neuron model with parallel ON and OFF pathways and spike frequency adaptation. Neural Netw. **106**, 127–143 (2018)

25. Fujisawa, R., Dobata, S., Sugawara, K., Matsuno, F.: Designing pheromone communication in swarm robotics: group foraging behavior mediated by chemical substance. Swarm Intell. **8**(3), 227–246 (2014)

26. Garnier, S., Combe, M., Jost, C., Theraulaz, G.: Do ants need to estimate the geometrical properties of trail bifurcations to find an efficient route? A swarm robotics test bed. PLoS Comput. Biol. **9**(3), e1002903 (2013)

27. Griparić, K., Haus, T., Miklić, D., Polić, M., Bogdan, S.: A robotic system for researching social integration in honeybees. PLoS ONE **12**(8), e0181977 (2017)

28. Hamann, H.: Space-time continuous models of swarm robotics systems: supporting global-to-local programming. Ph.D. thesis, Department of Computer Science, University of Karlsruhe (2008)

29. Hamann, H.: Swarm Robotics: A Formal Approach. Springer, Cham (2018). https://doi.org/10.1007/978-3-319-74528-2

30. Hamann, H., Markarian, C., auf der Heide, F.M., Wahby, M.: Pick, pack, & survive: charging robots in a modern warehouse based on online connected dominating sets. In: 9th International Conference on Fun with Algorithms (FUN 2018), vol. 100, pp. 1–13 (2018)

31. Herianto, H., Sakakibara, T., Kurabayashi, D.: Artificial pheromone system using RFID for navigation of autonomous robots. J. Bionic Eng. **4**(4), 245–253 (2007)

32. Holland, O., Melhuish, C.: An interactive method for controlling group size in multiple mobile robot systems. In: ICAR, pp. 201–206 (1997)

33. Hu, C., Arvin, F., Xiong, C., Yue, S.: A bio-inspired embedded vision system for autonomous micro-robots: the LGMD case. IEEE Trans. Cogn. Dev. Syst. **9**(3), 241–254 (2017)

34. de Ipiña, D.L., Mendonça, P.R.S., Hopper, A.: TRIP: a low-cost vision-based location system for ubiquitous computing. Pers. Ubiquit. Comput. **6**(3), 206–219 (2002)

35. Ismail, A.R., Desia, R., Zuhri, M.F.R.: The initial investigation of the design and energy sharing algorithm using two-ways communication mechanism for swarm robotic systems. In: Phon-Amnuaisuk, S., Au, T.W. (eds.) Computational Intelligence in Information Systems. AISC, vol. 331, pp. 61–71. Springer, Cham (2015). https://doi.org/10.1007/978-3-319-13153-5_7

36. Karpelson, M., et al.: A wirelessly powered, biologically inspired ambulatory microrobot. In: ICRA, pp. 2384–2391. IEEE (2014)

37. Khaliq, A.A., Saffiotti, A.: Stigmergy at work: planning and navigation for a service robot on an RFID floor. In: IEEE International Conference on Robotics and Automation (ICRA), pp. 1085–1092 (2015)

38. Klingner, J., Kanakia, A., Farrow, N., Reishus, D., Correll, N.: A stick-slip omnidirectional drive-train for low-cost swarm robotics: mechanism, calibration, and control. In: IROS, pp. 846–851 (2014)

39. Krajník, T., et al.: A practical multirobot localization system. J. Intell. Rob. Syst. **76**(3–4), 539–562 (2014)

40. Krajník, T., et al.: A practical multirobot localization system. J. Intell. Rob. Syst. **76**(3–4), 539–562 (2014)

41. Krajnik, T., Nitsche, M., Faigl, J., Duckett, T., Mejail, M., Preucil, L.: External localization system for mobile robotics. In: 2013 16th International Conference on Advanced Robotics (ICAR), pp. 1–6. IEEE (2013)
42. Krajník, T., Santos, J.M., Duckett, T.: Life-long spatio-temporal exploration of dynamic environments. In: 2015 European Conference on Mobile Robots (ECMR), pp. 1–8. IEEE (2015)
43. Krieger, M.J., Billeter, J.B., Keller, L.: Ant-like task allocation and recruitment in cooperative robots. Nature **406**(6799), 992–995 (2000)
44. Kulich, M., Chudoba, J., Košnar, K., Krajník, T., Faigl, J., Přeučil, L.: Syrotek - distance teaching of mobile robotics. IEEE Trans. Educ. **56**(1), 18–23 (2013)
45. Lightbody, P., Krajník, T., Hanheide, M.: An efficient visual fiducial localisation system. SIGAPP Appl. Comput. Rev. **17**(3), 28–37 (2017). https://doi.org/10.1145/3161534.3161537
46. Lightbody, P., Krajník, T., Hanheide, M.: A versatile high-performance visual fiducial marker detection system with scalable identity encoding. In: Proceedings of the Symposium on Applied Computing, SAC 2017, pp. 276–282. ACM, New York (2017). https://doi.org/10.1145/3019612.3019709
47. Martinoli, A., Ijspeert, A., Mondada, F.: Understanding collective aggregation mechanisms: from probabilistic modelling to experiments with real robots. Rob. Auton. Syst. **29**(1), 51–63 (1999)
48. Mayet, R., Roberz, J., Schmickl, T., Crailsheim, K.: Antbots: a feasible visual emulation of pheromone trails for swarm robots. In: Dorigo, M., et al. (eds.) ANTS 2010. LNCS, vol. 6234, pp. 84–94. Springer, Heidelberg (2010). https://doi.org/10.1007/978-3-642-15461-4_8
49. McLurkin, J., Smith, J., Frankel, J., Sotkowitz, D., Blau, D., Schmidt, B.: Speaking swarmish: human-robot interface design for large swarms of autonomous mobile robots. In: AAAI Spring Symposium (2006)
50. Mintchev, S., Ranzani, R., Fabiani, F., Stefanini, C.: Towards docking for small scale underwater robots. Auton. Robots **38**(3), 283–299 (2015)
51. Olson, E.: AprilTag: a robust and flexible visual fiducial system. In: ICRA, pp. 3400–3407. IEEE, May 2011
52. Pedre, S., Krajník, T., Todorovich, E., Borensztejn, P.: Hardware/software co-design for real time embedded image processing: a case study. In: Alvarez, L., Mejail, M., Gomez, L., Jacobo, J. (eds.) CIARP 2012. LNCS, vol. 7441, pp. 599–606. Springer, Heidelberg (2012). https://doi.org/10.1007/978-3-642-33275-3_74
53. Phoenix: Phoenix 3D motion capture. http://www.ptiphoenix.com/. Accessed 18 Aug 2018
54. Purnamadjaja, A.H., Russell, R.A.: Bi-directional pheromone communication between robots. Robotica **28**(01), 69–79 (2010)
55. Rekleitis, I., Meger, D., Dudek, G.: Simultaneous planning, localization, and mapping in a camera sensor network. Rob. Auton. Syst. **54**(11), 921–932 (2006)
56. Rezeck, P.A., Azpurua, H., Chaimowicz, L.: HeRo: an open platform for robotics research and education. In: Latin American Robotics Symposium (LARS) and Brazilian Symposium on Robotics (SBR), pp. 1–6 (2017)
57. Romero-Ramirez, F.J., Muñoz-Salinas, R., Medina-Carnicer, R.: Speeded up detection of squared fiducial markers. Image Vis. Comput. **76**, 38–47 (2018)
58. Russell, R.A.: Ant trails-an example for robots to follow? In: IEEE International Conference on Robotics and Automation, vol. 4, pp. 2698–2703 (1999)
59. Russell, R.A.: Air vortex ring communication between mobile robots. Rob. Auton. Syst. **59**(2), 65–73 (2011)

60. Santos, J.M., Krajník, T., Fentanes, J.P., Duckett, T.: Lifelong information-driven exploration to complete and refine 4-D spatio-temporal maps. IEEE Rob. Autom. Lett. 1(2), 684–691 (2016)
61. Schmickl, T., Hamann, H., Worn, H., Crailsheim, K.: Two different approaches to a macroscopic model of a bio-inspired robotic swarm. Rob. Auton. Syst. 57(9), 913–921 (2009)
62. Schmickl, T., et al.: Get in touch: cooperative decision making based on robot-to-robot collisions. Auton. Agents Multi-Agent Syst. 18(1), 133–155 (2009)
63. Soysal, O., Şahin, E.: A macroscopic model for self-organized aggregation in swarm robotic systems. In: Şahin, E., Spears, W.M., Winfield, A.F.T. (eds.) SR 2006. LNCS, vol. 4433, pp. 27–42. Springer, Heidelberg (2007). https://doi.org/10.1007/978-3-540-71541-2_3
64. Stump, E., Kumar, V., Grocholsky, B., Shiroma, P.M.: Control for localization of targets using range-only sensors. Int. J. Rob. Res. 28(6), 743–757 (2009)
65. Sugawara, K., Kazama, T., Watanabe, T.: Foraging behavior of interacting robots with virtual pheromone. In: IEEE/RSJ International Conference on Intelligent Robots and Systems (IROS), pp. 3074–3079 (2004)
66. Takaya, Y.U., Arita, T.: Situated and embodied evolution in collective evolutionary robotics. In: International Symposium on Artificial Life and Robotics (2003)
67. Turgut, A.E., Çelikkanat, H., Gökçe, F., Şahin, E.: Self-organized flocking in mobile robot swarms. Swarm Intell. 2(2), 97–120 (2008)
68. Valentini, G., et al.: Kilogrid: a novel experimental environment for the kilobot robot. Swarm Intell. 12, 1–22 (2018)
69. Vaussard, F., Rétornaz, P., Roelofsen, S., Bonani, M., Rey, F., Mondada, F.: Towards long-term collective experiments. In: Lee, S., Cho, H., Yoon, K.J., Lee, J. (eds.) Intelligent Autonomous Systems 12. AISC, vol. 194, pp. 683–692. Springer, Heidelberg (2013). https://doi.org/10.1007/978-3-642-33932-5_64
70. Vicon: Vicon MX Systems. http://www.vicon.com/products/viconmx.html. Accessed 12 July 2013
71. Wagner, D., Schmalstieg, D.: ARToolKitPlus for pose tracking on mobile devices. In: 12th Computer Vision Winter Workshop (CVWW), pp. 139–146 (2007)
72. Walter, V., Saska, M., Franchi, A.: Fast mutual relative localization of UAVs using ultraviolet LED markers. In: 2018 International Conference on Unmanned Aircraft Systems (2018)
73. Watson, R.A., Ficiei, S., Pollack, J.B.: Embodied evolution: embodying an evolutionary algorithm in a population of robots. In: Congress on Evolutionary Computation (1999)
74. West, A., Arvin, F., H. Martin, S.W., Lennox, B.: ROS Integration for Miniature Mobile Robots. In: Towards Autonomous Robotic Systems (TAROS) (2018)
75. Winfield, A.F., Nembrini, J.: Emergent swarm morphology control of wireless networked mobile robots. In: Doursat, R., Sayama, H., Michel, O. (eds.) Morphogenetic Engineering, pp. 239–271. Springer, Heidelberg (2012). https://doi.org/10.1007/978-3-642-33902-8_10
76. Yamamoto, Y., et al.: Optical sensing for robot perception and localization. In: IEEE Workshop on Advanced Robotics and its Social Impacts, pp. 14–17 (2005)
77. Yang, S., Scherer, S., Zell, A.: An onboard monocular vision system for autonomous takeoff, hovering and landing of a micro aerial vehicle. J. Intell. Rob. Syst. 69(1–4), 499–515 (2013)
78. Zhang, Z., Xu, X., Li, B., Deng, B.: An energy-encrypted contactless charging system for swarm robots. In: Magnetics Conference (INTERMAG) (2015)

Route Planning for Teams of Unmanned Aerial Vehicles Using Dubins Vehicle Model with Budget Constraint

David Zahrádka[✉], Robert Pěnička, and Martin Saska

Faculty of Electrical Engineering, Czech Technical University,
Technicka 2, 166 27 Prague, Czech Republic
{zahrada2,penicrob,saskam1}@fel.cvut.cz

Abstract. In this paper, we propose Greedy Randomized Adaptive Search Procedure (GRASP) with Path Relinking extension for a solution of a novel problem formulation, the Dubins Team Orienteering Problem with Neighborhoods (DTOPN). The DTOPN is a variant of the Orienteering Problem (OP). The goal is to maximize collected reward from a close vicinity of given target locations, each with predefined reward, using multiple curvature-constrained vehicles, such as fixed-wing aircraft or VTOL UAVs with constant forward speed, each limited by route length. This makes it a very useful routing problem for scenarios using multiple UAVs for data collection, mapping, surveillance, and reconnaissance. The proposed method is verified on existing benchmark instances and by real experiments with a group of three fully-autonomous hexarotor UAVs that were used to compare the DTOPN with similar problem formulations and show the benefit of the introduced DTOPN.

Keywords: Route planning ·
Dubins team orienteering problem with neighbourhoods · DTOPN ·
Unmanned Aerial Vehicles · Mapping · Data collection · Inspection ·
Reconnaissance · Surveillance

1 Introduction

Due to the recent increase in usage of autonomous systems, such as Unmanned Aerial Vehicles (UAVs) for search and rescue, information gathering and even military applications [22–24], the demand for effective control rises proportionally. This includes route planning algorithms, which can increase the productivity of autonomous vehicle deployment when effective. Since optimization problems such as the studied multi-goal planning are computationally complex, various algorithms and heuristics surface and compete in which is closer to an optimal solution and in the required computational time. One of the problem formulations used for route planning optimization over multiple target location is the Orienteering Problem (OP) [19].

© Springer Nature Switzerland AG 2019
J. Mazal (Ed.): MESAS 2018, LNCS 11472, pp. 365–389, 2019.
https://doi.org/10.1007/978-3-030-14984-0_27

The OP features a set of target locations each with an assigned reward for visiting its particular position. Both the starting and the ending locations are predefined. The travel budget used to visit the locations is limited by a given budget, and therefore only a subset of the locations can be visited. The objective of the OP optimization is thus to find a single route, limited by the travel budget, which maximizes the total collected reward while visiting each location at most once.

The studied OP can be described as a variant of the well-known Travelling Salesman Problem (TSP) [32]. However, compared to the TSP, in which the goal is to minimize the travel distance while visiting all locations, in the OP the goal is to maximize the score collected by visiting nodes within a limited travel budget. The OP is thus a better way to model situations where visiting all locations is impossible, for example in applications where the vehicles used have a limited amount of fuel or battery flight time.

While the OP is very useful for real-life applications, it is often possible to deploy multiple vehicles simultaneously. Utilizing multiple UAVs leads in many applications to greater collected score in the same amount of time. For these applications, the generalization of the OP, called the Team Orienteering Problem (TOP) [8], is very useful. An example field of application could be reconnaissance or visual inspection, where the deployment of multiple UAVs at the same time leads to faster inspection and information gathering [38]. Another example of practical application of the TOP can be planning of recruitment of athletes from high schools. A recruiter has to visit several schools in a given number of days. He visits the schools beforehand and rates their recruitment potential. As the recruiter's time is limited, he has to choose the schools to visit each day to maximize the recruiting potential.

All previously mentioned problem formulations utilize only holonomic vehicle models. This poses a problem for fixed-wing UAVs that are unable to make sharp turns, which the solutions of the OP and TSP often contain. Similar curvature constraints appear even with multirotor UAVs in scenarios where maintaining constant forward speed is required. This is addressed in the Dubins Orienteering Problem (DOP) [30]. It is a generalization of the OP that considers curvature-constrained vehicles called Dubins vehicles [13]. This allows planning routes for vehicles with limited turning radius, such as unmanned fixed-wing aircraft or certain Unmanned Ground Vehicles (UGVs).

Further increase of the collected reward can be achieved by utilizing locations neighborhoods. These represent an area around each of target locations that is sufficient to visit in order to claim the associated reward. By allowing to collect the reward from the close vicinity of target locations, the travel costs to visit locations are lowered, thus allowing to visit more locations and increase the total collected reward within the limited travel budget. This can, for example, translate into more data collected in one UAV flight. They can be used to model situations where the collection of sensory data is possible from the close vicinity, for example, from wireless sensor networks, or when visually inspecting locations using UAVs with onboard cameras [18]. The OP generalization that features neighborhoods is the Orienteering Problem with Neighborhoods (OPN) [15], also known as the Close Enough Orienteering Problem.

A problem formulation that combines the aforementioned generalizations is the Dubins Team Orienteering Problem with Neighborhoods (DTOPN), a novel problem formulation introduced in this paper. It combines the benefits of previously mentioned problem formulations and can be flexibly used in their place. An example of a solution of the DTOPN solution can be seen in Fig. 1. The DTOPN can thus be used for scenarios with multiple curvature-constrained Dubins vehicles with limited travel distance due to battery discharge and with the ability to collect data from within the neighborhoods of target locations. This makes it a very useful routing problem formulation for mapping, surveillance and reconnaissance scenarios where constant speed is crucial and where reducing speed in the particular areas of interest can be dangerous.

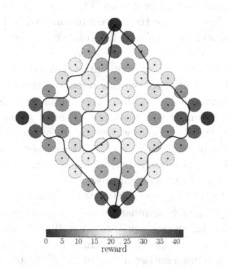

Fig. 1. An example of a DTOPN solution of the p6.3.j dataset by *Chao et al.*

A Greedy Randomized Adaptive Search Procedure [6] is introduced as a solution to the DTOPN. It is a flexible algorithm originally used to solve the TOP [36] extended using a sampling-based approach to consider the Dubins vehicle's minimal turning radius requirements and the locations respective neighborhoods. Using the known radius of the neighborhood, each locations neighborhood is sampled, resulting in a set of neighborhood locations. These are afterward used instead of the original location. By sampling the possible headings for each location and connecting them with curves respecting the minimal turning radius of the Dubins vehicle, the curvature constraints are enforced.

The proposed algorithm was tested in a real-life experiment which modeled a data-collection scenario using three Micro Aerial Vehicles (MAVs) initially developed for multi-robot applications [34]. The experimental verification was done for the regular TOP, and for the novel DTOP and DTOPN. The goal was to visit locations scattered on a field and visually inspect them with onboard cameras to collect their associated reward and thus show the benefits of the introduced DTOPN.

The rest of this work is organized in the following way. Related work is summarized in the next section. Section 3 contains formal definitions of the DTOPN and its subproblems. In Sect. 4, the algorithm proposed to solve the DTOPN is introduced. Section 5 contains the computational results and experiment verification in real-life conditions. Finally, Sect. 6 concludes the paper.

2   Related Work

The studied Dubins Team Orienteering Problem with Neighborhoods belongs to a large class of multi-goal routing problems. Therefore, the most relevant routing problems and approaches to solve them are summarized in this section.

The most classical routing problem formulation is the Travelling Salesman Problem. It is widely studied in many fields, including computer science, operational research, and robotics. The problem is defined as follows: A salesman wants to visit each of a set of cities exactly once and return to the starting city with minimal distance traveled [32]. Many algorithms were developed that solve it, since it is easy to formulate and, as an NP-hard problem [25], is difficult to solve. For example, the Ant Colony System as proposed by *Dorigo et al.* in [12] and further thoroughly described in [11]. It also serves as a basis for other problem definitions that generalize it by imposing other restrictions that extend its usefulness for many real-life applications.

The Orienteering Problem is a variation of the Travelling Salesman Problem that features a travel budget that must be met. It combines two NP-hard optimization problems, the Knapsack and the Travelling Salesman Problem. The similarity of the OP to the Knapsack problem is in the objective of maximizing collected reward by selecting a subset of target locations to be visited within the budget. This is combined with the TSP, in which the goal is to minimize the travel cost associated with visiting target locations on a route.

The OP, also known as the Selective TSP, was introduced by *Golden, Levy and Vohra* in [19]. Its first application was however mentioned by *Tsiligirides* in [39], using an example of a traveling salesman with not enough time to visit all possible cities.

It is often possible to deploy more than one salesman at a time. For these applications, the Multiple Traveling Salesman Problem (mTSP) [3] is very useful. The goal of the mTSP is to determine a set of routes for multiple salesmen whom all start from and return back to their home city.

Similar to the mTSP, the Team Orienteering Problem extends the classical OP by using multiple subsets of nodes without intersections to maximize the collected reward. The difference between the TOP and the mTSP is that the TOP uses a travel budget which originates from the Orienteering Problem. It was first introduced by *Chao et al.* in [8], even though a similar problem was already introduced by *Butt and Cavalier* in [5] called Multiple Tour Maximum Collection Problem.

Chao et al. also proposed a heuristic for TOP based on their five-step heuristic for the OP [7]. The key difference is that instead of selecting the best path, p best paths are selected. They also introduced benchmark instances for the TOP, which were partially based on datasets by *A. Tsiligirides* [39]. In [4], *Bouly et al.* proposed a Memetic Algorithm for the TOP. Other recent algorithms [21] include a branch-and-cut proposed in [9] by *Dang et al.*, which was based on a linear formulation with a polynomial number of binary variables, a Particle Swarm Optimization inspired algorithm by *Dang et al.* [10], a Multi-start Simulated Annealing by *Lin et al.* [29] and a Pareto Mimic Algorithm by *Ke et al.* [27].

Another interesting generalization of the TSP is the Close Enough Traveling Salesman Problem (CETSP), introduced in [20] by *Gulczynski et al.* In the CETSP, each city is associated with a neighborhood. The goal is to find the shortest tour that starts and ends in a given depot and travels within a given required radius of each city.

The Orienteering Problem with Neighborhoods (OPN), first proposed in [15] by *Faigl et al.*, is very similar to the CETSP. The OPN is a generalization of the Orienteering Problem in which reward collection within given non-zero communication range of target locations is possible. Unlike the OP, the OPN requires to determine the position within the circular neighborhoods of each target location and is, therefore, more computationally demanding due to the increased problem size. In order to be able to solve the OPN with reasonable computational power, the neighborhood can be sampled to a finite number of new locations, similar to headings in the DOP. In [15], *Faigl et al.* also proposed the use of Self Organizing Map-based (SOM) algorithm to solve the OPN.

As a problem formulation for vehicles with limited turning radius, such as fixed-wing UAVs or VTOL UAVs with constant forward speed, the Dubins Traveling Salesman Problem (DTSP) and Dubins Orienteering Problem (DOP) are useful. Both utilize the curvature-constrained Dubins vehicle model [13] with limited turning radius.

The DTSP was first proposed by *Savla et al.* in [35]. While the usage of Dubins vehicle model extends the usability of the DTSP, it raises the computational capacity requirements, since it is necessary to calculate the best way to connect different headings for each pair of visited cities. One of the approaches used to tackle this limitation is to use sampling-based algorithms.

The DOP is very similar to the DTSP. It was introduced by *Pěnička et al.* in [30] together with a sampling-based algorithm that solves it. The main difference between the DOP and the DTSP is that the DOP considers route limitations in the form of limited battery capacity, which is useful for real-life applications in robotics.

The algorithm proposed in [30] is based on the Variable Neighborhood Search algorithm, originally used on the OP. Its results were also verified in real-life testing scenarios. The algorithm iteratively performs shake and local search procedures. The shake routine randomly changes the best achieved solution to escape from a local maximum. The local search procedure then tries to improve the solution created by shake.

The Dubins Orienteering Problem with Neighborhoods (DOPN) combines both the limited curvature constraint of Dubins vehicle and the ability to measure data within a predefined circular neighborhood around each target location. It is also known as the Close Enough Orienteering Problem with Dubins Vehicle [16]. Because of the neighborhood extension, the collected reward is higher than in DOP, even though it uses the same Dubins vehicle model. It was proposed in by *Pěnička et al.* in [31] together with a VNS-based metaheuristic. It produced feasible results with a larger collected reward than in DOP by saving travel costs and even outperformed the only existing SOM-based approach for non-overlapping neighborhoods in the Euclidean Orienteering Problem with Neighborhoods.

Since great computational power is required to solve the DTOPN, a fast algorithm providing good results is required. Ideal candidates are algorithms able to solve the TOP since the Dubins and Neighborhoods extension mainly modifies the criteria of path creation.

According to a survey published by *Vansteenwegen* [42], the best scoring TOP heuristics are Slow VNS by *Archetti et al.* [1], which managed to have the lowest average gap between the generated and best-known solution, and Ant Colony Optimization by *Ke et al.* [26], which managed to find the highest number of best solutions.

The first to focus on obtaining good results in only a few seconds of computational time were *Vansteenwegen et al.* using a Guided Local Search approach [40] and VNS approach [41]. Their proposed Skewed Variable Neighbourhood Search (SVNS) [41] reduced the minimal achieved computational time from 63.6 seconds to 3.8, making it significantly faster than other heuristics while still managing to provide good results.

Very good results were, however, achieved by the Greedy Randomised Adaptive Search Procedure with Path Relinking extension by *Souffriau* [36]. It is capable of two modes of work, the Fast Path Relinking (FPR) and Slow Path Relinking (SPR). In the FPR mode, the algorithm ended up the second fastest with an average gap in between the fast performing SVNS by *Vansteenwegen et al.* [41] and the good results providing Ant Colony Optimization by *Ke et al.* [26].

In the SPR mode, it achieved lower average gap than VNS by *Archetti et al.* [1] and found more best solutions found than Ant Colony Optimization by *Ke et al.* [26] while maintaining faster execution times than both of them.

This makes it a very flexible algorithm that is well-suited for reliable and fast solving of the NP-hard DTOPN and therefore is proposed in this paper as a solution to the problem.

3   Problem Statement

In this section, the Team Orienteering Problem is defined as an optimization problem together with its constraints. The TOP is then extended to the Dubins

TOP and the DTOP with Neighborhoods. This shows the similarities and differences between these problem formulations and helps to explain the benefits of these extensions. The optimization problem formulation serves to identify the defining variables of each problem formulation and shows their influence on the resulting solution.

3.1   Team Orienteering Problem

The TOP is a combination of the EOP and mTSP. The goal is to determine P paths, each limited with T_{max}, that maximize the total collected reward. Each path is a subset $S_k^p \subset S$, where $S = \{s_1, ..., s_n\}$ is a set of target locations. The origin and ending locations are given and represented as s_1 and s_n. Each considered target location s_i is defined by its position denoted as $s_i \in \mathbb{R}^2$ and its reward r_i. The reward of both the starting and the ending location is assumed to be zero $r_1 = r_n = 0$ and is strictly positive for all other locations. The goal of the TOP is to determine for each of P paths, $p = (1, \ldots, P)$, a set of k target locations that define the subset $S_{k_p}^p$.

The sequence of their visits can be described as a permutation over k_p target location of path p using $\Sigma_{k_p}^P = (\sigma_1^p, \ldots, \sigma_{k_p}^p)$, with constraints $1 \leq \sigma_i^p \leq n$, $\sigma_i^l \neq \sigma_j^m$ for $i \neq j \wedge l \neq m$ and $\sigma_1^p = 1$, $\sigma_{k_p}^p = n$, where σ_i^p represents the target location index. These ensure that each node is visited by at most one path and at most once.

For the Euclidean distance $\mathcal{L}_e(s_{\sigma_i^p}, s_{\sigma_j^p})$ between two locations $s_{\sigma_i^p}$ and $s_{\sigma_j^p}$ both belonging to path p, the TOP can be formulated as the following optimization problem:

$$\underset{\Sigma_{k_p}^P, k_p}{Maximize\ R} = \sum_{p=1}^{P} \sum_{i=1}^{k_p} r_{\sigma_i^p}$$

$$s.t.\ \sum_{i=2}^{k_p} \mathcal{L}_e(s_{\sigma_{i-1}^p}, s_{\sigma_i^p}) \leq T_{max}\quad \forall p \in (1, \ldots, P)\,, \tag{1}$$

$$\sigma_i^l \neq \sigma_j^m \text{ for } i \neq j \wedge l \neq m\,,$$

$$\sigma_1^p = 1\,, \sigma_{k_p}^p = n\quad \forall p \in (1, \ldots, P)\,,$$

where R represents the total collected reward.

3.2   Dubins Team Orienteering Problem

In the DOP [30], the state of the Dubins vehicle $q = (x, y, \theta)$ consists of its position in plane $s = (x, y) \in \mathbb{R}^2$ and its heading ($\theta \in \mathbb{S}^1$), i.e., $q \in SE(2)$. The vehicle model is thus non-holonomic. The minimal turning radius ρ influences the length of the shortest path between two states. The kinematic model of Dubins vehicle with a constant forward velocity v and control input u can be described as:

$$\dot{q} = \begin{bmatrix} \dot{s} \\ \dot{\theta} \end{bmatrix} = \begin{bmatrix} \dot{x} \\ \dot{y} \\ \dot{\theta} \end{bmatrix} = v \begin{bmatrix} cos\theta \\ sin\theta \\ \frac{u}{\rho} \end{bmatrix}, u \in [-1, 1]. \tag{2}$$

For model (2), the shortest path between two states consists only of a straight line (L-segment) and arcs with the radius ρ (C-segment). The optimal path is then one of two possible maneuvers CCC, CLC, where C can be either a left-turning or right-turning arc [13], resulting in six possible combinations. These are further denoted as Dubins maneuvers.

While the Dubins maneuver for any two states q_i and q_j with its length $\mathcal{L}_d(q_i, q_j)$ can be determined analytically, in the DOP it is necessary to determine the particular headings θ_i and θ_j of the vehicle at corresponding locations s_i, s_j, respectively.

Each target location s_i is thus in the DOP considered as the state $q_i = (s_i, \theta_i)$ and in addition to the determination of the subset S_k of the k locations in the route and the permutation $\Sigma = (\sigma_1, ..., \sigma_k)$, the DOP intends to find the corresponding heading angles $\Theta = (\theta_{\sigma_1}, ..., \theta_{\sigma_k})$

The Dubins Orienteering Problem for the model can be then formulated as the optimization problem

$$\underset{k, S_k, \Sigma, \Theta}{Maximize} \; R = \sum_{i=1}^{k} r_{\sigma_i},$$

$$s.t. \sum_{i=2}^{k} \mathcal{L}_d(q_{\sigma_{i-1}}, q_{\sigma_i}) \leq T_{max}, \tag{3}$$

$$\sigma_1 = 1, \sigma_k = n.$$

In contrast to the Euclidean OP, the DOP considers the Dubins vehicle model and the path is constructed using the Dubins maneuvers between the adjacent target locations (states). The optimization problem is not only over all possible subsets and respective permutations of the target locations (k, Σ), but also over all possible heading angles Θ of the target locations. This makes the problem computationally challenging as the already NP-hard EOP is extended to optimize over heading angles.

For the DTOP, the described problem formulation must be modified to accommodate multiple paths. Each of the locations s_i^p is then considered as the state $q_i^p = (s_i^p, \theta_i^p)$ and the permutation becomes $\Sigma_{k_p} = (\sigma_1^p, ..., \sigma_{k_p}^p)$ with corresponding heading angles $\Theta_p = (\theta_{\sigma_1^p}, ..., \theta_{\sigma_{k_p}^p})$. The optimization problem is thus:

$$\underset{\Sigma_{k_p}^p, \Theta_p, k_p}{Maximize} \ R = \sum_{p=1}^{P} \sum_{i=1}^{k_p} r_{\sigma_i^p}$$

$$s.t. \ \sum_{i=2}^{k_p} \mathcal{L}_d(q_{\sigma_{i-1}^p}, q_{\sigma_i^p}) \leq T_{max} \quad \forall p \in (1, \ldots, P), \tag{4}$$

$$\sigma_i^l \neq \sigma_j^m \ \text{for} \ i \neq j \wedge l \neq m,$$

$$\sigma_1^p = 1, \ \sigma_k^p = n \quad \forall p \in (1, \ldots, P).$$

3.3 Dubins Team Orienteering Problem with Neighborhoods

In the Dubins Team Orienteering Problem with Neighborhoods, the reward can be collected by visiting a circular neighborhood around each location. The specific neighborhood is described by the neighborhood radius parameter δ that defines a δ-radius disk centered at the respective target location coordinates. It is expected for all target locations to have the same value of δ, except the starting location s_0 and the ending location s_k with zero neighborhood radius.

In contrast to the DTOP where k, Σ and Θ are determined, the DTOPN also requires determination of particular locations of the waypoints $W \subseteq \mathbb{R}^2$ at which the rewards are collected, where the waypoints are within δ distance from the respective target locations, i.e., $w_{\sigma_i} \in W, s_{\sigma_i} \in S_k$ and $|(w_{\sigma_i}, s_{\sigma_i})| \leq \delta$. The Dubins Orienteering Problem with Neighborhoods [31] can be then described as the optimization problem:

$$\underset{k, S_k, W, \Sigma, \Theta}{Maximize} \ R = \sum_{i=1}^{k} r_{\sigma_i},$$

$$s.t. \ \sum_{i=2}^{k} \mathcal{L}_d(q_{\sigma_{i-1}}, q_{\sigma_i}) \leq T_{max}, \tag{5}$$

$$q_{\sigma_i} = (p_q \sigma_i, \theta_{\sigma_i}, w_{\sigma_i} \in W_k, \theta_{\sigma_i} \in \Theta).$$

Four important variables must be determined to solve the DTOPN. These are, similar to the TOP, S_{k_p} and Σ_{k_p}, where S_{k_p} represents the locations in each route and thus influences the total collected rewards R, and the permutation Σ_{k_p} that defines the length of each path p over S_{k_p} constrained by the budget T_{max}. Furthermore, the DTOPN solution contains the sequence of heading angles Θ_{k_p} at the target locations that influence the length of each path because of the curvature constraints of the Dubins vehicle.

The final path length is also influenced by the neighborhoods of the respective target locations implied by $\|w_{\sigma_i}, s_{\sigma_i}\| < \delta$. This results in additional search among the waypoints $W_{k_p} = (w_{\sigma_1}, ..., w_{\sigma_{k_p}})$ and is the reason why the DTOPN is more challenging than the DTOP or TOP since it adds additional part of the continuous optimization for the locations of the waypoints in $\mathbb{R}^2$.

In the DTOPN, the same applies, only modified to accommodate multiple routes. The variables become $w_{\sigma_{k_p}^p} \in W_{k_p}^p = (w_{\sigma_1^p}, \ldots, w_{\sigma_{k_p}^p}), s_{\sigma_i^p} \in S_{k_p}^p$ where $|(w_{\sigma_i^p}, s_{\sigma_i^p})| \leq \delta$. The DTOPN can be described as the optimization problem:

$$\underset{\Sigma_{k_p}^P, \Theta_p, W_{k_p}^P, k_p}{Maximize} \quad R = \sum_{p=1}^{P} \sum_{i=1}^{k_p} r_{\sigma_i^p}$$

$$s.t. \quad \sum_{i=2}^{k_p} \mathcal{L}_d(q_{\sigma_{i-1}^p}, q_{\sigma_i^p}) \leq T_{max} \quad \forall p \in (1, \ldots, P) \ ,$$

$$q_{\sigma_i^p} = (w_{\sigma_i^p}, \theta_{\sigma_i^p}), w_{\sigma_i^p} \in W_{k_p}, \theta_{\sigma_i^p} \in \Theta_{k_p} \forall p \in (1, \ldots, P), \quad (6)$$

$$\|w_{\sigma_i^p}, s_{\sigma_i^p}\| \leq \delta \ \forall i \in (2, k-1) \forall p \in (1, \ldots, P),$$

$$\|w_{\sigma_1^p}, s_{\sigma_1^p}\| = 0, \|w_{\sigma_k^p}, s_{\sigma_k^p}\| = 0,$$

$$\sigma_i^l \neq \sigma_j^m \text{ for } i \neq j \wedge l \neq m \ ,$$

$$\sigma_1^p = 1 \ , \ \sigma_k^p = n \quad \forall p \in (1, \ldots, P) \ .$$

4   Proposed GRASP with Path Relinking Method for the DTOPN

The proposed algorithm to solve the DTOPN is based on the Greedy Randomised Adaptive Search Procedure (GRASP) with Path Relinking [36]. It consists of four procedures that are repeated indefinitely until the termination condition is met. The termination condition, in this case, is a number of X of iterations without improvement. The GRASP with Path Relinking has two modes of operation, the Fast Path Relinking (FPR) and the Slow Path Relinking (SPR). The key difference is in the number of iterations without improvement. The FPR has significantly lower X and thus produces faster results, while the SPR has higher, which increases the quality of solutions.

The four procedures are Construct, Local Search, Link to Elites and Update Elites. Construct and Local Search create and subsequently improve an independent solution, producing a locally optimal solution. This solution is then improved using the best solutions previously found, called elites, in the Link to Elites procedure. Update Elites then refreshes the set of best previously found solutions, called elite pool, depending on the number of elites currently present. If the pool is not full, the currently generated solution is inserted. Otherwise, if there are elites with a lower score than the current solution, the worst-performing elite is replaced. After the termination condition is met, the algorithm returns the best feasible solution found. The GRASP-PR algorithm for DTOPN is summarized in Algorithm 1.

4.1   Construct Procedure

The Construct procedure is responsible for generating new solutions that are subsequently improved. It is characterized by Greediness, a parameter randomly

---

Algorithm 1. GRASP-PR

1: **while** Nr of iterations without improvement **do**
2:     Construct
3:     Local Search
4:     Link to Elites
5:     Update Elites
6: **end while**

---

drawn from uniform distribution $<0, 1>$. Greediness describes the exact ratio between greediness and randomness of the currently constructed solution. Its random nature helps in generating various distinctive solutions and exploring new possible means to achieve a higher reward.

First, all infeasible nodes are discarded. Feasibility is checked by calculating the price of each node l, which is equal to $t_l = t_{il} + t_{lj}$ where i represents the starting node and j represents the ending node and t_l is the travel distance to visit the node. If $t_l < budget$, the node is feasible and inserted into a candidate list. Otherwise, it is discarded.

Afterward, each node's heuristic value is calculated: $h_l = \frac{r_l}{t_l}$, and threshold is set as

$$h_{threshold} = (h_{max} - h_{min}) \cdot greediness. \tag{7}$$

The candidates that fulfill the requirement $h_l > h_{threshold}$ are inserted into a restricted candidate list.

Then, all P paths of the current solution are populated by randomly selected nodes from the restricted candidate list so that the total length of a path p with k_p nodes

$$t^p_{total} = \sum_{i=2}^{k_p} t_{i-1,i} \tag{8}$$

does not violate the travel budget. The procedure ends when there are no feasible nodes left to include and returns P feasible paths of the current solution.

4.2   Local Search Procedure

The Local Search procedure summarized in Algorithm 2 tries to improve the solution using four different ways until finding the local optimum. It alternates between reducing the total price of the P solution paths and increasing the value by considering non-included nodes.

First, a 2-opt operator is applied. This ensures that the total price of each of the P paths is minimal.

Next, the neighborhood between paths of the solution is explored in such a way that minimizes the total price. If an improvement in the total price can be achieved by swapping two locations already present in the solution, they are swapped. This continues until no possible price improvement can be achieved by exchanging nodes between routes.

Algorithm 2. Local Search

1: **while** Improvement **do**
2:     Simplified 2-opt
3:     Swap
4:     Replace
5:     Insert
6: **end while**

Afterward, Replace tries to improve the total claimed score by replacing nodes currently in the solution with feasible nodes not included. The algorithm considers all non-included feasible nodes and finds their cheapest insertion place in the current solution. If it would violate the travel budget, the node with the lowest score that could be feasibly replaced is found and replaced. The Replace procedure executes moves that result in the best collected score increase, and, in case of a tie, the best collected score increase and the lowest solution price increase.

Finally, Insert tries to insert non-included nodes into the current solution in a way that would not increase the total price over the travel budget.

The procedure ends when no further improvements in score or solution length can be found.

4.3   Link to Elites Procedure

The Link to Elites procedure described in Algorithm 3 serves as a long-term memory component which ensures that the complete independence between solutions generated by Construct and Local Search procedures is avoided.

The procedure takes two solutions as an argument, a starting solution and a guiding solution, and explores the neighborhood between them. It is executed for the current solution as a starting solution and subsequently all members of the elite pool as a guiding solution and vice versa. Every time the procedure uses an elite as a guiding solution, it's age increments by 1. This ensures that the algorithm does not get stuck in the local optimum and explores as many feasible solutions as possible. The procedure returns the best solution found.

Algorithm 3. Link to Elites

1: Set intersections = common nodes of starting and guiding solutions
2: Set locationsToAdd = guiding solution minus intersections
3: **while** locationsToAdd is not empty **do**
4:     Insert locations, allow infeasibility
5:     Remove locations to restore feasibility
6:     Local Search
7: **end while**

The first step is creating a list of intersections between starting and guiding solutions, which contains all the nodes both solutions have in common. Then, a list of nodes to add is found by removing all intersection nodes from the guiding solution. This prevents adding duplicate locations into the solution.

Furthermore, the similarity between the starting solution and the guiding solution is calculated. If the similarity is above a certain threshold, the procedure does not continue. This helps to ensure that the resulting solutions improved by the Link to Elites procedure are new and different from existing elite solutions, which is necessary for discovering new and better solutions. It also helps to speed up the algorithm by minimizing the possibility of generating a solution identical to one already generated.

While the list of nodes to add is not empty, the procedure considers inserting nodes from the list of nodes to add to feasible paths. This differs from the way the Construct procedure works because even moves that would violate the travel budget are considered.

When all P paths are infeasible, the algorithm restores feasibility by removing nodes. For each node, it's heuristic value $h^{-1} = \frac{t_i}{r_i}$ is calculated. The node with the highest h^{-1} is then removed. This continues until all paths are feasible again. The algorithm iterates through all routes of the solution and removes a node from each infeasible route. The Local Search procedure is then called to optimize the solution price after iterating through all routes, which helps in preventing unnecessary location removal. The procedure ends when the list of nodes to add is empty and returns a list of elite candidate solutions.

4.4 Update Elites Procedure

The Update Elites procedure takes the best solution found by Link to Elites and compares it to recently best found solutions called Elites. There is a maximum of 10 Elites which are stored in the Elite Pool.

If any Elite solution reached the age of

$$max(10, \frac{X}{10})$$

where X is the number of iterations without improvement, it is removed. If the solution found is equal to an existing elite solution, the elite's age is reset to 0. If not, and the elite pool has not reached the maximum number of members (10), the solution is added. Otherwise, if the solution's collected score is higher than the worst elite's, it is replaced. This populates the Elite Pool, which is necessary for the solutions in order to be at least partially non-random.

4.5 GRASP Extensions for the DTOPN

To use the GRASP algorithm for the DTOPN, the algorithm must first be extended in order to consider the Dubins Vehicle constraints.

The procedure uses a sampling-based approach to create a discrete set of headings for each location by proportionally sampling possible heading angles with the provided headings resolution R_D. This is necessary because otherwise, the set of different heading angles would be infinite. The exact number of samples generated this way is given by the Dubins resolution parameter R_D. Next, the method used to calculate distances between each location is extended to calculate the distance between every sampled heading of each location using Dubins maneuvers with the given minimal turning radius as a parameter. This ensures that locations requiring sharp turns to visit are now penalized since curved routes have larger travel cost than straight lines.

The method responsible for adding locations to routes is modified to use the set of distances created by the previously mentioned procedure. Furthermore, when adding new locations, the optimal heading sample is calculated based on the travel distance and based on the entry heading of location. This ensures that the travel cost associated with Dubins maneuvers is minimized.

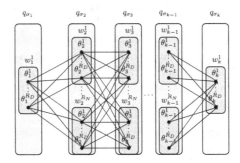

Fig. 2. Search graph with R_D uniformly sampled headings and with R_N uniformly sampled neighborhoods at each target location $q_{\sigma_i}, 0 \leq i \leq k$. Graph search algorithm over all heading and neighborhood sample combinations is utilized to find the path with minimal length connecting specified target locations $(q_{\sigma_1}, \ldots, q_{\sigma_k})$. This is applied to each of P paths of the solution.

Finding the optimal headings that minimize the travel cost can be described as a Dubins Touring Problem [17]. To minimize the length of a path, a graph search algorithm is used to determine the sequence of headings that produce a path with minimal travel cost. The algorithm is developed to utilize a dynamic programming technique [30] to store distances $||(q_{\sigma_1}, q_{\sigma_i})||$ and $||(q_{\sigma_i}, q_{\sigma_k})||$ for every location i to decrease the computational requirements needed to re-calculate distances when adding, exchanging or removing locations. The graph is visualized in Fig. 2.

Afterward, the GRASP algorithm must be further modified in order to be able to work with the neighborhoods around each location. Similar to the Dubins extension, the neighborhoods extension utilizes a sampling-based operation to reduce the number of samples for each vertex. This is necessary to reduce the number of neighborhood samples since it is otherwise infinite.

Using a given neighborhood resolution parameter R_N, the circular radius of location neighborhoods is equidistantly sampled into a finite number of points.

This results in a set of new locations, called neighborhood locations, in a circle of diameter δ around each of target locations. The method responsible for calculating distances between locations is then modified to iterate through every newly created neighborhood location and use their coordinates to calculate distances in addition to iterating through all locations and all heading samples.

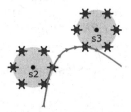

Fig. 3. An example showing how heading and neighborhood samples are connected using curves with radius ρ.

The method responsible for adding locations is then modified to find the neighborhood location and heading with lowest price increase when adding a new location. This ensures that every constructed solution uses routes with minimized travel costs (Fig. 3).

To find the optimal neighborhood sample, a similar graph search algorithm used to find optimal headings is used (Fig. 2). The information about which neighborhood location was visited is then stored in the route data, since it is necessary for further optimization and to eliminate the need to calculate the best headings and neighbors for every location when only one has changed.

5   Results

In this chapter, the testing instances are presented, and the results of the application of the GRASP-PR algorithm on the TOP, DTOP and DTOPN are compared, together with the influence of specific parameters on provided solutions. Furthermore, the real-life experiment used to verify the performance is presented, and its results are discussed.

All presented results were computed using Intel Xeon 6130 Gold CPUs (2.10 GHz). If not specified otherwise, parameters for DTOPN are neighborhood radius $\delta = 0.5$, Dubins radius $\rho = 0.5$, neighborhood resolution $R_N = 4$ and Dubins resolution $R_D = 4$. Values for Fast Path Relinking (FPR) and Slow Path Relinking (SPR) correspond to the parameter number of iterations without improvement, which is set as 10 for FPR and 300 for SPR. These values are based on the results in [36] where they were found as the best-performing pair. The dependency of the collected reward and required execution time on specific parameters was computed as an average of multiple runs using SPR. Datasets used for testing are Chao 4–7 [8].

5.1 Algorithm Performance

In order to establish a common ground for the results of the DTOPN solutions and other problem formulations, the results of this implementation of the GRASP-PR algorithm are compared to the existing computational results of GRASP-PR for the TOP. Table 1 contains results of both a single run of the algorithm and the best result of 10 runs, both using a sample of datasets from Chao 4–7 [8], compared to the results of GRASP-PR presented in [36] which are denoted as *SPR10 Original* and *FPR10 Original*.

Table 1. Reward collected by the GRASP algorithm used in this thesis compared to results by *Souffriau et al.* [36].

Dataset	FPR10 original	SPR10 original	FPR	SPR	FPR10	SPR10
p4.2.b	341	341	247	252	252	252
p4.3.j	858	861	748	761	748	777
p4.4.j	732	732	448	507	523	525
p5.2.n	925	925	910	900	910	910
p5.3.n	1070	1070	715	735	735	735
P5.4.p	760	760	700	740	705	755
p6.2.j	942	948	894	900	912	930
p6.3.l	1002	1002	858	972	948	978
p6.4.k	528	528	438	480	468	480
p7.2.j	632	646	559	605	615	623
p7.3.i	480	485	364	428	422	436
p7.4.i	364	366	293	291	295	302

The single-run results in Table 1 have a significant gap between its collected reward and the reference. However, in situations where it is crucial to produce solutions in a small amount of time, the results of a single run could be acceptable when exchanged for faster execution time.

The maximal results obtained after 10 runs of SPR (SPR10) and 10 runs of FPR (FPR10) still present a noticeable gap, albeit lower, when compared to the best achieved GRASP results. This may be caused by possible small differences in the implementation of the algorithm. Furthermore, due to the randomized nature of the algorithm, it can not be guaranteed that even the FPR10 and SPR10 will present the best solution that this algorithm is capable of providing.

5.2 Computational Results

Results of DTOPN compared to DTOP and TOP are shown in Table 2 containing the collected reward and execution times for a sample of the aforementioned

Table 2. Collected reward and execution times in seconds in TOP, DTOP and DTOPN instances

Dataset	FPR Reward	TOP Time	FPR Reward	DTOP Time	FPR Reward	DTOPN Time	SPR Reward	TOP Time	SPR Reward	DTOP Time	SPR Reward	DTOPN Time
p4.2.b	244	0.14	238	1.06	247	4.87	252	1.35	238	2.03	259	30.56
p4.3.j	713	0.95	200	0.9	694	12.25	785	18.30	718	32.38	812	128.52
p4.4.j	482	0.03	15	0.28	505	2.78	534	1.52	488	2.34	563	11.58
p5.2.n	830	0.62	391	1.37	920	54.12	900	18.82	850	53.94	950	468.16
p5.3.n	725	0.80	690	2.79	820	12.50	735	5.60	690	16.46	840	121.47
p5.4.p	675	0.41	655	0.80	745	2.18	740	2.21	700	6.90	845	37.20
p6.2.j	912	4.37	906	18.54	1086	39.42	924	23.76	870	43.26	1164	674.95
p6.3.l	894	0.89	900	7.82	978	4.10	978	15.38	888	33.82	1098	61.08
p6.4.j	306	0.02	282	0.42	420	1.28	330	0.31	312	0.60	462	8.55
p7.2.j	559	1.50	582	11.89	543	10.85	628	35.17	607	43.73	656	524.67
p7.3.i	394	0.35	377	0.93	321	2.65	435	2.04	419	3.86	461	30.56
p7.4.i	308	0.05	273	0.57	309	2.85	303	0.42	273	0.53	311	4.88
Average	586.8	0.84	459.1	3.95	632.3	12.32	628.7	10.41	587.5	19.99	701.8	175.18

datasets. Example TOP trajectory is shown in Fig. 4a, DTOP in Fig. 4b and DTOPN in Fig. 4c.

As shown in Table 2, the average collected reward in the DTOP instances was lower than in the TOP. This is due to the curvature constraints of the Dubins vehicle which increase the length of arcs connecting the locations, thus increasing the travel costs. In the DTOPN instances, the average collected reward was highest, since the neighborhoods extension lowers the travel budget required to visit locations.

The execution times presented in Table 2 show that the execution time required to solve the TOP, DTOP and DTOPN instances varies. This is due to the randomized nature of the algorithm. However, even with the variations, a trend of increasing execution time with each extension is present. While the DTOPN can save the travel costs expended to visit locations, thus increasing the collected reward, the execution time required increases as well, due to the additional neighborhood samples.

5.3   Algorithm Parameters Influence

Dubins Resolution Parameter R_D specifies how many heading samples are created for each location. It affects the total collected reward of the DTOPN solution since, with a higher number of samples, more precise results are computed. This, however, also influences the computational time required, because the higher number of samples increases the total number of possible combinations of locations and headings. The graph in Fig. 5 shows that for higher resolutions, larger average rewards are collected, but with increasing execution time.

Neighborhood Resolution Parameter similarly to the Dubins resolution, the neighborhood resolution parameter R_N influences the preciseness of neighborhood sampling. Figure 6 shows that higher resolution increases the execution time, but also tends to increase the average collected reward.

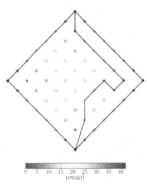

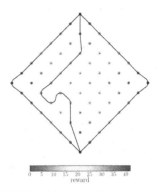

(a) TOP trajectory with total collected reward 804, $\rho = 0$, $\delta = 0$, $R_D = R_N = 1$

(b) DTOP trajectory with total collected reward 762, $\rho = 0.5$, $\delta = 0$, $R_D = 8$, $R_N = 1$

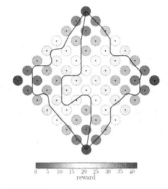

(c) DTOPN trajectory with total collected reward 1008, $\rho = 0.5$, $\delta = 0.5$, $R_D = R_N = 8$.

Fig. 4. Example TOP, DTOP and DTOPN trajectories generated by the proposed GRASP-PR algorithm on p6.3.j dataset

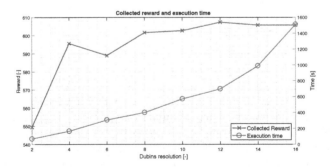

Fig. 5. The dependency of the average collected reward and execution time of the SPR DTOPN solution on the Dubins resolution parameter R_D.

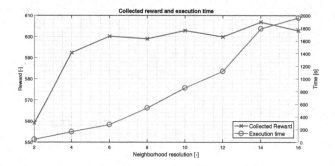

Fig. 6. The dependency of the average collected reward and execution time of the SPR DTOPN solution on the neighborhood resolution parameter R_N.

Dubins Radius Parameter ρ is one of the important parameters that influence the results of the algorithm. It represents the minimal turning radius of the Dubins vehicle and is usually determined by the parameters of the vehicle itself. For larger Dubins radii, the average collected reward is significantly lower, as seen in Fig. 7. However, the Dubins radius parameter does not seem to have a significant impact on the execution time.

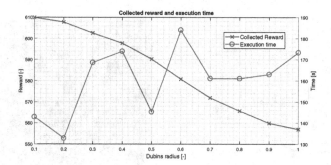

Fig. 7. The dependency of the average collected reward and execution time of the SPR DTOPN solution on the Dubins radius parameter ρ.

Neighborhood Radius Parameter δ similarly to the Dubins radius has a large impact on the quality of produced results. It acts opposite to the Dubins radius in the sense that the higher the radius, the higher the average reward collected, as seen on Fig. 8. It represents the radius (size) of the area around each location that must be visited in order to claim the associated reward.

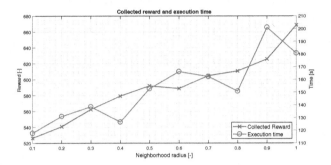

Fig. 8. The dependency of the average collected reward and execution time of the SPR DTOPN solution on the neighborhood radius parameter δ.

5.4   Experimental Verification with Micro Aerial Vehicles

The GRASP-PR algorithm for the DTOPN was tested in real-life conditions using a group of three unmanned hexarotor MAVs, which were initially developed for multi-robot applications [33, 34, 37]. The MAVs were equipped with relative localization system, Real Time Kinematic (RTK) GPS, onboard PC, autopilot and down-facing camera. Figure 9a shows the onboard equipment.

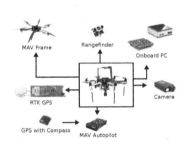

(a) Visualisation of hexarotor UAV hardware used in the real-life experiment.

(b) *UAV 1* hovering above its starting location (marked by a black block on the ground), taken by onboard camera of *UAV 2*.

Fig. 9. UAV hardware components and flying UAV above the starting position

The relative localization system is used by the onboard Collision Avoidance System (CAS) to prevent mid-air collisions within the group [14, 28]. They are also equipped with an onboard Model Predictive Controller (MPC) used for stabilization using only onboard sensors [2]. RTK GPS provides an accurate reading of UAVs position, which is necessary for precision of pre-planned missions.

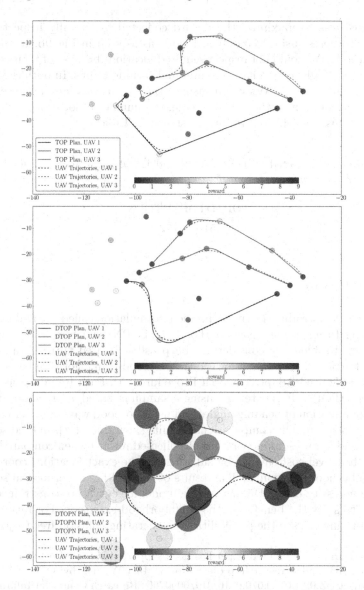

Fig. 10. Three pre-planned trajectories for the TOP, DTOP and DTOPN with a visualization of MAV trajectories flown in real-life experiment based on data from onboard positioning systems. The axes are shown in meters.

In the experiment, three different trajectories were prepared (TOP, DTOP and DTOPN), as shown in Fig. 10. The locations are represented by colored blocks as seen in Fig. 9b, each with its assigned reward, which were scattered on a field with an area of approximately 100×50 m. The utilized T_{max} budget was 80 and it represented maximal travel distance for each UAV in meters.

The goal was to maximize the reward collected by visually inspecting the locations of interest using the onboard cameras as seen in Fig. 9b. In the TOP instance, the MAVs could not inspect required locations because of their constant forward speed resulting in a non-zero minimal turning radius. In both the DTOP and the DTOPN, the MAVs could inspect the target locations because of the Dubins extension, but in the DTOPN case, the number of locations visited was higher and thus the collected reward as shown in Table 3.

Table 3. Reward collected in TOP, DTOP and DTOPN instances using FPR with parameters $T_{max} = 90$, $\rho = 6.15\ m$, $\delta = 4\ m$, $a_{max} = 2.6\ m \cdot s^{-2}$, $v_c = 4\ m \cdot s^{-1}$.

TOP	DTOP	DTOPN
55	51	72

6   Conclusion

In this paper we introduce the novel problem formulation called the Dubins Team Orienteering Problem with Neighborhoods (DTOPN) which extends the Team Orienteering Problem by considering the possible limited turning radius and remote data collection at target locations. We propose an algorithm based on the Greedy Randomized Adaptive Search Procedure with Path Relinking as a solution to the DTOPN. It is extended using a sampling-based approach to find the optimal combination of heading angles and neighborhood waypoints for each visited target location. The computational results show that the proposed solution is feasible and features an increase in the collected reward when compared to the TOP due to travel cost savings by substituting the exact locations coordinates with neighborhood waypoints. The results of the real-life experimental scenario verify the necessity of the Dubins vehicle model for curvature-constrained vehicles and also show the benefits of the neighborhood extension. For future work, we intend to investigate the possibilities of generating collision-free trajectories.

Acknowledgment. Presented paper has been supported by the Czech Science Foundation (GAČR) under research project No. 17-16900Y and by OP VVV MEYS funded project CZ.02.1.01/0.0/0.0/16_019/0000765 "Research Center for Informatics". Access to computing and storage facilities owned by parties and projects contributing to the National Grid Infrastructure MetaCentrum, provided under the programme "Projects of Large Research, Development, and Innovations Infrastructures" (CESNET LM2015042), is greatly appreciated. Support of the Grant Agency of the Czech Technical University in Prague No. SGS17/187/OHK3/3T/13 is gratefully acknowledged.

References

1. Archetti, C., Hertz, A., Speranza, M.G.: Metaheuristics for the team orienteering problem. J. Heuristics **13**(1), 49–76 (2007)
2. Baca, T., Loianno, G., Saska, M.: Embedded model predictive control of unmanned micro aerial vehicles. In: 2016 21st International Conference on Methods and Models in Automation and Robotics (MMAR), pp. 992–997. IEEE (2016)
3. Bektas, T.: The multiple traveling salesman problem: an overview of formulations and solution procedures. Omega **34**(3), 209–219 (2006)
4. Bouly, H., Dang, D.C., Moukrim, A.: A memetic algorithm for the team orienteering problem. 4or **8**(1), 49–70 (2010)
5. Butt, S.E., Cavalier, T.M.: A heuristic for the multiple tour maximum collection problem. Comput. Oper. Res. **21**(1), 101–111 (1994)
6. Campos, V., Martí, R., Sánchez-Oro, J., Duarte, A.: Grasp with path relinking for the orienteering problem. J. Oper. Res. Soc. **65**(12), 1800–1813 (2014)
7. Chao, I.M., Golden, B.L., Wasil, E.A.: A fast and effective heuristic for the orienteering problem. Eur. J. Oper. Res. **88**(3), 475–489 (1996)
8. Chao, I.M., Golden, B.L., Wasil, E.A.: The team orienteering problem. Eur. J. Oper. Res. **88**(3), 464–474 (1996). https://doi.org/10.1016/0377-2217(94)00289-4. http://www.sciencedirect.com/science/article/pii/0377221794002894
9. Dang, D.-C., El-Hajj, R., Moukrim, A.: A branch-and-cut algorithm for solving the team orienteering problem. In: Gomes, C., Sellmann, M. (eds.) CPAIOR 2013. LNCS, vol. 7874, pp. 332–339. Springer, Heidelberg (2013). https://doi.org/10.1007/978-3-642-38171-3_23
10. Dang, D.C., Guibadj, R.N., Moukrim, A.: An effective pso-inspired algorithm for the team orienteering problem. Eur. J. Oper. Res. **229**(2), 332–344 (2013)
11. Dorigo, M., Di Caro, G.: Ant colony optimization: a new meta-heuristic. In: Proceedings of the 1999 Congress on Evolutionary Computation, CEC 1999, vol. 2, pp. 1470–1477. IEEE (1999)
12. Dorigo, M., Gambardella, L.M.: Ant colony system: a cooperative learning approach to the traveling salesman problem. IEEE Trans. Evol. Comput. **1**(1), 53–66 (1997)
13. Dubins, L.E.: On curves of minimal length with a constraint on average curvature, and with prescribed initial and terminal positions and tangents. Am. J. Math. **79**(3), 497–516 (1957)
14. Faigl, J., Krajník, T., Chudoba, J., Přeučil, L., Saska, M.: Low-cost embedded system for relative localization in robotic swarms. In: 2013 IEEE International Conference on Robotics and Automation (ICRA), pp. 993–998. IEEE (2013)
15. Faigl, J., Pěnička, R., Best, G.: Self-organizing map-based solution for the orienteering problem with neighborhoods. In: 2016 IEEE International Conference on Systems, Man, and Cybernetics (SMC), pp. 001315–001321. IEEE (2016)
16. Faigl, J., Pěnička, R.: On close enough orienteering problem with dubins vehicle. In: IEEE/RSJ International Conference on Intelligent Robots and Systems (IROS), pp. 5646–5652 (2017)
17. Faigl, J., Váňa, P., Saska, M., Báča, T., Spurný, V.: On solution of the dubins touring problem. In: 2017 European Conference on Mobile Robots (ECMR), pp. 1–6. IEEE (2017)
18. Galceran, E., Carreras, M.: A survey on coverage path planning for robotics. Rob. Auton. Syst. **61**(12), 1258–1276 (2013)

19. Golden, B.L., Levy, L., Vohra, R.: The orienteering problem. Naval Res. Logist. **34**(3), 307–318 (1987)
20. Gulczynski, D.J., Heath, J.W., Price, C.C.: The close enough traveling salesman problem: a discussion of several heuristics. In: Alt, F.B., Fu, M.C., Golden, B.L. (eds.) Perspectives in Operations Research. Operations Research/Computer Science Interfaces Series, vol. 36, pp. 271–283. Springer, Boston (2006). https://doi.org/10.1007/978-0-387-39934-8_16
21. Gunawan, A., Lau, H.C., Vansteenwegen, P.: Orienteering problem: a survey of recent variants, solution approaches and applications. Eur. J. Oper. Res. **255**(2), 315–332 (2016)
22. Hodicky, J.: Autonomous systems operationalization gaps overcome by modelling and simulation. In: Hodicky, J. (ed.) MESAS 2016. LNCS, vol. 9991, pp. 40–47. Springer, Cham (2016). https://doi.org/10.1007/978-3-319-47605-6_4
23. Hodicky, J.: Standards to support military autonomous system life cycle. In: Březina, T., Jabłoński, R. (eds.) MECHATRONICS 2017. AISC, vol. 644, pp. 671–678. Springer, Cham (2018). https://doi.org/10.1007/978-3-319-65960-2_83
24. Hodicky, J., Prochazka, D.: Challenges in the implementation of autonomous systems into the battlefield. In: 2017 International Conference on Military Technologies (ICMT), pp. 743–747. IEEE (2017)
25. Karp, R.M.: Reducibility among combinatorial problems. In: Miller, R.E., Thatcher, J.W., Bohlinger, J.D. (eds.) Complexity of Computer Computations, pp. 85–103. Springer, Boston (1972). https://doi.org/10.1007/978-1-4684-2001-2_9
26. Ke, L., Archetti, C., Feng, Z.: Ants can solve the team orienteering problem. Comput. Ind. Eng. **54**(3), 648–665 (2008)
27. Ke, L., Zhai, L., Li, J., Chan, F.T.: Pareto mimic algorithm: an approach to the team orienteering problem. Omega **61**, 155–166 (2016)
28. Krajník, T., et al.: A practical multirobot localization system. J. Intell. Rob. Syst. **76**(3–4), 539–562 (2014)
29. Lin, S.W.: Solving the team orienteering problem using effective multi-start simulated annealing. Appl. Soft Comput. **13**(2), 1064–1073 (2013)
30. Pěnička, R., Faigl, J., Váňa, P., Saska, M.: Dubins orienteering problem. IEEE Rob. Autom. Lett. **2**(2), 1210–1217 (2017). https://doi.org/10.1109/LRA.2017.2666261. http://mrs.felk.cvut.cz/icra17dop
31. Pěnička, R., Faigl, J., Váňa, P., Saska, M.: Dubins orienteering problem with neighborhoods. In: 2017 International Conference on Unmanned Aircraft Systems (ICUAS), pp. 1555–1562, June 2017
32. Robinson, J.: On the hamiltonian game (a traveling salesman problem). Technical report, Rand Project Air Force, Arlington, VA (1949)
33. Saska, M., et al.: System for deployment of groups of unmanned micro aerial vehicles in GPS-denied environments using onboard visual relative localization. Auton. Robots **41**(4), 919–944 (2017)
34. Saska, M., Vonásek, V., Krajník, T., Přeučil, L.: Coordination and navigation of heterogeneous mav-ugv formations localized by a 'hawk-eye'-like approach under a model predictive control scheme. Int. J. Rob. Res. **33**(10), 1393–1412 (2014)
35. Savla, K., Frazzoli, E., Bullo, F.: On the point-to-point and traveling salesperson problems for dubins' vehicle. In: Proceedings of the 2005 American Control Conference, pp. 786–791. IEEE (2005)
36. Souffriau, W., Vansteenwegen, P., Berghe, G.V., Oudheusden, D.: A greedy randomised adaptive search procedure for the team orienteering problem. In: EU/MEeting, pp. 23–24 (2008)

37. Spurny, V., Baca, T., Saska, M.: Complex manoeuvres of heterogeneous MAV-UGV formations using a model predictive control. In: 2016 21st International Conference on Methods and Models in Automation and Robotics (MMAR), pp. 998–1003. IEEE (2016)
38. Thakur, D., et al.: Planning for opportunistic surveillance with multiple robots. In: 2013 IEEE/RSJ International Conference on Intelligent Robots and Systems (IROS), pp. 5750–5757. IEEE (2013)
39. Tsiligirides, T.: Heuristic methods applied to orienteering. J. Oper. Res. Soc. **35**(9), 797–809 (1984)
40. Vansteenwegen, P., Souffriau, W., Berghe, G.V., Van Oudheusden, D.: A guided local search metaheuristic for the team orienteering problem. Eur. J. Oper. Res. **196**(1), 118–127 (2009)
41. Vansteenwegen, P., Souffriau, W., Berghe, G.V., Van Oudheusden, D.: Metaheuristics for tourist trip planning. In: Sörensen, K., Sevaux, M., Habenicht, W., Geiger, M. (eds.) Metaheuristics in the Service Industry. LNE, vol. 624, pp. 15–31. Springer, Heidelberg (2009). https://doi.org/10.1007/978-3-642-00939-6_2
42. Vansteenwegen, P., Souffriau, W., Van Oudheusden, D.: The orienteering problem: a survey. Eur. J. Oper. Res. **209**(1), 1–10 (2011)

M&S of Intelligent Systems - AI, R&D and Application

Information Exchange Diagrams for Information Systems and Artificial Intelligence in the Context of Decision Support Systems

Sebastian Jahnen[(⊠)] and Stefan Pickl

Fakultät für Informatik, Universität der Bundeswehr,
85577 Neubiberg, Germany
{Sebastian.Jahnen,Stefan.Pickl}@unibw.de

Abstract. Nowadays we face the Information Age and nothing evolves more rapid than the evolution of Information Sciences (IS). Related to building networks between different Communities of Interests (COI) with the help of Information Systems, the major challenge is sharing information and giving information at the right place, in the right time, to the right COI. Basis for successful implementations in information systems and further basis for decision support in artificial intelligence (AI) systems are information exchange diagrams. The present paper deals with an automatic extraction from information exchanges in operational process models. For this purpose, a modelling method was specific designed and used. Based on graph theory approaches the modelling method gives the possibility with a specially developed software to extract the information exchanges automatically and produce needed diagrams. Our approach is a new concept in the field of information systems and AI in the context of decision support systems. We present theoretical foundations and first experimental designs.

Keywords: Information exchange · AI · Information system

1 Introduction

Basis for the exchange of information is communication. In communication there are some rules defined for more than forty years now. In the 70's Paul Grice drafted guidelines for communication which are still the basic policies in communication for enterprises and the military. The Grice's Maxims [1] especially the maxim of quantity ("Make your contribution as informative as it is required but do not make it more informative than it is required".) is one of the first rules in the exchange of information. Based on these principles the information exchange structure in military processes, especially military operations, is designed. Followed the goal of the so called *Organizational Transparency* (the organization provides relevant, timely, and reliable information to stakeholder demands) [2] information systems depend on given information exchange structures. Also, decision support systems are following these structures, providing information.

© Springer Nature Switzerland AG 2019
J. Mazal (Ed.): MESAS 2018, LNCS 11472, pp. 393–401, 2019.
https://doi.org/10.1007/978-3-030-14984-0_28

This paper deals with a method for automated generating information exchange diagrams from processes modelled for military operations, following the goal of optimize the information exchange structure while in the case of losing some communication ways in the battlefield during an operation.

2   Information Exchange

As mentioned above information exchange belongs to communication and follows the same principles and policies. The structure for exchange information in military operations is derived from the operation planning and procedure modelled as an operational process. For including an operational process in an enterprise architecture the NATO Architecture Framework (NAF v. 3.1) provides tools and methods [3].

The NAF is divided in views and sub views which represent all parts of an armed force. The operational view contains the sub views for the operational process (NATO Operational View 5, NOV-5) and the related information exchange diagram (NATO Operational View 2, NOV-2). With These two sub views it is possible to provide all necessary information to reach the mentioned goals:

- Operational Process
  - Actions
  - Information Flow
  - Information Element
- Information Exchange
  - Node/Role
  - Location
  - Information (Type/Element)
  - Information Flow

Shown in Fig. 1, there is an operational process (reconnaissance patrol) modelled as NOV-5. Related to this operational process Fig. 2 shows the NOV-2 information exchange diagram.

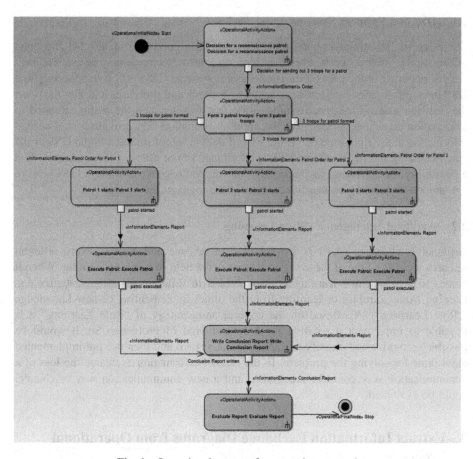

Fig. 1. Operational process for executing a patrol

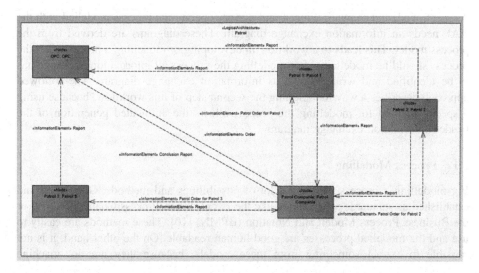

Fig. 2. Information exchange diagram related to Fig. 1

2.1   Decision Support Systems

The essential basic for fact-based decision making is good information [4] and good information is what decision support systems also need. Because many decision makers are disappointed in the past by decision support systems (DSS) [5] it is necessary that DSS have information very close to actual operations and processes and don't support the decision making by generic information that won't match the reality. Related to information exchange DSS need the input information for providing the needed information at the right time etc. If the way of delivering information to the DSS or the way of sharing information from DSS is interrupted or not usable it is necessary to find a possibility for delivering or sharing the information because nowadays DSS need to get and share information nearly real-time at the moment they occur [6].

2.2   Artificial Intelligence - Rote Learning

Artificial Intelligence (A.I.) is a wide field in science and include many different research areas [7]. One of these research areas is the field of machine learning. When it comes to A.I. machine learning is also divided in different learning strategies e.g. learning from examples or learning from the direct implementing of new knowledge ("Rote Learning") [8]. Based on the concept and strategy of "Rote Learning" it is possible to implement a solution for a special kind of problems. So, it would be possible recognize and classify problems by the system and then use pre-implemented algorithms for solving the problem. In the case of information exchange the loss of a communication way could be recognized and a new communication way or channel could be evaluated.

3   Extract Information Exchange Diagrams from Operational Processes

As mentioned, and shown in Figs. 1 and 2 every operational process modelled with the NAF needs an information exchange diagram. These diagrams are derived from the process model. This leads to a workflow in developing or modelling a process, first the process should be modelled. After modelling the process all information exchange has to be identified and written down as information exchange diagram. The followed approach describes a way for skipping the second step of this workflow, because using a special method for modelling the process allows the automated generation of the needed information exchange diagram.

3.1   Process Modelling

For modelling a process there are a lot of possibilities and methods. Often used and established as standards are methods like the Event-driven Process Chain (EPC) [9] or the Business Process Modell and Notation (BPMN) [10]. These methods are easily to use and the modelled processes are good human readable. On the other hand, it is not possible to extract something from these models, because they are not machine readable. For this purpose, a modelling method is developed which combines the advantage of easy use but is also machine readable.

The method is based on "easy to use methods" like the Bildkartenmethode (BKM) [11] and combines it with graph theoretical approaches and uses some characteristics from petri nets [12]. Where the nodes contain all information about an action (role/entity and task/action), the edges convey the information elements and the state, which is necessary to reach the next step in the process.

Figure 4 shows the result of this modelling, represented as GraphML-File [13] modelled with yEd[1]. The process is represented as a graph, where the vertices represent the operational steps and the "workflow" is represented by edges, which realize also the state of a step in the process and the transmitted information (see Fig. 3). All vertices convey the needed information about the task/action and the involved participants. Using the modelling method combined with this software allows further graph operations on the model. The process shown in Fig. 1 is based on a model which was modelled with this method and transformed by the algorithm explained in 3.2. In addition to all information given by the model shown in Fig. 4 the model in Fig. 1 automatically contains all required stereotypes given by the metamodel of the NAF (e.g. "OperationalActivityAction" for every process step, "InformationFlow" for every edge, etc.).

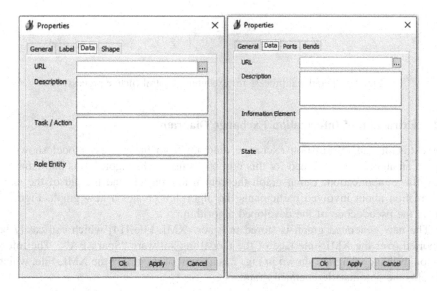

Fig. 3. Property of a node (left) and an edge (right)

---

[1] yEd is a graph drawing software; https://www.yworks.com/products/yed.

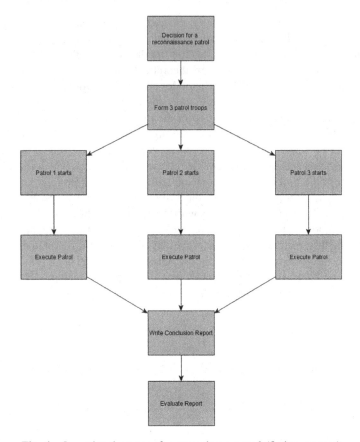

Fig. 4. Operational process for executing a patrol (fictive process)

3.2   Extraction of Information Exchange Diagrams

For extracting an information exchange diagram from the process model shown in
Fig. 3, an algorithm developed for this purpose is used. The algorithm is implemented
as a "Java" application. Using graph theoretical approaches and the aid of the extra
information about involved participants the algorithm creates a new graph. Figure 5
shows the pseudocode of the developed algorithm.

The new generated graph is stored in a new XML-File [14] which can easily be
imported over the XML-interface of the modelling software "Sparx EA"[2]. The information exchange diagram shown in Fig. 2 is a result of importing the XML-File, which
contains the automated generated graph.

---

[2] "Sparx Enterprise Architect is a comprehensive UML analysis and design tool for UML, SysML,
BPMN and many other technologies"; https://www.sparxsystems.eu/start/home/.

```
1    Generate Information Exchange Diagram from Process
2    Input:  Graph G  (represents the Process)
3    Output: Graph IE (represents Information Exchange Diagram)
4
5    get Graph G{
6         VM_IE = new Map for Vertices
7         EM_IE = new Map for Edges
8
9         VM_G = get Map with all Vertices from Graph G
10
11        for (from int i = 0; to end of VM_G; i++){
12             Vertex v = get Vertex i from VM_G
13             array Actors = get Actors from v
14
15                 for(from int j = 0; to end of Actors; j++){
16                     Vertex v_ie = new Vertex(ID := Actors[j])
17                     if(VM_IE doesn't contain v_ie) add v_ie to VM_IE
18                 }
19        }
20
21        EM_G = get Map with all Edges from Graph G
22        for (from int i = 0; to end of VM_E; i++){
23             Edge e = get Edge i from EM_G
24
25             Vertex from = get Vertex FROM from e
26             Vertex to   = get Vertex TO from e
27
28             array Actors_From = get Actors from Vertex from
29             array Actors_To = get Actors from Vertex to
30
31             for(int j = 0; to end of Actors_From; j++){
32                 for(int k = 0; to end of Actors_To; k++){
33                     Edge e_ie = new Edge (From := Actors_From[j]; To := Actors_To[k])
34                     if(EM_IE doesn't contain e_ie) add e_ie to EM_IE
35                 }
36             }
37        }
38
39        Graph IE = new Graph (Vertices := VM_IE; Edges := EM_IE)
40
41        return IE
42   }
```

Fig. 5. Pseudocode of the developed algorithm

4 Benchmark Evaluation and Further Applications

There are a lot of application possibilities of automated generated information exchange diagrams. As mentioned before fact-based decision making is related to information, so DSS are too. In military operations on the operational level there a many communication exchange ways and information which have to be shared, what is shown in the information exchange diagram related to the short process in Fig. 1.

For example, if the connection between the "OPC" and the "Patrol Company" is lost it is possible to build a connection between these two vertices over one of the three vertices "Patrol". For this small example it is easy to recognize which alternatives are available if one of the connections is lost. In large operations and communication networks the help of additionally systems is necessary. On this point it is possible to use A.I. in DSS.

As described in Sect. 2.2 the approach for "Rote Learning" could be implemented as A.I. in DSS. Shortest Path algorithms like Dijkstra [15] give the A.I. the possibility to determine new communication ways or connections based on the information exchange diagram. Further in addition with other information about the communication ways, e.g. which hardware, software, tool, etc. is used DSS can provide a list of alternative ways of communication.

In military operations this would support the commander's level, related to getting information for DSS and also provide a continuously communication for the subordinate level. Especially at the commander's level DSS are necessary and often used in software for common operational pictures (COP). The COP has a connection to the military data link network LINK 16, which provides the information of all participants in an operation. A common way for representing information in DSS is provide a graphical layer which supports the commander's level [16]. In addition alternative ways in communication exchange (e.g. in case of losing a communication way) could be provided as graphical layer or as model for decision support [17]. Followed, the principle that not every information has to be shared with everyone, an automated generated information exchange diagram could be useful in the planning of networks. Reducing the vertices which have to be connected supports an A.I. in automated network planning [18].

5 Conclusion

Giving information in the needed quality to the right time to the right COI (e.g. commander's level or on the level of subordinates) is the main discipline of DSS. The field of information quality is excluded in this paper, but information exchange diagrams address the problem of addressing the right COI. The automated generation of information exchange diagrams from process models guarantee the relation between both parts. The diagrams can provide the basics which are necessary for DSS to share and get the right information. These information exchange diagrams (extracted from the operational process) are the basis for an information exchange matrix, which shows what information has to be shared, transmitter and receiver of the information. This matrix could be used as NOV-3 and provides the possibility for further work in operation planning, for example, a prioritization of information (e.g. urgent information high-ranged etc.). Further, alternative communication ways can be provided with the help of A.I. implemented in the DSS, which use the information exchange diagrams.

References

1. Grice, H.P.: Logic and conversation. In: Cole, P., Morgan, J.L. (eds.) Syntax and Semantics. Speech Acts, vol. 3, pp. 41–58. Academic Press, New York (1975)
2. Williams, C.C.: Trust diffusion: the effect of interpersonal trust on structure, function, and organizational transparency. Bus. Soc. **44**(3), 357–368 (2005)
3. Matthes, D.: Enterprise Architecture Frameworks Kompendium: Über 50 Rahmenwerke für das IT-Management, 1st edn. Springer, Heidelberg (2011). https://doi.org/10.1007/978-3-642-12955-1

4. Power, D.J.: Decision Support Systems: Concepts and Resources for Managers, 1st edn. Greenwood Publishing Group, London (2002)
5. Drucker, P.F.: The next information revolution. Forbes **162**(4), 47–54 (1998)
6. Shim, J.P., Warkentin, M., Courtney, J.F., Power, D.J., Sharda, R., Carlsson, C.: Past, present, and future of decision support technology. Decis. Support Syst. **33**(2), 111–126 (2002)
7. Russell, S.J., Norvig, P.: Artificial Intelligence: A Modern Approach, 1st edn. Pearson Education Limited, Malaysia (2016)
8. Michalski, R.S., Carbonell, J.G., Mitchell, T.M. (eds.): Machine Learning: An Artificial Intelligence Approach, 1st edn. Springer, Heidelberg (1983). https://doi.org/10.1007/978-3-662-12405-5
9. Keller, G., Scheer, A.W., Nüttgens, M.: Semantische Prozeßmodellierung auf der Grundlage "Ereignisgesteuerter Prozeßketten (EPK)". In: Scheer, A.-W. (Hrsg.): Veroeffentlichungen des Instituts fuer Wirtschaftsinformatik, Heft 89, University of Saarland, Saarbruecken (1992)
10. White, S.A.: Introduction to BPMN (2004). http://www.bpmn.org. Accessed 17 Aug 2018
11. Gappmaier, M., Gappmaier, C.: Alles Prozess?!: Einfach wirksame Prozessoptimierung in jeder Situation mit der Bildkartenmethode (BKM), 3rd edn. Books on Demand, Nordstedt (2011)
12. van der Aalst, W.M.P.: Making work flow: on the application of petri nets to business process management. In: Esparza, J., Lakos, C. (eds.) ICATPN 2002. LNCS, vol. 2360, pp. 1–22. Springer, Heidelberg (2002). https://doi.org/10.1007/3-540-48068-4_1
13. Brandes, U., Eiglsperger, M., Lerner, J., Pich, C.: Graph markup language (GraphML). In: Tamassia, R. (ed.) Handbook of Graph Drawing Visualization, pp. 517–541. Taylor & Francis, London (2013)
14. Bray, T., Paoli, J., Sperberg-McQueen, C.M., Maler, E., Yergeau, F.: Extensible markup language (XML). World Wide Web J. **2**(4), 27–66 (1997)
15. Dijkstra, E.W.: A note on two problems in connexion with graphs. Numer. Math. **1**(1), 269–271 (1959)
16. Hodicky, J., Frantis, P.: Decision support system for a commander at the operational level. In: KEOD 2009, Proceedings of the 1st International Conference on Knowledge Engineering and Ontology Development, pp. 359–362 (2009)
17. Stodola, P., Mazal, J.: Tactical decision support system to aid commanders in their decision-making. In: Hodicky, J. (ed.) MESAS 2016. LNCS, vol. 9991, pp. 396–406. Springer, Cham (2016). https://doi.org/10.1007/978-3-319-47605-6_32
18. Mata, J., et al.: Artificial intelligence (AI) methods in optical networks: a comprehensive survey. Opt. Switch. Netw. **28**, 43–57 (2018)

Visual Data Simulation for Deep Learning in Robot Manipulation Tasks

Miroslav Surák[1], Karel Košnar[2(✉)], Miroslav Kulich[2], Viktor Kozák[2], and Libor Přeučil[2]

[1] Faculty of Electrical Engineering, Czech Technical University in Prague, Prague, Czech Republic
surakmir@fel.cvut.cz
[2] Czech Institute of Informatics, Robotics and Cybernetics, Czech Technical University in Prague, Prague, Czech Republic
{karel.kosnar,miroslav.kulich,viktor.kozak,libor.preucil}@cvut.cz
http://imr.ciirc.cvut.cz

Abstract. This paper introduces the usage of simulated images for training convolutional neural networks for object recognition and localization in the task of random bin picking. For machine learning applications, a limited amount of real world image data that can be captured and labeled for training and testing purposes is a big issue. In this paper, we focus on the use of realistic simulation of image data for training convolutional neural networks to be able to estimate the pose of an object. We can systematically generate varying camera viewpoint datasets with a various pose of an object and lighting conditions. After successful training and testing the neural network, we compare the performance of network trained on simulated images and images from a real camera capturing the physical object. The usage of the simulated data can speed up the complex and time-consuming task of gathering training data as well as increase robustness of object recognition by generating a bigger amount of data.

Keywords: Deep learning · CNN · Simulation · Ray-tracing · Robot manipulation · Random bin picking

1 Introduction

Robotic manipulators are used in an industrial automation for decades. Typical use-cases vary from welding to pick-and-place tasks. Nowadays, the cooperative robots share the workspace with humans and therefore traditional approaches, relying on precise predefined positions of items in robots workspace, are not working anymore. The robot needs to sense its working space with different sensors and adapt its actions according to the actual situation. With recent progress in deep learning, there start attempts to solve situations, where the robot needs to grasp an object in a random position with end-to-end neural networks trained from large training datasets.

© Springer Nature Switzerland AG 2019
J. Mazal (Ed.): MESAS 2018, LNCS 11472, pp. 402–411, 2019.
https://doi.org/10.1007/978-3-030-14984-0_29

The deep convolutional neural networks (CNNs), especially when working with images, need a huge amount of labeled data to train. Getting data with proper labels from the real world is usually time-consuming, and often a manual task. For example, this end-to-end network approach [10] makes use of RGB-D sensor and more than 50 thousands of grasping trials and needs 700 h of robot labor. Therefore, there is a need to speed up and automation of data collecting and labeling.

One possible way is to use simulated images for the training of the CNN. However, training from synthetic images can lead to overfitting of the network to 'unrealistic' details only present in synthetic images and failing to generalize well on the real images. The use of a simulator as realistic as possible is a way presented in this paper.

1.1 Related Works

Grasping movement is typically planned directly from RGB or RGB-D image of target objects. Analytic approaches register actual data to the database of 3D models of known objects with precomputed grasping points [1,6,14]. A registration often involves many intermittent steps like image segmentation, classification and pose estimation, where each step typically depends on multiple parameters, that are difficult to tune.

Very good results with utilizing simulated data of 3D point clouds achieves the approach described in [7] for defining grasping points. It achieves better results than analytic approaches. [8] introduces the extension of previous for suction cups grasping.

Alternative approaches are making use of deep learning to estimate the 3D pose of the object directly from intensity image and/or 3D point cloud [5,15]. As there is a need for a large number of training data, a new approach is to train the network on simulated images [13] and to adapt the representation to real data [12]. The work [3] improves the precision of recognition by adding synthetic noise to synthetic training images. Recent research suggests that in some cases it may be sufficient to train on datasets generated using perturbations to the parameters of the simulator [4].

2 Problem Definition

The problem solved in this paper is motivated by the real-world problem of picking of specific metallic parts of a single type from a transportation package and feed these to an automated industrial assembly line. As this task is highly repetitive and the motion performed by the human worker is tedious and one-sided, there is a request for automation.

Parts are not fully randomly distributed in the package, as they are originally organized in columns, but get scattered during the transport. It is expected by the end-user, that manipulator can pick more than 80% of the object from the package. The assembly line needs one part every 60 s. Another request is

Fig. 1. Image of the transportation package with objects and feeder of the automatic assembly line

flexibility of the solution, as there are many different types of parts manually feed to automated assembly lines. Therefore, modification of the solution for the new part should be as easy as possible (Fig. 1).

As the existing solution described in the following section is based on convolutional neural network, it needs a huge amount of training data. Therefore, this paper focuses on the generation of simulated training data and evaluation of the usage of this data in the described solution.

3   Solution

Pose estimation of the objects for picking is described in details in diploma thesis [11]. Pose estimation of the object position is divided into three steps: segmentation of the image and detection of regions that contain a single object, raw estimation of the position and accuracy improvement.

3.1   Segmentation

The segmentation of the object is base on the Histogram of Oriented Gradients (HOG) approach [2] with the sliding window. This segmentation method was used because is easy to train and performs well under different conditions, e.g. change of light. The parameters of the HOG detector are: block size 16×16, cell size 8×8, image patch size 64×64. A simple SVM classifier is used for classification if the window contains an object or not. Image patches detected as containing object by HOG are used in later steps of the algorithm.

3.2   Raw Estimation of Pose

The size and the position of the center of the patch segmented in the previous step by HOG are used as a first estimation of the distance and position of the object respectively. This first estimation is not accurate enough for reliable picking the object from the box. Therefore next step is necessary to improve the estimation of the position to the level, where the gripper can reliably pick the object. Moreover, the orientation (normal vector) of the object is necessary to estimate to allow successful picking of the object.

3.3   Accuracy Improvement Using CNN

For the further improvement of the object position accuracy, the deep convolution network (CNN) is used. For the needs of the CNN, the previously detected patches are resized to the unified size of 64 × 64 pixels. As the image patches are resized to unified size, it is not possible for the CNN to directly estimate the position and distance of the object and only multiplicative coefficient of the position in x-y plane and distance from the previous step are trained.

The input of the network is an image patch of size 64 × 64 pixels. It is followed by four convolutional layers with ReLU activation function followed by max pool layers. Usage of max pool layers effectively decreases the number of parameters of the model, because of the sparsity of data. The last layer of the network is a fully-connected layer with 3 neurons, whose output are predicted position coefficients. The network is learned by the back-propagation approach.

4   Gathering of Training Dataset

The gathering of training data is a semi-autonomous process. At first, the precise position of the learned part is determined by manually placing the gripper on the part. Then, the gripper with the camera is automatically placed into pre-defined positions in different distances and angles. As the position of the part in the transportation package, e.g. at the bottom, on the top or near the package wall, influence the appearance of the part, this procedure is repeated with part placed in the different configuration in the transportation package. For each configuration are gathered hundreds of images.

These images were then processed with HOG detector and only image patches that contains the part are used in later steps. Also, the relative position between the camera and the part was calculated in this step from the original position of gripper placed on the part and actual position of the gripper with the camera.

The training data consist from: (1) truth relative position between the camera and the object and (2) gray-scale image patch.

4.1   Synthetic Training Dataset

To be able to train CNN from synthetic training data, we need to obtain the same data in the same format. The most crucial part is the gray-scale image.

As the technical drawing of the part is available, it was easy to get the 3D model of the part in question. Now, the realistic gray-scale image of the model with proper lighting, shading, and reflection is necessary to simulate. As the most promising approach seems to use ray-tracing software. This software can realistically simulate all the complicated reflections and lighting of 3D models with different materials and textures. Our choice is to use the Persistence of Vision Raytracer (PoV-Ray) [9] (see Fig. 2 for example of the result) as it is open-source and authors are familiar with the usage of this software.

Fig. 2. Example of the result of PoV-Ray.

The real placement of the camera is in the center of the gripper head with circular light around the camera. See Fig. 3 depicting gripper head with the camera, sucking cups and circular light. Therefore, it was necessary to simulate the camera with the same field of view and the same light source around the camera, to get the same reflections on the surface of the parts. The object and the camera was placed in the same position as gathered by real manipulator with real part. So the synthetic dataset is as near to real one as possible.

The next task was to find the correct material and texture of the model, that is as near to appearance of the real part as possible. The similarity was evaluated by human eyes and improved in an iterative way to achieve the results depicted in Fig. 4.

5   Experiments Description and Evaluation

In the experiments, we compare the errors of the estimated position of the parts. We create two training datasets of the same size of 1000 images. The first dataset

Fig. 3. Configuration of the gripper head with centered camera, suction cups and circular light.

Fig. 4. Example of the real and corresponding simulated images.

was collected with the real camera placed on the real manipulator. The second dataset was generated in the PoV-Ray software. Both datasets contain the same items, the images taken from same positions with the same lighting.

Also, we create a testing dataset with 200 images. The testing dataset was collected with the real camera on the real manipulator.

Two networks were trained in the supervised-learning fashion using the Mean Squared Error. Adam optimizer with learning rate 0.001 was used to find the optimal weights. The training required 5000000 iterations, the dropout rate of 0.5 was used. The first network was trained on the real dataset and the second network was trained on the synthetic dataset.

To get the reference performance, the first network trained on real training dataset was run on the real testing dataset (see Fig. 5). The achieved errors where used as a reference point for the comparison. The performance of the second network ran on the real testing dataset (see Fig. 6) is compared with the first one.

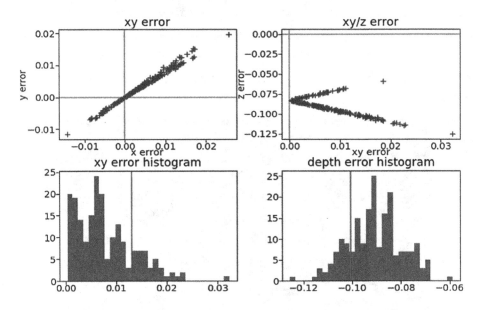

Fig. 5. Errors in position for network trained and tested on real dataset.

The results of the network trained on the synthetic dataset are slightly worse than the original network trained on the real images. The difference between the two networks are less than 10% and that is in the tolerance for the deployment into the real process. The precision of the position determination is in average 7% worse. The precision of the depth determination is in average 3.5% worse. The variance of the position and depth errors are not significantly worse.

The time needed for the collection of the real dataset with 1000 images is around 1.5 h. The synthetic dataset of the same size can be generated on the MetaCentrum Grid Infrastructure in order of minutes.

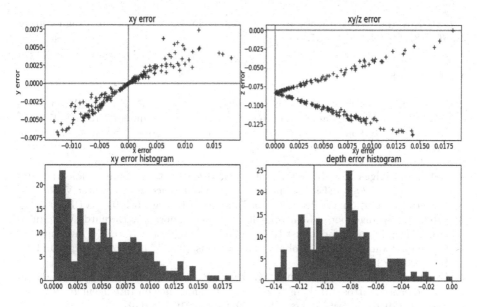

Fig. 6. Errors in position for network trained on synthetic dataset and tested on real dataset.

6   Conclusion and Further Work

The performance of the network trained on the synthetic dataset is slightly worse than the network trained on the real dataset, but the difference is in the tolerance, so the network trained on the synthetic dataset can be deployed with the real manipulator.

Now used part is quite simple and rotational symmetrical, so we can use a quite a small dataset for training. As we plan to use this system for more complex parts, there will be a need for the much bigger dataset and then the time savings will be more significant.

For further improvement, we plan to combine the real and synthetic data together to improve the performance of the network. Also, we plan to replace the manual tuning of the material parameters in the ray-tracing software with automated process of learning the parameters from the performance of the network. As the material parameters significantly influence the light reflections, it is expected, that with the better estimation of material parameters, the simulated images will be more realistic.

Acknowledgement. Access to computing and storage facilities owned by parties and projects contributing to the National Grid Infrastructure MetaCentrum provided under the programme "Projects of Large Research, Development, and Innovations Infrastructures" (CESNET LM2015042), is greatly appreciated. This work has been supported from the European Union's Horizon 2020 research and innovation programme under grant agreement No 688117 (SafeLog). The work was also supported by the Grant Agency of the Czech Technical University in Prague, grant No. SGS18/206/OHK3/3T/37.

References

1. Ciocarlie, M., Hsiao, K., Jones, E.G., Chitta, S., Rusu, R.B., Şucan, I.A.: Towards reliable grasping and manipulation in household environments. In: Khatib, O., Kumar, V., Sukhatme, G. (eds.) Experimental Robotics. Springer Tracts in Advanced Robotics, vol. 79. Springer, Heidelberg (2014). https://doi.org/10.1007/978-3-642-28572-1_17
2. Dalal, N., Triggs, B.: Histograms of oriented gradients for human detection. In: Proceedings of 2005 IEEE Computer Society Conference on Computer Vision and Pattern Recognition, CVPR 2005 (2005). https://doi.org/10.1109/CVPR.2005.177
3. Eitel, A., Springenberg, J.T., Spinello, L., Riedmiller, M., Burgard, W.: Multimodal deep learning for robust RGB-D object recognition. In: IEEE International Conference on Intelligent Robots and Systems (2015). https://doi.org/10.1109/IROS.2015.7353446
4. Gualtieri, M., Pas, A.T., Saenko, K., Platt, R.: High precision grasp pose detection in dense clutter. In: IEEE International Conference on Intelligent Robots and Systems (2016). https://doi.org/10.1109/IROS.2016.7759114
5. Gupta, S., Arbeláez, P., Girshick, R., Malik, J.: Aligning 3D models to RGB-D images of cluttered scenes. In: Proceedings of the IEEE Computer Society Conference on Computer Vision and Pattern Recognition (2015). https://doi.org/10.1109/CVPR.2015.7299105
6. Hinterstoisser, S., et al.: Multimodal templates for real-time detection of textureless objects in heavily cluttered scenes. In: Proceedings of the IEEE International Conference on Computer Vision (2011). https://doi.org/10.1109/ICCV.2011.6126326
7. Mahler, J., et al.: Dex-Net 2.0: deep learning to plan robust grasps with synthetic point clouds and analytic grasp metrics. In: Robotics: Science and Systems (2017). https://doi.org/10.15607/RSS.2017.XIII.058, http://arxiv.org/abs/1703.09312
8. Mahler, J., Matl, M., Liu, X., Li, A., Gealy, D., Goldberg, K.: Dex-Net 3.0: computing robust robot vacuum suction grasp targets in point clouds using a new analytic model and deep learning. In: 2018 IEEE International Conference on Robotics and Automation (IRCA) (2018). https://doi.org/10.1109/ICRA.2018.8460887
9. Persistence of Vision Pty. Ltd.: POV-Ray - the persistence of vision raytracer (2004). http://www.povray.org/
10. Pinto, L., Gupta, A.: Supersizing self-supervision: learning to grasp from 50K tries and 700 robot hours. In: Proceedings of IEEE International Conference on Robotics and Automation, pp. 3406–3413, June 2016. https://doi.org/10.1109/ICRA.2016.7487517
11. Sushkov, R.: Detection and pose determination of a part for bin picking. Master thesis, Czech Technical University in Prague, June 2017. https://dspace.cvut.cz/handle/10467/68468?show=full
12. Tzeng, E., et al.: Adapting deep visuomotor representations with weak pairwise constraints. In: The 12th International Workshop on the Algorithmic Foundations of Robotics (2016)
13. Varley, J., Weisz, J., Weiss, J., Allen, P.: Generating multi-fingered robotic grasps via deep learning. In: IEEE International Conference on Intelligent Robots and Systems (2015). https://doi.org/10.1109/IROS.2015.7354004

14. Xie, Z., Singh, A., Uang, J., Narayan, K.S., Abbeel, P.: Multimodal blending for high-accuracy instance recognition. In: IEEE International Conference on Intelligent Robots and Systems, pp. 2214–2221 (2013). https://doi.org/10.1109/IROS.2013.6696666

15. Zeng, A., et al.: Multi-view self-supervised deep learning for 6D pose estimation in the Amazon Picking Challenge. In: Proceedings of IEEE International Conference on Robotics and Automation (2017). https://doi.org/10.1109/ICRA.2017.7989165

Incremental Learning of Traversability Cost for Aerial Reconnaissance Support to Ground Units

Miloš Prágr$^{(\boxtimes)}$ ⓘ, Petr Čížek ⓘ, and Jan Faigl ⓘ

Faculty of Electrical Engineering, Czech Technical University in Prague,
Technicka 2, 166 27 Prague, Czech Republic
{pragrmil,cizekpe6,faiglj}@fel.cvut.cz
https://comrob.fel.cvut.cz

Abstract. In this paper, we address traversability cost estimation using exteroceptive and proprioceptive data collected by a team of aerial and ground vehicles. The main idea of the proposed approach is to estimate the terrain traversability cost based on the real experience of the multilegged walking robot with traversing different terrain types. We propose to combine visual features with the real measured traversability cost based on proprioceptive signals of the utilized hexapod walking robot as a ground unit. The estimated traversability cost is augmented by extracted visual features from the onboard robot camera, and the features are utilized to extrapolate the learned traversability model for an aerial scan of new environments to assess their traversability cost. The extrapolated traversability cost can be utilized in the high-level mission planning to avoid areas that are difficult to traverse but not visited by the ground units. The proposed approach has been experimentally verified with a real hexapod walking robot in indoor and outdoor scenarios.

1 Introduction

Robots deployed in remote missions such as extraterrestrial exploration [4] have to deal with challenges related to the traversability of the operational area to efficiently accomplish the mission and avoid hard-to-traverse regions that might impose risks to the robotic platform. Various robotic systems suitable for different terrain types are emerging, with multi-legged walking robots being already deployed in several scenarios, e.g., search and rescue tasks [17], inspection of the failed Fukushima nuclear power plant [5], and underwater marine operations [9]. Therefore, we consider a hexapod walking robot as a suitable ground robot that is capable of traversing complex terrains, especially when aided by an efficient locomotion control such as [12]. However, it is still beneficial to further increase mission efficiency by planning to avoid hard-to-traverse areas that can be assessed using a cost of transport characterizing the terrain traversability.

In autonomous missions, terrains may not be a priory known, and therefore, the robots have to learn the terrain traversability model incrementally to

© Springer Nature Switzerland AG 2019
J. Mazal (Ed.): MESAS 2018, LNCS 11472, pp. 412–421, 2019.
https://doi.org/10.1007/978-3-030-14984-0_30

immediately assess the traversed terrain and extrapolate the assessment for the forthcoming areas. Such a model can be based on the terrain classification into discrete classes [14], but it can also be a continuous measure based on exteroceptive [19] or proprioceptive data [10]. The fundamental requirement of the traversability model suitable for a high-level mission planning is the ability to extrapolate the learned traversability estimation to not visited but already seen areas, e.g., by utilizing Unmanned Aerial Vehicle (UAV) to efficiently explore the environment and evaluate its cost of transport using a ground walking robot. Such a system of cooperative navigation of a UAV and a legged robot is proposed in [6]; however, the authors based the traversability estimation on fixed local terrain characteristics, which deny the robot exploitation of newly collected information about the terrain traversability.

In this paper, we report on the deployment of learning framework [16] for the cost of transport regression from RGB-D data that allows combining numerous learning algorithms and terrain characterization features to learn and infer the traversability cost. However, we consider the Fast Incremental Gaussian Mixture Network (F-IGMN) [15], an online incremental Gaussian mixture algorithm, and a set of computationally efficient point cloud-based terrain descriptors for the herein presented results of the practical deployment of the framework. The descriptors are robust to viewpoint changes, and they can be captured both from the robot near-to-ground viewpoint and also from the aerial viewpoint.

The rest of the paper is organized as follows. The most related existing approaches and the utilized cost of transport are presented in the next section. The addressed problem is described in Sect. 3 and a brief overview of the used learning framework for the regression of the cost of transport is presented in Sect. 4. The main part of the paper is a report on the experimental deployment and achieved results for both indoor and outdoor scenarios that are reported in Sect. 5. Concluding remarks are dedicated to Sect. 6.

2   Related Work

Approaches that tackle the problem of the terrain traversability characterization can be divided into two main categories. The first category consists of methods utilizing classification to a discrete set of classes. These classes can be either human-observed terrain types [3,11,14] (e.g., grass, rock, sand, etc.), or simple binary distinction between passable/impassable terrains [19]. Three terrain classes for the simulated Martian terrain are considered in [14] but the passable-impassable distinction is utilized in [19]. The authors of [3,11] combine broader sets of classes with an impassable or otherwise undesirable obstacle class.

The second category characterizes the traversed terrains with a continuous function. Approaches that use proprioceptive [10], or exteroceptive [18,19] data can be further distinguished. Continuous traversability assessment functions can be utilized as a motion cost in planning algorithms. In [18] and [19], assessment functions based directly on locally observed properties are presented. Two-level discrete-continuous functions using three hazard criteria (i.e., slope, roughness, and step height) are proposed in [19]. A set of general global features is used to define the terrain traversability for a large wheeled vehicle in [18]. The utilized descriptors are the position, density, and point cloud distributions of the sensed obstacles. The main disadvantage of these approaches is that the traversability value assigned to a specific terrain cannot be directly updated based on the new robot experience.

In proprioceptive approaches such as the Cost of Transport (CoT) [20], the terrain traversability property is inferred from the terrain traversal experience. A particular value of the CoT is proportional to the energy consumed for traversing the respective terrain and to the inverse of the traversal speed. The CoT for battery powered robots is defined in [10] as

$$\text{CoT} = \frac{1}{m \cdot g} \cdot \frac{P_{\text{inst}}}{v}, \tag{1}$$

where m is the robot weight, g is the gravitational acceleration, v is the robot speed, and P_{inst} is the instantaneous power consumption computed from the battery voltage V and the instantaneous current I_{inst} drawn from the battery

$$P_{\text{inst}} = V \cdot I_{\text{inst}}. \tag{2}$$

Proprioceptive-based approaches and discrete terrain classification approaches need to be paired with terrain characterization features to enable terrain evaluation from a distance. Simple color space features as a part of the respective terrain descriptors are used in [1,2,14]. Color- and reflectance-based vegetation indexes have been used in [3,21]. The Gabor filter response combined with the HSV and grayscale descriptors are used as features on aerial images in [18].

Among the various shape and geometric features reported in [2,8,11,19], the most notable is the fact that all approaches use some height-based descriptor. It requires a selection of the canonical vertical orientation, which may be difficult, e.g., when dealing with sloped mountain terrains. The Principal Component Analysis (PCA) to characterize the terrain shape is utilized in [22] that is further combined with the normal vector based statistics in [11].

A simple feature set consisting of the height feature and histograms in the HSV color space is proposed in [2]. A set of seven geometric features extracted from a point cloud constructed using a stereo camera is utilized to characterize the terrain roughness and step height in [8]. Voxel, point cloud, and color features are used for terrain and vegetation classification by the authors of [3]. Their descriptors include scan line features, a vegetation index, and the PCA.

Based on the literature review briefly reported in this section, and also based on our results [16], we have selected the CoT terrain traversability measure [10] as the traversability assessment function that is combined with the computationally inexpensive color-geometric features.

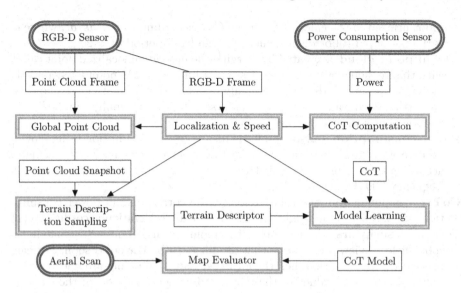

Fig. 1. Individual building blocks of the CoT regression framework [16].

3   Problem Statement

The problem addressed in this paper is a remote estimation of the CoT defined by (1). The requested model of the CoT is learned using our incremental learning framework [16] for the CoT regression of a priory unknown terrains by a hexapod walking robot that collects real measurements of the terrain traversability using RGB-D images and robot power consumption readings.

The learned traversability model is utilized for the evaluation of environment scans, which can be captured from a viewpoint that differs from the robot near-to-ground position, e.g., by a UAV. Hence, we aim for a two-agent setup, where the ground walking robot learns and creates the terrain traversability model and the UAV can evaluate the environment terrains remotely. Moreover, we consider that the ground robot incrementally learns the terrain traversability property and provides continuously improving traversability assessment.

4   Incremental Traversability Learning Framework

The herein presented results leverage on the CoT regression framework introduced in [16], which is schematically shown in Fig. 1. The framework can run online, and its building blocks are as follows.

RGB-D sensor provides RGB and depth images with 30 frames per second. For each RGB-D frame, an individual colored point cloud is created.

Localization and Speed localizes the robot and the captured RGB-D frames using the ORB-SLAM [13]. Besides, it allows to compute the speed of the robot utilized in the calculation of the CoT according to $v = ds/dt$, where ds is the measured robot displacement for the fixed period $dt = 20$ s.

Global point cloud is created by merging the individual localized point clouds with the frequency approx. 1 Hz. Its size is limited to 2M points, whereas the individual point clouds consist roughly of 300k points. Point cloud snapshots are further down-sampled to 5–15k points. Note, the externally supplied aerial environment scan is constructed similarly.

Power consumption sensor provides mean power consumption using moving average with 10 s long window of instantaneous power readings calculated according to (2) from the raw battery voltage and current measured with the frequency 500 Hz.

CoT is computed from the robot velocity, the corresponding power consumption, and known robot weight and gravitational acceleration according to (1).

Terrain descriptors are computed at randomly sampled interest points in the robot field of view, i.e., points that are likely to be traversed soon. For each processed robot position, i.e., the localized frame, 30 points are selected. The Lab channel mean values in the 0.3 m radius neighborhood of the sampled point and the PCA shape feature [11], based on the 0.2 m radius neighborhood, are computed using the point cloud snapshot.

CoT regression model is learned from the CoT of the traversed terrain and terrain descriptors in the robot field of view. Therefore, the descriptors must be stored until the robot passes the respective locations. The stored descriptors are randomly pruned, and only the locations in front of the robot are kept for an extended period. An online incremental Gaussian mixture model called the Fast Incremental Gaussian Mixture Network (F-IGMN) [15] is utilized for learning and regression. In the F-IGMN, a single data point is processed in $\mathcal{O}(kd^2)$ for k components and d dimensions. The particular parametrization of the F-IGMN is $k = 10$ components, grace period $v_{\min} = 100$, minimal accumulated posterior $sp_{\min} = 3$, and scaling factor $\delta = 1$ with $d = 7 + 1$ dimensional CoT annotated terrain descriptor.

Map evaluator evaluates an externally provided point cloud using the CoT regression model. The point cloud is centered to $[0, 0, 0]$, and the global ground plane is fitted using the RANSAC scheme [7]. In such a way, the 0.1 m spaced grid of the 2D cost map can be constructed, which is then utilized for the grid-based path planning.

5   Experimental Results

The incremental learning framework [16] has been deployed and experimentally verified in three testing scenarios using battery-powered hexapod robot shown in Fig. 2a. The robot has six legs, each with three joints actuated by the Dynamixel AX-12A servomotors. Thus the robot has 18 controllable actuators, and it is capable of traversing irregular terrains using an adaptive motion gait [12], which utilizes only the position feedback from the servomotors. The robot is equipped with the Intel RealSense D435 RGB-D camera mounted approx. 15 cm above the ground and slightly skewed towards the ground to collect exteroceptive measurements and support localization and map building.

The regression framework overviewed in Sect. 4 is employed in the traversability cost assessment that is used in the grid-based path planning. Moreover, the extrapolation of the learned CoT using features from exteroceptive measurements enables path planning for terrains not visited by the ground robot, but for which an aerial scan is available. In the experimental evaluation, an initial model of the real robot CoT is learned in an indoor setup first, and it is then evaluated in a testing setup that is reported in Sect. 5.1 and Sect. 5.2, respectively. In Sect. 5.3, we report on the performed outdoor deployment.

(a) Robot   (b) Indoor Learning   (c) Indoor Testing   (d) Outdoor

Fig. 2. The (a) hexapod walking robot and the (b–d) three experimental setups.

5.1 Indoor Learning Setup

In the first evaluation scenario, the robot is remotely guided over three artificial indoor terrains, see Fig. 2b. First, the robot traverses over variable height and slope 10×10 cm wooden bricks covered by an artificial turf, which is firm enough to support the robot even if some of its legs are not directly supported by the bricks. After leaving the artificial turf, the robot continues over flat PVC flooring where it crawls in one direction, and then, it turns around and returns. Finally, the last terrain type is a flat ground covered by slippery black fabric.

The robot is guided by a human operator over the individual terrains who adjusts the robot speed to prevent it from getting stuck. Therefore, based on the experience from [16], the operator chooses the most efficient motion, the CoT is minimized, and we consider its measured values to be the ground truth. In this setup, the artificial turf is the hardest terrain to traverse, and the robot is guided slowly. The fastest locomotion is over the flat PVC flooring, but the overall speed is hindered by the turn. The flat ground covered by the black fabric allows medium speed because the fabric is slippery.

The aerial scan of the track is captured by manually scanning the track from 2 m above the terrain. The scan is evaluated using the CoT model altogether four times: (1) at the beginning of the experiment; after traversing (2) the artificial turf; (3) flat PVC flooring; and (4) the black fabric. Each time the model encompasses the robot experience up to the given instant. For each case, the CoT annotation of the scan is created, and a path over the experimental track is computed using the 8-neighborhood A*. The edge cost is defined as the mean

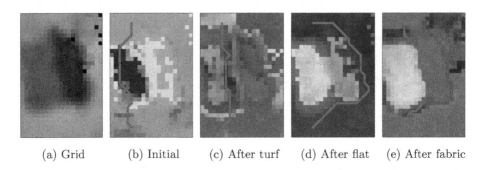

(a) Grid   (b) Initial   (c) After turf   (d) After flat   (e) After fabric

Fig. 3. Indoor learning scenario: the (a) 0.1 m sized cell grid over an aerial scan and its (b–e) CoT annotations with the planned path over the track. The used annotations correspond to four models sampled (b) at the beginning of the run; after traversing (c) the artificial turf; then after (d) the flat PVC flooring; (e) and finally the black fabric.

CoT of the edge incident nodes, and the minimal observed CoT is utilized as a multiplier of the Euclidean heuristic. The start and goal locations are situated in such a way that the shortest path leads over the hard-to-traverse turf, the longest over the flat ground, and the mid-length path over the black fabric. The aerial scan annotations and the respective paths are shown in Fig. 3.

Naturally, the robot learns the traversability after it traverses the terrain. Therefore, the shortest path over the rough terrain is computed first and then it is updated after observing both the rough and flat terrains. Besides, it can be noticed that the path planner avoids the black areas before traversing them.

5.2 Indoor Testing Setup

The model learned in the indoor learning setup has been utilized on a different indoor track shown in Fig. 2c. The track consists of the turf-covered bricks, flat PVC floor, and black fabric covered floor that all have been used in the learning setup. Besides, three additional (not previously utilized) terrains are used: the turf-covered floor, fabric covered bricks, and exposed bricks.

The same evaluation methodology as in the previous setup has been utilized, and individual evaluations of the aerial scan are visualized in Fig. 4. We found out that the terrains not presented in the learning phase appear to be evaluated correctly, and the turf over the bricks is considered to be the most costly terrain type in the final evaluation. Further, the flat terrain is learned to be the cheapest terrain, and the fabric and bare cubes have a medium value of the CoT. Even though there is only a limited difference between the CoT over the fabric covered flooring and covered bricks, the shortest path is found over the flat ground.

5.3 Outdoor Setup

Two terrain types, which are less distinctive than the indoor terrains, have been considered in the outdoor experimental setup, see Fig. 2d. In this setup, the robot

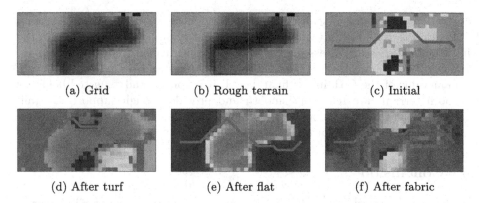

(a) Grid   (b) Rough terrain   (c) Initial

(d) After turf   (e) After flat   (f) After fabric

Fig. 4. Indoor testing scenario: the (a) 0.1 m sized cell grid over an aerial scan with the (b) rough terrain highlighted and its (c–f) CoT annotations with the planned path over the track. The used annotations correspond to the four models sampled (c) at the beginning of the run; after traversing (d) the artificial turf; (e) the flat PVC flooring; (f) and the black fabric.

traverses over grass first, where the area is relatively scorched, and thus being a mix of green grass, yellow grass, and brown dirt. Then, the robot traverses stone brick pavement. Due to nearby construction work, the gray pavement is partially covered with dirt.

(a) Grid   (b) Initial   (c) After grass   (d) After road   (e) Indoor model

Fig. 5. Outdoor scenario: the (a) 0.1 m sized cell grid over an aerial scan and its (b–e) CoT annotations. The used annotations correspond to three models sampled (b) at the beginning of the run; after traversing (c) the grass and (d) the pavement. The annotation of the outdoor track using the (e) indoor model from Sect. 5.1. A different behavior of the individual models can be observed when encountering areas not represented in the scan that are shown as the black areas in the grid image.

Similarly to the indoor learning setup, the CoT model is learned based on the measurements collected by the robot while traversing the outdoor track. Moreover, the final indoor model is also tested on the outdoor track. Individual evaluations of the aerial scan are depicted in Fig. 5. Again, the correct terrain evaluation is available only after the respective terrain type is traversed. In the initial model, the dirt/grass area is sharply distinguished from the pavement. However, the robot observes that the scorched grass and the pavement have a similar value of the CoT in the final model. It is correct and expected behavior as both terrains are relatively smooth and provide enough support. Overall, we found out that the indoor learned model performed surprisingly well when utilized on the outdoor dataset.

6   Conclusion

In this paper, we report on the deployment of the terrain traversability regression framework in practical indoor and outdoor setups. The framework infers a model of the CoT that can be extrapolated for aerial scans to build a cost map that is then utilized in path planning. Based on the reported results, the framework has been successfully evaluated, and it is a vital building block for further development. In particular, we aim to introduce an active perception into the presented framework to deliberately plan paths that will improve the model of the terrain traversability property.

Acknowledgements. This work has been supported by the Czech Science Foundation (GAČR) under research Project No. 18-18858S.

References

1. Bartoszyk, S., Kasprzak, P., Belter, D.: Terrain-aware motion planning for a walking robot. In: RoMoCo, pp. 29–34 (2017). https://doi.org/10.1109/RoMoCo.2017.8003889
2. Belter, D., Wietrzykowski, J., Skrzypczynski, P.: Employing natural terrain semantics in motion planning for a multi-legged robot. J. Intell. Robot. Syst., 1–21 (2018). https://doi.org/10.1007/s10846-018-0865-x
3. Bradley, D.M., Chang, J.K., Silver, D., Powers, M., Herman, H., Rander, P., Stentz, A.: Scene understanding for a high-mobility walking robot. In: IEEE/RSJ International Conference on Intelligent Robots and Systems (IROS), pp. 1144–1151 (2015). https://doi.org/10.1109/IROS.2015.7353514
4. Brown, D., Webster, G.: Now a stationary research platform, NASA's mars rover spirit starts a new chapter in red planet scientific studies. NASA Press Release (2010)
5. Falconer, J.: Toshiba unveils four-legged nuclear plant inspection robot. Innovation Toronto (2012). http://www.innovationtoronto.com/2012/11/toshiba-unveils-four-legged-nuclear-plant-inspection-robot/. Accessed 10 April 2018
6. Fankhauser, P., et al.: Collaborative navigation for flying and walking robots. In: IEEE/RSJ International Conference on Intelligent Robots and Systems (IROS), pp. 2859–2866 (2016). https://doi.org/10.1109/IROS.2016.7759443
7. Fischler, M.A., Bolles, R.C.: Random sample consensus: a paradigm for model fitting with applications to image analysis and automated cartography. Commun. ACM **24**(6), 381–395 (1981). https://doi.org/10.1145/358669.358692

8. Homberger, T., Bjelonic, M., Kottege, N., Borges, P.V.K.: Terrain-dependant control of hexapod robots using vision. In: Kulić, D., Nakamura, Y., Khatib, O., Venture, G. (eds.) ISER 2016. SPAR, vol. 1, pp. 92–102. Springer, Cham (2017). https://doi.org/10.1007/978-3-319-50115-4_9

9. Jun, B.H., Shim, H., Kim, B., Park, J.Y., Baek, H., Yoo, S., Lee, P.M.: Development of seabed walking robot CR200. In: OCEANS MTS/IEEE Bergen, pp. 1–5 (2013). https://doi.org/10.1109/OCEANS-Bergen.2013.6608164

10. Kottege, N., Parkinson, C., Moghadam, P., Elfes, A., Singh, S.P.N.: Energetics-informed hexapod gait transitions across terrains. In: IEEE International Conference on Robotics and Automation (ICRA), pp. 5140–5147 (2015). https://doi.org/10.1109/ICRA.2015.7139915

11. Kragh, M., Jørgensen, R.N., Pedersen, H.: Object detection and terrain classification in agricultural fields using 3D lidar data. In: Nalpantidis, L., Krüger, V., Eklundh, J.-O., Gasteratos, A. (eds.) ICVS 2015. LNCS, vol. 9163, pp. 188–197. Springer, Cham (2015). https://doi.org/10.1007/978-3-319-20904-3_18

12. Mrva, J., Faigl, J.: Tactile sensing with servo drives feedback only for blind hexapod walking robot. In: RoMoCo, pp. 240–245 (2015). https://doi.org/10.1109/RoMoCo.2015.7219742

13. Mur-Artal, R., Tardós, J.D.: ORB-SLAM2: an open-source SLAM system for monocular, stereo, and RGB-D cameras. IEEE Trans. Robot. **33**(5), 1255–1262 (2017). https://doi.org/10.1109/TRO.2017.2705103

14. Otsu, K., Ono, M., Fuchs, T.J., Baldwin, I., Kubota, T.: Autonomous terrain classification with co- and self-training approach. Robot. Autom. Lett. **1**(2), 814–819 (2016). https://doi.org/10.1109/LRA.2016.2525040

15. Pinto, R.C., Engel, P.M.: A fast incremental gaussian mixture model. PLoS One **10**(10), e0139931 (2015). https://doi.org/10.1371/journal.pone.0139931

16. Prágr, M., Čížek, P., Faigl, J.: Cost of transport estimation for legged robot based on terrain features inference from aerial scan. In: IEEE/RSJ International Conference on Intelligent Robots and Systems (IROS), pp. 1745–1750 (2018). https://doi.org/10.1109/IROS.2018.8593374

17. Roennau, A., Heppner, G., Nowicki, M., Dillmann, R.: LAURON V: a versatile six-legged walking robot with advanced maneuverability. In: AIM, pp. 82–87 (2014). https://doi.org/10.1109/AIM.2014.6878051

18. Sofman, B., Lin, E., Bagnell, J.A., Cole, J., Vandapel, N., Stentz, A.: Improving robot navigation through self-supervised online learning. J. Field Robot. **23**(11–12), 1059–1075 (2006). https://doi.org/10.1002/rob.20169

19. Stelzer, A., Hirschmüller, H., Görner, M.: Stereo-vision-based navigation of a six-legged walking robot in unknown rough terrain. Int. J. Robot. Res. **31**(4), 381–402 (2012). https://doi.org/10.1177/0278364911435161

20. Tucker, V.A.: The energetic cost of moving about: walking and running are extremely inefficient forms of locomotion. Much greater efficiency is achieved by birds, fish—and bicyclists. Am. Sci. **63**(4), 413–419 (1975)

21. Ünsalan, C., Boyer, K.L.: Linearized vegetation indices based on a formal statistical framework. IEEE Trans. Geosci. Remote Sens. **42**(7), 1575–1585 (2004). https://doi.org/10.1109/TGRS.2004.826787

22. Wellington, C., Stentz, A.: Online adaptive rough-terrain navigation in vegetation. In: IEEE International Conference on Robotics and Automation (ICRA), pp. 96–101 (2004). https://doi.org/10.1109/ROBOT.2004.1307135

Quantifying the Effects of Environmental Conditions on Autonomy Algorithms for Unmanned Ground Vehicles

Phillip J. Durst[(✉)] and Justin Carrillo

U.S. Army Engineer Research and Development Center,
3909 Halls Ferry Road, Vicksburg, MS 39180, USA
{phillip.j.durst,justin.t.carrillo}@usace.army.mil
http://www.erdc.usace.army.mil

Abstract. Autonomy for commercial applications is developing at a rapid pace; however, autonomous navigation of unmanned ground vehicles (UGVs) for military applications has been deployed to a limited extent. Delaying the use of autonomy for military applications is the environment in which military UGVs must operate. Military operations take place in unstructured environments under adverse environmental conditions. Military UGVs are infrequently tested harsh conditions; therefore, there exists a lack of understanding in how autonomy reacts to challenging environmental conditions. Using high-fidelity modeling and simulation (M&S), autonomy algorithms can be exercised quickly and inexpensively in realistic operational conditions. The presented research introduces the M&S tools available for simulating adverse environmental conditions. Simulated camera images generated using these M&S tools are run through two typical autonomy algorithms, road lane detection and object classification, to assess the impact environmental conditions have on autonomous operations. Furthermore, the presented research proposes a methodology for quantifying these environmental effects.

Keywords: Autonomy · Autonomous ground vehicles ·
Environment effects · Perception

1 Introduction

Autonomy for commercial ground vehicle applications is enjoying a great deal of success in the field, such as the Google car or Tesla [1]. However, autonomy for military ground vehicles has lagged far behind industry. While several factors have led to this lag, one primary obstacle to military autonomy applications is the operational environment. While commercial applications have the luxury of operating in stable, benign environments, military applications take place in harsh and unpredictable conditions. Autonomy algorithms, which are by nature often shown to be susceptible to the operational environment [2], often fail in harsh conditions, such as rain, snow, and fog. Moreover, little research has been

© Springer Nature Switzerland AG 2019
J. Mazal (Ed.): MESAS 2018, LNCS 11472, pp. 422–432, 2019.
https://doi.org/10.1007/978-3-030-14984-0_31

given over to understanding exactly how such environmental conditions impact autonomous operations. Rather, autonomous ground vehicles remain tested and fielded primarily in the case of predictable, known, on-road conditions.

The effects of adverse weather on autonomy are not understood or, to date, measured in a quantitative fashion. Human driver response to rain, dust, soft soil, etc. can be empirically measured and modeled. However, sensor/autonomy responses to these conditions require complex data acquisition that is difficult and expensive to recreate in the field. The goal of this paper is to demonstrate how high-fidelity simulations can be used in lieu of field testing and that the controlled and repeatable conditions achieved in simulation can enable the development of quantitative metrics for algorithm performance.

The paper is laid out as follows. Section 2 gives a brief introduction to the simulation software used in this study, Sect. 3 provides example evaluations of algorithm performance as a function of environment. Specifically, performance is assessed for both a lane detection algorithm operating in ideal and heavy rain situations and an object classification computational neural network (CNN) algorithm at varying times of day. Lastly, Sect. 4 provides concluding thoughts and recommendations for future work.

2   VANE: High-Fidelity Modeling and Simulation for Autonomous UGVs

The Virtual Autonomous Navigation Environment (VANE) is a high-fidelity, physics-based simulator for autonomous UGVs. The VANE began development primarily as a tool for developing autonomy algorithms for unmanned ground vehicles (UGVs) [3,4]. VANE originated as a tool to simulate an autonomous UGV performing a given mission in a given environment. As VANE has developed, it has shifted towards primarily being a tool for simulating sensor-environment interactions [5,6]. As such, VANE is a micro-scale, mission-level simulation tool for performance assessment and algorithm development for autonomous UGVs.

There are several key components to VANE, the most important of which is the sensor and environment models that are used in the development and testing of autonomy algorithms. On top of the sensor modeling are the mobility and vehicle dynamics models. These mobility and sensor models together recreate the physical behaviors of the UGV as it reacts to its autonomy algorithms. This study focuses on sensor-environment interactions as they impact autonomy algorithms, and therefore VANE's mobility models are not discussed further in this paper. Further details of VANE's sensor-environment models are given below.

The core of VANE is its sensor and environment models. Sensor outputs drive the autonomy algorithms used by UGVs. By providing high-fidelity, physics-based simulated sensor outputs, VANE can more accurately simulate autonomous UGV behaviors. The mechanism by which the VANE generates high-fidelity sensor outputs is its ray tracer.

The VANE ray-tracer (VRT) uses high-performance computing to simulate the radiative transfer of energy through the environment. The VRT is a full spectral simulation that calculates spectral reflectance properties using either the cosine lobe model or the He-Torrance-Sillion-Greenberg [7] bidirectional reflectance distribution function (BRDF) model for surface reflectance. The atmosphere in VANE is also modeled using a physics-first approach and implemented using numerical methods found within literature [8,9].

To obtain high-fidelity sensor outputs, a high-fidelity environment is required. The simulation environment itself must contain physical data to stimulate the sensors. The environment must contain not just the geometry of each object but also critical physical information, such as spectral reflectance. Moreover, the environment will "look" different to different sensor models. The modeled BRDF is critical to LIDAR and camera models, but not to GPS, which is more concerned with geometry. For the UGV mobility platform, the environment should contain the soil strength of the ground surface. Figure 1 shows an example VANE geo-environment. Coupled to the environment model are the sensor models. The VANE contains models for the sensors most commonly used by autonomous UGVs, e.g. camera, LIDAR, and GPS sensors, and Fig. 2 shows an example simulated sensor output for a CCD camera.

3   Algorithm Performance Under Varying Environmental Conditions

While it is well known qualitatively that adverse environmental conditions negatively impact autonomy algorithm performance, this degradation of performance has yet to be studied in a quantifiable way. Common sense shows that image processing algorithms will be less effective in the rain, and it has been shown in literature that rain reduces the performance of image processing algorithms [11,15]. Similarly, the difficulties image processing algorithms suffer when operating at dusk or dawn are well-known qualitatively [12]. The goal of this paper is to take these phenomena and leverage simulation to quantify performance as a function of environment for two popular navigation algorithms: object classification and lane detection.

3.1   Object Classification via CNN

CNNs can achieve reasonable performance on hard visual recognition tasks, sometimes matching or even exceeding human performance in some domains. Inception-v3 is a CNN that was trained for the ImageNet Large Visual Recognition Challenge. The CNN was developed using TensorFlow, which is an open-source software library for machine learning developed by the Google Brain team. To compare models in the ImageNet Large Visual Recognition Challenge, it is common practice to examine how often the model fails to predict the correct answer as one of its top five guesses - termed "top-5 error "rate" [13].

Fig. 1. An example VANE environment of a forested area.

Fig. 2. An output image from a Sony CCD camera compared with the real object being modeled within the geo-environment.

Inception-v3 [13] reached a 3.46% top-5 error rate for the ImageNet Large Visual Recognition Challenge using the data from 2012 while AlexNet achieved a 15.3%, Inception [14] achieved a 6.67%, and BN-Inception-v2 achieved a 4.9%. Since a pre-trained CNN was used in this study, the number of nodes was constrained to 90,000; therefore, the images had a size limit of 300 by 300.

Modern object recognition models contain a large number of parameters that can take long periods of time to fully train. This study uses a transfer learning technique to shortcut the training time by taking the fully-trained Inception-v3 model and retraining from the existing weights for new classes. The final layer is retrained from scratch, while leaving all the other layers untouched. The premise behind this idea is that the natural phenomenon of interest tends to have a hierarchical structure that deep neural networks naturally capture. At the lowest layers of the neural network, simple things such as lines and edges are detected and fed to higher layers for detecting more complex things, such as part shapes. The last layer of the neural network combines all parts together to detect objects of interest.

Several methods, such as deforming, cropping, or brightening the training inputs, were available and utilized for improving the results of image training in random ways. These methods have the advantage of expanding the effective size of the training data due to all of the possible variations of the same image and tends to help the network learn to cope with all the distortions that will occur in real-life uses of the classifier [13].

For this particular study, the CNN was used as a binary classifier for detecting a Heavy Expanded Mobility Tactical Truck (HEMTT). The VANE-generated imagery was used for the simulated data. The simulated data consisted of 1,871 images in an urban environment and 1,630 images in a rural environment with half of the images in a clear condition and the other half of the images in a hazy condition. The time of day was varied from 0900 to 1700. The camera views of the vehicle were randomly chosen for a 360° horizontal view of the vehicle with a maximum angle of 45° from the ground.

To evaluate the performance of the CNN, the neural network was given simulated data from a camera in a forest environment under varying rainfall rates and at different times of day, which could be set using VANE's simulation environment. The Fig. 3 below shows the impact of the rainfall rate on the CNN's ability to detect the HEMTT vehicle in the form of confidence of detection vs rainfall rate. Figure 4 shows a sample simulated image with rainfall. Probability of detection was not performed because probability of detection requires a large diverse data set for each rainfall rate to get a good measurement of the CNN's probability of detecting the vehicle. In this work, a single data set consisting of images with increasing rainfall rate was evaluated for proof of concept of the degradation of performance of the CNN's confidence in detection with respect to rainfall rate using simulated data.

Similarly, Fig. 5 shows the impact of time of day on detection confidence. The same image set of a HEMTT in a forested setting was used for training. The trained CNN was then tested on images over the course of a 24 h day. It this study, the HEMTT was aligned east to west, so that at dawn the sun was directly behind the vehicle and at dusk the sun was shining directly unto the vehicle. Figure 6 shows a sample simulated image at dawn. Figures 3 and 5

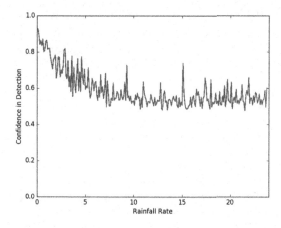

Fig. 3. The confidence of detection vs rainfall rate for the HEMTT in rain.

Fig. 4. An example simulated camera image of the HEMTT in rain.

show, and quantify, the environmental affects long known to hinder autonomy. Furthermore, by showing these effects quantitatively, objective performance of algorithms can be undertaken and algorithms can be compared rigorously for optimal performance.

3.2   Lane Detection

A key task that must be accomplished for on-road autonomous navigation is lane detection. For a ground vehicle to successfully follow a road, it must know the physical characteristics of said road, namely the width and curvature. Therefore, much research has been given to developing these algorithms, which are in use

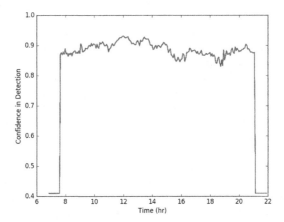

Fig. 5. The confidence of detection vs time of day for the HEMTT.

already for many commercial applications. However, little effort has been given over to quantifying these algorithms performance in adverse environmental conditions, and therefore these algorithms are typically not used in rain/snow/fog etc.

This is not to say that lane detection algorithms are never deployed in these challenging conditions. Several filters and other image processing techniques can be used to reduce the noise created by rain for lane detection algorithms, and details on these methods can be readily found in [15]. However, for the purposes of this study, a simple algorithm with no noise reduction was chosen. The reason for this choice is that this study does not aim to test a specific algorithm but rather to benchmark this particular type of algorithm's performance. The quantities measured below reflect a "base-line" performance and define a general operational envelope as a function of environmental conditions.

The chosen lane detection algorithm was tested against two sequences of simulated images: one under clear sky conditions and one in "heavy" rain condition of a rainfall rate of 20. This study quantifies lane detection performance in two ways. The first is percent detection. This metric is the number of frames with lanes detected divided by the total number of frames analyzed by the algorithm. For this study, 37 images were simulated, and the accuracy of the algorithm in clear and raining conditions can be found in Table 1. Table 1 also shows the average accuracy of the detections calculated as given below.

Accuracy of detection was taken as the magnitude of the differences between the ground truth slopes of the lanes in the image and the slopes of the detected lanes. As Table 1 and Figs. 7 and 8 show, a significant drop in accuracy was observed in the raining condition.

Figure 8 shows the output of the lane detection algorithm on an image taken by the camera sensor in the "heavy" rain condition. On the other hand, Fig. 7 shows the output for the same image on a clear day. Qualitatively, the algorithm

Fig. 6. An example simulated camera image of the HEMTT at dawn.

Table 1. Lane detection algorithm performance comparison between clear and heavy rain conditions.

Condition	Percent of frames with detection	Average detection accuracy
Heavy rain	81.1%	72.5%
Clear	91.2%	94.7%

Fig. 7. An example simulated camera image with lanes detected under clear sky condition.

clearly fails in the rain; however, using the methods described above, this failure can be quantified. In so doing, the accuracy of the "heavy" rain image is only 43.6%, whereas the accuracy of the "clear" image is 98.7%.

Fig. 8. An example simulated camera image with lanes detected under "heavy" rain condition.

An alternative to the benchmarking metrics used in this study is [16], which proposes a much more thorough standard of measurement. The "state-of-the-art" metrics proposed by this work include: Pixel-level correctness, Boundary position and deviation, Unoccupied lane length, and Corridor width. These metrics were not chosen for the current work due primarily to the fact that they are derived using experimental data and include metrics leveraging UGV behavior during field testing. Future work in assessing UGV performance using M&S should expand to include such metrics and change to leveraging these more informative metrics.

4   Conclusions

The future military and commercial vehicle fleets will almost certainly have some form of autonomy-enabled operations. Even manned vehicles are likely to have on-board algorithms to enable operators to perform at a higher level. To date, unmanned ground vehicles have not been fielded for military applications beyond experimental systems and limited testing. This lack of application is due in large part to the complexity of the operational environment for UGVs. Systems must be able to operate not only in unstructured and unpredictable environments but also operate in these environments in adverse environmental conditions. To push the envelope of operations for UGVs, a more comprehensive understanding of the impact environment has on autonomy is necessary.

To meet this need, this paper proposed and presented a method for quantitatively assessing autonomy performance as a function of environmental conditions. This study took two test cases: rainfall and time of day, and two algorithms: object classification and lane detection. Using high-fidelity M&S, the environmental conditions were fully controllable, which enabled consistency in experiments and well-defined environmental conditions. Using these benefits of M&S, data was collected in the form of simulated camera images, and algorithm performance was studied.

The findings of this study reflect the qualitative knowledge already contained in the autonomy community: rainfall degraded algorithm performance for both object classification and lane detection, and time of day degraded performance of object classification. The major impact of this study was to quantitatively measure these impacts. In so doing, algorithm developers will now have a means for objectively assessing algorithm performance, which will in turn enable more robust algorithm development. Furthermore, these quantitative assessments will enable better comparisons between algorithms to determine which algorithm is best suited to which environment. Leveraging M&S for these types of studies and expanding test cases to additional sensors, environments, and algorithms will allow for faster and more robust development of future autonomy for military applications.

Acknowledgments. Permission to publish was granted by Director, Geotechnical and Structures Laboratory.

References

1. Wright, A.: Automotive autonomy. Commun. ACM **54**(7), 16–18 (2011)
2. Luettel, T., et al.: Autonomous ground vehicles-concepts and a path to the future. Proc. IEEE **100**(Centennial Issue), 1831–1839 (2012)
3. Jones, R., et al.: Virtual autonomous navigation environment (VANE). In: Earth & Space 2008: Engineering, Science, Construction, and Operations in Challenging Environments, pp. 1–8 (2008)
4. Cummins, C.: Virtual autonomous navigation environment. DTIC Document (2008)
5. Goodin, C., et al.: Sensor modeling for the virtual autonomous navigation environment. In: IEEE Sensors, pp. 1588–1592 (2009)
6. Goodin, C., Durst, P.J., Gates, B., Cummins, C., Priddy, J.: High fidelity sensor simulations for the virtual autonomous navigation environment. In: Ando, N., Balakirsky, S., Hemker, T., Reggiani, M., von Stryk, O. (eds.) SIMPAR 2010. LNCS (LNAI), vol. 6472, pp. 75–86. Springer, Heidelberg (2010). https://doi.org/10.1007/978-3-642-17319-6_10
7. He, X.D., et al.: A comprehensive physical model for light reflection. In: ACM SIGGRAPH Comput. Graphics, vol. 25. no. 4 (1991)
8. Hosek, L., Wilkie, A.: An analytic model for full spectral sky-dome radiance. ACM Trans. Graph. (TOG) **31**, 95 (2012)
9. Bird, R.E., Riordan, C.: Simple solar spectral model for direct and diffuse irradiance on horizontal and tilted planes at the earth's surface for cloudless atmospheres. J. Clim. Appl. Meteorol. **25**(1), 87–97 (1986)
10. Narasimhan, S.G., Nayar, S.K.: Contrast restoration of weather degraded images. IEEE Trans. Pattern Anal. Mach. Intell. **25**(6), 713–724 (2003)
11. Kang, L.-W., Lin, C.-W., Fu, Y.-H.: Automatic single-image-based rain streaks removal via image decomposition. IEEE Trans. Image Process. **21**(4), 1742 (2012)

12. Bertozzi, M., Broggi, A., Fascioli, A.: Vision-based intelligent vehicles: state of the art and perspectives. Robot. Auton. Syst. **32**(1), 1–16 (2000)
13. Abadi, M., et al.: Tensorflow: a system for large-scale machine learning. In: OSDI, vol. 16 (2016)
14. Szegedy, C., Liu, W., Jia, Y., et al.: Going deeper with convolutions. In: Proceedings of the IEEE Conference on Computer Vision and Pattern Recognition, pp. 1–9 (2015)
15. Hillel, A.B., et al.: Recent progress in road and lane detection: a survey. Mach. Vis. Appl. **25**(3), 727–745 (2014)
16. Fritsch, J., et al.: A new performance measure and evaluation benchmark for road detection algorithms. In: 16th International IEEE Conference on Intelligent Transportation Systems. IEEE (2013)

Introducing Intelligence and Autonomy into Industrial Robots to Address Operations into Dangerous Area

Agostino G. Bruzzone[1]([⊠]), Marina Massei[1], Riccardo Di Matteo[2], and Libor Kutej[3]

[1] Simulation Team, DIME, University of Genoa, Genoa, Italy
{agostino,massei}@itim.unige.it
[2] Simulation Team, SIM4Future, Genoa, Italy
dimatteo@simulationteam.com
[3] University of Defence in Brno, Brno, Czech Republic
libor.kutej@unob.cz

Abstract. The paper addresses the issue to use new generation robotic systems inside industrial facilities in order to complete operations in dangerous area. The new robotic systems are currently adopting the autonomous approach already in use in military sector; however, in this context the intensity of operations and the necessity to interact with high productivity systems introduce different challenges. Despite the problems, it is evident that this approach could provide very interesting improvements in terms of safety for humans especially in relations to dangerous area. For instance, in confined spaces, Oil & Gas or Hot Metal Industry these new autonomous systems could reduce the number of injures and casualties. In addition, these systems could increase the operation efficiency in this complex frameworks as well as the possibility to carry out inspections systematically; in this sense, this could result in improving the overall reliability, productivity and safety of the whole Industrial Plant. Therefore, it is important to consider that these systems could be used to address also security aspects such as access control, however they could result vulnerable to new threats such as the cyber ones and need to be properly designed in terms of single entities, algorithms, infrastructure and architecture. From this point of view, it is evident that Modeling and Simulation represent the main approach to design properly these new systems. In this paper, the authors present the use of autonomous systems introducing advanced capabilities supported by Artificial Intelligence to deal with complex operations in dangerous industrial frameworks. The proposed examples in oil and gas and hot metal industry confirm the potential of these systems and demonstrate as simulation supports their introduction in terms of engineering, testing, installation, ramp up and training.

Keywords: Artificial Intelligence · Autonomous systems · Safety ·
Industrial plants · Security · Modeling and Simulation

© Springer Nature Switzerland AG 2019
J. Mazal (Ed.): MESAS 2018, LNCS 11472, pp. 433–444, 2019.
https://doi.org/10.1007/978-3-030-14984-0_32

1 Introduction

With the term UxVs (Unmanned Vehicles) are indicate all the remote vehicles operating in different environment (aerial, ground, surface, underwater) with different levels of autonomy (piloted remotely, fully autonomy). These new technologies are one of the focuses of Industry 4.0 due their characteristics that make them well adapt to substitute human for many danger operations [4, 53].

Indeed, due their agility, speed and expandable nature UxVs can reach easily place dangerous and/or difficult to access increasing safety in industrial plants. For these reasons, Autonomous Systems popularity is increasing exponentially in the last period and researchers are studying possible uses in different application field [23, 42]. In addition, the authority and agencies of security and prevention have manifested their interest in UxVs studying possible uses of them in many applications [13, 14]. In Italy for example, the Department of Technological Innovation (DIT) is promoting the diffusion of Autonomous System in dangerous operations substituting humans. In this way the want increase safety in industrial plants reducing the exposure of people to potential dangerous environments for addressing difficult tasks [47].

By the way, the introduction of UxVs in industrial Plants, if from one side increase the safety, from other side introduce new problems and risks that require to be investigated. Simulation can give great efforts in the investigation of these possible problems finding them soon and searching their solution (Fig. 1).

Fig. 1. Simulator operating autonomous systems for emergencies in industrial plants

2 Application Cases

In this paper are proposed different scenario where is investigate the potential use of different Autonomous Systems in Industrial Plants for increasing safety and security. These cases show various opportunities, risks and challenges derives from the uses of these technologies in the industrial sector and underline the need of creating training programs and procedures for the use UxVs. Many different kinds of UxV operate over multiple domains: land (UGV, Unmanned Ground Vehicles), air (UAV, unmanned Aerial Vehicles), surface (USV, Unmanned Surface Vehicles) and underwater (AUV, Autonomous Underwater Vehicles) [21, 24, 33, 45, 46, 49] (Fig. 2).

Fig. 2. UAV overview within critical infrastructure protection simulator

3 UxV and Safety

The use of UxVs in industrial plants is studied with the main goal of increasing safety; the versatility, the operational cost and the expandable nature of autonomous systems makes them perfect in substitute the human in those tasks that are particular difficult and/or in dangerous environment [1, 2, 25, 32, 34]. Unfortunately, the contexts of these operations are very complex and so results very complex the implementations of UxVs too. In order to investigate UxV problems, M&S (Modelling and Simulation) results the methodology most promising by its ability to well reproduce complex scenario [6].

The projects about using UxV for Industry and major Agencies along last years are several and one of the most active entity in this field is the Simulation Team of the University of Genoa that has acquires a large experience in autonomous systems. The Simulation Team is an International Network working on simulation technologies and is studying the use of autonomous systems in different application from several year, including the collaborative multi domain cases [7, 9].

In Italy INAIL-DIT is studying how to integrate Autonomous and Remote Pilot Systems in the actual Industrial plants paying particular attention to all new safety aspects that the use of them create in these scenarios. However, the safety is only one of the multiple topics that will need to take in account before the effective use of UxVs in working area; for example, could be useful study the development of new advanced equipment, or study new materials suitable for drones for protect them or people working in the same area [15, 43, 55]. It is also very important that regulation and tecnologies standard follow all the innovative processes within Industry 4.0 including autonomous systems and remote pilot systems [26]. If, for example, we analyze a typical example of the possible use of UxVs as could be the use of a remote piloting system, we immediately understand the complexity of the context. First, a person drives the robotic system and in the area of the UxV mission, there may be present human personnel working that introduce issues about safety (Fig. 3).

Fig. 3. Protection of port framework by USV and other autonomous systems

This implies the necessity of a training for the staff as well as the drafting of rules and regulations on the use of this technology to protect the safety of workers and for the transposition of the applicable product. From this context, it also emerges that the authorities involved are different; in fact, in addition to the European commission, state departments, ministries and public administrations are also involved. In the case of the example in Italy, they would be involved: the Ministry of Labour and Social Policies, the Health and Safety of Workers, the Ministry of Labour Infrastructure and Transport, for Aviation Security or Navigation and the National Agency for Civil Aviation [18]. The UxVs can be used in many application fields that different each other in terms of environment (e.g. open spaces or closed) and actions to be performed (e.g. monitoring, valve closures, taking samples) depending on both the type of drone that its equipment. For better understand how many and different can be the possible usages of UxVs, in the following are presented a set of possible fields of application. The UxVs could easily replace the humans in these tasks reducing risks and increasing the safety of workers no longer forced to expose themselves daily to certain dangers. Another possible field of application is the use of drones in confined spaces where there may be dispersion of hazardous materials: in these environments, in fact, accidents are still frequent maybe due to an incorrect analysis of risks [29, 36, 48]. The drones could access these areas with adequate sensors before human and warn of the presence of these possible substances [19, 55]. Obviously, before of using drone in these scenarios an accurate analysis of the boundary conditions should be performed in order to use UxVs in a safety mode avoiding possible risks of accidents (presence of liquid, electromagnetic fields, vapors etc.). The UxV could also be used to make environmental surveys with high resolution cameras or to monitor the infrastructures with the most modern sensors such as GPR (Ground Penetrating Radar) which allow to check the conditions of underground infrastructures (water leaks, cracks etc.). Furthermore, UxVs have the capability to operate in swarm exchanging instantaneously each other data increasing drastically the performance on monitoring plants [28] (Fig. 4).

Fig. 4. UGV with arm to perform task in industrial plant under development by simulation team

In this field, there are many studies because it is considered one of the most promising direction in the future use of autonomous systems [11].

Data acquired from swarm could be exchange also with a master control unit an advise instantaneously a human if is approaching a dangerous areas; in this way it is supposed to developed a smart guiding support that provide automatic corrective actions on vehicles, such as find an alternative way or reduce the speed [17, 27]. Another possible use of UxVs results to be the plant monitoring as well as evaluating flue gas emissions in different areas or by different machines [20]. The benefits from this application arise because often machinery, for construction restrictions, can't mount directly on themselves the sensors for monitoring their emissions and the environment near them doesn't permit to add fixed sensor [3, 22]. For these reasons, UxVs could be very good candidate to make these monitoring operations adding them the sensors needed for getting the measurements and, in some case, also tailored arm for reach the best position where take them. Another possible usage of UxVs that has been studied is in the field of irrigations machines in herbaceous and tree crops [39]. In this cases, UxVs can be used to evaluate drift of fertilizers equipping UAV with special high definition (HD) camera and recording the machineries from different point of view [38]. The tracked drift motion, thanks to an a red powdered food mixed to the liquid, can be replicated on aerial photo through GIS Support. In this way making more test with different weather condition it can be possible analyse the different drift situations. UxVs can make aerial reconnaissance for having direct view of the situation in some critical scenarios helping in this way in the management of emergency relief activities [16]. Indeed, the Image captured from AUV/ROV, are usable to understands the more critical area or which area is safe and which no etc. Therefore, the application fields are many and different and can involves different kind of drones. The cases presented are only some of the possible employments of UxVs and all of them should be analysed fine in order to finding the critical issues and procedures to be adopted. M&S is a powerful and useful instrument for address these problems [9] (Fig. 5).

Fig. 5. Man on the loop: supervising underwater oil operations within the SPIDER special CAVE

4   Safety in off-Shore Operations

The authors developed different cases related to oil & gas industries operations on offshore facilities where the combined use of AUV, ROV (Remoted Operated Vehicles), USV and UAV were devoted to conduct operations and improve safety and security [4]. Among these R&D (Research and Development) projects it could be mentioned the case of SO2UCI (Simulation for Off-Shore, On-Shore & Underwater Critical Infrastructure), CRIPEM (CRitical Infrastructure Protection in Extended Maritime framework) and Megacity for Wind Farms at sea [5, 9]. Similar projects addressed also industrial plants such as power plants and desalination facilities in on-shore installation as in T-Rex (Threat network simulation for REactive eXperience) and IDRASS (Immersive Disaster Relief and Autonomous System Simulation) to improve safety and security [7, 8]. In all these cases it was fundamental to develop collaboration capability among different UxV on different domains [10]. These aspects required to develop Artificial Intelligence (AI) modules for the coordination as well as for scenario awareness used also to assess situations and priorities; in this sense the authors adopted a combined use of data fusion among fuzzy logic allocation matrices defining fuzzy rules, intelligent agents (IA) controlling the platforms and artificial neural networks (ANN) for correlating complex aspects [5]. The use of SPIDER (Simulation Practical Immersive Dynamic Environment for Reengineering) solution developed and applied by the author to supervise multiple UxV collaborating is an example of how new generation CAVE (Cave Automatic Virtual Environment) could be useful in this field [4, 6].

5   UGV Inside a Plant

Another interesting case is related to the following example on the utilization of M&S as virtual prototyping approach to develop an innovative UGV in industrial plants; this was sometime combined with use of small UAV for indoor operations as in IDRASS cases or applied on oil & gas facilities as well as hot metal industry [4].

Hereafter the focus is just on ground operations that due to the complexity of plant environment require to introduce AI effective module for task execution and move-ments. The aim of this research is to evaluate the possibility to use mobile robots for performing dangerous operations inside a plant and thus replacing human activities. In facts in this case the UGV in this way reduce exposition and frequency of human presence in dangerous areas, especially during most risky operations, allowing to increase the safety; therefore in addition to prevent accident to humans, such flexible robotic system could also represent an useful resource to carry out operations during crises for assessing damages, carrying out safe procedures and even look for people blocked by the accident.

In facts this requires to develop special and complex capabilities in robotic systems such as the possibility to complete a complex mission in articulated environments considering the characteristics of the plant processes as well as the high density in such facilities of equipment, cables, machines and components that could introduce diffi-culties in moving and operating. It should be considered also that in case of accidents, many of these facilities result immediately pretty dangerous for humans and requires to carry out specific tasks respect industrial processes (e.g. shut down machines, close valves and apply safety procedures) to verify the possibility of accident escalation, addressing a proper emergency planning and actions for injured personnel such as triage [12, 30, 35, 37]. Due to these reasons the use of autonomous systems to act as "first responders" assessing the situation and to support relief operations are very promising especially if integrated with legacy systems already available on the field.

The research addressed the following topics:

- Goals and Expectations for a new industrial UGV
- Definition of Hypotheses and Constraints related to Operations, Environment and Boundary Conditions
- Survey on Existing Technologies, Autonomous System Configurations, Platforms
- Review of Examples & Experiences, carried out using Autonomous Systems, interesting for the case study
- Identification of platforms potentially usable
- UGV Operator Requirements and Regulations
- Definition of Tailored Solutions for the Case Study
- Feasibility, General Performance, Flexibility, Extensibility to other operations/plants/markets
- Capability Assessment, UGV Operational Modes
- Roadmap for a new UGV solution able to demonstrate the concepts by a preliminary virtual and physical prototype, setting a pilot project and proceeding in the Development Preliminary Analysis on costs and times to put the new UGV system into operations
- Risk Reduction & UGV Advantages and Criticalities
- Summary of Overall Benefits, Open Issues and Gaps

The authors have been involved in these studies to develop a simulator used to test the possibility of the UGV in dangerous industrial environment in which it has to operate, even in early phases, virtually.

Fig. 6. UGV performing indoor virtual operations

Furthermore the research led to the identification of the main autonomous platform characteristics and the expected degree of autonomy to face industrial plant challenges. The simulator gives good benefit to the research, indeed thanks to it, it has been possible identify multiple aspects such as the effective maximum dimension for the UGV, the degree of Freedom for the arm of the UGV for performing the tasks assigned to him, the possible configurations of his equipment etc…

The research is expected to provide a guideline and roadmap for future development of such UGV solution for creating an industrial UGV (Fig. 6).

6   Conclusions

The example propose examples interesting for the potential use of UxVs within Industrial Plants and Processes in order to increase safety. Indeed it is evident that many other different cases are possible and only a few parts of them has been already investigated. The possibilities for using UxVs are so many that it is very challenging to even list them all and it should expect a large development of these applications in near future.

Therefore, it's evident that the best promising technique able to test a large number of complex scenario in a fast way is M&S that is expected to become crucial for supporting the diffusion of UxVs in Industry. The virtual environment is also the ideal approach to test the IA and AI solutions to be used to enable collaborative missions and autonomy of these new robotic systems [31, 40, 41, 44, 50–52, 54]. From this point of view the authors are working with other research team to find solutions that could be flexible enough to be applicable in different industrial sectors.

References

1. Altawy, R., Youssef, A.M.: Security, privacy, and safety aspects of civilian drones: a survey. ACM Trans. Cyber-Phys. Syst. 1(2), 7 (2016)
2. Apvrille, L., Roudier, Y., Tanzi, T.J.: Autonomous drones for disasters management: safety and security verifications. In: Proceedings of 1st IEEE URSI Atlantic, May 2015
3. Bass, T.: Intrusion detection systems and multisensor data fusion. Commun. ACM 43(4), 99–105 (2000)
4. Bruzzone, A.G., et al.: Autonomous systems & safety issues: the roadmap to enable new advances in industrial applications. In: Proceedings of I3 M, Barcellona, Spain, September 2017
5. Bruzzone, A.G.: Information security: threats and opportunities in a safegurading perspective. Lectio Magistralis as Keynote Speech at World Engineering Forum, Rome, November 2017
6. Bruzzone, A., et al.: Disasters and emergency management in chemical and industrial plants: drones simulation for education and training. In: Hodicky, J. (ed.) MESAS 2016. LNCS, vol. 9991, pp. 301–308. Springer, Cham (2016). https://doi.org/10.1007/978-3-319-47605-6_25
7. Bruzzone, A.G., et al.: Simulation models for hybrid warfare and population simulation. In: Proceedings of NATO Symposium on Ready for the Predictable, Prepared for the Unexpected, M&S for Collective Defence in Hybrid Environments and Hybrid Conflicts, Bucharest, Romania, 17–21 October 2016

8. Bruzzone, A.G., Massei, M., Maglione, G.L., Di Matteo, R., Franzinetti, G.: Simulation of manned & autonomous systems for critical infrastructure protection. In: Proceedings of I3 M, Larnaca, Cyprus, September 2016
9. Bruzzone, A.G., Massei, M., Agresta, M., Poggi, S., Camponeschi, F., Camponesch, M.: Addressing strategic challenges on mega cities through MS2G. In: Proceedings of MAS, Bordeaux, France, 12–14 September (2014)
10. Bruzzone, A.G., et al.: Virtual framework for testing/experiencing potential of collaborative autonomous systems. In: Proceedings of I/ITSEC, Orlando, FL, USA (2013)
11. Bürkle, A., Segor, F., Kollmann, M.: Towards autonomous micro uav swarms. J. Intell. Robot. Syst. **61**(1–4), 339–353 (2011)
12. Cárdenas, A.A., Amin, S., Lin, Z.S., Huang, Y.L., Huang, C.Y., Sastry, S.: Attacks against process control systems: risk assessment, detection, and response. In: Proceedings of the 6th ACM Symposium on Information, Computer and Communications Security, March, pp. 355–366 (2011)
13. Clarke, R., Moses, L.B.: The regulation of civilian drones' impacts on public safety. Comput. Law Secur. Rev. **30**(3), 263–285 (2014)
14. Di Donato, L.: Intelligent systems for safety of industrial operators, the role of machines & equipment laboratories. In: SISOM Workshop, Rome (2017)
15. Djellal, F., Gallouj, F.: Services and the search for relevant innovation indicators: a review of national and international surveys. Sci. Public Policy **26**(4), 218–232 (1999)
16. Doherty, P., Rudol, P.: A UAV search and rescue scenario with human body detection and geolocalization. In: Orgun, M.A., Thornton, J. (eds.) AI 2007. LNCS (LNAI), vol. 4830, pp. 1–13. Springer, Heidelberg (2007). https://doi.org/10.1007/978-3-540-76928-6_1
17. Feddema, J.T., Lewis, C., Schoenwald, D.A.: Decentralized control of cooperative robotic vehicles: theory and application. IEEE Trans. Robot. Autom. **18**(5), 852–864 (2002)
18. Ferrandez, J.M., De Lope, H., De la Paz, F.: Social and collaborative robotics. Int. J. Robot. Auton. Syst. **61** (2013)
19. Floreano, D., Wood, R.J.: Science, technology and the future of small autonomous drones. Nature **521**(7553), 460 (2015)
20. Gardi, A., Sabatini, R., Ramasamy, S.: Stand-off measurement of industrial air pollutant emissions from unmanned aircraft. In: Proceedings of IEEE International Conference on Unmanned Aircraft Systems, June, pp. 1162–1171 (2016)
21. Grocholsky, B., Keller, J., Kumar, V., Pappas, G.: Cooperative air and ground surveillance. Robot. Autom. Mag. **13**(3), 16–25 (2006)
22. Ishiki, T., Kumon, M.: A microphone array configuration for an auditory quadrotor helicopter system. In: Proceedings IEEE International Symposium on Safety, Security, and Rescue Robotics, pp. 1–6 (2014)
23. Jans, W., Nissen, I., Gerdes, F., Sangfelt, E., Solberg, C.E., van Walree, P.: UUV covert acoustic communications – preliminary results of the first sea experiment. In: Techniques and technologies for unmanned autonomous underwater vehicles – a dual use view, RTO Workshop SCI-182/RWS-016, Eckernförde, Germany (2006)
24. Jones, D.: Power line inspection-a UAV concept. In: Proceedings of the IEEE Forum on Autonomous Systems, Ref. No. 11271, November 2005
25. Kastek, M., et al.: Multisensor system for the protection of critical infrastructure of seaport. In: Proceedings of SPIE, vol. 8288, May 2012
26. Kehoe, B., Patil, S., Abbeel, P., Goldberg, K.: A survey of research on cloud robotics and automation. IEEE Trans. Autom. Sci. Eng. **12**(2), 398–409 (2015)

27. Kim, D.H., Kwon, S.W., Jung, S.W., Park, S., Park, J.W., Seo, J.W.: A study on generation of 3D model and mesh image of excavation work using UAV. In: Proceedings of the International Symposium on Automation and Robotics in Construction, Vilnius, vol. 32, January 2015

28. Kovacevic, M.S., Gavin, K., Oslakovic, I.S., Bacic, M.: A new methodology for assessment of railway infrastructure condition. Transp. Res. Procedia **14**, 1930–1939 (2016)

29. Leão, D.T., Santos, M.B.G., Mello, M.C.A., Morais, S.F.A.: Consideration of occupational risks in construction confined spaces in a brewery. In: Occupational Safety & Hygiene III, vol. 343 (2015)

30. Magrassi, C.: Education and training: delivering cost effective readiness for tomorrow's operations. ITEC Keynote Speech, Rome, May 2013

31. Maravall, D., de Lope, J., Domíngueza, R.: Coordination of communication in robot teams by reinforcement learning. Robot. Auton. Syst. **61**, 661–666 (2013)

32. McCurry, J.: Dying robots and failing hope: fukushima clean-up falters six years after Tsunami. The Guardian, 9 March 2017

33. Merabti, M., Kennedy, M., Hurst, W.: Critical infrastructure protection: A 21 st century challenge. In: Proceedings of IEEE International Conference on Communications and Information Technology, ICCIT, March, pp. 1–6 (2011)

34. Merwaday, A., Guvenc, I.: UAV assisted heterogeneous networks for public safety communications. In: Proceedings of IEEE Wireless Communications and Networking Conference Workshops, March, pp. 329–334 (2015)

35. Mobley, R.K.: Plant Engineer's Handbook. Butterworth-Heinemann, Oxford (2001)

36. Nano, G., Derudi, M.: A critical analysis of techniques for the reconstruction of workers accidents. Chem. Eng. **31** (2013)

37. Palazzi, E., Caviglione, C., Reverberi, A.P., Fabiano, B.: A short-cut analytical model of hydrocarbon pool fire of different geometries, with enhanced view factor evaluation. Process Saf. Environ. Prot. **110**, 89–101 (2017)

38. Pizzella, L.A.E.: Contributions to the Configuration of Fleets of Robots for Precision Agriculture. Thesis, Universidad Complutense, Madrid, Spain, May 2014

39. Pulina, G., Canalis, C., Manni, C., Casula, A., Carta, L.A., Camarda, I.: Using a GIS technology to plan an agroforestry sustainable system in Sardinia. J. Agric. Eng. **47**(s1), 23 (2016)

40. Richards, A., Bellingham, J., Tillerson, M., How, J.: Co-ordination and control of multiple UAVs. In: Proceedings of the AIAA Guidance, Navigation, and Control Conference, Monterey, CA, August 2002

41. Ross, S., Jacques, D., Pachter, M., Raquet, J.: A close formation flight test for automated air refueling. In: Proceedings of ION GNSS-2006, Fort Worth, TX, September 2006

42. Salvini, P.: Urban robotics: towards responsible innovations for our cities. Robot. Auton. Syst. **100**, 278–286 (2017)

43. Sanchez-Lopez, J.L., Pestana, J., de la Puente, P., Campoy, P.: A reliable open-source system architecture for the fast designing and prototyping of autonomous multi-uav systems: Simulation and experimentation. J. Intell. Robot. Syst. **84**(1–4), 779–797 (2016)

44. Shafer, A.J., Benjamin, M.R., Leonard, J.J., Curcio, J.: Autonomous cooperation of heterogeneous platforms for sea-based search tasks. Oceans, 15–18 September, pp. 1–10 (2008)

45. Shkurti, F., et al.: Multi-domain monitoring of marine environments using a heterogeneous robot team. In: Proceedings of IEEE Intelligent Robots and Systems (IROS), pp. 1747–1753, 7–12 October 2012

46. Siebert, S., Teizer, J.: Mobile 3D mapping for surveying earthwork projects using an Unmanned Aerial Vehicle (UAV) system. Autom. Constr. **41**, 1–14 (2014)

47. Spanu, S., et al.: Feasibility study of an Augmented Reality application to enhance the operators' safety in the usage of a fruit extractor. In: Proceedings of FoodOPS, Larnaca, Cyprus, 26–28 September 2016
48. Spillane, J.P., Oyedele, L.O., Von Meding, J.: Confined site construction: an empirical analysis of factors impacting health and safety management. J. Eng. Des. Technol. **10**(3), 397–420 (2012)
49. Stilwell, D.J., Gadre, A.S., Sylvester, C.A., Cannell, C.J.: Design elements of a small low-cost autonomous underwater vehicle for field experiments in multi-vehicle coordination. In: Proceedings of the IEEE/OES Autonomous Underwater Vehicles, June, pp. 1–6 (2004)
50. Sujit, P.B., Sousa, J., Pereira, F.L.: UAV and AUVs coordination for ocean exploration. In: Oceans - EUROPE, 11–14 May, pp. 1–7 (2009)
51. Tanner, H.G.: Switched UAV-UGV cooperation scheme for target detection. In: IEEE International Conference on Robotics and Automation, Roma, Italy, April, pp. 3457–3462 (2007)
52. Tanner, H.G., Christodoulakis, D.K.: Decentralized cooperative control of heterogeneous vehicle groups. Robot. Auton. Syst. **55**, 811–823 (2007)
53. Tether, T.: Darpa Strategic Plan. Technical Report DARPA, May 2009
54. Vail, D., Veloso, M.: Dynamic multi-robot coordination. In: Multi-Robot Systems: From Swarms to Intelligent Automata, vol. II, pp. 87–100 (2003)
55. Valavanis, K.P., Vachtsevanos, G.J.: Handbook of Unmanned Aerial Vehicles. Springer, New York (2014). https://doi.org/10.1007/978-90-481-9707-1

Using Physics-Based M&S for Training and Testing Machine Learning Algorithms

Justin Carrillo[(⊠)], Burhman Gates, Gabe Monroe, Brent Newell,
and Phillip Durst

U.S. Army Engineer Research and Development Center, Vicksburg, MS, USA
{Justin.T.Carrillo,Burhman.Q.Gates,John.G.Monroe,
Brent.S.Newell,Phillip.J.Durst}@erdc.dren.mil

Abstract. Machine learning algorithms have been used to successfully solve many complex and diverse problems especially in the domain of unmanned vehicle systems. However, machine learning algorithms require training data that contain extensive variations in specimens. These variations include variations of the sensor settings, terrain conditions, environmental conditions, and variants of the objects of interest themselves. Capturing training specimens that span these variants is time consuming, expensive, and in some cases impossible. Training data in a narrow range of variations lead to decreases in performance such as overfitting. Therefore, collecting training data is often the limiting factor in developing robust machine learning algorithms for applications such as object detection and classification. Another time-consuming task is labeling training specimens with metadata needed in some training approaches. In this paper, we demonstrate using a physics-based modeling and simulation (M&S) capability to generate simulated training data spanning variations in sensor settings, terrain conditions, and environmental conditions that include a versatile automated labeling process. The product of prior efforts, the Virtual Autonomous Navigation Environment (VANE), is a high-fidelity physics-based M&S tool for simulating sensors commonly used on unmanned ground vehicles (UGVs) and unmanned aerial vehicles (UAVs).

Keywords: VANE · Modeling · Simulation · UGV · UAV · High-Fidelity · Physics-Based

1 Introduction

1.1 Robotic Simulation Tools

There are several robotic/sensor simulation tools currently available for industry, academia, and the Department of Defense (DoD). These simulation tools include Gazebo, Army Research Laboratory's (ARL's) Robotic Interactive Visualization and Experimentation Technology (RIVET), U.S. Army Engineer Research and Development Center's (ERDC's) Autonomous Navigation Environment Laboratory (ANVEL), Syncity, Synchrono, PBRT, ERDC's Computational Test Bed (CTB), and the Virtual Autonomous Navigation Environment (VANE). The most obvious difference in these tools is the level of detail and fidelity in the description of the environment and the

© Springer Nature Switzerland AG 2019
J. Mazal (Ed.): MESAS 2018, LNCS 11472, pp. 445–455, 2019.
https://doi.org/10.1007/978-3-030-14984-0_33

impact of the environment on sensor performance. This paper describes and utilizes the VANE, but this section briefly describes other robotic simulation tools that are currently available and the significant differences between them and the VANE. No one tool is capable of meeting all the requirements for researchers in every technology area. Recommending particular tools for each domain is difficult because a recommendation will depend on the fidelity requirements, whether there is a real-time constrain, and what level of system modeling must be supported [1]. Gazebo, RIVET, ANVEL, and Syncity take a real-time approach that utilizes lower-fidelity video-game technology while Synchrono, PBRT, ERDC's CTB, and ERDC's VANE use high-fidelity physics-based modeling and simulation.

Gazebo. Gazebo is a simulation tool originally developed in 2002 by Howard and Koenig [2]. Gazebo is designed to create a 3D dynamic multi-robot environment aimed at recreating the complex world that would be encountered by mobile robots. Gazebo is a tool that was described in Michal and Etzkorn [3] as a Robotic Development Environment (RDE). One of the key aspects of an RDE is to support simulations so that experimentation and debugging can be done without access to actual robot hardware. Gazebo is primarily for developers that want to debug their systems and not assess the impact of the environment on that systems performance.

RIVET. RIVET [4] is a computer-based simulation system that was developed by the Army Research Laboratory to merge game-based technologies with current and next-generation robotic development. RIVET provides a highly detailed, capable environment for sensor and algorithm development, integration, and assessment using common off-the-shelf computers. Just like Gazebo, RIVET requires compromises in model fidelity to support real-time and complete system simulation; therefore, RIVET is also primarily for developers that want to debug their systems and not assess the impact of the environment on that systems performance.

ANVEL. The ANVEL, developed by ERDC, is an unmanned ground vehicle (UGV) simulator that uses video-game technology to provide an interactive simulation environment. The ANVEL provides real-time interaction between vehicle models, their sensors, and their environment. The ANVEL can be used to evaluate the performance of a UGV in a mission in a relevant environment, but the level of detail in the environmental physics is limited to what can be simulated in real-time on typical laptop or desktop computers. The ANVEL also supports pre- and post-processing of VANE simulations.

Syncity. Syncity, a commercial software, is a real-world simulator for autonomous applications. Syncity constructs different types of real-world scenarios that otherwise would be difficult to record due to challenging and potentially hazardous conditions for the purposes of training and testing autonomous applications. It is a hyper realistic simulator generating synthetic data robust enough to train and validate algorithms for autonomous applications. Although the key features and sensors of Syncity are compelling, Syncity is a commercial software with very little to no documentation on the actual sensor and environment physics used. Syncity appears to utilize video-game technology rather than physics-based modeling and simulation.

Synchrono. Synchrono, developed by the University of Wisconsin-Madison, is a framework in which dynamic multi-agent simulations can be conducted to understand agent interplay and develop control algorithms in a safe and flexible environment. Unlike simulation tools such as Gazebo that simplify the noise models by assuming a Gaussian distribution, Synchrono's sensor modules are responsible for generating and recording data representing the data accumulated by various sensors [5]. Synchrono is built on top of project Chrono, which is a simulation platform created by Professor Tasora in 1997 and has since been used and developed by the Simulation Based Engineering Lab at the University of Wisconsin-Madison. Although Synchrono's framework is aimed towards greater fidelity for sensor-environment interactions and vehicle-terrain interactions, Synchrono is still in full development, and the sensor and environment models are still simplistic representations.

PBRT. PBRT [6] is a physics-based rendering engine that focuses exclusively on photorealistic rendering, which can be defined variously as the task of generating images that are indistinguishable from those that a camera would capture in a photograph or as the task of generating images that evoke the same response from a human observer as looking at the actual scene. Although PBRT is focused towards entertainment applications such as the movie special-effects industry, PBRT provides high-fidelity physics-based simulations of cameras. Much of the physics that are in PBRT are also in the VANE, but PBRT does not simulate sensors such as GPS, LiDARs, and IMUs and is not integrated with multi-body dynamics engines for dynamic interactions.

ERDC's CTB. The ERDC developed a near-surface computational test bed (CTB) to help understand the effects of geophysical phenomena on signatures sensed by various sensors operating in the electro-magnetic (EM) spectrum [7]. The CTB produces 3-D, physics-based, high-fidelity numerical modeling simulations of the geo-environment. This suite of physics-based models in CTB include the Adaptive Hydrology (ADH) soil model, the vegetation model that computes radiative transfer in plants, and the EO/IR sensor model. This modeling capability can be used to predict and improve the performance of current and future sensor systems for surface and near-surface anomaly detection amid highly heterogeneous and complex environments.

1.2 Virtual Autonomous Navigation Environment

The VANE is a simulation software for predicting the performance of unmanned and autonomous ground and aerial vehicles. The VANE integrates high-fidelity, physics-based sensor simulations with realistic vehicle dynamics and terrain and environment simulations to provide a complete picture of the factors influencing the performance of unmanned systems. As discussed in the previous section, while several robotics simulators exist, most lack the physics fidelity to accurately capture the influence of environmental conditions on the performance of the mobility platform and the sensors that enable autonomous operations [8]. The VANE uses the most realistic physics simulations for the physical processes impacting the robot, ensuring that the simulation is both realistic and predictive. Although VANE utilizes high-performance computing resources, a major misconception about the VANE is that high-performance computing resources are required for using the VANE. VANE simulations have been successfully run on PCs (Windows, Mac, and Linux) and workstations as well as supercomputers.

Sensors in the VANE include cameras, LiDAR, GPS, and IMUs. The VANE can be used to simulate digital cameras in the optical and near-infrared wavelengths that use CCD or CMOS sensors at the focal plane array. Non-ideal lensing systems can be simulated as well as electrical properties such as the gain and gamma compression of the sensor. For systems in the optical and near-infrared wavelengths, the properties of the sensing system can be modeling using radial and tangential distortion coefficients. These features allow the VANE to replicate distortions exhibited in real cameras. There are several important sources of error in LiDAR measurements. In time-of-flight (TOF) LiDAR systems, which are the predominant type used on outdoor robots, there are intrinsic errors introduced by the accuracy and precision of the time system and electronics. For well-calibrated commercial systems, these errors are typically on the order of a few centimeters and can be modeled as Gaussian noise. However, errors introduced by environmental conditions, including target reflectance and geometry, weather, and beam scattering, can be significantly higher. An example of results from the camera and LiDAR sensors are shown in Fig. 1.

Fig. 1. VANE simulated camera image (left); VANE simulated LiDAR point cloud (right).

Because the 3D representation of an environment must have detail and accuracy commensurate with the simulation goals and because the VANE can leverage HPC assets, very large and detailed scenes and environments are used for VANE simulations. Additionally, dynamic actors and animations like humans and vehicles can be scripted as input into the VANE in order to evaluate dynamic interactions and sensor responses and algorithms. Snow and dust are simulated in the VANE using a particle system, while rain is simulated using a random mask generator. Both methods interact using physics-based models, resulting in realistic signatures for both the camera and LiDAR sensors. An example of rain being simulated in the VANE is shown in Fig. 2. For LiDAR sensors, predicting the influence of dust and smoke is more complex. This is because the LiDAR acts as both the source and receiver, and the two-way radiative transfer must be calculated. The VANE employs an empirical model based on field and laboratory measurements for calculating the probability of LiDAR returns from dust [10]. More details on the VANE can be found in [8].

Fig. 2. VANE rain simulations (left to right: light, medium, heavy).

2 Using Simulation for Machine Learning Support

Need for Simulation. Machine learning tools have been used for a variety of problems such as assessment of planetary gear health, assessment of fatigue and cracking in vibrating load tests, automated 3-D tissue/organ segmentation from CT scans for soldier protection, armor mechanics problems, automated optical, thermal, and acoustic monitoring of the additive manufacturing process, automated first-pass analysis of video streaming data, and evaluation of human-annotated maintenance reports toward sensor-based anomaly detection in vehicles [11]. The current and future operational applications include military intelligence, natural language processing, data mining, anomaly detection, automated target recognition, robotics, self-healing, ethics, cybersecurity, prognostic and structural health monitoring, sequence mining, and medical diagnosis [11]. Military operations are expected to have higher than usual density of static and dynamic objects in a chaotic environment. Capturing training specimens that span these variants is time consuming, expensive, and in some cases impossible. Training data in a narrow range of variation lead to decreases in performance such as overfitting. Therefore, collecting training data is often the limiting factor in developing robust machine learning algorithms for applications such as perception applications. Combining simulation data with field-collected data provides a robust data set for training, testing, and validating machine learning algorithms for autonomous applications. The VANE provides diverse high-fidelity physics-based datasets that maximize machine learning by including high variation and randomization within the simulation to increase entropy within the dataset.

Machine Learning Support. Recently, VANE was upgraded to provide simultaneous ground truth within the sensor data for rapid training of machine learning-based perception algorithms. Developing labeled training specimens with certain meta-data can be time-consuming, limited in accuracy, and sometimes impossible in real-world data collects but requires little effort and is highly accurate using simulation tools such as the VANE. The VANE can provide metadata for the sensor, sensor position, target of interest, and environmental conditions. An example of a labeling-file with metadata for a training specimen is shown in Fig. 3. The VANE is able to provide information such as altitude, sensor orientation, bounding box of target within the image, ground-sampling-distance (GSD), target-sampling distance (TSD), pixels-on-target (PoT), and environment conditions such as visibility and rainfall rate.

```
"Version": 2.0
},

"Sensor": {
    "source": "SIMULATION",
    "type": "IMAGER",
    "parameters": {
        "mode": "RGB",
        "gamma": 0.75
    },

    "pixels": [4608,3456],
    "focal_length (m)": 0.00397
},

"Sensor Position": {
    "height_above_ground (m)": 43,
    "LookFrom (scene)" : [62.2128,52.5503,54.4752],
    "LookTo" : [0,0,-1],
    "LookUp" : [-0.207912,-0.978148,0]
},

"Objects": [
    {
        "name" : "HMMWV.obj",
        "bounding_box" : [2136,1637,2471,1819],
        "average TSD (width) (m/pixel)" : 0.0166191,
        "average TSD (height) (m/pixel)" : 0.018183,
        "average TSD (m/pixel)" : 0.0173999,
        "average TSD (viewing plane) (m/pixel)" : 0.0140944,
        "average GSD (width) (m/pixel)" : 0.0332608,
        "average GSD (height) (m/pixel)" : 0.0373623,
        "average GSD (m/pixel)" : 0.0353115,
        "average GSD (viewing plane) (m/pixel)" : 0.0297893,
        "Pixels on Target (PoT)" : 152247,
        "Horizontal PoT" : 50566,
        "Vertical PoT" : 50413,
        "location (scene) (x,y,z)" : [62.2128,52.5503,11.4752]
    }
],

"Scene": {
    "name": "CampLejeune",
    "position_orientation_matrix": [1,0,0,-90.8753, 0,1,0,88, 0,0,1,0, 0,0,0,1],
    "location_origin (lat_ddeg,lon_ddeg,elev_m)": [-90.8753,88,0]
},

"Environment": {
    "Sun Position": [-0.503217,0.860509,0.0793477],
    "Sun theta" : 1.49137,
    "time_zone": -5,
    "datetime": [2014,249,23,0,0],
    "turbidity": 8,
    "albedo": 0.5,
    "temperature (Celsius)": 30,
    "pressure (hPA/mbars)": 1006,
    "water_vapor (cm)": 5,
    "tau500": 0.4,
    "rainfall_rate (mm/h)": 25.4,
    "rainfall_density (num/m^3)": 350.422,
    "drop_size (m - diameter)": 0.00147937,
    "rain_velocity (m/sec)": -5.37981,
    "wind_speed (m/sec)": [0,0,0]
}
}
```

Fig. 3. JSON file containing image metadata.

For many applications, detection and classification of multiple objects and their relation to one another is required. For example, determining mobility obstacles requires detecting and classifying objects that could possibly be obstacles and then relating their relative positions onto road networks. The VANE provides ground truth data such as road segmentation and object location on a per-pixel basis as shown in Fig. 4. The image to the left is from a UAV at 247 m above the ground. The image to the right shows the ground truth data for the per-pixel labeling of the image. The white pixels represent the road segmentation and the gray pixels represent a HMMWV.

Fig. 4. VANE generated image (left). Per pixel labeling of image (right).

3 Analysis

UGV Systems. To highlight VANE's unique capability to support machine learning through simulated training data and testing environments, the VANE was used for training and testing a convolutional neural network for UGV systems [12]. In this study, high-fidelity vehicle dynamics using ERDC's Computational Research and Engineering Acquisition Tools and Environments – Ground Vehicles (CREATE^{TM} – GV) was coupled with VANE's high-fidelity sensor simulation to simulate a camera mounted on a vehicle. The objective of the study was to evaluate a convolutional neural networks ability to detect and classify a HEMTT vehicle from a camera mounted on a HMMWV. As the machine learning algorithm, the study implemented the Inception-v3 CNN using TensorFlow. The VANE was used to generate the simulated training and testing data, and physical experimental data were collected for both the training and testing scenarios. Receiver operating characteristics (ROCs) curves and precision-recall curves were used from comparing the various CNN testing conditions. Highlights from the work are shown in Fig. 5.

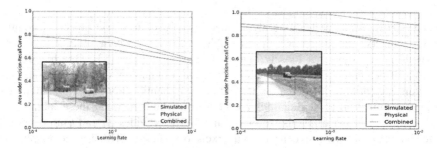

Fig. 5. UGV study results using CNN.

Although the study focused more towards comparing classifier performance based on training data and the ability of simulation to test and evaluate classifier performance, additional analysis such as observing and evaluating classifier performance with respect to rain fall rate and time of day can be utilized using the VANE. Figures 6 and 7 show the impact of rainfall rate and time of day on the same machine learning algorithm that was used in the study.

Fig. 6. VANE image with rain (left). Rain impact on classifier (right)

Fig. 7. VANE image early morning (left). Time of day impact on classifier (right)

UAS Systems. Machine learning algorithms can be used to detect objects, recognize specific objects, and determine the probable class of unknown objects. The term "recognition" is somewhat loosely defined in common usage [13]. One hierarchical ordering of perceptual processes include detection, orientation, clutter rejection, classification, recognition, identification (friend-or-foe), identification, discrimination, and intent discrimination [13]. Relations are quite often built between these hierarchical levels and the number of pixels required. The VANE provides ground truth data such as the number of successive horizontal pixels on a target (HPoT), the number of successive vertical pixels on a target (VPoT), and the total number of pixels on a target (PoT). Machine learning algorithms can be evaluated to determine the minimum number of pixels required for each hierarchical level which in turn can provide flight requirements for UAVs. For example, there is a relationship between the number of pixels on an object and the target sampling distance (TSD) in the viewing plane. The TSD is a function of flight height and sensor settings. An estimate derived from the VANE is provided in Eq. 1, where S_w is the sensor width (mm), S_h is the sensor height (mm), h is the flight height (m), f is the focal length (mm), I_w is the image width (pixels), and I_h is the image height (pixels). Figures 8 and 9 shows verification results of Eq. 1 using two different camera sensors. The camera parameters are shown in Table 1.

$$PoT_{HMMWV} = g(f(\textit{flight height, sensor settings}), \textit{target})$$

$$= 32.383 \left(\frac{h}{2f} \left(\frac{S_w}{I_W} + \frac{S_h}{I_h} \right) \right)^{-2.0}, \ 50 \leq h \leq 300 \tag{1}$$

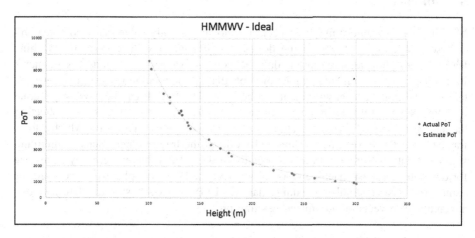

Fig. 8. Pixels on target with respect to height for HMMWV using Ideal camera.

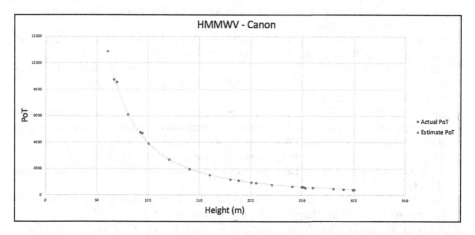

Fig. 9. Pixels on target with respect to height for HMMWV using Canon camera.

Table 1. Camera parameters

Parameters	Ideal camera	Canon camera
Resolution (pixels)	1920 × 1080	1920 × 1080
Sensor size (mm)	(1.894, 1.894)	(22.3, 14.9)
Focal length (mm)	3.5	18
Gamma	0.75	1.20
Gain	10.0	8.0

4 Conclusions

Machine learning algorithms have been used to successfully solve many complex and diverse problems especially in the domain of unmanned vehicle systems. Machine learning algorithms require training data that contain extensive variations in specimens. In this paper, we presented a simulation tools that provide variations in sensor settings, terrain conditions, and environmental conditions. The VANE can be used to generate training specimens that span these variants efficiently and without the cost of collecting the data in real-world scenarios. The VANE is completely physics-based, which allows for the closest representation of the real world in the training specimens. Combining simulated data with real-world data provides a dataset that has the entropy necessary for developing robust machine learning algorithms. Two examples were provided for using the VANE's simulated training data for UGV and UAV systems that also take advantage of a versatile automated labeling process.

References

1. Fields, M., et al.: Simulation tools for robotics research and assessment. In: Proceedings of the SPIE, vol. 9837 (2016)
2. Koenig, N., Howard, A.: Design and use paradigms for gazebo, and open-source multi-robot simulator. In: IROS, vol. 3, pp. 2149–2154. IEEE, USA (2004)
3. Michal, D.S., Etzkorn, L.: A comparison of player/stage/gazebo and microsoft robotics developer studio. In: Proceedings of the 49th Annual Southeast Regional Conference, pp. 60–66. ACM, USA (2011)
4. Brewer, R., Schaefer, K.E., Avery, E.: Robotic interactive visualization experimentation technology (RIVET): game-based simulation for human-robot interaction research. In: 2015 Winter Simulation Conference (WSC), pp. 3224–3225 (2015)
5. Elmquist, A., Negrut, D.: Virtual sensing for autonomous vehicle simulation in chrono. Simulation-Based Engineering Lab University of Wisconsin-Madison, Technical report TR-2016-13 (2017)
6. Pharr, M., Jakob, W., Humphreys, G.: Physically Based Rendering: From Theory to Implementation, 3rd edn. Morgan Kaufmann, Burlington (2017)
7. Nagaraju, K., et al.: High-Fidelity Modeling to Support Route Clearance. North Atlantic Treaty Organization (NATO) Science and Technology Organization (STO) STO-MP-MSG-143 (2016)

8. Goodin, C., et al.: Unmanned ground vehicle simulation with the virtual autonomous navigation environment. In: International Conference on Military Technologies (2017)
9. Durst, P., Goodin, C.: High fidelity modelling and simulation of inertial sensors commonly used by autonomous mobile robots. World J. Modell. Simul. **8**, 172–184 (2012)
10. Goodin, C., Durst, P., Prevost, Z., Compton, P.: A probabilistic model for simulating the effect of airborne dust on ground-based LiDAR. In: SPIE Defense, Security, and Sensing. International Society for Optics and Photonics (2013)
11. Lee, M., Valisetty, R., Breuer, A., Kirt, K., Panneton, B., Brown, S.: Current and future applications of machine learning for the US Army. US Army Research Laboratory, Technical report ARL-TR-8345 (2018)
12. Carrillo, J., Davis, J., Osorio, J., Goodin, C., Durst, J.: High-fidelity physics-based modeling and simulation for training and testing convolutional neural networks for UGV systems. J. Unmanned Veh. Syst. (in review)
13. Harney, R.: Combat Systems. Volume 1. Sensor Elements. Part I. Sensor Functional Characteristics (2004)

Adaptive Image Processing Methods for Outdoor Autonomous Vehicles

Lucie Halodová[1], Eliška Dvořáková[1], Filip Majer[1], Jiří Ulrich[1],
Tomáš Vintr[1], Keerthy Kusumam[2], and Tomáš Krajník[1(✉)]

[1] Artificial Intelligence Center, FEE, Czech Technical University,
Prague, Czech Republic
tomas.krajnik@fel.cvut.cz
[2] University of Nottingham, Nottingham, UK

Abstract. This paper concerns adaptive image processing for visual teach-and-repeat navigation systems of autonomous vehicles operating outdoors. The robustness and the accuracy of these systems rely on their ability to extract relevant information from the on-board camera images, which is then used for the autonomous navigation and the map building. In this paper, we present methods that allow an image-based navigation system to adapt to a varying appearance of outdoor environments caused by dynamic illumination conditions and naturally occurring environment changes. In the performed experiments, we demonstrate that the adaptive and the learning methods for camera parameter control, image feature extraction and environment map refinement allow autonomous vehicles to operate in real, changing world for extended periods of time.

1 Introduction

Digital cameras are gradually becoming one of the most important sensors used in mobile robotics that are deployed in natural environments. Their popularity can be attributed to the low-price, small size and the fact that they can provide large amounts of data in real time. However, while a typical camera generates a significant amount of data over relatively short time periods, most of this data is not relevant to the tasks assigned to a robot. Thus, a crucial problem of vision-based systems is to tackle the extraction, storage and management of relevant information contained in the image streams. Ideally, a vision-guided robot should not only be able to extract information rich enough to perform the tasks at hand, but also to build upon and gradually improve its knowledge of the environment to perform its tasks efficiently.

To achieve that, the images acquired by the camera should contain a sufficient number of identifiable, salient elements that are pertinent to the task at hand. These elements have not only to be extracted but, most importantly, correctly interpreted and utilised for the required task. Interpretation of images becomes more efficient and reliable if the robot can anticipate the information it searches

The work has been supported by the project 17-27006Y.

© Springer Nature Switzerland AG 2019
J. Mazal (Ed.): MESAS 2018, LNCS 11472, pp. 456–476, 2019.
https://doi.org/10.1007/978-3-030-14984-0_34

for and adapt the aforementioned phases, the extraction and the interpretation, accordingly. This requires the use of a knowledge base, which provides the robot with the description of salient image elements that the robot might encounter within different spatial and temporal contexts. An intelligent system should be able to build and refine this knowledge base during the course of its routine operation.

In the context of the visual navigation, a robot must ensure that the images have a sufficient contrast to identify an adequate number of environmental features necessary for position estimation of the robot relative to its goal. This not only means that the robot must adapt the camera settings and feature extraction parameters properly but also carefully choose which of the obtained features qualify to store for the later use and when to retrieve or discard them from its memory. In other words, the robot not only needs to make sure that it perceives the information relevant to its environment model but also that it can update the environment model based on the perceived information. The ability to adapt the settings of the robot perception subsystems, as well as its knowledge bases, is especially crucial in environments which exhibit appearance and structural changes. The causes for these changes include short-term factors like varying illumination and unpredictable weather and long-term, partially predictable seasonal processes.

This paper hence concerns adaptive methods, which are specifically aimed at ensuring a reliable long-term operation of visually-navigated robots in outdoor environments. In particular, we will discuss methods which *(i)* control camera parameters to ensure sufficient quality of obtained images, *(ii)* adapt image preprocessing modules to extract a suitable number of salient image elements for self-localisation and mapping, *(iii)* gradually adapt the environment map, so that it stays up-to-date with the environment variations and *(iv)* interprets the temporal behaviour of the changes in the map and predict which elements of the map will be visible at a particular time and location. Apart from a detailed description of the adaptive methods, we demonstrate that their integration into a visual navigation pipeline significantly improves the efficiency of the long-term mobile robot operation in outdoor, unstructured environments.

2 Related Work

Visual navigation systems can be divided according to their working principle to map-less, map-based and map-building based [11]. Map-less systems such as [9] assume that the environment contains traversable structures such as highway lanes, pathways or roads. Thus, these systems attempt to identify the specific structures and control the autonomous vehicle in a way to keep it on them. Map-based systems do not assume that traversable structures in the robot's environments will be easy to recognise. Instead, these systems navigate and localise the robots based on maps, which are known a priori [18]. Map-building-based systems are able to build these maps themselves typically by utilising the Simultaneous Localisation and Mapping (SLAM) [13,17,36] technology. These metric maps are then used to determine the position of the robot relative to its goal so that the robot can be driven along the intended trajectory. An alternative line of map-building approaches does not perform metric position estimation of the robot within the created maps. Instead, these systems use principles of a visual servoing, which allow the robot to repeat the path it was taught by a human operator during a teleoperated drive [3,6,34,42,43]. While these 'teach-and-repeat' systems are somewhat less versatile, because the robot can only reach positions it was driven to before, they were reported to be successfully deployed for long periods of time [20,39,45]. Regardless of their principle, most of the aforementioned systems are based on a similar processing pipeline: they capture images, process them to extract salient features, match those features to environment representations they created beforehand and determine how to steer the robot so that it stays on the intended path. Some of these systems are able to gradually adapt their environment representations [39] or perception systems [21] to the changes they observe as they repeat the taught paths.

The first stage of the visual processing pipeline is the image acquisition, which determines if the images contain information relevant for further processing. This is mainly influenced by the robot camera exposure settings, which have to be adapted to varying illumination conditions. A fully autonomous system needs to adapt the camera exposure constantly to cope with varying illumination. However, typical built-in auto-exposure methods are not aimed at visual navigation and thus fail to set the exposure properly [44,47]. Furthermore, while the impact of settings of subsequent processing stages can be analysed offline based on the gathered images, compensating for incorrect exposure setting is difficult.

Because of the aforementioned issues, several researchers proposed different methods of exposure setting. For example, Lu et al. [30] measure colour image quality by information-theoretic methods and propose to set the exposure to maximise the Shannon entropy of the RGB image and optimise the camera parameters by achieving similar values of exposure time and camera gain. Neves et al. [38] build a histogram of pixel intensities and set the camera exposure based on the mean sample value of that histogram. Furthemore, they take into account the size of underexposed and overexposed areas in the image. Shim et al. [44] base their exposure settings on the sum of gradients of individual pixel intensities in the image, which they aim to maximise. Zhang et al. [47] compare four gradient-based metrics that indicate the quality of images and conclude that Shim's method could be improved by the use of more advanced statistics and the photometric response function compensation [10]. To assess the quality of the images, Zhang uses the number of extracted FAST [31] features and tries to obtain as many FAST keypoints as possible. As the FAST features are used in many visual navigation systems, including the one which we use in this work [20], Zhang's method is highly relevant for the work presented herein.

Once an image is acquired successfully, the positions of its salient features (or keypoints) are extracted in a process called keypoint detection. A typical keypoint detection method, such as [1,27,31], assigns each pixel a given measure of saliency and reports pixels with saliencies which are locally maximal and exceed a pre-set threshold. For example, a SURF [1] keypoint detection is based on a Hessian matrix determinant, which retrieves points in the image with sufficient contrast that makes them easy to localise and track. Since the aforementioned threshold is set prior to the keypoint detection process, it is hard to predict how many keypoints are going to be detected in the processed image. However, the number of detected features strongly influences the ability of the robot to successfully establish the relation of the current image to its environment model and thus, to navigate reliably. As shown in [21], setting the threshold too high and acquiring a small number of relevant features negatively impacts navigation accuracy. On the other hand, too low threshold produces an excessive number of features which do not contribute to the quality of navigation but hamper the ability of the system to match them to the map in real time. Thus, similarly to the camera exposure time, the saliency threshold needs to be continuously adapted to the images that are being processed.

Once the salient keypoints are detected, another set of methods, called feature descriptors, are applied to the vicinity of these keypoints. The keypoints' descriptors, which are typically engineered [1,4,27], are generally meant to be invariant to contrast, scale, rotation and viewpoint changes. However, these invariances are often not needed for autonomous navigation and other properties such as robustness to illumination and seasonal changes, are desired. This led to the research of learning methods, which can generate feature description algorithms robust to illumination and seasonal changes [5,21]. In their latest work, [46] even demonstrated that adaptation of the feature descriptors during routine autonomous operation of the robots improves their ability for long-term operation. However, the feature description adaptation is planned to be included in our system in the future and is not subject to the evaluation in this work.

The extracted features, along with their positions and descriptions, are used either to build an environment map (teaching phase), or they are matched to the map to determine the robot's position relative to the intended path (repeat phase). However, as the environment changes, the features stored in the initially-taught map disappear and other features become visible. After a sufficiently long time, an environment might change its appearance so much, that the map becomes completely obsolete and irrelevant. Thus, long-term operation requires that the map is adapted to the changes during the robot operation. One of the most popular frameworks for map maintenance is based on 'experiences' [7], which allow representing the same location with several appearances or 'experiences', which depend on particular environmental conditions such as day/night or weather. During navigation, a robot retrieves several 'experiences' tied to a given location and tries to associate them with its current view. The failed association indicates that the appearance of the particular location changed drastically and the perceived appearance is added to the set of experiences associated with a given location. The experience-based approach allows not only smart management of the environment maps [14,26], but it was successfully integrated into teach-and-repeat systems [39]. Another popular approach, inspired by the interplay of long- and short-term memory, is described in [2,8], who gradually add newly detected features and discard features which are no longer visible. A similar approach, called Summary Map [35], ranks all the map features based on their past visibility and uses the rank to remove or add the elements to the current map. Inspired by [2,8,35], we implemented a similar scheme, which gradually adapts the map during the routing operation of the robot. Unlike in [2,8,35], where the experimental evaluation is performed off-line, our navigation system updates the maps during its operation.

Once the navigation system can cope with the environment changes it observes during routine operation, its ability to operate for longer time periods is greatly improved. Robust, long-term operation opens the possibility of repeated re-observation of the same locations, which capture the long-term temporal behaviour of the environment variations. These observations allow the robot to create not only spatial but also spatiotemporal models of its operational environment. These models can predict, how a given location will look like at the time of robot operation. For example, Lowry et al. [28] use consecutive observations to distinguish between time variable and time-invariant image components. Other works [25,37,40] attempt to predict how a winter scene will look based on its appearance captured during summer and vice versa. Visibility of image features in a period of time is based on a temporal model, which is recomputed while repeatedly watching the same or similar places. Rosen et al. [41] assume that persistence of features is limited and they employ the survivability theory to create environmental models that predict, which elements are not likely to be visible anymore. Similarly, Krajník et al. [23] propose the use of spectral analysis to capture the periodic behaviour of feature visibility caused by day/night and seasonal cycles. Our navigation system employs the FreMEn method proposed in [22,23], to model the temporal behaviour of each feature.

Since the problem of long-term, reliable operation gradually comes into focus of the robotics research community, the aforementioned list of papers is by no means exhaustive. Thus, we refer to a comprehensive review of approaches for long-term visual localisation in [29] and a general review of AI methods for long-term autonomy in [24].

3 Adaptive Navigation System

The navigation paradigm which we base our system on belongs to the group of teach-and-replay navigation systems. These systems allow mobile robots to autonomously traverse paths, through which they have been previously driven by human operators. In the teaching phase, a person guides the robot along the intended path and the robot creates an environment map. Once the map is created, the robot can navigate along the path autonomously, which is referred to as 'replay'. The underlying principle of the autonomous replay depends on the robot's sensory and computational equipment and on the way it represents the environment.

Fig. 1. Navigation method overview: During a teleoperated drive or 'teaching', a robot creates a sequence of local maps at regular intervals (blue circles). Each map contains the image captured at the given spot and the features extracted from the image. During the replay phase, a robot at a given distance from the path start (white circle) selects the closest map (red circle) and establishes correspondences between the visible and mapped features. Difference between the horizontal coordinates of the feature pairs, which corresponds to the horizontal shift of the images, determines the robot steering velocity (shown as red arrows at the bottom). (Color figure online)

3.1 Navigation System Core

In our case, the robot is equipped with a monocular camera and odometry. During the teaching phase, the robot uses the odometry to measure the travelled distance and once every certain distance it saves the latest captured image along with its features into a local map. Thus, at the end of the teaching phase, the taught path is represented as a sequence of local maps indexed by their distance from path start, see Fig. 1. During the autonomous navigation or 'replay', the robot retrieves the local map according to its distance from the path start and matches the features extracted from its current camera image to the ones from the local map. Then, it subtracts the horizontal image coordinates of the corresponding pairs and uses a histogram voting scheme to determine the most prevalent difference δ. This process, referred to as image registration, aims to recover the horizontal shift δ between the image which is currently seen and the image stored in the local map. The value of δ indicates not only the difference of the robot's current heading from the heading it had during the teaching but also its lateral displacement from the path. Thus, using a simple regulator to steer the robot in a way, which would keep the difference between the horizontal coordinates of the corresponding features close to zero, causes the robot to follow the taught path. The works [20,33,45] show that if a robot traverses a closed path repeatedly, the aforementioned steering correction scheme efficiently suppresses both lateral and longitudinal errors of the robot position and keeps the robot on the taught path. The aforementioned navigation process is illustrated in Fig. 1, and also in several videos available from http://bearnav.eu.

The processing stages of the navigation pipeline along with the information flows are shown in Fig. 2. A classic, non-adaptive visual navigation pipeline would compose only of the modules drawn in black, which acquire an image, extract its features, establish their correspondences with the map and calculate the robot steering. However, the outdoor visual navigation requires that the navigation deals with changing illumination and environment variations. Thus, our system extends the classic pipeline by several components responsible for adaptation with the aforementioned variations – these modules are shown in Fig. 2 in blue colour. These modules *control camera exposure* to ensure sufficient quality of captured images, *adapt feature extraction* to obtain a suitable number of image features, gradually *adapt maps* of the environment by removing obsolete features and adding new ones, and *predict* which features are going to be visible at a particular time and location. Since this paper is aimed at the evaluation of the adaptative methods on the quality of visual navigation, we provided only a coarse overview of the navigation method. For further details, please refer to [20, 32, 33]. In the following sections, we will explain how the individual modules (shown in blue in Fig. 2) handle the adaptation of the core modules of the visual navigation.

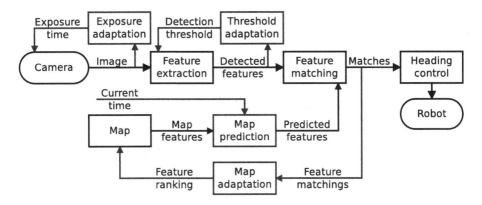

Fig. 2. Navigation system modules: the standard, core processing modules are drawn in black, and the modules responsible for adaptation to the environmental conditions are shown in blue. (Color figure online)

3.2   Camera Exposure Control

The quality of the image features directly depends on the quality of the input image stream, which, in outdoor environments, is affected by varying illumination. Therefore, we developed a simple method, which compensates unstable illumination by adaptively setting exposure of the robot on-board camera, which, in our case, does not provide us with a built-in automatic exposure setting. To do so, we had to edit the driver of e-Con TARA camera, which is used on most of our robots. Since the robot on-board camera is aimed forwards, the bottom halves of most images contain the ground, which does not provide information useful for navigation. Thus, our method attempts to control the camera exposure to keep the mean brightness of the top half of the images at a certain value.

Each time an ith image is captured, we calculate the mean brightness b_i of its top half. Then we compare the desired brightness b_d to the actual one (b_i) and calculate the next exposure setting e_{i+1} according to:

$$e_{i+1} = e_i + c_e\, e_i \left(\frac{b_d}{b_i} - 1\right), \tag{1}$$

where e_i is current exposure setting, $c_e \in {<}0,1{>}$ is the exposure control gain, b_d is the desired brightness and b_i is current brightness of the image top half. Since the exposure setting of the camera takes some time to take effect, we perform this calculation once per five frames obtained.

The Eq. 1 has two parameters which have to be set: the desired image brightness b_d and the exposure control gain c_e. To ensure a quick response to the changes in illumination, while avoiding possible oscillations of the image brightness, we can set c_e to or just below the value of one. As the paper [16] showed that in challenging lighting conditions, slightly underexposed images are more likely to be registered correctly, we set the desired mean brightness b_d in the range of 0.4–0.5. While this exposure setting scheme is rather simple, the experiments shown in Sect. 4.2 indicate that the images obtained are more suitable than if the exposure was controlled automatically in the traditional way, i.e. according to the brightness of the entire image.

3.3   Feature Detection Adaptation

The accuracy and reliability of the visual navigation is affected by the number of features the detector extracted from the on-board camera image. However, the relation between the number of features and the quality of the navigation is not straightforward. While the higher number of extracted features typically results in more accurate registration, exceeding a certain number of elements does not bring significant improvement [21]. Moreover, a high number of features takes more time to extract and match, which slows down the response of the robot to its position perturbations. Furthermore, a high number of features increases the chance of obtaining incorrect correspondences, which also negatively impacts the navigation accuracy. From the experiments performed in [21], it seems that the optimal number of features depends on the particular environment. However, the results in [19, 21] show that it is better to process the images in a way, which produces a certain, fixed number of features per image.

In the case of the SURF detector, which is used in our experiments, the number of the extracted features depends on the local image contrast and the value of the Hessian matrix threshold. Since the images the robot perceives along the path differ in contrast, setting a fixed threshold would produce a different number of features for each image. This often results in a deficiency in the features required for accurate image registration or in detection slowdown due to the excessive number of extracted features. Thus, one needs to adapt the Hessian threshold on the fly during the robot routine operation.

To achieve that, we adapt the Hessian threshold according to the number of features obtained in the last image. Let us assume that we want to obtain f_d features for further processing. Since we can always erase features that are in excess, but too many detected features would slow down the detection, we attempt to set the Hessian threshold to obtain a slightly higher number of features f'_d, where $f'_d = f_d (1 + o)$, where o stands for an 'overshoot' factor, which we set to 0.3 in our experiments. Assume that the last i^{th} image with a Hessian threshold of t_i provided us with f_i features and we need to calculate the next threshold t_{i+1}. To do so, we order the detected features according to their response, i.e. the value of their Hessian matrix determinant, obtaining a sequence $\mathcal{H}$, where $h(1)$ is the Hessian response of the most prominent feature and $h(f_i)$ is the Hessian threshold of the least prominent detected feature. If the number of features f_i is higher than f'_d, then we simply set the t_{i+1} to the response of the feature, which is at position $(f'_d + 1)$ in the aforementioned set, i.e. $t_{i+1} = h(f'_d + 1)$. In case that $f_i < f'_d$, we take the 10 last values of $\mathcal{H}$ and use linear extrapolation to determine t_{i+1}. In particular, we use the values of $h(f_i - 10), h(f_i - 9) \ldots h(f_i)$ to estimated a linear function $h'(n)$ and then set t_{i+1} to $h'(f'_d)$.

As mentioned in the previous section, features that are most useful for heading correction typically appear in the upper half of the image. Thus, the detector is set up to extract the features from this upper half. The impact of the aforementioned Hessian threshold adaptation on the quality of visual navigation is evaluated in Sect. 4.4, which shows that extracting the features from the upper half of the image results in more accurate image registration.

3.4   Map Adaptation

Outdoor, natural environment is subject to gradual, but perpetual change, which affects both its structure and appearance. As the environment changes, the map which was created during the teaching phase becomes gradually obsolete. Thus, to achieve long-term operation, a mobile robot should be able to adapt its environment representation to the changes it observes. Ideally, the adaptation of the map should be performed automatically during the routine robot operation, and it should not require human intervention.

Inspired by the methods presented in [2,8,35], we implemented a method, which allows to refine and update the environment maps during the routine robot navigation. The core idea is to evaluate the utility of the elements in the maps, keep the features which are useful, remove the ones which are often matched incorrectly, and add features which were not visible before. To do so, we developed a system, which ranks the features according to their usefulness for navigation by continuously monitoring if they provide correct information about the robot heading. As mentioned in Sect. 3.1, our navigation system continuously retrieves the image features from the local maps in the robot vicinity, matches these features to the ones extracted from the camera image, and uses a histogram voting method to determine the most frequent difference in the horizontal coordinate of the feature pairs.

Thus, the system can assess the correctness of the established correspondences by comparing the horizontal difference of the feature pairs to the result provided by the histogram voting method. The Fig. 3 illustrates the results of this assessment – the feature pairs, whose horizontal displacement is in consensus (which is established by the histogram voting), and are considered to be correctly established, are drawn in green. The feature pairs, whose horizontal coordinates differ from the consensus, are considered to be incorrectly established (these are drawn in red colour in Fig. 3). Whenever the system establishes the correspondences and divides them into correct and incorrect, it increases the rating of the map features, which were correctly matched and decreases the rating of the map features, which produced incorrect correspondences. The score of unmatched features is not changed. In this way, features that contribute to the correct estimation of the robot heading gradually improve their rating, while map features that are often mismatched end up with the low rating.

Fig. 3. Feature-based image registration results: the green correspondences are consistent with the results of the histogram voting, and the map features that belong to them will have their ranking increased. On the other hand, the correspondences in red are considered to be incorrect, and ranking of their map features will be decreased. Features with low ranking will be eventually removed from the map. (Color figure online)

The system also rates unmatched 'view' features, i.e. features extracted from the current camera image, according to the distance of their descriptors to the features of the map. In particular, the rank of each view feature equals its distance (in descriptor space) to the nearest feature in the local map. Such a rating corresponds to the feature saliency or uniqueness. Highly-rated view features, which are unlikely to be mismatched and paired incorrectly, are good candidates to be added to the map.

Thus, to adapt the map, the system selects n map features with the lowest ranking (i.e. the ones which are often mismatched) and substitutes them with

n view features with the highest ranking. One of the important questions is the choice of n, which defines how quickly the map adapts to the changes. If the n is low, the map cannot adapt to fast environmental changes, but it's robust to occasional glitches of the image registration, which might result in wrong features being added to the map. An extreme case is $n = 0$, which indicates that the map does not adapt to the environment change at all. If the n is high, the map can adapt to rapid changes. However, since the navigation algorithm does not establish the robot heading with perfect accuracy, the positions of the new features tend to drift, and the map gradually deteriorates. Moreover, any failure of the image registration populates the map with features at wrong positions. An extreme case is $n = f_i$, which means that the system discards all features and creates a completely new local map. The thesis [15] demonstrates that both extremes of $n = 0$ and $n = f_i$ are not suitable for long term navigation.

In our experiments, we set n to the half of the number of correctly-established correspondences, i.e. to the half of the value of the highest bin of the histogram voting. This value ensures that the map gradually adapts to the changes observed. Since n is much lower than the number of features that constitute the highest histogram bin, it makes the map update also prone to occasional failures of the navigation. If the navigation method fails and the features are added to the wrong positions, these wrong features do not cause the method to fail in the subsequent navigation run. Rather, the system will start to decrease their rank, because their positions will not conform to the results of the histogram voting, and these features will be eventually removed from the map. To ensure that the map always adapts to the changes, we set the minimal value of n to 10.

3.5 Map Prediction

The previously presented map adaptation is capable of dealing with gradual environment changes. If the appearance change is significant and abrupt because the robot observes a given location after a large hiatus, no reliable correspondences will be established and the navigation method will fail. However, the visibilities of the features are not random, but they are strongly influenced by processes that exhibit certain temporal properties. For example, day-night, seasonal and vegetation cycles cause certain features to disappear and appear again with known periodicities, and tree growth rate decides the feature persistence. Since the adaptive map already provides the robot with the ability to operate for long time periods, the navigation system has the opportunity to learn about the influence of time on the visibility of the image features in its maps, i.e. to find the aforementioned cycles and map decay rates. In other words, the adaptive map provides sufficient information to create a spatiotemporal representation of the operational environment which allows predicting which features will be visible at which time. Thus, a robot can predict the environment appearance at night even if it would radically differ from the appearance observed during the last robot operation.

To build such a predictive model, we capture and store the visibility information of each map feature along with the time it was (un-)detected. Once we have

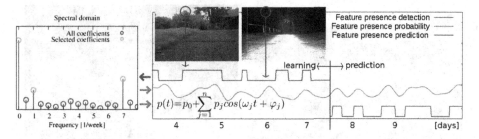

Fig. 4. FreMEn feature map for visual localisation and navigation: The observations of image feature visibility (centre, red) are transferred to the spectral domain (left). The most prominent components of the model (left, green) constitute an analytic expression (centre, bottom) that represents the probability of the feature being visible at a given time (green). This allows predicting the feature visibility at a time when the robot performs visual navigation (blue). Courtesy of [22]. (Color figure online)

enough information, we can start to reason about the feature's persistence (i.e. how likely the feature will be visible in the future) [41], or periodicity [22,23]. In particular, our system employs the concept of the Frequency Map Enhancement (FreMEn) [23], which retrieves the periodicities of the feature visibility through a Fourier transformation-based spectral analysis. The resulting model can predict the likelihood of each feature visibility in the future, which is especially beneficial if the robot visits a given area after a long time [12,23]. The core principle of the FreMEn method is illustrated in the Fig. 4 and at www.fremen.uk.

Thus, every time our robot loads the environment maps for navigation, it creates temporal models of all features and calculates the likelihood of their visibility at the time of its operation. The most likely-to-be-seen features (selected by a greedy or a Monte-Carlo scheme), then constitute the map for a given navigation trial. As explained in [12], the Monte-Carlo scheme is applied to avoid situations, where some features are not detected simply because the system does not place them in the environment map. The main benefit of the FreMEn-based predictions comes from the fact that the FreMEn model captures the cyclic behaviour of the feature visibility caused by the changing position of the main outdoor illuminant, the Sun.

4   Experimental Evaluation

To evaluate how the proposed adaptive methods affect the performance of the visual navigation, we tested them on a CAMELEON ECA tracked robot for difficult terrain. The robot was modified by installing an aluminium superstructure, which contains several mounts for cameras, Intel i3 control laptop and other equipment. In the experiments presented here, we used 768×480 px monochrome images from the left camera of the e-Con TARA stereo device. For night experiments, we installed the Fenix 4000 lumen torch. The robot configuration used in our experiments is shown on Fig. 5. The data gathering took place at Hostibejk

Fig. 5. Cameleon robot configuration during the experimental trials.

Hill in Kralupy nad Vltavou, Czechia, which is a small forest park with one building, see Fig. 5. Near this building, the robot was taught a closed trajectory, and then it traversed the taught path autonomously more than 100 times over the course of one month in various environmental conditions ranging from cloudless days, overcast, light rain, sunset and night. Each time the robot used a local map to determine its steering, it saved its current image and associated it with the given local map. Since the taught path is represented by ∼80 local maps, and the robot traversed the path autonomously more than 100 times, the resulting dataset is composed of more than 8000 images. Approximately one-third of the images contain a small building otherwise, the images associated with the local maps contain trees, shrubs and other close and distant structures. To document the experiments, we provide datasets, codes and videos from the experimental site at http://bearnav.eu.

4.1   Evaluation Metrics

The purpose of the image processing in our system is to register the images from the local maps to the images captured by the robot camera. The horizontal shift between the mapped and perceived images, which is recovered by establishing correspondences between these image features, is used to correct the robot heading. Thus, the more accurate and robust the image registration is, the more precise and reliable visual-based navigation would be. Therefore, the criterion to evaluate the visual navigation pipeline is the accuracy of the image registration.

To do so, we use the dataset images captured during the autonomous drive to simulate the robot movement by 'replaying' them to the robot navigation system. The navigation system uses the histogram voting to calculate the horizontal shift between the dataset images and the images in the local maps in the same way as during normal operation. After that, we compare the calculated shift to the ground truth, obtained by manual annotation. The difference between the shift calculated by the system and the human-generated one is the measure of the method accuracy. Since during each simulated drive, the system registers several hundreds of images, we denote the result of the ith image registration as δ_i (see Sect. 3.1). By comparing the calculated δ_i to the values obtained by human annotation γ_i, we obtain a sequence $\epsilon_i = |\delta_i - \gamma_i|$ which corresponds to the accuracy of the image registration for i^{th} image pair.

Thus, the evaluation of each method or its setting produces another error sequence of ϵ_i. To determine which method or which settings produce more accurate navigation, we compare the values ϵ_i produced by the procedure mentioned above statistically and qualitatively. For statistical evaluation, we apply Student's paired sample test, which is able to determine if an error sequence of ϵ_i^A generated by method A is statistically significantly smaller than another error sequence ϵ_i^B, generated by method B. For qualitative evaluation, we use the values of ϵ_i to calculate a function $p(\epsilon_t)$, which indicates the probability that the registration error was lower than a given threshold, i.e. $p(\epsilon) = P(\epsilon_i \leq \epsilon_t)$. While the first test clearly indicates if one method performs better than the other, displaying $p(\epsilon_t)$ of two different methods inside of the same graph provides one with a better insight how much these methods actually differ.

4.2   Camera Exposure Control

To evaluate the impact of camera exposure adaptation, described in Sect. 3.2, we let the robot traverse the taught path several times with three different exposure settings. To provide a fair experimental setting, we performed the trial during the late afternoon, when the illumination was quite stable. The first, 'standard' setting, used in our system by default, controls the camera exposure to keep the brightness of the upper image half at 0.5. The second, 'full' setting, controls the camera exposure to keep the brightness of the entire image at 0.5. The third, 'fixed' setting, does not perform adaptation. Rather, the operator sets the camera exposure at the beginning of the autonomous drive to a value, which causes the overall contrast of the image to be equal to 0.5. During each traversal, the robot collected 658 images, which were subsequently used for evaluation as described in Sect. 4.1. For examples of the images gathered, see Fig. 6.

While the graph shown in Fig. 7 seems to indicate only the small difference in the performance of the methods, the statistical tests confirmed that the 'standard' setting achieves statistically significantly lower registration errors compared to the 'full' and 'fixed' exposure settings. Later on, we performed the same test during sunset. Here, the robot was not able to navigate properly neither with 'full' nor 'fixed' exposure setting, but it worked with the 'standard' one. In this case, the navigation performance itself proved the superiority of the 'standard' exposure adaptation scheme, and it was not necessary to perform the statistical or qualitative evaluation.

4.3   Feature Detection Adaptation

To evaluate the impact of feature number adaptation, proposed in Sect. 3.3 we used the dataset gathered with the 'standard' exposure setting. We replayed the dataset images through the navigation system as described in Sect. 4.1 using three different schemes of feature detection. The first, 'standard' setting, used

Fig. 6. Different exposure settings: the first image is captured with fixed exposure, the second with exposure control applied to the complete image and the last image was captured with exposure controlled according to the top half part of the images.

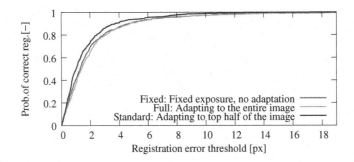

Fig. 7. The probability of registration error being smaller than a given number of pixels for different exposure adaptation schemes during sunset.

in our system by default, controls the Hessian threshold as described in Sect. 3.3 in a way to extract 500 image features from the top image half. The second, 'full' setting, controls the Hessian threshold as described in Sect. 3.3 in a way to extract 500 image features from the entire image. The third, 'fixed' setting, does not adapt the Hessian threshold. Rather, the Hessian threshold is set to hand at the beginning of the simulated autonomous drive to obtain approximately 500 features from the first image. Examples of images with different keypoint detection strategy are shown in Fig. 8.

While the graph, shown in Fig. 9, seems to indicate only a small difference in the performance of the methods, the statistical tests confirmed that the 'standard' Hessian adaptation achieves statistically significantly lower registration errors compared to the 'full' case where the Hessian threshold is adapted to extract features from the entire image as well as to the 'fixed' case, where the Hessian threshold does not adapt but is fixed. Our system used the combination of the upright SURF [1] detector and the BRIEF [4] descriptor. A typical time to extract features from the 768 × 480 pixel images captured by the robot's camera was 33 ms on an Intel i3 laptop, which is used as the robot control PC.

Fig. 8. The different setting of the keypoints detection: in the first image, 500 features are detected only from the top half. In the second image, the same amount is extracted from the whole image, and the last image has a fixed Hessian threshold, which results in feature deficiency.

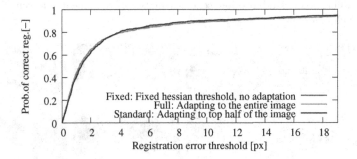

Fig. 9. The probability of registration error being smaller than a given number of pixels for different Hessian threshold adaptation schemes.

4.4   Map Adaptation

To evaluate the effect of map adaptation, we used the same dataset as in Sect. 4.3, i.e. the dataset gathered with the 'standard' exposure setting. Again, we replayed the dataset images through the navigation system as described in Sect. 4.1 using three different map adaptation schemes. The first, 'Adaptive map' adaptation scheme, used in our system by default, gradually exchanges the map features as described in Sect. 3.4. The second, 'Plastic' scheme, discards all features from the map and saves the ones from the current view. This scheme corresponds to setting the n parameter described in Sect. 3.4 to a very high value. The third, 'Static map', does not adapt the map during autonomous traversals, which corresponds to the n parameter being set to 0. The Fig. 10 indicates that the 'Adaptive map' achieves lower registration errors compared to the 'Plastic' adaptation scheme, which is caused by the feature position drift, see [15] for details. Since the testing dataset does not exhibit any significant appearance changes, the performance of the 'Static map' is comparable to the performance of the 'Adaptive' one, which was confirmed by the statistical tests.

To further test the ability of the adaptive mapping to deal with the changing environment appearance, we performed 12 autonomous traversals during sunset, where the robot started its traversals during daylight and ended at full dark. To allow the robot to navigate at night autonomously, we switched on its 4000 lm

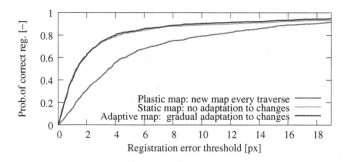

Fig. 10. The probability of registration error being smaller than a given number of pixels for different map adaptation schemes in the environment without significant appearance changes.eps

Fig. 11. Feature matching between the map and the current robot view: The first and the third image shows that the use of 'Adaptive map' results in more correspondences compared to the 'Static map' use, shown in the second and fourth image. The second image pair shows a situation, where the light changed from full day to full night, and the 'Static map', created during the day, becomes obsolete. The 'Adaptive map' has enough correct correspondences due to using information from 'Static map' augmented with features from later traversals.

torch, see Fig. 5. As shown in Fig. 11, the 'Adaptive map' provides more relevant features to the navigation system compared to the 'Static map'. As indicated by the Fig. 12 and the statistical tests performed, the 'Adaptive' and 'Plastic' maps also outperform the 'Static' one in terms of registration accuracy. Note, that the effects of the 'Plastic map' drift, observed in the previous experiment, are insignificant compared to the effects caused by the changes.

4.5   Map Prediction

To evaluate the effect of map prediction, we processed data from 87 autonomous traversals. We divided the dataset into a training set of length 57, which is used to build temporal models of the features and a testing set, which we use for evaluation as described in Sect. 4.1. To predict the features which are going to be visible, we used the FreMEn [23] temporal model to predict the 500 most

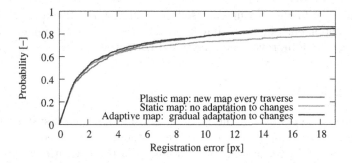

Fig. 12. The probability of registration error being smaller than a given number of pixels for different map adaptation schemes with significant appearance changes.

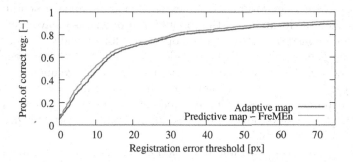

Fig. 13. The probability of registration error being smaller than a given number of pixels with standard and predicted maps.

Fig. 14. Image blur caused by deteriorating lighting during a sharp turn. The left image was captured during the day, while the other ones were captured after sunset.

likely visible features as specified in Sect. 3.5. Then, we simulated the robot drive using the testing dataset, which contains both day and night images.

The difference between the registration error of the system using and not using map prediction is shown in Fig. 13. Statistical testing, performed as specified in Sect. 4.1, confirmed that the map predicted by the FreMEn and Monte Carlo scheme achieved lower registration error.

In comparison with the previous results, a higher error is more probable which was caused by insufficient illumination of late testing drives. The lack of light caused the camera to set longer exposure which caused significant blur in places where the robot had to turn. Figure 14 demonstrates the gradual deterioration of image quality in the testing set used. The images are taken from the same part of the robot path at a different time. Despite the blurry images, the map prediction method improved the directional correction.

5 Conclusion

We have presented adaptative methods for a visual navigation system, intended for robots that are supposed to operate in outdoor environments for extended periods of time. The purpose of these methods is to deal with the appearance changes commonly occurring outdoors due to lighting variations and natural processes which affect the environment structure. In a series of experiments, performed on a real robot over the course of several weeks, we demonstrate that adaptation of camera exposure, feature extraction and map representation has a positive impact on the ability of the robots to autonomously navigate outdoors. Finally, we demonstrated that a robot operating for extended periods of time could acquire enough observations to understand how the environment changes over time and use this knowledge to predict the future environment appearance, which further improves the efficiency and reliability of its operation. To ensure the reproducibility of the research presented here, we provide the system's source codes as well as access to the datasets used for evaluation at https://github.com/gestom/stroll_bearnav/wiki.

One of the most surprising findings was that use of the plastic map adaptation scheme resulted in a fast deterioration of the map quality, see Sect. 4.4. We hypothesise that this effect was not reported before because other researches either evaluate their map adaptation methods without considering the fact that the vision is actually used to guide the vehicle, or do not traverse the same trajectory as many times as we did in our experimental evaluations.

We believe that the adaptive and predictive capabilities of environment representations are crucial to achieving the long-term autonomy of mobile robots. Thus, our future research will aim at gradual map adaptation and temporal models for prediction of the future environment appearance.

References

1. Bay, H., Ess, A., Tuytelaars, T., Van Gool, L.: Speeded-up robust features (SURF). Comput. Vis. Image Underst. **110**(3), 346–359 (2008)
2. Biber, P., Duckett, T.: Dynamic maps for long-term operation of mobile service robots. In: RSS (2005)
3. Blanc, G., Mezouar, Y., Martinet, P.: Indoor navigation of a wheeled mobile robot along visual routes. In: International Conference on Robotics and Automation (ICRA) (2005)
4. Calonder, M., Lepetit, V., Strecha, C., Fua, P.: BRIEF: binary robust independent elementary features. In: Proceedings of the ICCV (2010)
5. Carlevaris-Bianco, N., Eustice, R.M.: Learning visual feature descriptors for dynamic lighting conditions. In: IROS. IEEE (2014)

6. Chen, Z., Birchfield, S.T.: Qualitative vision-based path following. IEEE Trans. Rob. Autom. **25**(3), 749–754 (2009)
7. Churchill, W.S., Newman, P.: Experience-based navigation for long-term localisation. IJRR **32**(14), 1645–1661 (2013). https://doi.org/10.1177/0278364913499193
8. Dayoub, F., Duckett, T.: An adaptive appearance-based map for long-term topological localization of mobile robots. In: IROS (2008)
9. De Cristóforis, P., et al.: Real-time monocular image-based path detection. J. Real Time Image Process. **11**, 335–348 (2013)
10. Debevec, P.E., Malik, J.: Recovering high dynamic range radiance maps from photographs. In: SIGGRAPH. ACM (2008)
11. DeSouza, G.N., Kak, A.C.: Vision for mobile robot navigation: A survey. IEEE Trans. Pattern Anal. Mach. Intell. **24**(2), 237–267 (2002). https://doi.org/10.1109/34.982903
12. Dvořáková, E.: Temporal models for mobile robot visual navigation. B.S. thesis, Czech Technical Univerzity in Prague (2018)
13. Engel, J., Schöps, T., Cremers, D.: LSD-SLAM: large-scale direct monocular SLAM. In: Fleet, D., Pajdla, T., Schiele, B., Tuytelaars, T. (eds.) ECCV 2014. LNCS, vol. 8690, pp. 834–849. Springer, Cham (2014). https://doi.org/10.1007/978-3-319-10605-2_54
14. Gadd, M., Newman, P.: Checkout my map: Version control for fleetwide visual localisation. In: IROS. IEEE/RSJ (2016)
15. Halodová, L.: Map management for long-term navigation of mobile robots. Bachelor thesis, Czech Technical University, May 2018
16. Halodová, L., Krajník, T.: Exposure setting for visual navigation of mobile robots. In: Student Conference on Planning in AI and Robotics (PAIR) (2017)
17. Holmes, S., Klein, G., Murray, D.W.: A square root unscented Kalman filter for visual monoSLAM. In: International Conference on Robotics and Automation (ICRA) (2008)
18. Kosaka, A., Kak, A.C.: Fast vision-guided mobile robot navigation using model-based reasoning and prediction of uncertainties. CVGIP: Image Underst. **56**(3), 271–329 (1992)
19. Krajník, T., Cristóforis, P., Nitsche, M., Kusumam, K., Duckett, T.: Image features and seasons revisited. In: European Conference on Mobile Robots (ECMR) (2015)
20. Krajník, T., Majer, F., Halodová, L., Vintr, T.: Navigation without localisation: reliable teach and repeat based on the convergence theorem. In: IROS (2018)
21. Krajník, T., et al.: Image features for visual teach-and-repeat navigation in changing environments. Rob. Auton. Syst. **88**, 127–141 (2017)
22. Krajník, T., et al.: Long-term topological localization for service robots in dynamic environments using spectral maps. In: IROS (2014)
23. Krajník, T., et al.: FreMEN: frequency map enhancement for long-term mobile robot autonomy in changing environments. IEEE Trans. Rob. **33**(4), 964–977 (2017)
24. Kunze, L., Hawes, N., Duckett, T., Hanheide, M., Krajnik, T.: Artificial intelligence for long-term robot autonomy: a survey. IEEE RAL **3**(4), 4023–4030 (2018). https://doi.org/10.1109/LRA.2018.2860628
25. Latif, Y., Garg, R., Milford, M., Reid, I.: Addressing challenging place recognition tasks using generative adversarial networks. In: ICRA (2018)
26. Linegar, C., Churchill, W., Newman, P.: Work smart, not hard: recalling relevant experiences for vast-scale but time-constrained localisation. In: ICRA (2015)
27. Lowe, D.G.: Distinctive image features from scale-invariant keypoints. Int. J. Comput. Vis. **60**(2), 91–110 (2004)

28. Lowry, S., Milford, M.J.: Supervised and unsupervised linear learning techniques for visual place recognition in changing environments. IEEE T-RO **32**(3), 600–613 (2016)
29. Lowry, S., et al.: Visual place recognition: a survey. IEEE T-RO **32**(1), 1–19 (2016)
30. Lu, H., Zhang, H., Yang, S., Zheng, Z.: Camera parameters auto-adjusting technique for robust robot vision. In: ICRA. IEEE (2010)
31. Mair, E., Hager, G.D., Burschka, D., Suppa, M., Hirzinger, G.: Adaptive and generic corner detection based on the accelerated segment test. In: Daniilidis, K., Maragos, P., Paragios, N. (eds.) ECCV 2010. LNCS, vol. 6312, pp. 183–196. Springer, Heidelberg (2010). https://doi.org/10.1007/978-3-642-15552-9_14
32. Majer, F., Halodová, L., Krajník, T.: Source codes: bearing-only navigation. http://bearnav.eu
33. Majer, F., et al.: A versatile visual navigation system for outdoor autonomous vehicles. In: Modeling and Simulation for Autonomous Systems (2018, in review)
34. Matsumoto, Y., Inaba, M., Inoue, H.: Visual navigation using view-sequenced route representation. In: International Conference on Robotics and Automation (ICRA) (1996)
35. Mühlfellner, P., Bürki, M., Bosse, M., Derendarz, W., Philippsen, R., Furgale, P.: Summary maps for lifelong visual localization. J. Field Rob. **33**(5), 561–590 (2016)
36. Mur-Artal, R., Montiel, J.M.M., Tardós, J.D.: ORB-SLAM: a versatile and accurate monocular SLAM system. IEEE Trans. Rob. **31**(5), 1147–1163 (2015). https://doi.org/10.1109/TRO.2015.2463671
37. Neubert, P., Sunderhauf, N., Protzel, P.: Appearance change prediction for long-term navigation across seasons. In: European Conference on Mobile Robotics (2013)
38. Neves, A.J.R., Cunha, B., Pinho, A.J., Pinheiro, I.: Autonomous configuration of parameters in robotic digital cameras. In: Araujo, H., Mendonça, A.M., Pinho, A.J., Torres, M.I. (eds.) IbPRIA 2009. LNCS, vol. 5524, pp. 80–87. Springer, Heidelberg (2009). https://doi.org/10.1007/978-3-642-02172-5_12
39. Paton, M., MacTavish, K., Berczi, L.-P., van Es, S.K., Barfoot, T.D.: I can see for miles and miles: an extended field test of visual teach and repeat 2.0. In: Hutter, M., Siegwart, R. (eds.) Field and Service Robotics. SPAR, vol. 5, pp. 415–431. Springer, Cham (2018). https://doi.org/10.1007/978-3-319-67361-5_27
40. Porav, H., Maddern, W., Newman, P.: Adversarial training for adverse conditions: robust metric localisation using appearance transfer. In: ICRA (2018)
41. Rosen, D.M., Mason, J., Leonard, J.J.: Towards lifelong feature-based mapping in semi-static environments. In: ICRA. IEEE (2016)
42. Royer, E., Lhuillier, M., Dhome, M., Lavest, J.M.: Monocular vision for mobile robot localization and autonomous navigation. Int. J. Comput. Vis. **74**(3), 237–260 (2007)
43. Segvic, S., Remazeilles, A., Diosi, A., Chaumette, F.: Large scale vision based navigation without an accurate global reconstruction. In: CVPR (2007)
44. Shim, I., Lee, J.Y., Kweon, I.S.: Auto-adjusting camera exposure for outdoor robotics using gradient information. In: IROS. IEEE/RSJ (2014)
45. Krajník, T., Faigl, J., Vonásek, V., et al.: Simple, yet stable bearing-only navigation. J. Field Rob. **27**(5), 511–533 (2010)
46. Zhang, N., Warren, M., Barfoot, T.: Learning place-and-time-dependent binary descriptors for long-term visual localization. In: ICRA. IEEE (2016)
47. Zhang, Z., Forster, C., Scaramuzza, D.: Active exposure control for robust visual odometry in HDR environments. In: ICRA (2017)

Interaction with Collaborative Robot Using 2D and TOF Camera

Aleš Vysocký$^{(\boxtimes)}$, Robert Pastor, and Petr Novák

Faculty of Mechanical Engineering, Department of Robotics,
VŠB-Technical University Ostrava,
17. listopadu 15, 708 33 Ostrava, Czech Republic
{ales.vysocky, robert.pastor, petr.novak}@vsb.cz

Abstract. With increasing count of applications with collaborative robots it is important to be able to set up conditions for controlling the robot and its interaction with human operator. This article describes a quick response system detecting hands of the operator using 2D and TOF camera in the shared operator-robot workspace. The technology provides data about position of hands and gestures which is used to instant reactions of the robot on presence of the operator and to control the robot. Interaction with robot with gestures is for the operator more intuitive with no need of expertise knowledge. Operator can use both hands, so the operation is more efficient and faster. This system was tested on prototype workplace and results are evaluated in this article.

Keywords: Cobot · HMI · TOF camera · Gesture control · HRC · Industrial robot

1 Introduction

With rapidly growing market of robots used in industrial applications there is also a great increase of deployed cobots. Prognoses and forecasts reports even bigger expansion of robot technology [1]. According to reference projects of robotic companies [2], collaborative robots are mostly used in manipulation tasks in coexistence manner of human-robot cooperation [3]. Typical applications are machine tending workplaces with a cobot. In this kind of application have cobots advantages during pre-production phase, because pick and place positions can be programmed with hand guiding of the robot. In the production phase are utilized sensitive abilities of the robot for adaptation on actual conditions [4] (inaccuracy of the object of manipulation or wear and tear of the clamping device). There is no need of cooperation with human in production phase. This is analogous in quality inspection tasks with a camera mounted to the robot, technological tasks like polishing, glue dispensing applications or bin picking.

This paper focuses on applications where robot and operator share the workplace and are working on the same component in the time of production. Example of such an application is assembly, packing operations or positioning and handling. Cooperation workplaces are not that common in industry and one of the reasons might be a lack of devices supporting effective interaction between robot and the operator. In some

© Springer Nature Switzerland AG 2019
J. Mazal (Ed.): MESAS 2018, LNCS 11472, pp. 477–489, 2019.
https://doi.org/10.1007/978-3-030-14984-0_35

existing applications the operator is waiting till the robot finish its operation, then he carries out his task and enable the robot to continue in its cycle. Robot application is limited to simple commands for stopping the robot (emergency stop button) and resume in program.

More effective control interfaces could be set up with modern detection and display technologies. The most essential way of control by an operator is when he does not have to use any device. Using gestures and voice commands [5] is natural for a human operator. Gesture control becomes widely used not only in gaming applications, but car infotainment can be controlled by gestures, devices to control computer applications [6] are available with sophisticated SDKs. With this kind of devices could be developed various applications for navigation in 3D space [7] or 3D scene scanning and reconstruction [8].

3D sensors used in this article are based on TOF (Time of Flight) technology. Which has a higher accuracy [9] and results are less influenced with environment conditions than technologies based on structured light projection and triangulation.

Several techniques of hand gesture recognition were proposed and evaluated [10]. Gestures are detected from image – 2D, depth data [11], marker position data, or are recognized on non-image basis. As a source image can be captured the scene and detected hands in the environment or special cases like wrist camera [12]. Gestures are detected from the image based on object finding and processing or based on neural network [13, 14]. Object recognition begins with blob analysis [15], shape detection and classification [16]. Hands are found in the image using skin color recognition [17]. When hands are separated from the scene, processing can be done. OpenCV image processing library is using contour finding which is input for convex hull definition [18]. Fingertips are then edge points of the convex hull and points most distant from the hull are inner points of the hand. With combination of relative positions of detected points, several gestures can be recognized.

Recognition based on neural networks provides robust results [19] in gesture classification. Deep learning can be useful in working with larger set of gestures and also for detecting not only hands but other features of environment with reasonable in operation computational time.

Gestures are static where the actual state of the hand is detected and evaluated. Dynamic gestures [20] can provide more information and except the static data we have information about direction of the gesture, speed or change of gesture in time. For dynamic gesture recognition and prediction Hidden Markov Models, Dynamic Bayesian networks and Dynamic Time Warping are used.

Non-image gesture recognition is based on electromyography [21], where electric signals are based on muscle activity. Other technologies relate to special gloves provided with different sensors (resistive, shape memory, accelerometers, strain gauge [22]). Advantage of non-image recognition is working environment conditions independence, wider space usability (camera angle of view independent), but additional accessory can limit the operator in his work and can be uncomfortable.

2   Experimental Workplace

Application is developed on a simple collaborative workplace which is used for teaching at our department. This station consists of table with mounted robot and working desk. The robot is Universal Robots UR3 [23] with maximal reach of 500 mm and payload 3 kg. This robot is extended with force-torque sensor for more precise sensitivity and AFAG EU20 electric gripper with 3D printed jaws designed for manipulation with wooden cubes. 5 cubes in a foam stack represent five objects of manipulation.

2.1   Interaction with Buttons

On the right side of the working desk is situated an emergency stop button which is an essential element of the workplace. With this button operator immediately stops the manipulator and all peripheral devices. In the left corner is placed a universal button connected to digital input of the robot.

With a single button operator can confirm that he finished his part of an operation and robot can start/resume its program. Except pressing a button this kind of simple interaction could be also done by abandoning the work zone or by touching the robot body in a specific direction. Solution of this simplicity limits possible intervention to the production process. In a simple application like this could the input device consist of 5 buttons for every cube and robot would move to preprogrammed positions. This solution is strongly dedicated to specific operation and scene.

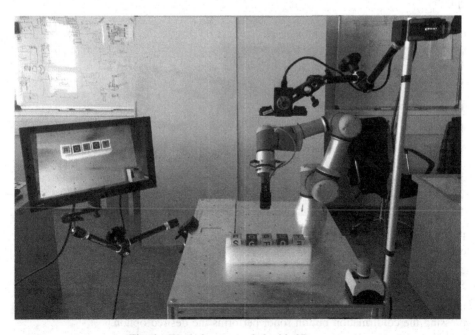

Fig. 1. Workplace extended with 3D camera

Simple robot workplace was during experiments extended with several user interface. For scenarios with mechanical devices were added buttons and joystick. For camera mounting, the mast with universal holder was attached to the table. Touch screen for either interaction and feedback was attached also with universal holder.

Scenarios with buttons and 2D industrial camera worked just with robot's control unit. For 3D image processing and gesture recognition the workplace had to be extended with external computer.

2.2 Scene Detection

For interaction with robot in a changing scene it is necessary to scan and recognize the scene. The workplace is extended with 2D camera which is situated above the work-place and is focused on working desk. The camera is a monochrome COGNEX In-Sight 8000 Vision system [24]. In-sight vision system is capable of wide range of 2D image processing functions optimized for industrial applications. Localization functions like edge detection, blob location, pattern matching and inspection functions like measurement, text/codes reading, flaw detection and many other are available directly in the camera and there is no need for any additional hardware.

Processed scene is projected on a touch screen monitor. Because of testing different types of user input, camera software In-sight built-in functions are not used and the image from the camera is processed in an application running on external computer. In the acquired image desired features are detected with OpenCV library functions. For this basic example square shaped objects are found in the scene and a yellow border is drawn around them (Fig. 2). The middle cube is highlighted with blue color as a default option.

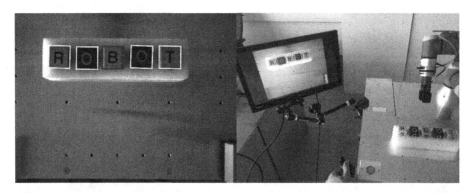

Fig. 2. Cubes detected in 2D image (left), cube selection with joystick (right) (Color figure online)

For object selection a joystick device is placed to the right corner next to confir-mation button. With this device operator can choose from detected objects and after pressing the confirmation button robot performs the desired option.

In this simple example with 5 exactly positioned objects joystick selection is user friendly, with increasing complexity of the scene and number and variants of selectable components this scenario becomes insufficient. More irregular disposition of detected objects was tested with targeting mounting holes on the working desk. Navigation with joystick is comfortable in one dimension or a regular matrix, with irregular disposition is joystick less effective. Satisfactory option is direct selection on the touch screen.

If objects do not have defined position or the number of positions is higher, programming of pick-positions of the robot becomes time-consuming. 2D image can provide X and Y coordinates of the object on the working desk which are provided to controller of the robot and this can manipulate its gripper over the object. To synchronize positions generated by recognition software with the robot, calibration must be done. There are several types of camera-robot calibration. With the marker connected to the robot gripper, this could be found in camera image and coordinates can be synchronized [25]. OpenCV offers camera calibration using a chessboard [26]. This method provides position calibration and elimination of camera distortion. Chessboard is precisely placed on the working desk with specific offset from mounting position of the robot.

Recognition application in the example application is tuned for basic shape finding based on contours detection. For real applications this will be extended with pattern matching functions for standard shapes like screw heads. After specification of the production operation detection of requested objects can be done by deep learning and use of convolutional neural network [27].

3   Gesture Control

Gestures and voice commands allow operator to control the robot in the most intuitive way [28]. Operator can give commands with empty hands and be ready for his task in the process. With more operations dependent on optical technologies must be its limitations taken in account:

- Steady light conditions/possibility of calibration and adaptation
- Reflectivity of materials
- Vibrations and rigid fastening of the camera
- Focal distance and focus capability
- Dust, mist/smoke and pollution effecting the lens.

3.1   3D Sensor

Hand localization and gesture recognition is possible using 2D image [29], but more robust is using a combination of 2D and 3D data from ToF camera [30]. Experimental workplace is extended for gesture recognition with 3D sensor. For 3D scanning of the scene the sensor must provide the depth data. Typical technologies used in available sensors are IR-light grid projection and processing (LEAP, Kinect) or Time of Flight (Intel RealSense, Kinect v2).

Experimental workplace was at the first time equipped with Kinect v2 which supports body and hand recognition. The system is optimized for Xbox console and tracking the whole body. After some testing the workplace was equipped with Creative BlasterX SENZ3D sensor (Fig. 1) based on Intel RealSense technology. This sensor provides 720p (1280 × 720) color image and VGA (640 × 480) depth data at 60FPS. Sound can be captured with dual microphone. Advantage of the sensor is SDK for implementing functions in application and functions directly prepared for hand tracking and gesture recognition [31].

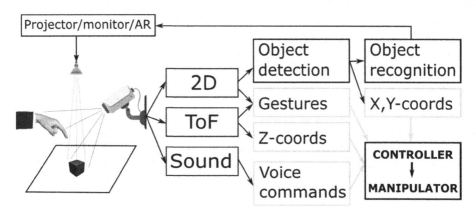

Fig. 3. Interaction diagram (Color figure online)

Interaction diagram provides basic workplace concept (Fig. 3). Color stream from the camera is processed similarly like in previous scenario with 2D camera. Objects are detected, recognized and visualized to the operator as a feedback. Information can be visualized on the monitor or can be projected back to scene. Projection can bring a recurrent problem with object detection in next scene capture because the scene is affected by projected information. Next option is using of virtual reality which can be used to visualize scanned scene and feedback information to the head mounted display [32, 33]. Virtual reality in industrial environment could be dangerous because an operator must be aware of his surroundings. Better solution can be use of augmented reality [34] which can add the feedback information to the operator's field of vision.

2D image is used also in combination with depth data to detect hands of an operator and detect static and dynamic gestures. In this application are recognized and processed 3 basic gestures. Pointing gesture for selection (Fig. 4), thumbs up gesture for confirmation (Fig. 6) and fist gesture to soft stop of the robot movement (Fig. 7). Depth map data in the form of point cloud can be used for pose recognition, 3D model matching and other functions which are not yet implemented in the application.

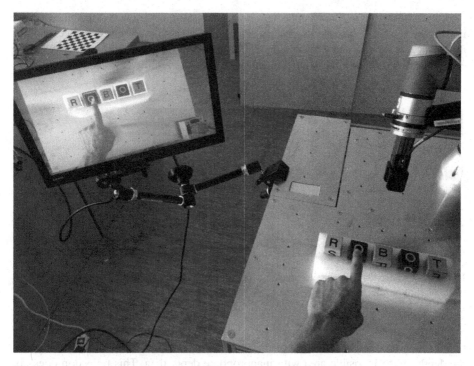

Fig. 4. Selection with pointing gesture

If the operator needs to use both hands for his task, gesture demonstration can limit him. In this case the system has a speech recognition subsystem to enable voice commands [35]. In this experimental workplace only "STOP" command is implemented which has the same function as fist gesture.

3.2   Selection of Option

When the application starts, cubes in the scene are recognized and yellow borders are drawn to the image. Yellow boxes are representing selectable options. When the end enters the space above the working desk and is calibrated (SDK supports calibration of spread fingers palm), application is waiting for pointing finger gesture. When the gesture is detected a green dot is drown to the extremity point of the gesture which is located at the end of the finger. When the point enters the selectable area, border is highlighted with red color.

In the simple application was not counted with multiselect option. For selecting multiple objects, it is possible to confirm every selection or select item with longer presence of the finger above the selection border.

Processed image with borders around cubes is displayed on the monitor for operator, gesture is recognized from color image and depth image (Fig. 5). Depth image is grayscale image in its raw form, in the figure colorization of depth layers is used for better clarity.

During tests two problems appeared. Emergency stop button was detected as a hand in fist gesture. This is caused by gesture recognition algorithm. Solution is disruption of the shape. This was done by adding 2 black stripes on the button which optically changed the button and the software did not detect the button as a hand.

Fig. 5. 2D image with detected cubes and pointing finger (left), colored depth image (right) (Color figure online)

Second problem is with glossy surface of the working desk. Desk is made of polished aluminum and reflection of the surface is problematic for ToF technology. In the depth image is visible area with inappropriate depth data. This reflection does not affect gesture recognition but influence depth information for future processing.

3.3 Confirmation Gesture

Selection procedure is finished when thumbs-up gesture is detected. On the display the selected item highlight is switched from red to blue and the information about the selected object is sent to the robot. With confirmation gesture movement of the robot is started.

Fig. 6. Thumbs up gesture for confirmation (Color figure online)

Application is communicating with robot via socket communication. Requested target data is sent to the robot and after confirmation the robot program is started. If the operator must stop or suspend the robot movement, except of pressing safety emergency button which turns the robot to emergency stop (robot drives are disconnected from power and breaks are activated, operator must confirm the state and restart the robot) application provides soft stop. Soft stop is activated either with fist gesture or with voice command "STOP". This command is sent to dashboard server of the robot.

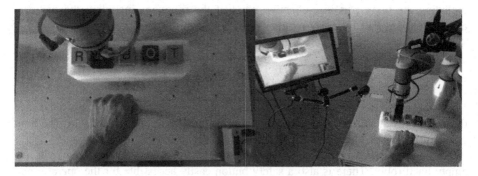

Fig. 7. Fist gesture to stop the robot

Communication with robot is settled over Ethernet TCP/IP protocol. Universal Robots controller provides different types of client access:

- **Primary/secondary interface** – provides and receives robot state data and commands with 10 Hz update rate
- **Realtime interface** – like primary/secondary interface but update rate is 125 Hz
- **Dashboard server** – simple commands to control the robot control application (GUI)
- **Socket communication** – exchange of various messages via socket communication
- **XML-RPC – Remote Procedure call using XML**
- **RTDE (Real-Time Data Exchange)** – transmit of custom state data with 125 Hz update rate.

4 Safety

This workplace permits direct contact of robot arm and the operator during robot motion. The motion must fulfill safety restrictions emerging from the safety standard ISO TS 15066 [36] which has recommendations for safe using of collaborative robots. This workplace is designed for front-side access of the worker. Operator can have arms and fingers in the shared workspace during robot motion. In the standard are values of maximal allowable transient contact conditions. The speed of the robot must be limited so that the maximal relative speed between robot and the operator does not exceed the $v_{rel,max}$ (1).

$$v_{rel,max} = \frac{F_{max}}{\sqrt{\mu \cdot k}} = \frac{p_{max} \cdot A}{\sqrt{\mu \cdot k}} \qquad (1)$$

where F_{max} is a maximal contact force for specific body region. Contact force depends on maximal permissible pressure which is based on research of comfortable collisions and the contact area. Formula (2) gives μ which is a reduced mass of two-body system including the body part m_H, effective mass of the robot m_R including the payload. Spring constant k of the body region can be found in the standard.

$$\mu = \left(\frac{1}{m_H} + \frac{1}{m_R} \right)^{-1} \qquad (2)$$

In the tested scenario operator's hands could be outside the robot-operated area during its movement or rested in monitored predefined positions on the table. The limit speed of the robot towards the rested operator is than directly the calculated $v_{rel,max}$. According to these equations different limits are set for robot carrying a payload and empty hand robot. There is also a safety button easily accessible for the operator and robot has preset movement boundaries so that it cannot carry out unpredictable movement to the operator or into the working desk.

5 Conclusions

Experimental workplace was developed for testing human robot interaction in gradually more universal and intuitive scenarios. Basic interaction with simple input device such as button provides limited possibilities for workplace control. For more complex task like assembly or disassembly of various objects or orientation in indefinite environment user must have an option to adapt the process. Adaptation relates to recognition of the scene and objects of interest. With recognized scene user can control the robot according to actual situation at the workplace. This workplace robotic workplace was designed for education of robot control and technologies testing. Results will be processed into learning scripts and used in lectures.

Gesture control was tested with 3D camera and its SDK in combination with application for robot communication and control. Using gestures to interpret simple commands and for navigation in workspace appeared intuitive for the operator. With using of optical technologies many limiting and influencing factors come to account and the application must be optimized for individual scene to reach desired robustness.

Future Work. This is a prototype workplace prepared for further testing and experiments. The focus in next research will be to make the workplace robust for different environment and conditions. Statistical analysis of application behavior and long-time human robot interaction will be done in next part of the project. Simple cube objects representing detectable entities will be substituted for demanded shapes with use of advanced detection and recognition tools.

Acknowledgements. This article has been elaborated in the research project Research Centre of Advanced Mechatronic Systems, reg. no. CZ.02.1.01/0.0/0.0/16_019/0000867 in the frame of the Operational Program Research, Development and Education, project "Coboty – vývoj periferií" HS3541801 in cooperation with Moravskoslezský automobilový klastr, z.s. This article has been supported by specific research project SP2018/86 and financed by the state budget of the Czech Republic.

References

1. Allied Market Research: Robotics Technology Market by Type (Industrial Robots, Service Robots, Mobile Robots and Others) and Application (Defense and Security, Aerospace, Automotive, Domestic and Electronics) Global Opportunity Analysis and Industry Forecast, 2013–2020 (2019). https://www.alliedmarketresearch.com/robotics-technology-market. Accessed 23 Jan 2019
2. Universal-robots.com: Applications for collaborative robot arms—Universal Robots (2019). https://www.universal-robots.com/applications/. Accessed 23 Jan 2019
3. Vysocky, A., Novak, P.: Human – robot collaboration in industry. MM Sci. J. **2016**(02), 903–906 (2016). https://doi.org/10.17973/mmsj.2016_06_201611
4. Francis, S.: Collaborative robotic system makes 'monotonous and physically demanding tasks' at BMW easier. Robotics & Automation News (2019). https://roboticsandautomationnews.com/2017/06/15/collaborative-robotic-system-makes-monotonous-and-physically-demanding-tasks-at-bmw-easier/12889/. Accessed 23 Jan 2019
5. Han, X., Rashid, M.: Gesture and voice control of Internet of Things. In: 2016 IEEE 11th Conference on Industrial Electronics and Applications (ICIEA) (2016). https://doi.org/10.1109/iciea.2016.7603877
6. Marin, G., Dominio, F., Zanuttigh, P.: Hand gesture recognition with leap motion and kinect devices. In: 2014 IEEE International Conference on Image Processing (ICIP) (2014). https://doi.org/10.1109/icip.2014.7025313
7. Hodicky, J., Frantis, P.: Gesture and body movement recognition in the military decision support system. In: Proceedings of the 9th International Conference on Informatics in Control, Automation and Robotics, pp. 301–304 (2012). https://doi.org/10.5220/0003971903010304
8. Das, A., Murmann, D., Cohrn, K., Raskar, R.: A method for rapid 3D scanning and replication of large paleontological specimens. PLoS ONE **12**(7), e0179264 (2017). https://doi.org/10.1371/journal.pone.0179264
9. Wasenmüller, O., Stricker, D.: Comparison of kinect V1 and V2 depth images in terms of accuracy and precision. In: Chen, C.-S., Lu, J., Ma, K.-K. (eds.) ACCV 2016. LNCS, vol. 10117, pp. 34–45. Springer, Cham (2017). https://doi.org/10.1007/978-3-319-54427-4_3
10. Liu, H., Wang, L.: Gesture recognition for human-robot collaboration: a review. Int. J. Ind. Ergon. **68**, 355–367 (2018). https://doi.org/10.1016/j.ergon.2017.02.004
11. Xu, C., Cheng, L.: Efficient hand pose estimation from a single depth image. In: 2013 IEEE International Conference on Computer Vision, pp. 3456–3462 (2013). https://doi.org/10.1109/iccv.2013.429
12. Chen, F., Deng, J., Pang, Z., Baghaei Nejad, M., Yang, H., Yang, G.: Finger angle-based hand gesture recognition for smart infrastructure using wearable wrist-worn camera. Appl. Sci. **8**(3), 369 (2018). https://doi.org/10.3390/app8030369
13. Chaudhary, A., Raheja, J., Das, K., Raheja, S.: Intelligent approaches to interact with machines using hand gesture recognition in natural way: a survey. Int. J. Comput. Sci. Eng. Surv. **2**(1), 122–133 (2011). https://doi.org/10.5121/ijcses.2011.2109

14. Jalab, H.: Static hand gesture recognition for human computer interaction. Inf. Technol. J. **11** (9), 1265–1271 (2012). https://doi.org/10.3923/itj.2012.1265.1271
15. Ganokratanaa, T., Pumrin, S.: The vision-based hand gesture recognition using blob analysis. In: 2017 International Conference on Digital Arts, Media and Technology (ICDAMT), pp. 336–341 (2017). https://doi.org/10.1109/icdamt.2017.7904987
16. Nalepa, J., Kawulok, M.: Fast and accurate hand shape classification. In: Kozielski, S., Mrozek, D., Kasprowski, P., Małysiak-Mrozek, B., Kostrzewa, D. (eds.) BDAS 2014. CCIS, vol. 424, pp. 364–373. Springer, Cham (2014). https://doi.org/10.1007/978-3-319-06932-6_35
17. Kawulok, M., Kawulok, J., Nalepa, J., Smolka, B.: Self-adaptive algorithm for segmenting skin regions. EURASIP J. Adv. Signal Process. **2014**(1), 170 (2014). https://doi.org/10.1186/1687-6180-2014-170
18. Mesbahi, S., Mahraz, M., Riffi, J., Tairi, H.: Hand gesture recognition based on convexity approach and background subtraction. In: 2018 International Conference on Intelligent Systems and Computer Vision (ISCV), pp. 1–5 (2018). https://doi.org/10.1109/isacv.2018.8354074
19. Oyedotun, O., Khashman, A.: Deep learning in vision-based static hand gesture recognition. Neural Comput. Appl. **28**(12), 3941–3951 (2016). https://doi.org/10.1007/s00521-016-2294-8
20. Barros, P., Maciel-Junior, N., Fernandes, B., Bezerra, B., Fernandes, S.: A dynamic gesture recognition and prediction system using the convexity approach. Comput. Vis. Image Underst. **155**, 139–149 (2017). https://doi.org/10.1016/j.cviu.2016.10.006
21. Kerber, F., Puhl, M., Krüger, A.: User-independent real-time hand gesture recognition based on surface electromyography. In: Proceedings of the 19th International Conference on Human-Computer Interaction with Mobile Devices and Services - MobileHCI 2017, pp. 1–7 (2017). https://doi.org/10.1145/3098279.3098553
22. Su, Z., et al.: Microsphere-assisted robust epidermal strain gauge for static and dynamic gesture recognition. Small **13**(47), 1702108 (2017). https://doi.org/10.1002/smll.201702108
23. Universal-robots.com: UR3 collaborative table-top robot arm that automates almost anything (2019). https://www.universal-robots.com/products/ur3-robot/. Accessed 23 Jan 2019
24. Cognex.com: In-Sight 8000 Vision Systems (2019). https://www.cognex.com/products/machine-vision/2d-machine-vision-systems/in-sight-8000-series. Accessed 23 Jan 2019
25. Ilonen, J., Kyrki, V.: Robust robot-camera calibration. In: 2011 15th International Conference on Advanced Robotics (ICAR), pp. 67–74 (2011). https://doi.org/10.1109/icar.2011.6088553
26. opencv.org: Camera Calibration—OpenCV 3.0.0-dev documentation (2019). https://docs.opencv.org/3.0-beta/doc/py_tutorials/py_calib3d/py_calibration/py_calibration.html. Accessed 23 Jan 2019
27. Grover, P.: Evolution of Object Detection and Localization Algorithms. Towards Data Science (2019). https://towardsdatascience.com/evolution-of-object-detection-and-localization-algorithms-e241021d8bad. Accessed 23 Jan 2019
28. Rogalla, O., Ehrenmann, M., Zollner, R., Becher, R., Dillmann, R.: Using gesture and speech control for commanding a robot assistant. In: Proceedings 11th IEEE International Workshop on Robot and Human Interactive Communication, pp. 454–459 (2002). https://doi.org/10.1109/roman.2002.1045664
29. Malima, A., Ozgur, E., Cetin, M.: A fast algorithm for vision-based hand gesture recognition for robot control. In: 2006 IEEE 14th Signal Processing and Communications Applications, pp. 1–4 (2006). https://doi.org/10.1109/siu.2006.1659822

30. Arachchi, S., Hakim, N., Hsu, H., Klimenko, S., Shih, T.: Real-time static and dynamic gesture recognition using mixed space features for 3D virtual world's interactions. In: 2018 32nd International Conference on Advanced Information Networking and Applications Workshops (WAINA), pp. 627–632 (2018). https://doi.org/10.1109/waina.2018.00157

31. Software.intel.com: Intel® RealSense™ SDK, Hand Tracking Tutorial (2019). https://software.intel.com/sites/default/files/Hand_Tracking.pdf. Accessed 23 Jan 2019

32. Frantis, P., Hodicky, J.: Human machine interface in command and control system. In: 2010 IEEE International Conference on Virtual Environments, Human-Computer Interfaces and Measurement Systems, pp. 38–41 (2010). https://doi.org/10.1109/vecims.2010.5609345

33. Frantis, P., Hodicky, J.: Virtual reality in presentation layer of C3I system. In Proceedings of International Congress on Modelling and Simulation: Advances and Applications for Management and Decision Making, MODSIM, pp. 3045–3050 (2005)

34. Kot, T., Novák, P., Babjak, J.: Application of augmented reality in mobile robot teleoperation. In: Mazal, J. (ed.) MESAS 2017. LNCS, vol. 10756, pp. 223–236. Springer, Cham (2018). https://doi.org/10.1007/978-3-319-76072-8_16

35. Holada, M., Pelc, M.: The robot voice-control system with interactive learning. In: New Developments in Robotics Automation and Control (2008). https://doi.org/10.5772/6284

36. ISO/TS 15066:2016: Robots and robotic devices, collaborative robots (2016)

Analysis of Tensor-Based Image Segmentation Using Echo State Networks

Charles Donkor[1], Emmanuel Sam[2], and Sebastián Basterrech[3]([✉]) [iD]

[1] Department of Computer Science, University of Cape Coast, Cape Coast, Ghana
Cdonkor@knust.edu.gh
[2] School of Computing and Technology, Wisconsin International University College,
Accra, Ghana
elsam@wiuc-ghana.edu.gh
[3] Department of Computer Science, Faculty of Electrical Engineering,
Czech Technical University, Prague, Czech Republic
Sebastian.Basterrech@fel.cvut.cz

Abstract. In pixel classification-based segmentation both the quality of
the feature set and the pixel classification technique employed may influence the accuracy of the process. Motivated by the potential of Structural
Tensors in the extraction of hidden information about texture between
neighboring pixels and the computational capabilities of Echo State Network (ESN), this study proposes a tensor-based segmentation approach
using the standard ESN. The pixel features of the image are initially
extracted by incorporating Structural Tensors in order to enrich it with
information about image texture. Then the resulting feature set is fed
into ESN and the output is trained to classify unseen pixels from the
testing set. The effect of the two main parameters that impact the accuracy of ESN: reservoir size and spectral radius, was also evaluated. The
results are promising when compared to recent state-of-the-art segmentation approaches.

Keywords: Image segmentation · Image processing ·
Echo State Network · Support Vector Machine · Feature extraction

1 Introduction

An image segmentation technique consists of partitioning an image into different
meaningful regions such that each region is homogeneous [13]. It forms an essential component of image processing, as it seeks to delineate the boundaries of
various objects of interest in an image or parts of an image for further analysis.
The delineation is usually based on image features such as colour, shape, texture

This work has been supported by the Czech Science Foundation (GAČR) under
research project No. 18-18858S, and the authors acknowledge the support of the OP
VVV MEYS funded project CZ.02.1.01/0.0/0.0/16_019/0000765 "Research Center for
Informatics".

© Springer Nature Switzerland AG 2019
J. Mazal (Ed.): MESAS 2018, LNCS 11472, pp. 490–499, 2019.
https://doi.org/10.1007/978-3-030-14984-0_36

or a mixture of these. But the process is accompanied by complexities such as the presence of noise, overlap between intensities of different objects, variation of contrast and weak edges [15]. Therefore, even though a number of segmentation techniques have been proposed, no single technique can be considered good for all images. The quality of image segmentation based on features created by structural Tensor has been analyzed recently in [5]. Structural Tensor (ST) is capable of extracting information on texture, which is within the neighborhood of a pixel. In [5], structural information extended with information about color, intensity, or a mixture of these features was found to be useful in achieving optimal color image segmentation outcome. Therefore this study chooses to adopt this method of feature extraction to test the quality of image segmentation using Echo State Network (ESN) [6].

ESN and a closely related approach known as Liquid State Machine (LSM) [11] were introduced in the early 2000s as alternative approach to gradient-descent based approaches for training Recurrent Neural Networks (RNNs). The computational approach introduced by these two independent but simultaneous models, which has recently become known as Reservoir Computing, demonstrates that RNN can still perform significantly well even when only a subset of the network weights are trained. ESN consists of a RNN of fixed random weights known as reservoir, and a supervised learning model called the readout. When driven by an input signal, the reservoir improves the linear separability of the input data using a high dimensional feature map, and preserves the nonlinear transformation of the input history in its internal states. A classification or prediction problem can then be solved by training only the weights of the readout structure using the collected reservoir activation states. Therefore the training procedure is fast, and avoids the problem of vanishing exploding gradient [3] introduced by gradient-descent methods. However, the reservoir is influenced by a number of global parameters which in turn impacts the accuracy of the model. The most relevant of these parameters are sparsity, spectral radius of the reservoir matrix, and reservoir size or the dimension of the reservoir matrix [10]. Guidelines on how these parameters can be tuned to guarantee that the state of the reservoir is suitable for good predictions can be found in [6,9].

The internal reservoir activations provided by the dynamics of ESN reservoir has been sufficient in solving many benchmark problems(e.g. [6,7]) and practical problems such as time series predictions [2,9]. In [15], the potential of the ESN reservoir to refine pixel features for colour image segmentation and the influence of the above mentioned parameters on the results have been investigated. In this work, the readout for the classification was realized with a Multi-Layer Perceptron (MLP) consisting of two hidden layers with 15 neurons. Besides, a tensor-based supervised classification method that uses Tucker decomposition to approximate the outputs from the ESN reservoir was proposed in [14], and numerical experiments carried out with spatiotemporal data outperforms the traditional method based on linear output weight, in terms of classification accuracy.

However, this study explores the potential of the standard ESN model with linear regression readout for achieving good image segmentation results when applied on Tensor-based feature set. To test the resilience of ESN to redundant and noised pixel attributes, no feature reduction techniques were applied. Finally, we evaluate the influence of spectral radius and reservoir size on the accuracy of image segmentation and compare the best result with the accuracy attained with Support Vector Machine [4] (when applied on the same tensor-based feature set) as well as the accuracy of existing state of the art methods. The rest of this paper is organized as follows. Section 2.2 presents a description of Structural Tensor as well as ESN and its properties. Section 3 provides a description of the dataset and the experimental setup for this study. The results of the experiments and related discussions on how reservoir size and spectral radius impacts the quality of image segmentation are presented in Sect. 4. We conclude and present recommendation for future work in Sect. 5.

2   Background

This section consist of two parts. It presents a description of the concept of Structural Tensor, followed by a description of the ESN model and its properties.

2.1   Structural Tensor

Structural tensor allows for the extraction of valuable information on texture in addition to information on color and intensity [5]. For any given two dimensional image I, the structural tensor $\mathbf{ST}$ at a point $\mathbf{p_0}$ can be computed using the following formula:

$$\mathbf{ST}_{(\mathbf{p_0})} = G_{R_{(\mathbf{p_0})}}(\mathbf{DD}^T) \tag{1}$$

where $\mathbf{R}(\mathbf{p_0})$ is the compact nearest neighbourhood of $\mathbf{p_0}$, $G_{R_{(\mathbf{p_0})}}$ is an averaging operator in the region R, centered at a point $\mathbf{p_0}$, and D denotes an image gradient vector at each point $\mathbf{p}$ in R, computed as follows [5]:

$$\mathbf{D}(\mathbf{p}) = \begin{bmatrix} I_x(\mathbf{p}) \\ I_y(\mathbf{p}) \end{bmatrix} \tag{2}$$

where $I_x(\mathbf{p})$ and $I_y(\mathbf{p})$ are discrete spatial derivatives of I at point $\mathbf{p}$ in the x and y directions respectively. The averaging operator $G_{R_{(\mathbf{p_0})}}$, can simply be realized using a discrete binomial or Gaussian filter [8]. However, for more precise computations nonlinear anisotropic filter is used. Discussions on this can be found in [5]. After applying the filter G_R to average over a set of points in R, $\mathbf{ST}$ becomes a symmetric positive 2D matrix which elements describe average values of the gradient components in the neighborhood defined at the given point $\mathbf{p_0}$ [5]:

$$\mathbf{ST} = G_R \left(\begin{bmatrix} I_x \\ I_y \end{bmatrix} \begin{bmatrix} I_x I_y \end{bmatrix} \right) = \begin{bmatrix} I_x I_x & I_x I_y \\ I_y I_x & I_y I_y \end{bmatrix} = \begin{bmatrix} T_{xx} & T_{xy} \\ T_{xy} & T_{yy} \end{bmatrix} \tag{3}$$

The structural tensor $\mathbf{ST}$ does not only carry information on signal changes at a single point $\mathbf{p_0}$ but also at all points in the nearest neighborhood of $\mathbf{p_0}$. Therefore it conveys information on overlapping regions (i.e. image texture and local curvature) around $\mathbf{p}$. To include information on colour/intensity, the 2D vector $\mathbf{D}(\mathbf{p})$ can be extended to get the 3D vector $\mathbf{E}$:

$$\mathbf{E}^{\mathbf{T}}(\mathbf{p}) = \left[\mathbf{D}^{\mathbf{T}}\ \mathbf{I}(\mathbf{p})\right]^{\mathbf{T}} = \left[I_x\ I_y\ I\right]^{\mathbf{T}} \tag{4}$$

By substituting 4 into 1, the Extended Structural Tensor(EST) is obtained as follows:

$$\mathbf{EST} = G_R(EE^T) = G_R\left(\begin{bmatrix} I_x \\ I_y \\ I \end{bmatrix}\left[I_x I_y I\right]\right) = \begin{bmatrix} I_x^2 & I_x I_y & I_x I \\ I_y I_x & I_y^2 & I_y I \\ I_x I & I_y I & I^2 \end{bmatrix} \tag{5}$$

Now $\mathbf{EST}$ contains average components of the gradient, besides average squared intensity signal, as well as mixed products of the gradient component and intensity [5]. To account for the components I_R, I_G, I_B of a colour image, $\mathbf{EST}$ can be obtained by replacing $\mathbf{D}$ in 1 with the following vector: the following vector $\mathbf{F}$:

$$\mathbf{F}^{\mathbf{T}}(\mathbf{p}) = \left[\mathbf{D}^{\mathbf{T}}\ \mathbf{I}(\mathbf{p})\right]^{\mathbf{T}} = \left[I_x\ I_y\ I_R\ I_G\ I_B\right]^{\mathbf{T}} \tag{6}$$

The result is a positive definite symmetrical matrix, which contains fifteen components.

2.2 Description of Echo State Network

As shown in 1, the architecture of ESN consist of an input layer with p neurons connected to d hidden neurons, and an output layer with o neurons. The connections between these layers form two main structures: a reservoir structure and a readout structure. The reservoir (RNN) structure is defined by the tuple $(\mathbf{W}^{\text{in}}, \mathbf{W}^{\text{r}})$, where $\mathbf{W}^{\text{in}}$ and $\mathbf{W}^{\text{r}}$ are randomly generated input connection (input-to-hidden) and recurrent connection (hidden-to-hidden) weight matrices with dimensions $d \times (1 + p)$ and $d \times d$ respectively, and the readout consist of a hidden-to-output weight matrix $\mathbf{W}^{\text{out}}$ with dimensions $o \times (1 + p + d)$, which are usually trained with linear regression model. The 1 accounts for the dimension of the first row of $\mathbf{W}^{\text{in}}$ and $\mathbf{W}^{\text{out}}$ which usually contains $1s$ corresponding to bias terms. When p is driven by an input signal $\mathbf{s}(t)$ from an input space $\mathbb{R}^p$ at any time t, the reservoir uses $\mathbf{W}^{\text{in}}$ as a high dimensional feature map to transform the signal into a larger space $\mathbb{R}^d$ with $p \ll d$, and then memorizes the nonlinear transformations of the input history in its internal states.

Given a training set composed of input signal $\mathbf{s}(t) \in \mathbb{R}^p$, the reservoir updates its internal states $\mathbf{x}(t) = (x_1(t), ..., x_N(t))$ according to the following formula:

$$\mathbf{x}(t) = g_h(\mathbf{s}(t), \mathbf{x}(t-1), \mathbf{W}^{\text{in}}, \mathbf{W}^{\text{r}}). \tag{7}$$

Next, the parametric function shown below uses the actual reservoir states to execute the model output:

$$\hat{\mathbf{y}}(t) = g_o(\mathbf{x}(t), \mathbf{W}^{\text{out}}), \tag{8}$$

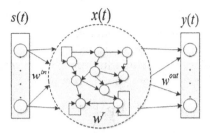

$s(t)$ $\quad$ $x(t)$ $\quad\quad$ $y(t)$

Fig. 1. Architecture of a standard ESN model.

where $g_o(\cdot)$ is an activation function with parameters in $\mathbf{W}^{\text{out}}$.

Although in the standard ESN model there are no connections between the input and readouts neurons [6], another readout form is the following [9]:

$$\hat{\mathbf{y}}(t) = g_o(\mathbf{s}(t), \mathbf{x}(t), \mathbf{W}^{\text{out}}). \tag{9}$$

In this study, we used hyperbolic tangent $\tanh(\cdot)$ as the activation function $g_h(\cdot)$, and so the dynamics was computed as:

$$\mathbf{x}(t) = \tanh(\mathbf{W}^{\text{in}}\mathbf{s}(t) + \mathbf{W}^{\text{r}}\mathbf{x}(t-1)). \tag{10}$$

The output of the model, $\hat{\mathbf{y}}(t)$ at a given time t is computed as follows:

$$\hat{\mathbf{y}}(t) = \mathbf{W}^{\text{out}}[\mathbf{s}(t); \mathbf{x}(t)], \tag{11}$$

where $[\cdot; \cdot]$ denotes a vector concatenation operation.

3   Methodology

In this section, we explain the dataset and the experimental setup used to evaluate the usefulness and effectiveness of ESN for tensor-based colour image segmentation.

3.1   Data Description

The datasets employed in this study to analyze the quality of tensor-based image segmentation using ESN were the original datasets extracted by a proposed Tensor-Based Image Segmentation Algorithm (TBISA) in [5] for a similar study using other classifiers. The datasets were based on Berkley Segmentation Benchmark images [12], which consisted of seven images identified as '35058', '41033', '66053', '69040', '134052', '161062', and '326038'. These images consisted of masks for two and three class detection problems. However, this study concentrated on the masks for two class detection problems. The masks were defined by [5] based on the original segmentation contour published on source page.

The EST feature set is made up of fifteen features which include tensor information on a pixel and its neighborhood, as well as mixed products of these. Besides, it has a corresponding class label that indicates whether the pixel belongs to an object or a background. Detailed description of the extraction process, and the Tensor-Based Image Segmentation Algorithm (TBISA) employed can be found in [5].

3.2   Experimental Setup

First of all, the EST datasets for all the seven images were combined into one dataset. The values of attributes which form each feature vector were then normalized to lie between 0 and 1, and the pixels were randomized. The resulting data was divided into two subsets: 80% was used to train the ESN and the remaining 20% was used to test it. We experimented with two different classifiers: a standard ESN where $\mathbf{W}^{in}$ and $\mathbf{W}^{r}$ are randomly initialized with uniformly distributed weights in the range $[-0.5, 0.5]$, and SVM classifier which uses a Gaussian radial basis (RBF) kernel function. In order to test the effect of spectral radius and reservoir size on the effectiveness of ESN as a classification algorithm, the reservoir was configured with different combinations of the following reservoir sizes d and spectral radius ρ: $d = \{150, 200, 300, 400, 500, 600\}$ and $\rho = \{0.1, 0.3, 0.5, 0.99\}$. The selection of these range of values were informed by previous researches (e.g. [1,15]) which considered the influence these parameters on the performance of ESN. The reservoir states were updated with a leaking rate of 0.3 and the output weights were estimated by setting the regularization factor in the linear regression model to 1×10^{-2}. The classification accuracy resulting from the selected pair of parameters (d, ρ), was estimated using Accuracy. The experiment for each combination of parameters was repeated five times and the average accuracy was recorded.

4   Results and Discussion

The mean and standard deviation of all the accuracies that resulted from the various combinations of d and ρ were 0.955431026 and 0.000026323 respectively. Table 1 below shows the maximum and minimum accuracies as well as the values of d and ρ that led to each accuracy. These results imply that, all things being equal, changes in spectral radius and reservoir size leads to a fairly significant change in accuracy. The trend for the changes realized is depicted (in both 3D and 2D format) in Fig. 2 below. It can be observed that, in most cases, an increase in the value of ρ caused a slight corresponding improvement in classification accuracy. Besides, for each spectral radius, an increase in d leads to a positive change in accuracy in virtually all cases. Hence the best accuracy, as shown in Table 1, was attained with a spectral radius of 0.1 and reservoir size of 600. This confirms the observation of [15] about spectral radius in a similar study where ESN was used for feature selection and MLP was used as readout for classification, and a spectral radius of 2.0 or less was proposed as the best choice for good quality colour image segmentation. Our finding for the best reservoir size is, however, contrary to what was suggested in [15], as $d > 400$ rather appear to be a good choice of reservoir dimension for a good image segmentation quality. These findings served as motivation for further experiments to ascertain if classification accuracy can get any better when d is increased beyond 600 neurons and ρ is decreased below 0.1. Therefore combinations of the following values of d and ρ were tested: $d = 700, 800, 900, 1000$ and $\rho = 0.01, 0.03, 0.05, 0.09$. As depicted in Fig. 3 below, a steady increase in accuracy was realized as d increases. The effect of ρ, however, became unpredictable and did not have as much effect as d. Based on this, we selected 0.9565224 attained with $d = 1000$ and $\rho = 0.01$ as the best performance of ESN for color image segmentation as far as this study is concerned.

Table 1. Classification performance of ESN when applied on Extended Structural Tensor feature set.

	Accuracy	Reservoir size (d)	Spectral radius (ρ)
Maximum	0.955488667	600	0.1
Minimum	0.955397455	400	0.3

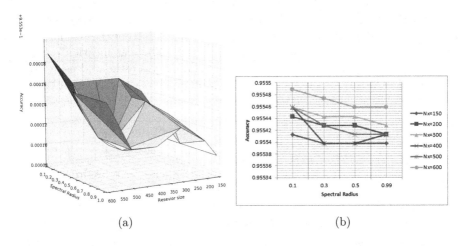

(a)   (b)

Fig. 2. Effect of a range of d and ρ on the Accuracy of Tensor-based Image Classification using ESN. d ranges from 150–600 and ρ ranges from 0.1–0.99.

4.1   Comparing ESN and Other Techniques

Table 2 below shows the best performance of ESN obtained through the experiments explained above; the accuracy of SVM when applied on the same EST feature set; and the results obtained in related studies using other approaches.

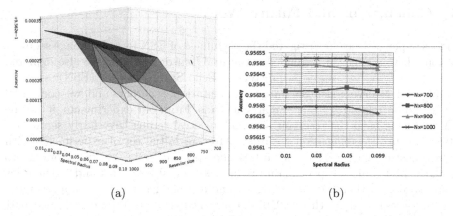

(a)   (b)

Fig. 3. Effect of a range of d and ρ on the Accuracy of Tensor-based Image Classification using ESN. d ranges from 700–1000 and ρ ranges from 0.01–0.099.

It can be seen that the segmentation accuracy of ESN is relatively better than the average performance of SVM as far as this study is concerned. Although our tensor-based approach did not involve any form of feature selection, as was done in some related studies (e.g. [14, 15]), the performance of ESN was promising. This demonstrates the resilience of ESN to redundant and noised data. It is also worth mentioning that in [5], the accuracy was estimated over each image dataset; the pixels used for training were also used for testing. But this study took a different approach in order to test the ability of ESN to classify unseen pixels and also to make the model insusceptible to over-fitting: pixels used for testing were different from those used for training. When compared to the classification result attained by the tensor-based method proposed in [14], it can be concluded that the result of approximating the reservoir outputs using a Tucker decomposition may be similar to that of using raw reservoir outputs based on EST feature set.

Table 2. Comparison of ESN to SVM and some existing State of the Art Classifiers.

Algorithm	Accuracy	Dataset
ESN	0.9565224	All images combined*
SVM	0.86106	All images combined*
Random Tree (RT) [5]	0.9322	'326038'*
MLP [5]	0.9938	'35058'*
MLP [15]	0.926	SSDS dataset [15]

*'35058', '41033', '66053', '69040', '134052', '161062', '326038'

5   Conclusions and Future Work

In this paper, we have investigated the usefulness of ESN for color image segmentation when applied on Extended Structural Tensor features. We also explored the effect of the two main parameters of Echo State Network (ESN): spectral radius and reservoir size on image segmentation accuracy, and we compared our result to related state of the art methods. From our experiments, it can be inferred that both the reservoir size and spectral radius have fairly significant effects on pixel classification accuracy. After an initial set of experiments with reservoir size ranging from 150 to 600 and spectral radius ranging from 0.1 to 0.99, the best choice of spectral radius and reservoir size were 0.1 and 600 neurons respectively. Further tests with $\rho < 0.1$ and $d > 600$ led to further improvement in accuracy. Based on this, 0.9565224 attained with $d = 1000$ and $\rho = 0.01$ was chosen as the best performance of ESN for color image segmentation as far as this study is concerned Tensor-based segmentation accuracy using ESN was comparatively better than SVM, and showed promising results when compared with recent state-of-the-art segmentation approaches. However, in future work, we plan to analyze the accuracy of other types of ESN models, and we expect to apply the NN and Structural Tensor feature set refined through various feature selection processes.

References

1. Basterrech, S.: Empirical analysis of the necessary and sufficient conditions of the echo state property. In: 2017 International Joint Conference on Neural Networks, IJCNN 2017, Anchorage, AK, USA, 14–19 May 2017, pp. 888–896 (2017). https://doi.org/10.1109/IJCNN.2017.7965946
2. Basterrech, S., Rubino, G.: Echo State Queueing Networks: a combination of reservoir computing and random neural networks. Probab. Eng. Inf. Sci. **31**, 457–476 (2017). https://doi.org/10.1017/S0269964817000110
3. Bengio, Y., Simard, P., Frasconi, P.: Learning long-term dependencies with gradient descent is difficult. IEEE Trans. Neural Netw. **5**(2), 157–166 (1994). https://doi.org/10.1109/72.279181
4. Drucker, H., Burges, C., Kaufman, L., Smola, A., Vapnik, V.: Support vector regression machines. In: Neural Information Processing Systems, no. 1, pp. 155–161 (1996). 10.1.1.10.4845
5. Jackowski, K., Cyganek, B.: A learning-based colour image segmentation with extended and compact structural tensor feature representation. Pattern Anal. Appl. **20**, 401–414 (2017)
6. Jaeger, H.: The "echo state" approach to analysing and training recurrent neural networks. Technical report 148, German National Research Center for Information Technology (2001)
7. Jaeger, H., Haas, H.: Harnessing nonlinearity: predicting chaotic systems and saving energy in wireless communication. Science **304**(5667), 78–80 (2004)
8. Lathauwer, L.D.: Signal Processing Based on Multilinear Algebra. Katholieke Universiteit Leuven Faculteit der Toegepaste Wetenschappen Department Elektrotechniek (1997)

9. Lukoševičius, M., Jaeger, H.: Reservoir computing approaches to recurrent neural network training. Comput. Sci. Rev. **3**, 127–149 (2009). https://doi.org/10.1016/j.cosrev2009.03.005
10. Lukoševičius, M.: A practical guide to applying echo state networks. In: Montavon, G., Orr, G., Müller, K.R. (eds.) Neural Networks: Tricks of the Trade. Lecture Notes in Computer Science, vol. 7700, pp. 659–686. Springer, Heidelberg (2012). https://doi.org/10.1007/978-3-642-35289-8_36
11. Maass, W., Natschläger, T., Markram, H.: Real-time computing without stable states: a new framework for a neural computation based on perturbations. Neural Comput. **14**, 2531–2560 (2002)
12. Martin, D., Fowlkes, C., Tal, D., Malik, J.: A database of human segmented natural images and its application to evaluating segmentation algorithms and measuring ecological statistics. In: Proceedings of 8th International Conference on Computer Vision, vol. 2, pp. 416–423, July 2001
13. Pal, N.R., Pal, S.K.: A review on image segmentation techniques. Pattern Recognit. **26**, 1274–1294 (1993)
14. Prater, A.: Classification via tensor decompositions of echo state networks. In: 2017 IEEE Symposium Series on Computational Intelligence (SSCI), pp. 1–8 (2017). https://doi.org/10.1109/SSCI.2017.8280968
15. Souahlia, A., Belatreche, A., Benyettou, A., Curran, K.: An experimental evaluation of echo state network for colour image segmentation. In: 2016 International Joint Conference on Neural Networks (IJCNN), pp. 1143–1150 (2016). https://doi.org/10.1109/IJCNN.2016.7727326, http://ieeexplore.ieee.org/document/7727326/

AxS in Context of Future Warfare and Security Environment (Concepts, Applications, Training, Interoperability, etc.)

LAWs: Latent Demand for Simulation of Lethal Autonomous Weapon Systems

Agostino G. Bruzzone[1($\boxtimes$)], Giulio Franzinetti[2], Marina Massei[3],
Riccardo Di Matteo[4], and Libor Kutej[5]

[1] Simulation Team, DIME, University of Genoa, Genoa, Italy
agostino@itim.unige.it
[2] Lio Tech Ltd., London, UK
giulio.franzinetti@liotech.eu
[3] Liophant Simulation, Savona, Italy
marina.massei@liophant.org
[4] SIM4Future srl, Genoa, Italy
riccardo.dimatteo@sim4future.com
[5] University of Defence in Brno, Brno, Czech Republic
libor.kutej@unob.cz

Abstract. This paper provides an overview on the Lethal Autonomous Systems (LAWs) and related critical issues caused by dynamically evolving context. Traditional approaches to evaluate new system development, software review, prototyping and testing are often not really efficient or even not applicable in this LAWs even due to the evolution in terms of social and operational scenarios. Vice versa, it is evident that simulation plays the key role to support evaluation of scenarios considering that is practically the only methodology able to develop and to conduct virtual tests on concepts, general principia, strategic decisions, technology impacts and related implications.

1 Introduction

The evolution of technology allowed us to experience a new world where autonomous vehicles, intelligent and robotic systems are a dynamically evolving reality; it is currently over a decade since such solutions have been used in operations in theaters (e.g. peacekeeping, peace enforcement, war, etc.), however, their current technological capabilities and quick evolving advances in this field make it very hard to define the boundaries of operational use and their future developments. Discussions about the impact of these systems are ongoing and have been for several years, therefore the authors in this paper propose some kind of wrap up of very consolidated concepts and an overview on the way ahead by defining M&S (Modeling and Simulation) as the key technology to be used as reference guideline.

The use of lethal force by Autonomous Systems is always twofold considering that for someone the autonomous systems are not very different from any other weapon while for others these have to be considered as new kinds of weapons based on "applicative" artificial intelligence. Indeed, many people consider responsibility of the human component to "use" or "activate" autonomous systems; however sometime, it

© Springer Nature Switzerland AG 2019
J. Mazal (Ed.): MESAS 2018, LNCS 11472, pp. 503–513, 2019.
https://doi.org/10.1007/978-3-030-14984-0_37

could seem that autonomous systems resemble to intelligent weapons that, in "some way", "self-direct" on the foe targets. Other people further stress the LAWs context by adopting a different point of view; they suggest that a robotic system could be considered not really different from a simple assault gun that shots bullets against enemies as soon as the trigger is activated by a soldier; the point is that obviously an autonomous intelligent system, contrary to an Automated Guided Missile (AGM), or even to a rifle, could have degrees of freedom in how to proceed based on his situation awareness and boundary conditions. In facts, even just to execute and order of direct fire on target, an UMS (Unmanned Systems), when remotely operated from far away, needs autonomy in executing orders just to guarantee an effective real-time control respect to communication latency and constraints. So, it is evident that such systems could, from a technological standpoint, already deal with engaging a target while patrolling or carrying out other tasks introducing some kind of robotic "weird free will".

2   Capabilities and Priorities

The new computational and operational capabilities of robotics systems introduce the main issue on this topic: it is evident that the word "autonomy" represents the crucial element of this equation and the problem is not limited only to liability for acts during operations, but involve many more considerations; therefore there is a way to address these issues and it deals with using simulation to create an independent validation and verification framework to check compliance of the UMS with RoE (Rules of Engagement), laws of war, international humanitarian law, etc.

These new capabilities of UMS are the major reasons why the new UAV (Unmanned Aerial Vehicle), thanks to their advanced technology supporting autonomy, are able to achieve valuable results almost impossible for the original generations of RPV (Remotely Piloted Vehicle) planned to be used in operations, half a century ago (e.g. AQM-34 V). Therefore there are still limits; for instance it is quite a while from the first very positive results on tests where a RPV interceptor engaged in dog-fight against an highly experienced pilot on a F-4 [13], however even today the use of UAV in these roles is still limited.

In facts, it has taken close to a decade from appearance of the new generations of UAV since their first operational use; in facts during first years of third millennium we record first cases of lethal force use by autonomous systems are recorded in combat over Asia, but we had to wait another decade to observe first UAV-to-UAV refueling between two Global Hawks of NASA.

This basic observation of existing situation makes the priorities clear, but also it outlines what it is the general trend in this sector and suggests that humans should address these issues as soon as possible and in proper way; simulation is the way to create virtual worlds where it is possible to investigate these elements and to evaluate alternatives and potential solutions [3].

Figure 1 proposes an example where the immersive interactive interoperable SPIDER (Simulation Practical Immersive Dynamic Environment for Reengineering) by SimulationTeam is used to supervise and review joint operations involving different UAV and UGV (i.e. fixed wing and rotary wing) in a CBRN Scenario, getting full understanding of the implications of different directions, RoE and policies [4]. Lethal force assigned to autonomous systems often introduces serious legal, ethical and practical issues that turn to be even more crucial challenges in relations to new generations of UMS such as UAV, UGV (Unmanned Ground Vehicle), USV (Unmanned Surface Vehicle), UUV (Unmanned Underwater Vehicle) or AUV (Autonomous Underwater Vehicle).

Fig. 1. Supervision of UMS in CBRN scenario

3 Evaluation and Development of LAWs

The evaluation and the development of Lethal Autonomous Weapons Systems (LAWs) affect all actors in conflict. LAWs generate a paradigm shift and not a simple evolution of existing weapon systems with limited impact within well defined perimeters. These toys play by themselves and simulation can be offered to developers, evaluators, opponents because the battle for and against LAWs has already started in several theatres. Legal development, political communications and technologies could be soon based on simulation more than on experience. Therefore, the market for simulation of autonomous systems, lethal or not, includes system developers, political communicators, legislators, lawyers, military and civilian trainers and last but not least gamers and new, interdisciplinary teams that have yet to be established.

"An autonomous system is capable of understanding higher-level intent and direction. From this understanding and its perception of its environment, such a system is able to take appropriate action to bring about a desired state. It is capable of deciding a course of action, from a number of alternatives, without depending on human oversight and control, although these may still be present. Although the overall activity of an autonomous unmanned aircraft will be predictable, individual actions may not be." [15].

4 Leaving Nothing to Chance

The decision-making process separates automatic from autonomous weapons. Indeed, decision making is a cognitive process and not a mechanical one while the choice is based on alternatives that have to be evaluated according to values and preferences. The process requires much more than engineering as ontological and epistemological issues that govern decision making process. The complexity of such processes was addressed in the early 90 s in Artificial Morality - Virtual Robots for Virtual Games [9, 10, 17] in which the author uses PROLOG to write simple decision-making programs.

Laws of War and Rules of Engagement define the principles of conduct and in engineering terms constitute higher and intermediate constrains, from an epistemological point of view they constitute the ethical or moral boundaries. Danielson combines morality with game theory to develop intelligent agents. "The constructive resources of artificial intelligence allow us to build most of the desired agents and manage the complexities that result from their interactions" [9].

Last, and most interesting, is the realisation that less constrained behaviour frameworks or simple maximisation ("SM") strategies underperform and might never outperform constrained maximisation ("CM") whose pursuit of simple self-interest can harm overall performance.

The use of the word morality evokes images and reactions from both civilian and military minds that are often unhelpful in this context. The reader is best served by accepting moral theory as rational choice theory.

In much of the academic literature on the subject of LAWs the problem posed by Isaac Asimov emerges defensively, apologetically for fear of quoting an icon of popular culture. Asimov, a scientist and academic and a military man himself, coined the term robotics. In his writings he illustrates the difficulty robots encounter when applying the laws. Other attempts to define rules that should define a robots' cognitive processes include those of the Engineering and Physical Research Council and the Arts and Humanities Research Council of Great Britain (https://epsrc.ukri.org). Sadya Nadella provided six basic rules of AI in an article in Slate magazine [16] later reviewed in The Verge [21].

There is currently a communication battle on media where opposition to killer robots creates headlines, legal issues arising from the use of any UMS. These elements are pretty complex it is almost impossible to evaluate risks and consequences without using simulation. In near future we will see this to be further extended to advanced technological developments of AI that could represent major opportunities or risks; in facts these considerations could evolve on media being viewed in emotionally charged terms.

5 A List of Problems/Opportunities to Offer Simulation as a Tool

The opportunity for the development of simulation starts with testing the hypotheses on the mission environment and conditions related to use of LAWs. As testing these devices incurs significant cost and risk layering simulation ahead of battlefield testing makes a lot of sense. You can test a tank in a range and just stand back... not so with many LAW systems. The contexts need to address multiple different issues that requires to be properly considered [18] as in www.cnas.org/publications/reports/robotics-on-the-battlefield-part-ii-the-coming-swarm.

Cognitive issues

(a) Discriminating between different agents
(b) Deciding how to behave with the different agents
(c) Self-interest of the robots vs humans
(d) Conflict of interest and conflict of preference
(e) The decision of which agents have to be treated morally and which ones should not

Technical issues

(a) Incomplete information
(b) Unexpected accelerated pace of interactions
(c) Unexpected interactions between adversarial systems
(d) Spoofing
(e) Behavioural hacking
(f) Technical hacking

Risk

(a) Strategic Mission Failure
(b) Tactical Mission Failure
(c) Failure to Respect Rules of Engagement and Laws ow War
(d) Civilian casualties
(e) Fratricide
(f) Unintended escalation

6 Modelling Situations as Games

The analytical reasoning process could be analysed using tools such as logic programming languages such as Prolog, ASP and Datalog in which clauses represent rules: these were developed initially in the 30 s and have allowed for the interaction of the skillsets of logicians, psychologists, ethicists, lawyers with the experience of operational military personnel.

Logic games provide the first testing ground for LAWs. Simple ones include Prisoner's dilemma, involving parameters such as fitness, egoism and utility; this evolves into the Iterated Prisoner's dilemma to add a level of complexity. Tit for tat, with its simple logic, but still requiring an evaluation of the symmetry of action and effect. Far more complex is the decision mechanism in the Chicken game, and so on.

The development of games starts by addressing basic challenges such as maximisation of interest, determining the optimal level of constraints on self-interests. Also, the level of cooperation between agents and what conditions should be imposed on it.

Testing the potential for endless loops given the logic. There is a long list, but the question that manufacturers will have to address is how they can act responsibly to the extent that limits their liability. Will the manufacturers expose the weaknesses in the internal logic of LAWs and in the interaction with other agents early enough to do something about it or will we have problems?

7 Major Challenges and Opportunities

"Strategic agents consider all alternatives and consider what the other agent can be expected to do in response to each of them", "That is what distinguishes sophisticated strategic thinking... from the simpler parametric situation of an agent" [9].

The intelligent agent, or agents, in a LAWs have to be able to constrain their actions for the sake of benefits shared by others. Due to these reasons the authors suggest the adoption of advanced solutions inspired by the IA-CGF created by SimulationTeam [3, 6, 7]. The Intelligent Agent Computer Generated Force have evolved along two decades and show ability to address complex scenarios in disaster relief [5], hybrid warfare [2], cyber warfare [8] and protection of critical infrastructure [4] with very interesting results and experimentations (www.liophant.org/projects). Indeed this necessity is over stressed when multiple autonomous systems interoperate as happen in JESSI Scenarios (Joint Environment for Serious Games and Simulation Interoperability) including UUV, USV, AUV and traditional assets; Fig. 2 illustrates the review of operations by a AUV that due to its nature (very limited communication capabilities) representing the most evident necessity to define criteria for operating autonomously in Joint Naval Operations [3]; another case is proposed by T-REX simulator in Fig. 3 related to a swarm attack by micro UAV on a critical infrastructure simulated jointly with cyber-attacks. In addition, a very interesting aspect is related to the dual use of these techniques for both military and civilian Applications [1].

8 The State of the Argument

United Nations and this Subject

Current discussions on conventions are taking place in Geneva on an annual basis within the framework created by the UN within the High Contracting Parties to the Convention on Prohibitions or Restrictions on the Use of Certain Conventional Weapons Which May be Deemed to be Excessively Injurious or to Have Indiscriminate Effects (CCW)... one could not try to make a more obscure name or setting as if to hide a debate rather than bring it to the forefront of attention. However, the list of

Fig. 2. Criteria analysis for AUV in virtual world

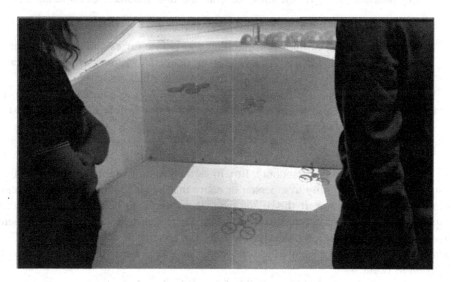

Fig. 3. Simulating swarm attack of suicide UAV

participants is endless including governments, human rights organizations other NGOs and prominent individuals including International Committee for Robot Arms Control, Human Rights Watch, Seguridad Humana en Latino América y el Caribe and the Campaign to Stop Killer Robots Need for meaningful human control [20].

Arguments against LAWs

a. that the use of AWS is either per se unethical or less moral than the use of currently available technology because of the exclusion of the human form specific decision processes dehumanise the target and the attacking side; and

b. that the decision-making process of AWS cannot currently match the ability of humans to respect the rules of engagement based on the Law of War and International Humanitarian Law and, according to some commentators but not all, that it is impossible for AWS (i.e. computers) to apply the rules of engagement set within the framework of the Laws of War and International Humanitarian Law, as well as a human being, would.

c. Hitherto, weapon technology has relied on the control exercised by humans in the loop, yet it allowed for the very existence of humanity has been threatened by errors that came close to causing nuclear war. Close Calls with Nuclear Weapons and the increasing reliance on autonomous systems may be viewed as irresponsible and unethical until such a time when it can be proved that AWS can be relied upon to act within the framework provided by the Rules of Engagement, based on the Laws of War and International Humanitarian Law.

Arguments supporting the development and deployment of LAWs

a. that the exclusion of human in the loop will reduce errors and violation of the rules of engagement due to mistakes or deliberate actions of insubordinate personnel, thereby producing a less dangerous conflict; and

b. that the reduction of the number of exposed military personnel and the more surgical and precise nature of military strike action will reduce civilian casualties, thereby making was less destructive.

Legal Arguments

The legal argument is a Pandora's Box in its own right. Criminal law requires the assignation of responsibility to a person or entity that has the moral agency to accept responsibility for a crime. If this is the case, there seems to be no solution for the responsibility gap if we accept this gap in our day-to-day lives. For example, when driving a car, we cannot predict with certainty the conditions of an entire road. The smallest anomaly, like a small shard of glass, can cause a tire to burst and a fatal collision to occur.

The driver cannot be held responsible for any death as he was not responsible for placing the shard of glass. The manufacturer who built the car with the highest standard of 'due care' cannot be held responsible because the car itself was not at fault. Finally, the workers who paved the road cannot be accountable as we could never know for sure when the shard of glass was introduced. The collision would be deemed an accident. We seem to have much higher standards when war is concerned, and a responsibility gap that allows for accidents and lack of responsibility seems unacceptable when it comes to taking a human life. If an AWS can be causally responsible for deaths seen in a war but cannot be legally and morally responsibility, then arguably the war in question is not a Just War nor a humanitarian one. It appears that for the time being, responsibility for the actions of AWS will remain with a human controller, until such a time when a fully autonomous system will possess the appropriate level and kind of autonomy to assume full legal and moral responsibility for its actions [11].

9   Stimulating Quotes

It is evident that this problem is pretty extended, and this paper proposes just elements to be considered, therefore the identification of Simulation as approach to address these issues is an important step forward. Indeed, the following quotes are proposed to stimulate discussion and to facilitate the summarizing of final conclusions:

"if we continue to develop our technology without wisdom or prudence, our servant may prove to be our executioner". (Gen. Omar Bradley)

"The only theoretical reason to take artificial intelligence more seriously than clockwork is the powerful suggestion that our minds work on computational principles" [12].

"the AI arms race is propelled by unstoppable forces: geopolitical competition, science pushing at the frontiers of knowledge, and profit-seeking technology businesses. So the question is whether and how some of its more disturbing aspects can be constrained. At its simplest, most people are appalled by the idea of thinking machines being allowed to make their own choices about killing human beings. And although the ultimate nightmare of a robot uprising in which machines take a genocidal dislike to the human race is still science fiction, other fears have substance" [19].

10   Conclusions

Digital simulations are already used to train, teach and evaluate soldiers in a range of operations from cultural interactions to language skills to weapons training and even the treatment of PTSD [14], but these simulations follow a relatively narrow set of parameters and the cost of any mistakes is limited.

The use Lethal Autonomous Weapons Systems (LAWs) poses serious legal, ethical and practical issues.

Precedent or convention do not yet define the liability for illegal acts resulting from the operation of LAWs in spite of the efforts of the United Nations Office for Disarmament Affairs. In the absence of an international convention, users and manufacturers will try to assign responsibility to each other. The only way both parties can reduce the burden of responsibility is by introducing independent verification of the compliance of the systems with the rules of engagement, with the laws of war and with the international humanitarian law.

An analytical approach involving the review of code and engineering specifications is neither practical nor likely to be effective. Hence, simulation is the only empirical approach that users and manufacturers of LAWs can employ to counter accusations of negligence and criminal lack of regard for the respect of codes of military conduct. The use of simulation will require exposure of the weaknesses and errors in the logic embedded in systems which may not be readily accepted by the supplier.

The need to match the enemy's use of LAWs and the power of the arms manufacturers may put pressure on governments to accept minimal standards of verification of the compliance of systems with rules of engagement and other legal requirements.

The emotional and political impact of the advent of the "killer robot" will contrast the cost/benefit analysis implied by calculating the number of violations of laws and rules of engagement due to human activity vs deployment of systems.

The requirement for simulation will probably emanate from the following:

1. Manufacturers of LAWs and or anti-LAW systems
2. Political communicators wishing to attack or defend the use of LAWs
3. Legislators who wish to evaluate cost/benefit and risk/reward profiles
4. Lawyers who are attempting to establish or reject liability for the use of a system
5. Game development companies in both training and recreational markets.

References

1. Bruzzone, A.G., Massei, M., Mazal, I., di Matteo, R., Agresta, M., Maglione, G.L.: Simulation of autonomous systems collaborating in industrial plants for multiple tasks. In: Proceedings of SESDE, Barcelona, September 2017
2. Bruzzone, A.G., Di Bella, P., Di Matteo, R., Massei, M., Reverberi, A., Milano, V.: Joint approach to model hybrid warfare to support multiple players. In: Proceedings of WAMS, Florence, September 2017
3. Bruzzone, A.G.: New challenges & missions for autonomous systems operating in multiple domains within cyber and hybrid warfare scenarios. Invited Speech at Future Forces, Prague, Czech Republic (2016)
4. Bruzzone, A.G., Massei, M., Maglione, G.L., Di Matteo, R., Franzinetti, G.: Simulation of manned & autonomous systems for critical infrastructure protection. In: Proceedings of DHSS, Larnaca, Cyprus, September 2016
5. Bruzzone, A.G., Massei, M., Di Matteo, R., Agresta, M., Franzinetti, G., Porro, P.: Modeling, interoperable simulation & serious games as an innovative approach for disaster relief. In: Proceedings of I3M, Larnaca, 26–28 September 2016
6. Bruzzone, A.G., et al.: Human behavior simulation for complex scenarios based on intelligent agents. In: Proceedings of the 2014 Annual Simulation Symposium, SCS, San Diego, April 2014
7. Bruzzone, A.G.: Intelligent agent-based simulation for supporting operational planning in country reconstruction. Int. J. Simul. Process Model. 8(2–3), 145–151 (2013)
8. Bruzzone, A.G., et al.: Virtual framework for testing/experiencing potential of collaborative autonomous systems. In: Proceedings of I/ITSEC, Orlando, FL, USA (2013)
9. Danielson, P.: Artificial Morality - Virtuous Robots for Virtual Games. Routlege, New York (1992)
10. Gauthier, D.P.: Morality, rational choice and semantic representation. Soc. Philos. Policy 5, 173–221 (1988)
11. Gocek, S.A.: What ethical concerns are raised by autonomous weapons systems? 15 October 2018. The Global Dispatches. www.theglobaldispatches.com. Accessed October 2018
12. Haugeland, J.: Artifical Intelligence, the Very Idea. MIT Press, Cambridge (1987)

13. Larm, D.: Expendable remotely piloted vehicles for strategic offensive airpower roles. Air University Maxwell AFB, AL, School of Advanced Airpower Studies (1996)
14. Mead, C.: War Play. Houghton Mifflin Harcourt Publishing Company, New York (2013)
15. Ministry of Defence: Joint Doctrine Publication 0-30.2 Unmanned Aircraft Systems. Ministry of Defence, The Development, Concepts and Doctrine Centre. UK Government, London (2017)
16. Nadella, S.: The Partnership of the Future, 28 June 2016. Slate. http://www.slate.com/articles/technology/future_tense/2016/06/microsoft_ceo_satya_nadella_humans_and_a_i_can_work_together_to_solve_society.html
17. Oldenquist, A.: The possibility of selfishness. Am. Philos. Q. **17**, 25–33 (1980)
18. Scharre, P.: Robotics on the Battlefield Part II: The Coming Swarm, 15 October 2014. CNAS. https://www.cnas.org/publications/reports/robotics-on-the-battlefield-part-ii-the-coming-swarm. Accessed October 2018
19. The Economist: Autonomous weapons are a game changer, 25 January 2015. The Economist. https://www.economist.com/special-report/2018/01/25/autonomous-weapons-are-a-game-changer. Accessed October 2018
20. United Nations Office at Geneva: 2018 Group of Governmental Experts on Lethal Autonomous Weapons Systems (LAWS), 9–13 April 2018. https://www.unog.ch/80256EE600585943/(httpPages)/7C335E71DFCB29D1C1258243003E8724. Accessed October 2018
21. Vincent, J.: Satya Nadella's rules for AI are more boring (and relevant) than Asimov's Three Laws, 29 June 2016. The Verge. https://www.theverge.com/2016/6/29/12057516/satya-nadella-ai-robot-laws

Using Unmanned Aerial Systems in Military Operations for Autonomous Reconnaissance

Petr Stodola$^{(\boxtimes)}$, Jaroslav Kozůbek, and Jan Drozd

University of Defence, Brno, Czech Republic
{petr.stodola,jaroslav.kozubek,jan.drozd}@unob.cz

Abstract. The article deals with modern technologies to support military operations in order to yield new innovations and development in the area of security and sustainability of military units. The article is divided into two key parts. The first part presents the model of the Autonomous Aerial Reconnaissance Problem (AARP). Firstly, the basic features and principles of the model are discussed. The AARP can be seen as a well-known Multi-Depot Vehicle Routing Problem (MDVRP); however, the different optimal criterion is used. Thus, the AARP is formulated for the first time as a new problem. Then, the basic aspects of the original metaheuristic solution proposed by the authors to the AARP is introduced. Finally, the algorithm is verified on the benchmark instances to show its effectiveness. The second key part of the article deals with the tactical aspects of the AARP; the use of the model is shown on the platoon level within a chosen tactical activity.

Keywords: Unmanned aerial systems · Reconnaissance operations ·
Meta-heuristic algorithms · Ant colony optimization ·
Multi-depot routing problem · Raid

1 Introduction

Decision support systems have become an integral part of the commander's decision-making process [1]. They are used as tools to support them with the gathering and sharing the information, analyses, evaluation of the possible action of the enemy, and even they can propose the variants of activity to fulfil the task of the commander optimally [2, 3].

Modern technologies, unmanned vehicles and autonomous robots have also their essential place on the modern battlefield. The use of decision support systems in combination with robotic devices can bring a great advantage over the enemy, especially in the context of asymmetric combat.

This article deals with one of the models of the Tactical Decision Support System (TDSS) being developed since 2006 at the University of Defence, Czech Republic. The model can be used for planning the optimal aerial reconnaissance of the area of interest by a fleet of unmanned aerial systems.

The article is organised as follows. Section 2 deals with the literature review. In Sect. 3, the model of the Autonomous Aerial Reconnaissance Problem is introduced. Firstly, the basic principles of the model are presented. Then, the AARP is

© Springer Nature Switzerland AG 2019
J. Mazal (Ed.): MESAS 2018, LNCS 11472, pp. 514–529, 2019.
https://doi.org/10.1007/978-3-030-14984-0_38

mathematically formulated; it is based on the similar Multi-Depot Vehicle Routing Problem but with the different optimal criterion. Next, the metaheuristic algorithm and its verification on the benchmark instances follows. Section 4 shows the tactical use of the model on a chosen tactical activity – raid.

2   Literature Review

This section presents the literature review of the topic both from the theoretical and practical points of view. The former looks at the model as a mathematical problem that can be solved by algorithms and methods. The latter reviews the literature dealing with similar problems or concepts.

2.1   Autonomous Reconnaissance as a Mathematical Problem

The Autonomous Aerial Reconnaissance model plans routes of Unmanned Aerial Systems (UAS) to explore the area of interest as discussed below in Sect. 3 in more detail. This task can be seen as the well-known Multi-Depot Vehicle Routing Problem (MDVRP). Since the time MDVRP was formulated by Dantzig and Ramser [4], many exact, heuristic, and meta-heuristic algorithms have been proposed.

MDVRP is an NP-hard problem [5], i.e. exact algorithms are time-consuming and hardly ever applicable to problems with more than 50 customers unless P = NP [6]. Therefore, heuristic and metaheuristic approaches have started to be popular for MDVRP. Some of the most recent heuristic methods were proposed in [7–10].

The principle of metaheuristic approaches consists in exploring the search space in a finite number of iterations. In each iteration, some heuristics as well as information obtained in the previous iteration are used to find a solution. Popular metaheuristic approaches are tabu search, genetic algorithms, simulated annealing and ant colony optimization. Examples of such methods applied to MDVRP include works in [11–15].

2.2   Trajectory Planning for Unnamed Aerial Systems

In the second part of this section, literature dealing with trajectory planning for UASs is considered. Geiger [16] in his dissertation evaluated several methods for trajectory optimization of UAVs including linear programming, dynamic programming, genetic algorithms, and neural networks. He also considered the UAV path planning in a surveillance problem. Wang et al. [17] proposed a method for planning the three-dimensional path for low-flying unmanned aerial vehicles in complex terrain based on an interfered fluid dynamical system.

Zhan et al. [18] introduced an improved A* algorithm for the real-time path planning of UAVs in a 3D large-scale battlefield environment aiming at high survival rates and low fuel consumption. Fu et al. [19] proposed a method based on genetic algorithm to generate path for UAV in the existence of unknown obstacle environment. Keller et al. [20] introduced algorithms for fixed-wing UAS integrating on-board sensor capabilities and vehicle manoeuvre constraints to reliably satisfy the objectives of persistent surveillance, path planning, and trajectory management.

Lu et al. [21] proposed a dynamic RRT algorithm applied to fixed-wing UAVs path planning in dynamic three-dimensional environment. Da Silva Arantes et al. [22] applied genetic algorithms to the path planning problem for UAVs during emergency landing. Ingersol et al. [23] modelled UAV path planning as a single objective optimization problem that utilizes a receding horizon approach where the path is constrained to avoid obstacle collision. Cekmez, Ozsiginan and Sahingoz [24] dealt with a problem of minimum time coverage of ground areas using a number of UAVs.

3   Autonomous Aerial Reconnaissance Problem

This section deals with the model of the Autonomous Aerial Reconnaissance Problem (AARP). The aim of the model is to plan the reconnaissance operation of the area of interest by a fleet of Unmanned Aerial Systems (UASs). The operation should be carried out as fast as possible.

3.1   Basic Features and Principles of the Model

The AARP model considers the area of interest to be explored as a polygon specified by the commander and deployed in the area of operations.
    The basic principles of the model are as follows:

- The area of interest is covered by a number of evenly distributed waypoints.
- The maximal allowed distance from any point in the area of interest to the nearest waypoint reflects the requirements of the commander and the parameters of UASs used (see below).
- The area of interest is assumed as explored when all waypoints are visited by at least one of the UASs.
- The model plans the routes of individual UASs between the waypoints in the optimal or near-optimal way according to the optimal criterion (see Sect. 3.3).

The principle described above is shown in Fig. 1 on an example situation. The green lines represent the area of interest. The green dots are waypoints to be visited (there are 104 waypoints in this case where maximum distance was set to 100 m). There are 3 UASs (labelled by letters A, B, and C) available for the reconnaissance operation.
    The value of the maximum distance between any point lying in the area of interest and the nearest waypoint D_{max} can be set according to:

- Commander's requirements such as minimum or maximum height of flight of UASs in the fleet.
- Parameters of sensorial systems of UASs such as the camera resolution and/or vertical and horizontal field of view.

For example, when both the height of flight above the ground level H_{AGL} and the horizontal and vertical fields of view A_{HFOV}, A_{VFOV} of UASs cameras are specified, the distance D_{max} is computed according to formula (1). The lower value from A_{HFOV} and

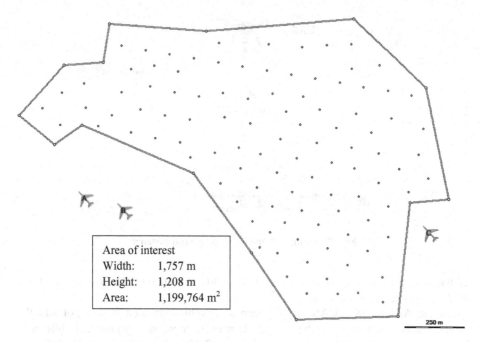

Area of interest
Width: 1,757 m
Height: 1,208 m
Area: 1,199,764 m²

250 m

Fig. 1. Example situation of the reconnaissance operation (Color figure online)

A_{VFOV} is used in the formula as trajectories of individual UASs are not know at the moment of deploying waypoints. It leads to the lower (i.e. worse) value of D_{max}.

$$D_{max} = 2 \cdot H_{AGL} \cdot \tan\left(\frac{\min(A_{HFOV}, A_{VFOV})}{2}\right),\tag{1}$$

where

D_{max} is the maximum distance,
H_{AGL} is flight height above the ground level,
A_{HFOV}, A_{VFOV} are horizontal and vertical fields of view.

Figure 2 shows the situation graphically. If the waypoints are deployed in the way that the D_{max} condition is met, then there is no space in the area of interest unobserved by sensorial systems of the unmanned vehicles.

3.2   AARP as a Multi-depot Vehicle Routing Problem

The problem of autonomous aerial reconnaissance can be seen as a problem similar to the well-known Multi-Depot Vehicle Routing Problem (MDVRP) which was formu-lated almost 60 years ago. The MDVRP is an NP–hard problem [5].

The MDVRP is composed of depots from which customers should be served. A fleet of vehicles is based at each depot. Each vehicle starts from one depot, services the customers assigned to that depot, and returns back to the same depot. The objective

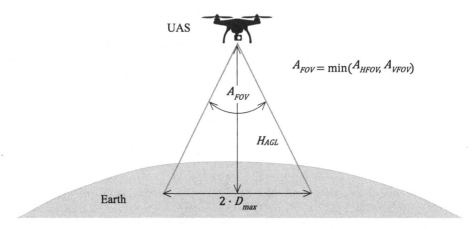

Fig. 2. Computation of the maximum distance

of the problem is to service all customers while minimizing the total sum of the distances travelled by all vehicles.

In the AARP, each UAS can be seen as a vehicle located in a depot and the waypoints are customers to be served. However, there is a significant difference between the objectives of the AARP and MDVRP. The objective of the MDVRP is to minimize the total travel distance whereas the objective of the AARP is to minimize the travel time of the longest route (see Sect. 3.3) as the goal is to conduct the reconnaissance operation as fast as possible.

Figure 3 presents the MDVRP and AARP solutions to the example situation depicted in Fig. 1. All UASs were considered as homogeneous with the average speed 10 m/s, $H_{AGL} = 60$ m, $A_{FOV} = 80°$, $D_{max} \cong 100$ m. The solutions were found via the meta-heuristic algorithm proposed by the authors (see Sect. 3.4). The difference in results is obvious as shown in Table 1.

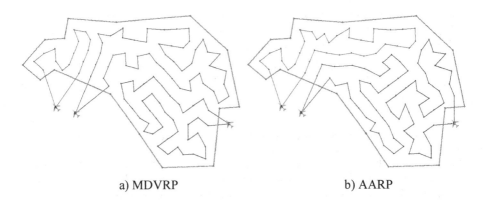

a) MDVRP   b) AARP

Fig. 3. The MDVRP and AARP solutions for the situation in Fig. 1

Table 1. The difference between the MDVRP and AARP solutions

Problem	Travel distance (km)				Operation duration (min:sec)			
	UAS A	UAS B	UAS C	Sol.	UAS A	UAS B	UAS C	Sol.
MDVRP	1.96	5.46	4.15	**11.57**	03:16	09:06	06:55	**09:06**
AARP	3.95	3.93	3.90	**11.78**	06:35	06:33	06:30	**06:35**

3.3 Problem Formulation

This section deals with the mathematical formulation of the AARP. Let $G = (V, E)$ be a graph where $V = U \cup W$ is a set of vertexes (nodes) composed of a set of vehicles $U = \{U_1, U_2, \ldots, U_M\}$ and depots $W = \{W_1, W_2, \ldots, W_N\}$ and E is a set of edges between all vertexes in the graph.

For every edge E_{V_i, V_j} between existing nodes V_i and V_j $(i \neq j)$, a non-negative cost c_{V_i, V_j} is associated which might be interpreted as the travel time. Each vehicle leaves from its original position with the purpose of visiting a sequence of waypoints. After the trip, the vehicle returns back to its original position.

A list of nodes to be visited in the correct order represents a route R_i for each vehicle $U_i \in U$. Route $R_i = \left(V_{(0)}^i, V_{(1)}^i, V_{(2)}^i, \ldots, V_{(K_i)}^i, V_{(K_i+1)}^i \right)$ where K_i is the number of waypoints in route R_i $(0 \leq K_i \leq N$ and $\sum_{i=1}^{M} K_i = N)$, $V_{(0)}^i = U_i$, $V_{(K_i+1)}^i = U_i$, $V_{(k)}^i = W_{(k)}^i$ for all $k = 1 \ldots K_i$.

Waypoint $W_{(k)}^i$ is k-th waypoint on the route R_i. No waypoint is visited more than once, i.e. $W_{(k)}^j \neq W_{(m)}^l$ for all $j, l = 1, 2, \ldots, M$ and $k = 1, 2, \ldots, K_j$ and $l = 1, 2, \ldots, K_l$ $(j \neq l \wedge k \neq m)$. In total, there is M routes, one for each vehicle, i.e. a set of routes $R = (R_1, R_2, \ldots, R_M)$.

The total cost C_i of route R_i is computed according to formula (2). The objective of the AARP is expressed in formula (3) which is to minimize the maximum cost of all routes R.

$$C_i = \sum_{j=0}^{K_i} c_{V_{(j)}^i, V_{(j+1)}^i} \tag{2}$$

$$\text{minimize}(\max(C_1, C_2, \ldots, C_M)) \tag{3}$$

3.4 Meta-heuristic Solution

The solution proposed by the authors for the AARP is based on the Ant Colony Optimization (ACO) theory. This section presents only basic aspects of the ACO algorithm; more details can be found in [25].

The algorithm is a probabilistic technique for solving computational problems. It belongs to the swarm intelligence methods. The principle is adopted from the natural world where ants explore their environment to find food. The exploring starts randomly at first; when an ant finds food, however, it lays down a pheromone trail along its route. When other ants find such a trail, they are likely to follow it (and lay down their own trail if successful).

Figure 4 shows the basic phases of the ACO algorithm. As the approach is metaheuristic, the search space is explored in a finite numbers of iterations. The algorithm works with the number of ants. In each iteration, a solution is found for each ant via the heuristic approach which involves the cost between nodes and the strength of pheromone trails (phase ①). In the next step, pheromone trails evaporate gradually (phase ②), thus reducing its strength; it has an effect of avoiding the convergence to a locally optimal solution. Then, pheromone trails are updated according to the best solution found in the current iteration (phase ③) and the next iteration starts unless the termination condition is met.

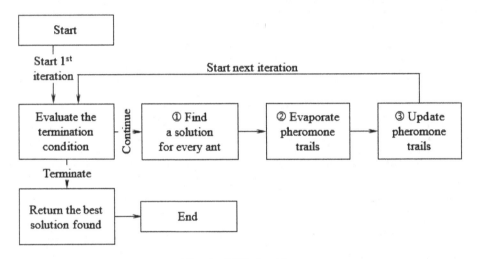

Fig. 4. ACO algorithm

The strength of the algorithm is that it can be used both for the MDVRP and AARP. The difference is only in the evaluation of the quality of solutions according to the given optimal criterion which can be either the sum of travel distances for the MDVRP or the maximum travel time for the AARP.

3.5   Experiments on Benchmark Instances

This section presents the experiments conducted on benchmark instances. As the AARP was formulated for the first time in this article, Cordeau's MDVRP benchmark instances were used. The goal was to show the difference between the optimal criteria for the MDVRP and AARP along with the impact on the real reconnaissance operation.

Table 2 records 10 benchmark instances p01 to p10; table also shows the number of nodes N and number of depots M for each instance. The best known solutions for the

MDVRP problems found in literature and taken from NEO Web [26] are recorded in column (1). The MDVRP optimal criterion was used. Next column (2) presents the same solution as in column (1) but reevaluated according to the AARP optimal criterion.

Table 2. Results for benchmark instances

Instance (N/M)	Best known solution		ACO solution		AARP error
	(1) MDVRP	(2) AARP	(3) AARP	(4) MDVRP	
p01 (50/4)	576.87	237.81	152.54	607.66	55.90%
p02 (50/4)	473.53	169.57	125.82	495.34	34.77%
p03 (75/5)	641.19	207.23	136.05	670.82	52.32%
p04 (100/2)	1001.59	532.24	511.41	1021.36	4.07%
p05 (100/2)	750.03	378.17	378.17	750.72	0.00%
p06 (100/3)	876.50	432.59	301.75	902.91	43.36%
p07 (100/4)	885.80	246.21	228.95	907.55	7.54%
p08 (249/2)	4437.68	2474.82	2224.85	4449.65	11.24%
p09 (249/3)	3900.22	1836.00	1364.70	4085.51	34.54%
p10 (249/4)	3663.02	1049.12	960.76	3825.73	9.20%

The next two columns of Table 2 shows solutions found by the ACO algorithm where the AARP optimal criterion was selected. Column (3) shows the results whereas column (4) records the same solution as in column (3) but reevaluated according to the MDVRP optimal criterion.

The last column in Table 2 presents the difference (error) between columns (2) and (3) in percent. This error can be imagined as the difference in results when planning the reconnaissance operation independently with two variants of optimal criteria. For example, in the case of instance p01, the reconnaissance operation would be more than 55% longer when planned according to the MDVRP optimal criterion instead of the AARP optimal criterion. The average error for all instances is 25.29%.

Figure 5 shows the results for instance p01. On the left, best known solution is shown. On the right, the solution found via the ACO algorithm with the AARP optimal criterion assigned is shown. As can be seen, Cordeau's benchmark instances incorporates two other conditions which cannot be violated. These conditions are (a) the maximum loop length and (b) the maximum capacity of deliveries in a loop. The loop is a part of the route in which the vehicle leaves the depot at the beginning of the loop and returns back at its end. Thus, the vehicle can return back to the depot more than once in a route to create more loops. The ACO algorithm supports both conditions. In Fig. 5b, the similar lengths of all 4 routes (blue, red, green, cyan) are obvious as the criterion is to conduct the operation as fast as possible.

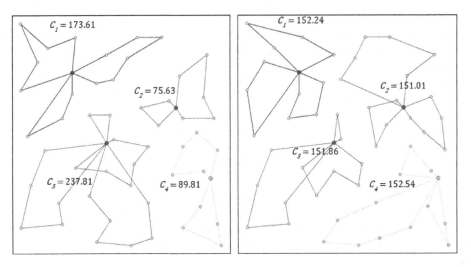

Fig. 5. Solutions for benchmark instance p01 (Color figure online)

4   Practical Use of the AARP on the Platoon Level

This section deals with the practical use of the AARP within a chosen tactical activity. In order to describe the possibilities of the AARP from mission planning phase up to the execution phase there was chosen one offensive tactical activity which is raid. The use of all kinds of UAVs is currently very topical particularly on the battalion level and above. However, there are a lot of possibilities to use small drones or compact UAVs on the tactical level too [27].

UAVs are especially valuable in environments where immediate information feedback is needed, manned aircraft are unavailable, or excessive risk or other conditions render use of manned aircraft less than deliberate. The UAVs can conduct day and night operations to support units.

The UAV's missions can include:

- Route, area, object and zone reconnaissance,
- Surveillance of named areas of interest (NAIs),
- Adjusting indirect fire weapons, close air support (CAS), and close-in fire support (CIFS),
- Support to:
  - Combat search and rescue (SAR),
  - Target acquisition (TA),
  - Battle damage assessment (BDA),
  - Rear area security,
  - Situation awareness (SA) development,
  - Intelligent preparation of battlefield (IPB),
  - Electronic warfare (EW),
  - Communications relay,

- Mine and chemical detection,
- Weather surveillance,
- Material and munition air resupply.

In case of that UAV is equipped by a weapon system (considered as a combat UAV) it can fulfill combat missions to destroy enemy in open terrain, in vehicles or covered in shelters or in buildings.

4.1 Use of the AARP for the Raid Mission

In order to understand whole outcome of the use of the AARP on platoon level it is necessary to briefly explain raid as chosen tactical activity.

It is vital to understand that raid is a part of the group called offensive military activities and it is conducted against a static objective. So the use of drones scripted with the AARP is in this case easier than in other dynamic activities.

There are countless definitions of raid; however all of them cover the basic of this activity. The NATO standardized document which clarifies and unifies meanings of military terms and definitions is NATO Glossary of Terms and Definitions.

According to NATO Glossary, the "raid" is considered as an operation, usually small scale, involving a swift penetration of hostile territory to secure information, confuse the enemy, or destroy his installations. It ends with a planned withdrawal upon completion of the assigned mission.

Usually the raid is conducted within the enemy territory, which means that the assigned friendly unit has to infiltrate into this territory and after the raid is completed withdrawal back into friendly territory. Both movements are extremely dangerous and required detailed planning.

The most important for success of a raid is to have very reliable information about the objective and enemy in there. Platoon leader has usually information only from company commander. Before the execution phase of the raid the platoon leader must organize the reconnaissance mission, which could provide up to date information. However, there is a significant risk of disclosure.

In this kinds of military activities, the utilization and usage of unmanned aerial systems as UAVs can significantly save time, decrease threats and mainly reduce own casualties.

There was carried out computer aided experiment utilizing environment originated by the tactical virtual simulator Virtual battlespace (VBS). The experiment provided valuable results. The tactical virtual simulator VBS is one of the most suitable tools for experimentation of military issues at the low tactical level [28].

4.2 Use of the AARP During Mission Planning

Mission planning is a crucial part of the army unit preparation process for assigned mission (operation). The army unit leaders on squad, platoon and company level apply for mission planning methodical procedure named Troop Leading Procedures.

Troop Leading Procedures (TLP) is dynamic process used to analyse a mission, develop a plan, and prepare for an operation. These procedures enable leaders to maximize available planning time while developing effective plans and adequately preparing their unit for an operation.

TLP consists of following eight steps:

1. Receipt of mission;
2. Issue a warning order;
3. Make a tentative plan (situation analysis);
4. Initiate movement;
5. **Conduct reconnaissance;**
6. Complete the plan;
7. Issue the order;
8. Supervise and refine.

The sequence of the steps of TLP is not rigid. Leaders modify the sequence to meet the mission, situation, and available time. For example, conducting of reconnaissance can be carried out after receipt of mission or during making of tentative plan or before movement or permanently during whole planning process. To carry out reconnaissance during raid execution is also valuable and essential.

It is highly possible and recommended to use the AARP always when platoon leader decides to carry out reconnaissance. The use of AARP can provide platoon leader almost online information not only about the objective but about the appropriate avenues of approach or people and vehicles moving in the vicinity of objective which can be evaluated as suspicious or enemy activities. During the raid mission planning, when we know the objective where the enemy is located, the most significant value of the AARP use should be saving the time.

After the platoon leader analyzed the received mission in the first step of TLP and issued his warning order in the second step, it is a time to make a tentative plan. In order to do so platoon leader has to create raw time calculation for the planning phase, movement and execution of received mission within the assigned time frame. The raw time calculation is the final part of the first step of TLP where platoon leader must incorporate all necessary consequent activities.

If the platoon leader is equipped by UAV or two or three compact UAVs and if he/she is using some of battle management system tools as the AARP, he/she can take situation over the enemy and launch raid much earlier to get momentum of surprise over the enemy which is a fundamental for success of the mission.

Platoon leader without the support of the AARP have to work only with the information given by company commander and information gathered during his own line of sight reconnaissance. However, that information could be old and not up to date and subsequently could lead to wrong platoon leader's decision.

Based on the mission analysis and time schedule the platoon leader could be able to define the basic requirement on the AARP in order to gain essential information.

4.3 Experiment of Used AARP During Raid Mission

The experiment was focused on comparing the offensive action supported by a drone and the same action without the drone. The situation in the experimental scenario was as follows:

Troops of enemy are formed by military and paramilitary units, which are experienced in conducting of battle in urban area and forest area. Within the platoon attack zone there were identified forwarded combat units in the GAULA village (see Figs. 6, 7 and 8).

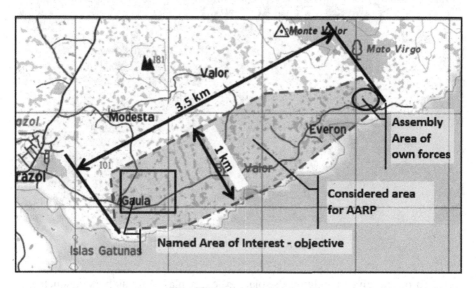

Fig. 6. Scheme of situation

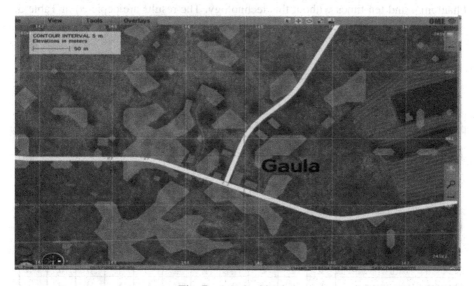

Fig. 7. Attack objective

There were identified two enemy troops with one civilian vehicle. Main equipment of enemy soldiers is 7.62 mm AR AK-47 and reactive antitank grenade launcher RPG-7. The 1st platoon was tasked to attack on order enemy forces in village Gaula as an objective 1 not later than H + 2 in order to secure the area GAULA.

Fig. 8. Attack objective in 3D

Based on the given condition the platoon leader played by authors conducted all steps of TLP and VBS 2 artificial intelligence conducted attack ten times using drone Phantom 4 and ten times without this technology. The results are depicted in Table 3.

Table 3. Experiment results

Followed values	Experiments without Phantom 4										
	EXP 1	EXP 2	EXP 3	EXP 4	EXP 5	EXP 6	EXP 7	EXP 8	EXP 9	EXP 10	Avg
Movement to LD duration (with obstacle)	15	16	16	20	14	15	18	15	16	17	16,2
Attack duration (from LD to ENY destruction	35	45	42	45	47	52	41	45	42	47	44,1
Overal duration (include area securing)	58	72	65	78	70	78	66	68	68	78	70,1
Losses of lives	2	3	2	1	3	2	4	2	1	2	2,2
Number of wounded	2	1	1	2	0	1	1	0	1	1	1
Material losses (BMP2)	1	0	1	1	1	0	0	1	1	1	0,7
TLP duration	25	24	23	27	21	26	25	24	22	23	24
Followed values	Experiments with Phantom 4										
	EXP 1	EXP 2	EXP 3	EXP 4	EXP 5	EXP 6	EXP 7	EXP 8	EXP 9	EXP 10	Avg
Movement to LD duration (with obstacle)	11	12	17	13	14	15	12	14	13	15	13,6
Attack duration (from LD to ENY destruction	35	42	40	38	39	38	37	32	37	34	37,2
Overal duration (include area securing)	45	67	64	59	60	59	56	54	55	54	57,3
Losses of lives	1	1	0	2	1	2	1	0	2	1	1,1
Number of wounded	1	0	1	1	1	1	0	1	1	1	0,8
Material losses (BMP2)	0	0	0	1	0	0	0	1	0	1	0,3
TLP duration	20	15	17	22	21	19	18	19	18	15	18,4

Thanks to the use of the AARP the platoon leader could have continuous knowledge about his/her activities in the objective and could immediately react to any changes. The results of experiment clearly show the advantages of using the AARP during offensive tactical activity – raid.

5 Conclusions

The article presents the model of the Autonomous Aerial Reconnaissance Problem. This problem has been formulated in this article for the first time. As shown in Sect. 3.5, the planning of the routes for individual UASs in the fleet according to the new model shortens the time to conduct the whole reconnaissance operation.

Using drones scripted with the AARP model significantly diminish this risk and provide even more accurate information about the objective and enemy there. Moreover, the soldiers who normally conduct the reconnaissance mission are not put in danger and could focus on execution phase of the raid.

There are countless possibilities to use the AARP in any type of tactical activities. The AARP is very promising model which could significantly influence whole approach to tactics on a level of platoon.

There is an open area for experimentation and verification of many new tactical solutions in the tactical virtual simulator VBS.

It is required and necessary to utilize the coherent application software into battle management systems in order to provide whole complex of the information support to leaders at the tactical level.

Designed model of the AARP can be one of the first swallow to be implemented into the Army of the Czech Republic Battle Management Vehicular Information System.

References

1. Hodicky, J., Prochazka, D., Prochazka, J.: Training with and of autonomous system – modelling and simulation approach. In: Mazal, J. (ed.) MESAS 2017. LNCS, vol. 10756, pp. 383–391. Springer, Cham (2018). https://doi.org/10.1007/978-3-319-76072-8_27
2. Stodola, P., Mazal, J.: Tactical decision support system to aid commanders in their decision-making. In: Hodicky, J. (ed.) MESAS 2016. LNCS, vol. 9991, pp. 396–406. Springer, Cham (2016). https://doi.org/10.1007/978-3-319-47605-6_32
3. Blaha, M., et al.: Perspective method for determination of fire for effect in tactical and technical control of artillery units. In: International Conference on Informatics in Control, Automation and Robotics, Lisboa, pp. 249–254 (2016)
4. Dantzig, G.B., Ramser, J.H.: The truck dispatching problem. Manag. Sci. 6(1), 80–91 (1959)
5. Ho, W., et al.: A hybrid genetic algorithm for the multi-depot vehicle routing problem. Eng. Appl. Artif. Intell. 21, 548–557 (2008)
6. Sharma, N., Monika, M.: A literature survey on multi-depot vehicle routing problem. Int. J. Sci. Res. Dev. 3(4), 1752–1757 (2015)

7. Gulczynski, D., Golden, B.L., Wasil, E.: The multi-depot split delivery vehicle routing problem: an integer programming-based heuristic, new test problems, and computational results. Comput. Ind. Eng. **61**(3), 794–804 (2011)

8. Imran, A.: A variable neighborhood search-based heuristic for the multi-depot vehicle routing problem. Jurnal Teknik Industri **15**(2), 95–102 (2011)

9. Reiners, Ch.: Constraint programming-based heuristics for the multi-depot vehicle routing problem with a rolling planning horizon. Ph.D. thesis, University of Duisburg-Essen, Duisburg (2015)

10. Bae, H., Moon, I.: Multi-depot vehicle routing problem with time windows considering delivery and installation vehicles. Appl. Math. Model. **40**(13–14), 6536–6549 (2016)

11. Maischberger, M., Cordeau, J.-F.: Solving variants of the vehicle routing problem with a simple parallel iterated tabu search. In: Pahl, J., Reiners, T., Voß, S. (eds.) INOC 2011. LNCS, vol. 6701, pp. 395–400. Springer, Heidelberg (2011). https://doi.org/10.1007/978-3-642-21527-8_44

12. Vidal, T., et al.: A hybrid genetic algorithm for multi-depot and periodic vehicle routing problems. Oper. Res. **60**(3), 611–624 (2012)

13. Yalian, T.: An improved ant colony optimization for multi-depot vehicle routing problem. Int. J. Eng. Technol. **8**(5), 385–388 (2016)

14. Gao, J.: Automobile chain maintenance parts delivery problem using an improved ant colony algorithm. Adv. Mech. Eng. **8**(9), 1–13 (2016)

15. Ma, Y., Han, J., Kang, K., Yan, F.: An improved ACO for the multi-depot vehicle routing problem with time windows. In: Xu, J., Hajiyev, A., Nickel, S., Gen, M. (eds.) Proceedings of the Tenth International Conference on Management Science and Engineering Management. AISC, vol. 502, pp. 1181–1189. Springer, Singapore (2017). https://doi.org/10.1007/978-981-10-1837-4_96

16. Geiger, B.: Unmanned aerial vehicle trajectory planning with direct methods. Ph.D. thesis, Pennsylvania State University, Old Maine (2009)

17. Wang, H., et al.: Three-dimensional path planning for unmanned aerial vehicle based on interfered fluid dynamical system. Chin. J. Aeronaut. **28**(1), 229–239 (2015)

18. Zhan, W., et al.: Efficient UAV path planning with multiconstraints in a 3D large battlefield environment. In: Mathematical Problems in Engineering, pp. 1–12 (2014)

19. Fu, S.-Y., et al.: Path planning for unmanned aerial vehicle based on genetic algorithm. In: International Conference on Cognitive Informatics & Cognitive Computing, pp. 140–144. IEEE (2012)

20. Keller, J.: Coordinated path planning for fixed-wing UAS conducting persistent surveillance missions. IEEE Trans. Autom. Sci. Eng. **14**(1), 17–24 (2017)

21. Lu, L., Zong, C., Lei, X., Chen, B., Zhao, P.: Fixed-wing UAV path planning in a dynamic environment via dynamic RRT algorithm. In: Zhang, X., Wang, N., Huang, Y. (eds.) ASIAN MMS 2016, CCMMS 2016. LNEE, vol. 408, pp. 271–282. Springer, Singapore (2016). https://doi.org/10.1007/978-981-10-2875-5_23

22. Da Silva Arantes, J., et al.: Heuristic and genetic algorithm approaches for UAV path planning under critical situation. Int. J. Artif. Intell. Tools **26**(1), 1–30 (2017)

23. Ingersol, B.T., et al.: UAV path-planning using bezier curves and a receding horizon approach. In: AIAA Modeling and Simulation Technologies Conference, Washington, D.C., 2016, pp. 1–14 (2016)

24. Cekmez, U., Ozsiginan, M., Sahingoz, O.K.: Multi-UAV path planning with parallel genetic algorithms on CUDA architecture. In: Genetic and Evolutionary Computation Conference Companion, Denver, pp. 1079–1086 (2016)

25. Stodola, P., Mazal, J.: Applying the ant colony optimization algorithm to the capacitated multi-depot vehicle routing problem. Int. J. Bio-Inspired Comput. **8**(4), 228–233 (2016)

26. NEO Web. Networking and Emerging Optimization. University of Malaga, Spain (2018). http://neo.lcc.uma.es/vrp/vrp-instances/multiple-depot-vrp-instances/. Accessed 7 July 2018
27. Drozd, J., Flasar, Z.: Unmanned aerial vehicle influence of troops leading procedure. In: Applied Technical Sciences and Advanced Military Technologies, Sibiu, Romania "Nikolae Balcescu" Land Forces Academy, pp. 155–162 (2017)
28. Kozubek, J., Flasar, Z.: Possibilities of verification the required capabilities according to NATO network enabled capabilities concept. Croatian J. Educ. **14**(1), 87–98 (2012)

Evaluating a Helicopter Pilot HMI for Rotor Strike Warning in a Simulated Environment

Markus Kaiser$^{(\boxtimes)}$ and Axel Schulte

Institute of Flight Systems, University of the Bundeswehr,
Werner-Heisenberg-Weg 39, 85577 Neubiberg, Germany
markus.kaiser@unibw.de

Abstract. Helicopter pilots often encounter unknown situations especially while in tactical flight or during rescue missions where they are demanded to react adequately to occurring hazards. This requires the pilots to continuously monitor their systems as well as their environment particularly in confined areas where there are numerous obstacles reducing free space for helicopter maneuvers. In those environments, pilots need to have very precise knowledge about the position and, if somehow possible, the shape of obstacles around them to be able to avoid accidents. Existing assistance systems only marginally support the pilot in this task. The R&D project "Human-Machine Interface (HMI) for Rotor Strike Warning & Hostile Fire Indication" conducted partially at our institute aims to improve those shortcomings by offering the pilots additional information. The project's goal is to design an HMI that is intuitively and efficiently usable by helicopter pilots. It uses aural as well as visual transmission of the required information. The paper describes this HMI concept as well as our helicopter simulator used for its verification. Additionally, based on the simulation environment VBS3 that is used in our simulator it explains which steps are necessary to provide the data required for the HMI concept evaluation. The paper concludes with an explanation of some possibilities for improvement, which are partly already planned for implementation.

Keywords: HMI · Rotor-Strike Warning · Simulated environment

1 Introduction

Today's helicopters are used for various purposes, which underlines their great flexibility and versatility. However, past air accidents also show their vulnerability to obstacles that are difficult or impossible to detect which may either be the case for small sized objects like wires or under DVE (brown-out, white-out, low/no daylight, etc.) and bad weather conditions [1].

Several attempts have been made to counteract such risks. For example multiple US Army and US Navy helicopters carry a "Wire Strike Protection System" (WSPS) which is intended to cut wires so that they cannot get caught in the main rotor [2].

In principle, however, the only way to reduce mentioned risks is to make the pilot aware of surrounding obstacles and narrows and thus reduce the probability of

© Springer Nature Switzerland AG 2019
J. Mazal (Ed.): MESAS 2018, LNCS 11472, pp. 530–543, 2019.
https://doi.org/10.1007/978-3-030-14984-0_39

collisions, meaning there needs to be some kind of Human-Machine-Interface (HMI) that transmits the necessary information to the pilot.

The project "HMI for Rotor Strike Warning and Hostile Fire Indication" conducted partly at the Institute of Flight Systems at the University of the Bundeswehr Munich aims to design such an HMI which is ideally also compatible to existing helicopter display concepts.

For the development and testing, the Institute of Flight Systems' own helicopter mission simulator is partly used. It was designed with various aspects in mind and expanded again for the project, which is described below as well as the basic HMI concept developed and its possibilities for expansion.

2   Related Work

In the recent past two obstacle warning systems consisting of the necessary sensors and an HMI have been developed and investigated, one by Airbus Defence and Space [3], the other by AgustaWestland (now Leonardo Helicopters, [4]). Both systems try to maintain the pilot's situational awareness by showing him the current helicopter environment as a 360° plan view. Additionally they use tones of different frequency and clocking to guide the pilot's attention towards the most critical obstacles. However, the aim of both projects was not to design a display concept but rather a holistic approach to the development of a complete warning system consisting of both display and the necessary sensory equipment.

Based on those projects Ebrecht et al. [5] recently developed and evaluated an HMSD that uses both a 3D conformal symbology and an additional 360° top view. Acoustic information or alarm tones were not employed, however.

Bronkhost et al. [6] and Simpson et al. [7] examined the effects of the use of spatial sound in the cockpit and whether its application supports the pilot. Both works show a significant improvement. Bronkhost et al. [6] also emphasize that the combination of 3D audio with visual systems brings even greater benefits. King et al. [8] investigate in their work which requirements have to be placed on the acoustic hardware in order to actually be able to use spatial sound.

3   Objective

Due to their quite simple architecture the Obstacle Warning Systems mentioned in the previous section could be evaluated in real flight with limited integration effort. Increasing system complexity, however, significantly reduces this possibility, meaning a huge part of future system development has to be done in simulated systems and based on results that are gathered from experiments conducted in those setups.

This is especially true if complex systems shall be evaluated in mission contexts. In such cases the use of full mission simulators like the one developed and used at the Institute of Flight Systems makes it possible to carry out examinations that are too expensive or dangerous if conducted in real flight. Several publications, e.g. Františ and Hodický [9], underline that so-called constructive simulation can be of great benefit for development and evaluation of future human machine systems and their respective interfaces.

As detailed in the next section there are several requirements for the subsystems forming the helicopter mission simulator. The evaluation of Rotor-Strike Warning Concepts is only possible if appropriate and accurate input data is generated by the simulated sensors. A huge number of mission simulation systems, however, does not provide interfaces to gather such sensor simulation data – after all, most of them focus on the correct mapping of interactions and correlations between people and items involved in a mission and not on the realistic representation of physical effects. The paper describes which adjustments and extensions to our simulator were required to combine both aspects and thereby allow the evaluation of our Rotor-Strike Warning Concept.

Additionally these changes are the foundation of several ideas outlined at the end of this paper, that aim to better support the pilot by taking into account the current workload and ongoing tasks, for which a deeper integration into the helicopter system is required.

4   Simulator Setup

The following section details the main subsystems of which our simulator is composed. All of them use defined interfaces based on a data distribution system to receive and transmit data which allows extensive logging as well as real-time and post processing.

4.1   Subsystems – Requirements and Design

Simulation Control Server
This component is the central node through which all clients involved in the mission are synchronized. Accordingly, the virtual sensors required for data acquisition are also connected here. In addition, it provides interfaces, mostly implemented via plug-ins, via which the mission process can be controlled or influenced.

Flight Simulation
An important part of every helicopter simulation setup is the calculation of the aircraft dynamics. It has to be sufficiently realistic to create the desired feeling of presence[1], although that does not necessarily mean that all aspects of the aircraft dynamics need to be represented perfectly. Since our simulator is constructed with focus upon missions and tasks, a COTS helicopter simulation that enables basic operations like take-off and landing, models helicopter instabilities and considers terrain information is sufficient for our purposes.

The only hard requirement that should be fulfilled is low-latency processing of pilot inputs. Otherwise controlling such an aircraft takes a huge part of the pilot's mental

---

[1] Presence and immersion are often confused; see e.g. the sidebar titled "Immersion and Presence" in [10] for a distinction.

resources and the results of experiments relying heavily on measured workload that are conducted in such a setup are prone to errors.

To maintain the flexibility of our helicopter mission simulator we added the requirement that an interface has to exist by which parameters like weather data, aircraft parameters or terrain data can be transmitted to the flight simulation. All requirements combined finally led to the decision for "X-Plane".

Out-of-the-Window View

According to Bradley and Abelson [13], visual information dominates all other human sensory perception in simulation setups. The most important part of the helicopter mission simulator therefore is our VBS3-based Out-of-the-Window image generation system. It consists of three separate, flat projection surfaces, each equipped with a rear-projection system which is controlled by its own graphics generation computer. Each surface shows a software horizontal field of view of approximately 60° and is designed to also span the same physical angle, so that the pilot is surrounded by the projection surfaces.

Tan et al. [11, 12] have found that large displays enable humans to use better, egocentric strategies to find their way in complex spatial situations because their feeling of presence heavily increases. This allows the assumption that the degree of immersion has previously increased, which is supported by Bradley and Abelson's statement [6] that due to the visuals surrounding the simulator cockpit, it "probably gives rise to very potent induced motion effects".

Besides the physical dimensions it is of great importance to ensure that new frames are rendered with a constant frame rate of more than 30 Hz. If that is not the case the pilot may lose the ability to predict further aircraft movement which could lead to pilot induced oscillations. Additionally, due to this unfamiliar behavior which the operator does not know from reality he could experience a loss of immersion.

We use distinct VBS3 clients to generate the images shown on each surface so that we can ensure we meet the above conditions. The image generation is based upon the flight dynamics data calculated by the respective component and uses real aerial imagery as terrain texture to minimize differences between simulation and reality.

Cockpit

As already mentioned our research concentrates on the investigation of complete missions using a generic helicopter. Accordingly, the exact reproduction of a specific helicopter cockpit is unnecessary. That does not mean that we would change the well-known configuration of a helicopter cockpit - in our setup the pilots still sit side by side and guide the helicopter using the usual flight controls, namely cyclic, collective and pedals – but instead of the classic displays equipped with line switch keys, we use touch-sensitive models that allow us to modify display and interaction concepts significantly faster.

Additionally our cockpit is equipped with an eye tracking system that provides us with data about where the pilot is looking. By applying statistical methods we are able to tell which information the pilots have not yet perceived and thus needs to be highlighted in order to ensure safe helicopter operations.

Virtual LIDAR-Like Distance Sensors

The aim of using virtual sensors is to obtain information about the distances between the helicopter and the surrounding obstacle scenery. Similar to the Out-of-the-Window views described above each of the available sensors uses a separate VBS3 instance which gets its data from the VBS3 mission control server and is attached to the simulated helicopter. The horizontal field of view is chosen to be

$$fov_h = \frac{360°}{n} \tag{1}$$

with n being the number of virtual sensors, so that merging all of the virtual sensors' data results in a point cloud that represents the entire helicopter vicinity.

The merging process, which will be described in detail in Sect. 4.2, requires the sensors to have minimum time offset. Ideally there should be some kind of synchronization between the sensors so that to-be-processed frames are taken at the same point in time. Many mission simulation systems, however, do not offer such an opportunity. Then, a lack of synchronicity can be compensated by increasing the framerate. With our VBS3-based setup tests have shown that above approximately 60 Hz the offsets between different sensors can be neglected and the resulting point cloud is sufficiently accurate for our purposes. Of course, this can be different if the data is intended for a different use.

4.2   Virtual Sensor Data Extraction and Processing

As stated before the virtual sensors are used to gather distance information to surrounding objects. The images rendered and then displayed by the VBS3 client instances, however, do not contain any type of distance information. Therefore a different approach is necessary.

Approach 1: Implementation of LIDAR Principle

The first possibility is to implement the principle of a LIDAR scanner using the API of the simulation system, i.e. guiding a visual beam around the helicopter at regular azimuth and elevation intervals and checking whether collisions with the environment occur. Our tests show that using this approach, 256 samples per second, i.e. 16 sampling intervals each in horizontal and vertical direction, are achieved at acceptable frame rates. According to equation

$$s = \sin\left(\frac{\frac{360°}{n}}{2}\right) * r \tag{2}$$

where r is the beam radius and n is the number of scans per level, at a distance of 50 m the spacing s between two sample points is as large as approx. 9.7 m. This resolution is neither sufficient in terms of time nor space.

Unfortunately increasing the resolution is impossible as well, as this would lead to an unacceptable reduction of the calculation frequency.

Approach 2: Using GPU Depth Data

To be able to decide which objects in a scene are hidden by other objects the GPU needs to know their positions. Accordingly the distance information for each object must be present as well. However, most of the available simulation systems like VBS3 do not offer this distance data natively over some kind of interface. Thus the only possibility is to inspect the rendering cycle and try to find calls to the graphics API which correspond to the appropriate operations.

At the time of writing VBS3 uses Direct3D9, for which multiple very specific conditions have to be met to be able to extract distance data. However, multiple graphics engines offer to apply a technique called Ambient Occlusion which enhances the rendered image's quality by using distance information[2]. This data resides in the GPU's working memory and is in general accessible via custom shaders.

Injecting the Dynamic Linking Library

In order to log the application's calls to the graphics framework and afterwards being able to copy data, the first step in our setup is to insert a dynamic linking library file into the application directory, named exactly like the one offering the system-wide interface to the graphics framework and containing all function declarations that are present in the original library file. This way, the library can act as a proxy and before passing the function calls on to the interfaces that are available in the original library modify or supplement them appropriately.

One of these supplements is the command to log all calls to functions that are related to the creation of graphics data buffers, also known as textures, and their given arguments so that a subsequent in-depth analysis is possible.

That is necessary due to the fact that there are only specific buffer formats[3] suitable to store distance data and at least one of them must be present in the log files to perform the next steps, otherwise it is impossible to implement the following steps.

As the log files reveal VBS3 uses multiple textures of the correct format. Using custom so-called shaders that are introduced via modification of the respective methods in the described proxy library, the buffer contents can be made visible. It turns out that only the largest buffer contains the desired data, which is also stored inversely proportional to the actual distance and must therefore be corrected using appropriate instructions in the shader. Figure 1 shows the original rendered image, Fig. 2 the corresponding not corrected depth data.

Applying further modifications to the mentioned shaders the data can be copied into a previously created, additional buffer in the Video RAM, which is explicitly marked as being CPU accessible. This way it is possible to transfer the depth data into system memory and afterwards process the data using conventional CPU based method.

Data Correction and Data Fusion

Figure 3 contains clearly visible distortions towards the edges, which result from the application of the pinhole camera principle and indicate the uneven resolution. Due to the direct correlation between the depth data and the pixels in the rendered image that

---

[2] For an in depth description and comparison of different Ambient occlusion techniques see [14].

[3] For D3D9 one example is D3DFMT_R32F, where depth data is stored only in the red channel.

Fig. 1. Example of an originally rendered image

Fig. 2. The corresponding depth data to Fig. 1

also means that the samples are unevenly distributed, which contradicts the behavior of a real LIDAR (Fig. 4).

Again, there are two possibilities to circumvent this problem. Either additional measuring points are inserted at the necessary positions, for example by interpolation, or, conversely, excess measuring points are removed. Since interpolation and similar methods effectively increase uncertainty, we have opted for the second option as shown in Fig. 5. If necessary the loss of information can be compensated by an increase of the initial render resolution.

The final step in providing an accurate picture of the helicopter environment is now to combine the individual sensor data. In cylindrical or spherical coordinates, this requires very little effort, because the data of the individual virtual sensors can simply be added sequentially.

5   Concept and Potential Improvements

The concept is based on the knowledge gained during the development of similar systems (cf. Sect. 2). It consists of three distinct components that provide the information by different means: a multi-function head down display (MFD), a head-mounted sight display (HMSD) and a 3D audio generator. The following sections describe their design and some special aspects that must be taken into account. Two areas are then identified in which improvements make sense or are planned.

Fig. 3. Original image of one of the virtual sensors; the distortion is clearly visible

Fig. 4. The corresponding, linear distance data to Fig. 3

Fig. 5. Screenshot of the debug application window showing the scan data after all corrections have been applied

5.1   Current Concept

Multi-function Head Down Display

Figure 6 shows the kind of presentation which is used on the multifunction display. In the center of the display is the helicopter symbol, which is surrounded by a safety circle with an offset of one meter to the helicopter. This is the absolute limit of proximity and should therefore never be reached. Concentrically around this circle further lines are

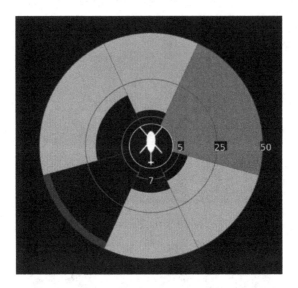

Fig. 6. Screenshot of the type of presentation chosen for the multi-function head-down display

visible, which indicate the different levels in the assessment of criticality. Depending on the concrete scenario and the helicopter model used, the distances may have to be varied in order to take into account the experience of the pilots.

If an obstacle enters the vicinity of the helicopter, the sector surrounding the obstacle is highlighted in the color corresponding to the warning level. Intentionally not all sectors have the same size. Instead, they are shaped so that the boundaries between the two lateral sectors and the three backward sectors coincide with the boundaries that mark the areas directly visible to the pilot.

The reason for our decision to use this type of display is, on the one hand, that military pilots in particular are used to sector-based displays due to other systems (e.g. Electronic Warfare System) and, on the other hand, because large areas that change color can attract the attention of the pilot much more easily than would be the case with more filigree variants.

HMSD

Especially in confined areas, helicopter pilots rarely look at their instruments. Instead, they look outside to identify obstacles and assess their own movement towards the environment. In such a situation they could therefore miss changes on the MFD.

To prevent this the HMSD is designed as a counterpart to the MFD. Because it always displays its information in the pilot's field of vision, the probability of perception is significantly higher than that of information displayed on the MFD.

It can be assumed that the MFD could also be completely dispensed with if a sufficiently detailed overview display is integrated into the HMSD. As already mentioned in Sect. 2, Ebrecht et al. [5] uses the example of maintenance missions on wind turbines to show a presentation concept designed in this way, in which apparently no further displays are used.

Nevertheless, we do not renounce the MFD. Often, significantly more data and graphics need to be displayed simultaneously in the HMSD than is the case in the images of Ebrecht's missions. If these data and graphics get too extensive, the pilot could be obstructed in his sight and lose the ability to safely guide the aircraft. Accordingly, it is of great importance to handle the available display area with restraint, i.e. use simple, slim graphics if possible.

In our case we have therefore opted for an abstract, two-stage representation of the obstacle backdrop, which does not contain any concrete distance information. If required, the pilot can still receive these via the MFD.

3D Audio

Even if critical situations are displayed on the HMSD or MFD, the pilot can miss them. To prevent this, synchronous audio warnings are output, consisting of a sequence of tones and a subsequent direction indication ("proximity one o'clock). In addition, the system has the ability to position sounds virtually in space, giving the impression that these sounds actually come from this direction.

Due to the possibility of being able to locate the initial sound sequence directly when a warning occurs, it should be possible for the pilot to locate the source of danger much faster. However, it is subject of forthcoming investigations whether this assumption actually applies.

5.2   Evaluation Results

Three helicopter crews of the German Armed Forces with experience on different helicopter types and total flight hours between 600 h and 5000 h with a median of 3100 h were invited for a basic evaluation of the described concept. The system was tested in various situations that were part of two fictional military missions.

In the first one the pilots were instructed to follow a specific flight path after take-off from their base and over the course of this flight practice several off-field landings in areas with heavy tree cover. On the one hand, this task served to familiarize the pilots with the helicopter simulator, on the other hand, the pilots were forced to actively enter situations in which the Rotor Strike Warning System could potentially prove itself.

The second mission then consisted of three different tasks with increasing difficulty. At the beginning they were supposed to land on a sports field, which was confined by high floodlight poles and partly blocked by construction machinery, subsequently, they should land on a sandbank limited on three sides by tree growth and, last but not least, they had to pick up a sling load in the immediate vicinity of a high lattice mast of a cableway.

Following each mission, a questionnaire survey consisting of 15 five-level Likert type items and multiple yes-no questions was used to find out how the pilots assessed the HMI in general and specific design aspects in those situations. One of the helicopter crews was only able to complete the first mission, which is why their ratings are not included in the figures below.

Figure 7 as well as Fig. 8 clearly show that the system is considered at least partly helpful by all pilots. In addition, there are no negative ratings, which indicates that also the basic concept of the system is considered appropriate. Also the general satisfaction with the system seems to increase from the first to the second mission, as nearly all

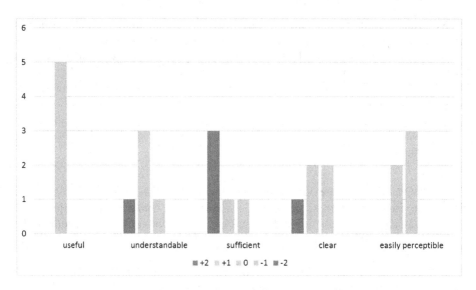

Fig. 7. Questionnaire results (ratings) after the first mission

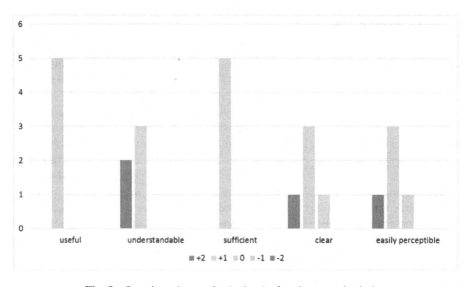

Fig. 8. Questionnaire results (ratings) after the second mission

ratings increase. This leads to the assumption that with further training, even better values can be achieved.

However, as the assessment of the benefits also shows, none of the pilots completely agrees with the statement that the system is helpful, i.e. rates it with +2, so there is obviously still room for improvement. This obviously also applies to the amount of information transmitted, as can be seen from a comparison of the corresponding ratings after both missions. It is reasonable to assume that the second mission required additional information due to a different task profile, the significance of which was not yet clear at the time of the first mission.

5.3   Intended Improvements and Outlook

The already existing Rotor Strike Warning systems described before are, like our existing warning concept, restrainedly designed. If critical situations occur, they are displayed or indicated with acoustic notices, but no further action is taken, although the available data basis also permits more advanced options.

Based on the performance data of the helicopter, the system could, for example, distinguish safe from unsafe areas and transmit these findings to the pilot in a suitable form. It would be conceivable, for example, to color hazard areas in the HMSD or to display alternative recommendations. Of course, the principle of using the available display area as sparingly as possible must also be applied here in order to avoid cluttering.

Ultimately, such a system could even cause the flight vector to change automatically, i.e. force the helicopter to retard in slow flight or alternatively change pitch and/or bank if the pilot does not react to occurring hazards.

The latter proposal should of course be treated with great caution, as automatic interventions can quickly lead to rejection of the system if they do not occur in a comprehensible or predictable manner or interfere with the pilot's intensions. Accordingly, a collision avoidance system should basically be configured so that it actually only intervenes if the collision can no longer be avoided by humans. Then there is a reasonable expectation that the system will actually be accepted by the pilots. Comparable systems called GCAS (Ground Collision Avoidance System) are already in use in fighter aircraft such as the F-16 [15] and have proven to be very effective.

If there is enough time before the collision, the pilot should in any case be informed of an imminent intervention of the system.

In the simplest case, this happens as soon as the time remaining until the collision falls below a certain value. This value should take into account the typical reaction time of the pilot as well as parameters depending on the current flight situation in order to give the pilot the possibility to intervene by himself, provided he is (still) capable of doing so.

The weak spot of this approach is the assumption that a pilot would react equally to warnings in any situation, provided he is able to do so in principle - in fact, his reactions differ quite significantly. If his workload is low, he may perceive an emerging hazard without any indication from the Rotor Strike Warning Systems at all. In such a situation warnings would probably even disturb the pilot. Conversely, more support is necessary in phases of high workload, which should also take place at other times. The support must therefore be context- and situation-dependent. Theißing et al. [16] and Honecker et al. [17] refer to this as "adaptive assistance".

Key element to its success is the ability to closely monitor the pilot's interactions, among other things for example also by means of eye-tracking, its mapping to specific tasks (see [18] for an example) and ultimately the comparison with models, which provide information about the tasks to be completed at this point in time. From this, appropriately adapted actions of the assistance system can then be derived.

This graduated form of warning could of course also be applied to the original warning concept, independently of the assistance functions described before. This would increase both the probability not to hinder the pilot in his tasks and the probability to give him the possibility to react in time in case of danger.

Ideally then the system should also consider Wickens' Multiple Resource Theory [19], according to which every human being has only limited mental resources that are divided into different channels. If a channel is fully occupied by certain tasks or actions, it represents a bottleneck for further information that is received via the same channel, even if other available mental resources are freely available. For example, if the pilot is communicating via radio, he may have problems processing spoken warnings that are issued in parallel. In such a situation, a system that works adaptively would use resources that are not already occupied by other tasks.

For the near future, the main focus is on the implementation of the automatic intervention capability. Experiments will show whether helicopter pilots can be convinced of such deep interventions in flight control.

References

1. Baker, S.P., Shanahan, D.F., Haaland, W., Brady, J.E., Li, G.: Helicopter crashes related to oil and gas operations in the Gulf of Mexico. Aviat. Space Environ. Med. **82**, 885–889 (2011). https://doi.org/10.3357/ASEM.3050.2011
2. Nagaraj, V.T., Chopra, I.: Safety Study of Wire Strike Devices Installed on Civil and Military Helicopters (2008)
3. Waanders, T., et al.: Helicopter rotorstrike alerting system. In: 41st European Rotorcraft Forum (2015)
4. Brunetti, M., Costa, C.: The guardian project: reasons, concept and advantages of a novel obstacle proximity lidar system. In: 40th European Rotorcraft Forum (2014)
5. Ebrecht, L., Ernst, J.M., Döhler, H.-U., Schmerwitz, S.: Integration of an exocentric orthogonal coplanar 360 degree top view in a head worn see-through display supporting obstacle awareness for helicopter operations. In: Yamamoto, S., Mori, H. (eds.) HIMI 2018. LNCS, vol. 10905, pp. 369–382. Springer, Cham (2018). https://doi.org/10.1007/978-3-319-92046-7_32
6. Bronkhorst, A.W., Veltman, J.A. (Hans), Van Breda, L.: Application of a three-dimensional auditory display in a flight task. Hum. Factors J. Hum. Factors Ergon. Soc. **38**, 23–33 (1996). https://doi.org/10.1518/001872096778940859
7. Simpson, B., et al.: 3D audio cueing for target identification in a simulated flight task. In: Proceedings of the Human Factors and Ergonomics Society Annual Meeting, vol. 48, pp. 1836–1840 (2004). https://doi.org/10.1177/154193120404801610
8. King, R.B., Oldfield, S.R.: The impact of signal bandwidth on auditory localization: implications for the design of three-dimensional audio displays. Hum. Factors J. Hum. Factors Ergon. Soc. **39**, 287–295 (1997). https://doi.org/10.1518/001872097778543895
9. Františ, P., Hodický, J.: Human machine interface in command and control system. In: 2010 IEEE International Conference on Virtual Environments Human-Computer Interfaces and Measurement Systems Proceedings, VECIMS 2010, pp. 38–41 (2010). https://doi.org/10.1109/vecims.2010.5609345

10. Bowman, D.A., McMahan, R.P.: Virtual reality: how much immersion is enough? Computer (Long. Beach. Calif) **40**, 36–43 (2007). https://doi.org/10.1109/mc.2007.257
11. Tan, D.S., Gergle, D., Scupelli, P., Pausch, R.: With similar visual angles, larger displays improve spatial performance. In: Proceedings of the Conference on Human Factors in Computing Systems, CHI 2003, p. 217 (2003). https://doi.org/10.1145/642647.642650
12. Tan, D.S., Gergle, D., Scupelli, P.G., Pausch, R.: Physically large displays improve path integration in 3D virtual navigation tasks. In: Proceedings of the 2004 Conference on Human Factors in Computing Systems, CHI 2004, vol. 6, pp. 439–446 (2004). https://doi.org/10.1145/985692.985748
13. Bradley, D.R., Abelson, S.B.: Desktop flight simulators: simulation fidelity and pilot performance. Behav. Res. Methods Instrum. Comput. **27**, 152–159 (1995)
14. Aalund, F.P.: A comparative study of screen-space ambient occlusion methods, vol. 1, pp. 1–59 (2013)
15. Swihart, D.E., et al.: Automatic ground collision avoidance system design, integration, & flight test. IEEE Aerosp. Electron. Syst. Mag. **26**, 4–11 (2011). https://doi.org/10.1109/MAES.2011.5871385
16. Theißing, N., Liegel, A., Schulte, A.: Verhindern von Pilotenfehlern durch ein zustandsadaptives Assistenzsystem. In: Grandt, M., Schmerwitz, S. (eds.) Kooperation und kooperative Systeme in der Fahrzeug- und Prozessführung (DGLR-Bericht 2015-01), pp. 97–113 (2015)
17. Honecker, F., Brand, Y., Schulte, A.: A task-centered approach for workload-adaptive pilot associate systems. In: Proceedings of the 32rd Conference on European Association for Aviation Psychology, Cascais, Port, pp. 485–507 (2016)
18. Honecker, F., Schulte, A.: Automated online determination of pilot activity under uncertainty by using evidential reasoning. In: Harris, D. (ed.) EPCE 2017. LNCS (LNAI), vol. 10276, pp. 231–250. Springer, Cham (2017). https://doi.org/10.1007/978-3-319-58475-1_18
19. Wickens, C.D.: Multiple resources and performance prediction. Theor. Issues Ergon. Sci. **3**, 159–177 (2002). https://doi.org/10.1080/14639220210123806

Experiment of the Tactical Decision Support System Within Company Defensive Operation

Jan Drozd[(⊠)]

University of Defence, Brno, Czech Republic
jan.drozd@unob.cz

Abstract. This paper presents experiment of the Tactical Decision Support System (TDSS) implementation within company defensive operation. TDSS system fully integrates model of aerial reconnaissance, which significantly influence decision-making of the company commander during the defensive operation. The experiment was conducted with support of simulation technology, particularly Steel Beast Pro software. As a part of various scenarios the paper presents results of UAS reconnaissance model use within different scenario on tactical level. The key part of this article defines parameters of the experiment, evaluates the results and confirms the purpose of the model in support of the decision-making process of a commander. The paper also deals with impact of usage modern technology on decision-making process.

Keywords: Unmanned aerial system · Aerial reconnaissance model ·
Tactical decision support systems · Experiments · Defensive operation ·
Static defense · Steel Beasts

1 Introduction

Military decision making process or in case of company level Troops leading procedure (TLP) is one of the crucial prerequisites for successful fulfilment of the given task. Effective TLP is essential in any tactical activities carried out by company commander and could, and definitely will significantly influence success probability. In order to support commander's TLP there are countless systems using. TDSS is a system used within Czech Army and its purpose is to automated decision making process of the commander on any level [1–3].

TDSS fully integrates UAS reconnaissance model, which is a planning model to reconnoiter area of interest of the commander. In the experiment described lower the model has been used to support decision making process of the company commander during the defensive operation. The focus has been put on the decision made during the execution phase where a lot of variables take place. The aim of the paper is not to fully and in complexity describe this model, but to verify its usage during the company defensive operation, particularly static defense and quantify some varieties.

© Springer Nature Switzerland AG 2019
J. Mazal (Ed.): MESAS 2018, LNCS 11472, pp. 544–552, 2019.
https://doi.org/10.1007/978-3-030-14984-0_40

2   UAS Reconnaissance Model

The model has been developed and implemented to the TDSS in the University of Defense in Czech Republic. The model was designed to reconnoiter area as fast as possible and allows using more than one drone or any other unmanned aerial vehicles. During the described experiment the model was used within the simulation software to cover specified area and only one drone has been used [3, 4].

The principle of the model is to set up several waypoints which cover designated area. The distance of the waypoints is given by the specification of the used drone or UAV. The main technical feature which has to be taken into the consideration is capacity of the battery, quality of camera, minimum and maximum height act.

The model plans the roads to scan whole area of commander's interest in order to minimize the maximum time of flight. The model is designed to take into the consideration unlimited number of drones or used UAV, however in our case only one drone was used. The mathematical formulation could be found in [1, 5–7]. Basically the formula of the model is:

$$\text{Minimize} \sum_{i=1}^{N} T_i \tag{1}$$

Where T_i is the distance travelled by i-th vehicle,
N is the total number of vehicles.

3   Experiments

The section deals with description of whole experiment as well as with short introduction of the simulation software used. The experiment was focused on the tactical activities define as a static defense on the company level.

3.1   Steel Beasts Pro Software

The software developed by the eSim games company Steel Beasts Pro (SB Pro v4.0) has been used in order to repeatedly simulate tactical activities on the company level. Currently Steel Beasts professional is used as a virtual constructive simulation by military personnel all over the world (Austria, Germany, Estonia, Finland, Sweden, Spain and others). Steel Beasts professional has wide training spectrum one of which is company level combined arms tactical training. Significant advantage of the Steel Beasts professional is a fact, that the artificial intelligence (AI) is on very high level and simulation could be conducted without additional operators. This aspects played significant role. Additional operator could bring unexpected behavior or decision making which would result in distorted results of the experiment.

Scenario for the experiment was prior completion consulted with the authors of the UAS model. In order to verify primarily above mentioned model, UAS reconnaissance model has been temporarily implemented to the simulation software. Subsequently during the creation of scenario this model has been used to plan all UAS waypoints as well as communication means with simulated company commander.

3.2   The Experiment Design

In order to verify impact of the UAS model used within TDSS during the defensive company operation, following condition of the experiment were set up:

- The organization structure of the infantry company is as depicted in Fig. 1;
- The organization structure of the enemy is in Fig. 2;
- Equipment and weaponry was set up the same for blue and red forces;
- The level of the training was set up the same for both parties;
- The initial deployment was set up prior simulation by the author Fig. 3;
- The simulation was conducted ten times with using of the UAS model and ten times without the model.

As it was mentioned the task organization of the friendly forces are depicted in the Fig. 1. For purpose of the experiment the weaponry and detailed structure of the friendly forces mechanized company is not important since the structure of the enemy forces are the same.

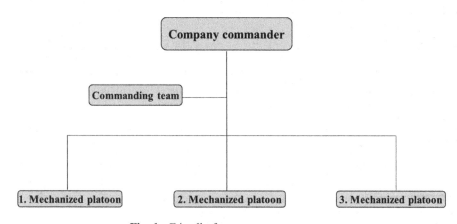

Fig. 1. Friendly forces company structure

Task organization of the enemy forces, particularly enemy mechanized battalion is depicted in the Fig. 2. The role of the commanding company as well as logistic company was not significant during the experiment since at the beginning of the battle all units were fully equipped by the necessary materials for more than two days battle.

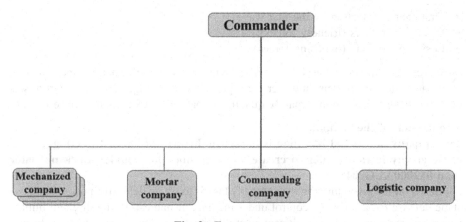

Fig. 2. Enemy structure

In the Fig. 3, there is an initial deployment of friendly forces before operational time of attack. The initial deployment of enemy is not depicted in the Fig, since it was not part of experiment and all enemy forces were steered by AI of the simulator.

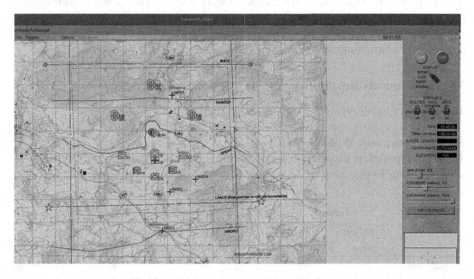

Fig. 3. Initial deployment of friendly forces

The aim of the experiment was to verify or falsify following hypothesis:

The usage of the UAS model implemented within TDSS, shortens the reaction time of blue forces as well as diminishes losses on the lives.

In order to verify or falsify above hypothesis there were set of varieties as follows:

- Reaction time of the cover forces – starting time of any movement, open fire or regrouping cause by enemy activities;

- First enemy detection by the first wave;
- Losses of soldiers (friendly forces);
- Losses of soldiers (opposing forces).

Of course the fact that UAV will provide accurate information and improve information flow is obvious; however the experiment was designed as a one part of sets of experiments which should mainly quantify impact of UAS reconnaissance model.

The mission of the company:
1st company of the 2nd battalion is to secure the area of interest (see Fig. 3) and cause enemy loses in order to create the conditions for counterattack not later than 012000AUG2018

Based on the given mission the company task is to set up defensive position and use static defense tactics to keep control under the area of interest. By the fire and limited movement maneuver cause significant enemy loses in order to create conditions to take initiative and by the offensive action defeat the enemy.

The simulation was conducted within the operation time starting by 0700 h in the morning. Varieties were related to this time. After the enemy cross the Front Edge of Battle Area (FEBA) the simulation was stopped.

Conditions for simulation were for all experiments the same. In order to have reliable results all experiments were conducted in four computers where four experiments were held simultaneously. Subsequently all computers were clean up (registers) and second part of experiments were conducted. The last two experiments were conducted just in two computers.

3.3   Experiments Results

The results of experiments are sum up in the Table 1.

3.3.1   Experiment Without UAS
The variety **reaction time** was counted mostly by the first movement or by the time when covering forces open fire on enemy. In all experiments both actions starts almost

Table 1. Results of experiments without UAS

	Reaction time	First enemy detection	Losses of soldiers (friendly forces)	Losses of soldiers (opposing forces)
Experiment 1	45	65	25	48
Experiment 2	42	78	19	43
Experiment 3	52	69	29	52
Experiment 4	63	89	25	39
Experiment 5	45	60	17	40
Experiment 6	51	78	26	41
Experiment 7	49	61	26	36
Experiment 8	39	58	20	39
Experiment 9	55	71	18	42
Experiment 10	37	60	23	46

in the same time or movement delay was not significant. The average time was 47.8 min. It is necessary mentioned that covering forces are in front of troops deployed within the first wave. And they will detect enemy first prior the enemy reach main battle area.

First enemy detection was counted by the time when the first soldier (in the first wave) of the friendly forces reported enemy activities in front of FEBA. By the detection the visual contact is meant. The average time of enemy detection was 68.9 min since simulation starts.

Losses of friendly and opposing forces were counted as a number of killed soldiers just by the opposing party. There were two cases of killed soldiers by the friendly fire, but these were not taken into the consideration. The average of losses in case of friendly forces was 22.8 soldiers and for enemy forces was 42.6 soldiers.

3.3.2 Experiment with UAS

Second set of experiments was conducted in the same conditions. Just the difference was that friendly forces were using UAS implemented within TDSS. Particularly the company commander had one UAS all the time observing area in front of covering forces (about 2–3 km in front of FEBA) and one UAS in front of FEBA (Table 2).

Table 2. Results of experiment with UAS

	Reaction time	First enemy detection	Losses of soldiers (friendly forces)	Losses of soldiers (opposing forces)
Experiment 1	30	61	15	50
Experiment 2	32	68	17	52
Experiment 3	24	55	12	56
Experiment 4	26	54	13	48
Experiment 5	29	49	15	52
Experiment 6	32	52	16	57
Experiment 7	33	51	18	53
Experiment 8	32	55	17	49
Experiment 9	28	54	15	47
Experiment 10	29	54	11	49

Average of the **reaction time** was 29.5 min. Compare to the first set of experiment without UAS there was significant reduction of time from 47.8 min to 29.5 min with represent 38.3%.

Average of the **first enemy detection** (visual) was in the second set of experiments 55.3 min witch represent compare to the first set of experiment (average 68.9) reduction for 19.7%.

Average of the friendly forces losses was 14.9 soldiers. In the first set of experiments it was 22.8 soldiers. There was reduction for 34.6%.

Average of the opposing forces losses was 51.3 soldiers which is compare to the first set of experiments increase for 20.4%.

3.4   Results Explanation

Based on the simulation there were some remarkable results.

Reaction Time of Covering Forces

The goal of the covering forces is primarily to reveal the enemy troops as soon as possible in order to provide enough time to the first wave of friendly forces to get ready for defense. Moreover the covering forces are to conduct deceives operation in order to disoriented enemy about the real position of FEBA.

As it was mentioned above by using UAS, particularly one UAS was used (simulated) for covering forces, there was reduction of the reaction time by 38.3%. Such reduction could be considered as a significant. However there was one more interesting result of the comparison. During the first set of experiments without UAS there was not substantial time between the first movement of the covering forces and fire opening. The covering forces at the beginning open the fire and subsequently start regrouping based on the enemy observation.

During the second set of experiments the movement or regrouping preceded opening fire. The reason for this is that covering forces firstly observed enemy movement thanks to the UAS which gave to the covering forces enough time to regroup in order to prepare effective fire. This fact has relation to another varieties enemy and friendly forces losses.

First Enemy Detection

This variety was, by UAS usage, diminished for about the 19.7%. The important fact is not only significant reduction of the first enemy observation but also the fact that enemy was firstly spotted by the UAS with attached HD cameras. This means that enemy was observed by friendly forces from out of the direct fire distance. This gave friendly forces more time to prepare for first contact with the enemy similarly to covering forces.

Losses of Soldiers (Friendly Forces)

Experiments proved that usage of UAS within TDSS lead to the significant reduction of soldier's losses. In the case of the above experiment the reduction was for 34.6%. The number corresponds to the experiment from (odkaz na minulý článek). During this experiment smaller unit was taken into the consideration. However the results of all experiments clearly show that usage of UAS within TDSS using above mentioned algorithm is effective.

Despite the numbers there was also another aspect which is necessary to mentioned. During the first set of experiments more than 50% of friendly forces losses were from the covering units and the second half from the first wave units on FEBA. During the second set of experiments only 25% was from covering units. This covering forces losses reduction was cause by the fact, that they were able to prepare action against first line of enemy in advance since they had almost online information about their position. The increase of the friendly forces losses on the FEBA was obviously cause by the superiority of the enemy.

Losses of Soldiers (Opposing Forces)
Average of the enemy losses increase from the first set of experiment for 20.5%. This number could be considered as a very significant. Similarly to the friendly forces losses there was another aspect of UAS usage. During the second set of experiments more than 60% of losses were cause by the contact with covering friendly forces. Very interesting is that in meantime losses of covering forces contrary to first set of experiments were diminish by 50%.

4   Conclusions

The experiments result revealed very interesting data. It is possible to state that usage of the UAS within TDSS during the defense operation on the company level has significant effect on the efficiency. The data proved that all varieties changed in a positive way for friendly forces. There are still a lot of variables which has to be taken into the consideration prior any decision. Most of the units are aware of the UAS or similar technologies and are taking countermeasures. However the results of the experiment and also the results of the experiment on the squad level [1], shows how important modern technology could be in a modern warfare.

The hypothesis for this experiment was defined as follows:

The usage of the UAS model implemented within TDSS, shorten the reaction time of blue forces as well as diminish losses on the lives.

There could be clearly stated that hypothesis has been verified. Moreover except the varieties which were part of the hypothesis another very interesting facts already mentioned above were revealed by the experiment.

Moreover UAS usage could bring significant change military tactics procedure. To have continuously updated data about the enemy position and movement could tremendously reduce time for preparation phase of any action or operation.

References

1. Drozd, J., Stodola, P., Křišťálová, D., Kozůbek, J.: Experiments with the UAS reconnaissance model in the real environment. In: Mazal, J. (ed.) MESAS 2017. LNCS, vol. 10756, pp. 340–349. Springer, Cham (2018). https://doi.org/10.1007/978-3-319-76072-8_24. ISBN 978-3-319-76071-1
2. Stodola, P., Mazal, J.: Tactical decision support system to aid commanders in their decision-making. In: Hodicky, J. (ed.) MESAS 2016. LNCS, vol. 9991, pp. 396–406. Springer, Cham (2016). https://doi.org/10.1007/978-3-319-47605-6_32
3. Hodický, J., Františ, P., Litvaj, O.: Validation of simulator supporting movement of small group or individuals in different terrains. In: 2012 Mechatronika, pp. 282–284. Czech Technical University in Prague, Prague (2012). ISBN 978-80-01-04985-3

4. Františ, P., Hodický, J.: Decision support system for a commander at the operational level. In: Proceedings of 1st International Conference on Knowledge Engineering and Ontology Development, INSTICC – Institute for Systems and Technologies of Information, Control and Communication, Madeira (2009). ISBN 978-989-674-012-2
5. Drozd, J., Flasar, Z., Stodola, P.: Use of modern technologies for combat units preparation and management, vol. 65, no. 3, pp. 602–616. Military Technical Courier (2017)
6. Šilinger, K., Blaha, M.: The new automated fire control system for artillery units based on interoperability and standards. In: 2017 Proceedings of the 14th International Conference on Informatics in Control, Automation and Robotics ICINCO, pp. 332-337. SciTePress, Setúbal (2017). ISBN 978-989758263-9
7. International Archives of the Photogrammetry, Remote Sensing and Spatial Information Sciences: ISPRS Archives Volume 42, Issue 2W6, pp. 101–107, 23 August 2017. 4th ISPRS International Conference on Unmanned Aerial Vehicles in Geomatics, UAV-g 2017, Bonn, Germany; 4–7 September 2017, Code 130171

Possibilities of Raster Mathematical Algorithmic Models Utilization as an Information Support of Military Decision Making Process

Jan Nohel[(✉)]

University of Defense, Brno, Czech Republic
jan.nohel@unob.cz

Abstract. The planning and decision-making process for conducting combat operations must take into account the development of the situation in a number of diverse areas that affect the action of forces and resources on the battlefield. An analysis of their impact can be done through computational programs and mathematical algorithmic models, using a raster representation of geographic and tactical data. Raster data layers in the mathematical model allow for a combination of surface effects, terrain elevation, and weather effects, as well as the forces and vehicles of both the enemy and your own units. Mathematical calculations result in a manoeuver route, optimized based on predefined criteria, such as speed and safety. The result can have a double use. In case of the manoeuver of enemy forces and vehicles, it is possible to simulate their probable future activity. The result of combining all effects of the situation on the battlefield is the optimal route of manoeuver of your own units designed in several seconds. Utilization of such software will provide commanders substantive independence and speed of decision making in the course of military operations. Sharing of designed route of manoeuver with all adjacent units and higher headquarters also enable to coordinate the activity of all superior task force.

Keywords: Decision-making process · Mathematical models · Algorithm ·
Raster data · Route of manoeuver

Every combatant or leader of an armed group and, above all, the commander of a military unit, must strive for the effective accomplishment of the intended combat task. The security of their own units, the protection of the civilian population, the time expenditure, the economy of forces and resources, and the maximum destructive or combat effect while maintaining their own losses at minimum are the most important factors in the successful conduct of combat operations. Detailed and comprehensive assessment and use of these factors is carried out by the crew of the military task force or unit commander during the decision process. The result is a decision on the activity of subordinate forces and resources, which is issued either in the planning and preparation process of the combat activity, or during the course of the combat activity. Especially in the latter case, great emphasis is placed on the speed of decision that can be created and realized in direct contact with the enemy. A delayed or ineffective

J. Mazal (Ed.): MESAS 2018, LNCS 11472, pp. 553–565, 2019.
https://doi.org/10.1007/978-3-030-14984-0_41

decision may in such a situation cause, for example, the encroachment and destruction of our own units, disproportionately high losses or failure to perform the specified operational task. To eliminate these risks, it is necessary to ensure a rapid, comprehensive and comprehensible communication and information support to the command and control process. Its purpose is to create a situation report, to understand the situation on the battlefield and to gain predominance over the enemy. To do this, it is necessary to continuously process multispectral information from all areas of the battlefield and to distribute them through Common Operational Picture (COP). Obtaining and evaluating a great deal of information from various situational factors, and using it to make a qualified decision, is a time-consuming process. The solution of these crisis situations is the use of digitized information, raster representation of data and mathematical algorithms of a tactical computer information system.

1 Obtaining Superiority on the Battlefield

Without gaining predominance over the enemy, they cannot be defeated. The superiority of the command and control process consists of several interconnected parts [1]. The information superiority over the enemy is based on the continuous collection of information, through a network of sensors from the full spectrum of operation in the space of operation, in real time. The information thus obtained must be converted into a format compatible with the other elements of the information system, so that it can be summarized and further processed. In order to achieve information superiority, it is therefore necessary to obtain, process into a uniform format, and distribute all information relevant to the orientation on the battlefield faster and more efficiently than the enemy.

Quickly accomplished complex orientation and understanding of the situation in the area of operation creates conditions for gaining the knowledge superiority over the enemy. The basic element for understanding the battlefield is the COP [2], which usually contains the current position of our own forces and means, the enemy, the civilian population, and the obstacles or impassable terrain displayed on the map background. Through the COP, the commander acquires at least a rough overview of the situation, but without a more detailed assessment of its development.

Means for creating COP are:

- Command and control communication support that provides data and information transmission between two physically separate locations,
- Information support for command and control, which is the process of collecting and retrieving data and information, processing, distributing, archiving and protecting it.

The next step is to gain knowledge, which means to have an overview not only of the kind and position of the enemy's forces and resources, but also of their ability to act and their intended activities.

Knowledge of the situation on the battlefield consists of merging two types of information:

- unambiguous information
  - describing, for example, the specification and position of forces and means, geographical characteristics, and weather forecasts,
  - does not need almost any additional interpretation,
  - a common overview of the situation is created by summarizing this type of information,
  - without further assessment, there may be a problem with their implication in a wider context,
- tacit information
  - require interpretation, and complement the findings,
  - include, for example, the capabilities of forces and resources, enemy's tactical procedures, local habits, and the probable enemy intent,
  - correct interpretation of the information depends on the knowledge and experience of the analyst who interprets the information.

Knowing the enemy's abilities and showing them at the COP promotes a comprehensive understanding of the battlefield situation that allows for more effective use of combat power. Thus, predominance in decision-making means comprehensively analysing and comparing possible alternatives to the operation of our own units more effectively and faster than the enemy. Rapid creation of the optimal variant of maneuver of our own forces and resources in a combat operation allows commanders to gain superiority over the enemy in decision making. Finally, if the commander has a decision-making priority, based on quick source analysis of all the important factors on the battlefield, he can flexibly and effectively use his maneuver units to conduct combat activities. The whole unit thus gains operational superiority over the opponent, which significantly increases its combat skills. But of course, the ability of communication and information means also to directly and comprehensibly distribute the information and intent of the commander to his subordinates so that it is ready for immediate use.

2 Automation of the Command and Control Process

Obtaining superiority over the enemy requires extensive automation of information support. Each soldier, unit, or weapon system in a military operation is a sensor where, as a source of information in its field of activity, it contributes to the creation of the COP. Sensors (information gathering elements) are crucial to the ability to independently search for and identify things or objects that affect the operation of our own forces and resources in the operating space. Acquired data is immediately sent to a single information database or to selected decision elements and means of action, which need them to carry out their tasks. By storing them in structured databases and displaying them on a computer screen or data projector, the unit creates a COP on the battlefield, accessible to all authorized users in an operation, based on unambiguous information. Based on this COP, commanders and staff create the knowledge base for making decisions about the activities of subordinate units. Defined decisions are then implemented by units as maneuvering elements.

Modern networked armed forces combine the activities of all these elements into an integrated, dynamic and flexible information management process. Its center is the COP, consisting of a map of the space of interest and multispectral information from various sources, converted into a uniform format. The networking of sensors with decision elements and means of action helps us to use the information obtained to quickly and efficiently operate our own maneuvering forces and means to achieve the objectives of the operation. Maneuvering resources can obtain the information needed for their activity either from a common overview and situation, or directly by requiring them from specific sensors in contact. Thus, all authorized elements on the battlefield can contribute to the common information awareness, sharing information obtained in the course of their own tasks.

Separate or automated sensor activity may include locating an object of interest ina designated area in the case of an observation means, identifying it, specifying, continuously monitoring, and distributing its own location and operation data. Subsequent distribution of the information thus obtained, along with its geographic location, contributes to the creation of the COP for all users in the joint network. The received data is displayed on the map background via its positional characteristics. Network sharing and display of information allows for a wide range of tactical information systems, such as the United States Army BFT (Blue Force Tracker) or the Czech Army's Operational Tactical System Command and Control of Ground Forces, see Fig. 1. Their use is easy to get the COP on the battlefield, usually without tacit information. By implementing appropriate mathematical algorithmic tools into tactical geographic information systems, their maneuverability and maneuverability can be calculated for the identified combat units and techniques. This implementation must be based on a fusion of visibility, range or combat effects and throughput in relation to the type of surface, the vegetation and the relief of the terrain, possibly affected by weather conditions. Automated data on the deployment of enemy forces and resources, complemented by a specification of their combat capabilities, will provide a general overview of the situation of the commanders and the combat force and the effectiveness of all units in the combat operation. Comprehensive knowledge of the battlefield supports autonomy, initiative, and consistent synchronization in performing tasks for all cooperating units in the military operation.

Nowadays, in every developed army, a battlefield digitization project, which is a network of interconnected computer stations, is being developed. Mutual sharing and exchange of information between computer stations is the core of this system. The purpose is to provide each commander with the maximum amount of information needed to navigate the area of operations and to make informed decisions. The US Army has developed the NCW (Network Centric Warfare) concept at the end of the last century. Its purpose is to collect information through sensors, to quickly process and analyze them, and then to distribute and share all levels of command and control to individual soldiers. NNEC (NATO Network Enabled Capability) is a less expensive European version of NCW. According to the United Kingdom Department of Defense, the NEC concept includes connecting sensors, decision elements, weapon systems, and supporting capabilities to achieve greater military effectiveness by using available information [4]. Networked sensors create a COP or a knowledge base for elements of the commander and staff, who decide on the effective assignment and action of networked maneuvering forces and resources. Maneuvering forces and resources carry out these tasks in order to achieve ultimate success in the military operation. In near real-time, the computer creates an overview of all available tactical geographic information on the battlefield [5]. The data and information obtained can be searched for in databases according to time, space, type and significance of their occurrence, for clear and immediate use. The networking of sensors and decision elements with the means of action facilitates and greatly accelerates the use of the information obtained to effectively use our own forces and means to achieve the objectives of the operation.

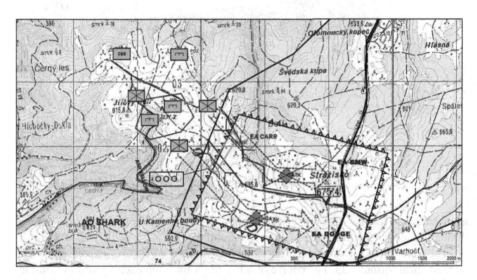

Fig. 1. COP provided by OTS VŘ PozS [3]

3 Model of the Optimal Maneuver Route

In 2015, the author of this article developed the model of the optimal maneuver route [6] as a part of his doctoral study at University of Defence, the top educational institution of the Army of the Czech Republic. The model evaluates and comprehensively summarizes the possibilities of the terrain throughput, focusing on five information areas. It has the form of a raster layer of HF (Horizontal Factors) or VF (Vertical Factors) throughput. The vegetation and the terrain surface are the basis of the Throughput Cost Surface ($P_{np1.i}$), which is bound to the influence of the terrain relief (VF_2), the weather (HF_3), and the effects of the combat potential of the enemy's forces and means (HF_4) with the combat potential of our own forces and means (HF_5). You can edit newly discovered information in each of the areas mentioned above. The Cumulative Throughput Cost Surface (SP_{np}) then evaluates the resulting raster layer that combines the influence of all layers of the model based on defined criteria and map algebra, see Fig. 2. The model distinguishes three basic types of moving elements, namely pedestrian, wheel and track vehicles. Each of these three modes of movement has different throughput characteristics. This is also linked to different

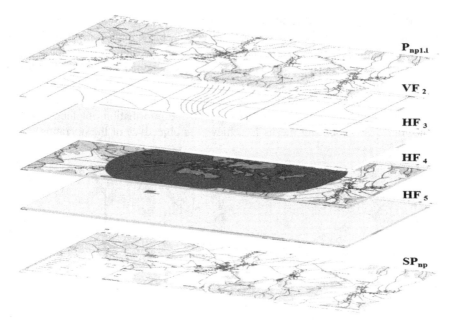

Fig. 2. Raster layers of model [7]

achievable speeds on specific surface types, different rate of climb, and limited throughput under the influence of snow and rainfall.

The vertical factor of the terrain slope for vehicles (VF_{2V}) is expressed by the mathematical formula (1). The course of VF_{2V} has a linear character given by constant value $K_{V2V} = -0.004$. The terrain slope affects are represented by (ω). The limiting passable terrain slope is set for tracked vehicles in the range of $-30°$ to $+30°$, and for wheeled vehicles of $-30°$ to $+20°$. Out of the range of these values, the terrain slope in the model is assessed as impassable, with $VF_2 = 0$.

$$VF_{2V} = K_{V2V} \cdot \omega + 1 \tag{1}$$

The vertical factor (VF_{2C}) of the terrain slope (ω) affects the dismounted movement downhill or uphill differently depending on the topography and the safe movement controllability (2). Its difficulty in walking uphill increases exponentially as the slope increases. When walking downhill, it drops down up to $20°$, when it is equal to the difficulty of movement on the flat ground. When walking downhill with the angle of slope of more than $20°$, the difficulty increases again with the increasing slope. Such a course is caused by a degree of gravity that facilitates the movement at first. However, when the terrain slope is more than $20°$, it forces the dismounted movement of individuals to brake in order to maintain a safe control over their movement. Its values have been borrowed from the thesis developed by Lenka Mezníková [8]. The limiting passable terrain slope for the dismounted movement is set in the range of $-50°$ to $+50°$.

$$VF_{2C} = \frac{1}{K_{V2C}} \tag{2}$$

The value of K_{V2C} is represented by the mathematical formula (3), which has been generated as regression equation of the third degree polynomial curve. Mentioned polynomial curve was put through curve of origin values [8]. The value of the regression equation reliability is 0.9875.

$$K_{V2C} = -0.0121\left(\frac{\omega}{10}\right)^3 + 0.0968\left(\frac{\omega}{10}\right)^2 + 0.3156\frac{\omega}{10} + 0.9933 \tag{3}$$

Horizontal factor of weathe (HF_3) represents affects od rainfall and snowfall. The model of the movement route for forces and equipment evaluates the effects of rainfall only for tracked and wheeled vehicles that move off the paved roads. For the movement of tracked and wheeled vehicles the derived mathematical formula is used (4). The coefficients of the snow layer influence in the field with a zero slope ($K_{3.1}$) related to particular types of moving elements were mathematically derived, as follows:

- For tracked vehicles: $K_{3.1P} = 0.01205$;
- For wheeled vehicles: $K_{3.1K} = 0.0192$;
- For the dismounted movement: $K_{3.1C} = 0.008$.

The combination of the influences of the terrain slope and the snow layer thickness ($L_{3.1}$) implements the coefficient of the snow-covered topography ($O_{3.1}$) in the calculation of $HF_{3.1}$, which is differentiated according to the type of a moving element. $O_{3.1}$ attains the following values:

- For tracked vehicles: $O_{3.1P} = 0.034$
- For wheeled vehicles: $O_{3.1K} = 0.0576$
- For the dismounted element: $O_{3.1C} = VF_{2C}$.

For the movement of tracked and wheeled vehicles the derived mathematical formula is used as follows (4):

$$HF_{3.1P/K} = 1 - (K_{3.1P/K} \cdot L_{3.1}) - (O_{3.1P/K} \cdot \omega) \tag{4}$$

For the movement of the dismounted element the derived mathematical formula is used as follows (5):

$$HF_{3.1C} = 1 - (K_{3.1C} \cdot L_{3.1}) - (1 - O_{3.1C}) \tag{5}$$

The limiting passable snow thickness for the dismounted element is set for 90 cm of snow since a thicker layer of snow is negotiable with great difficulties or even impassable for the dismounted element.

The model of the movement route for forces and equipment evaluates the effects of rainfall only for tracked and wheeled vehicles that move off the paved roads. The limitation of the terrain passability due to rainfall is generally assessed in four steps based on the precipitation amount for the purpose of creating a movement route in the model [7, 9], see Table 1.

Table 1. The value of $HF_{3.2}$ with different rainfall amounts in three days

$HF_{3.2}$	October–April	May–September
0 impassable	Larger than 80 mm	Larger than 120 mm
0.25	Larger than 60 mm	Larger than 90 mm
0.5	Larger than 40 mm	Larger than 60 mm
0.75	Larger than 20 mm	Larger than 30 mm
1 without an effect	Smaller than 20 mm	Smaller than 30 mm

The combat potential of the enemy's forces and means (HF_4) are defined in the model by their geographical position and the attributes of the tactical and technical characteristics of his weapons. From the location of their actual deployment the area that is visible within the effective range of enemy weapon systems is then evaluated as impassable. Similar principle was applied to destruction effect of land mines and IED (Improvised Explosive Devices).

The combat potential of our own forces and means (HF_5) expresses the degree of reduction in the impact of the enemy activity in terms of supporting the passage of the danger area of the task performance. The layer of friendly forces and equipment does not affect the passability of the area as a whole. It expresses only the ability or capabilities of friendly forces and equipment to support the manoeuvring element by eliminating security risks.

Movement speeds on specific types of terrain surface can be adjusted, depending on the type and technical condition of the vehicle, the driver and unit commander's experience, or the physical capabilities of the members of the pedestrian element. The forces and means of the enemy in the model represent an impassable space, defined by the possibilities of effective range of the main weapons and visibility from their position. In a similar way, the influence of our own forces and resources is also simplified, eliminating the impact of the enemy within the effective range of their main weapons and visibility. For specific forces and means of the enemy as well as own units, the range of the main weapon systems can be varied on the basis of the type of military equipment or the size and equipment of the unit. In the model, it is also possible to define impassable spaces due to the location of explosive and non-explosive deposits. This can be both the currently prepared roadblocks, and the locations of their historic discoveries or past attacks. Similarly, the model allows for the definition of other types of kinetic attacks or other offensive activities, as they may indicate the long-term effects and efforts of an enemy military unit or an organized group in the operation area. All areas of influence on throughput are specified in the model by mathematical algorithms, with critical assessments of their variant occurrence. The basis for the calculations of the implemented model is the raster matrix of the speed of overcoming different types of terrain surface. Its throughput cost surface is mathematically added to the speed limitation caused by the relief of the terrain, rain and snowfall, and the occurrence of enemy forces and means, see Eq. (6).

$$SP_{np} = \frac{P_{np1.i}}{VF_2 \cdot HF_3 \cdot \min(1, HF_4 \cdot HF_5)} \tag{6}$$

The mathematical operations of the implemented model and the search algorithm for the shortest paths result in a pathway in the operation area, optimized for safety and speed, see Fig. 3.

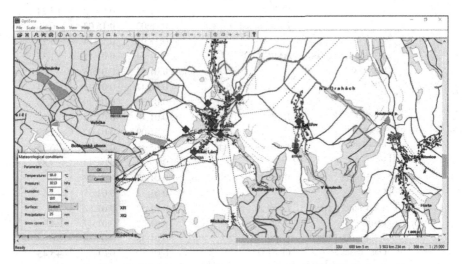

Fig. 3. Model of optimal maneuver route in TDSS [8]

The optimal maneuver route model was implemented by associate professor Ing. Petr Stodola, Ph.D. in the Tactical Decision Support System (TDSS), which was created at the University of Defence in 2006. The TDSS authors are associate professor Ing. Petr Stodola, Ph.D. and associate professor Ing. Jan Mazal, Ph.D. The system is an experimental platform for testing algorithmic mathematical models, using raster representation of digital tactical-geographic data and map algebra.

The functionality of the model has been verified by performing practical measurements on different surface types with different transfer route profiles, and using all three of the above-mentioned ways of movement. The discrepancy between the results of practical measurements and mathematic model calculations was 2.74%, on a smooth surface with medium thick vegetation in case of pedestrian movement, see Table 2. There was a greater variation when moving in areas with a significant occurrence of micro-relief forms. Of course, a significant factor in the speed of overcoming the transfer route is the experience and excessive caution of drivers or pedestrians, and weight or size of the transferred load.

Table 2. Results of practical measurements in the field

Segments of measurements (route)	Type of surface (layer)	Length of route (m)	Method of movement	Speed of the model $\left(V_i^h, \text{km/h}\right)$	Real time in the field $\left(t_i^h\right)$	Time in the model $\left(t_i^h\right)$	Deviation $\left(t_i^h\right)$ model/real (%)
1	1.3.5 other roads	1,484	BMP-2	36	0:02:18	0:02:18	0
					0:02:07	0:02:15	+6.29
			Lorry (T-815)	53	0:01:34	0:01:35	+1.06
					0:01:27	0:01:32	+5.74
			Dismounted	6.13	0:15:02	0:15:02	0
					0:13:05	0:12:24	−5.22
2	1.3.7 unpaved roads	606	BMP-2	28	0:01:14	0:01:15	+1.35
					0:01:11	0:01:13	+2.81
			Dismounted	5.85	0:06:42	0:06:43	+0.24
					0:05:36	0:05:20	−4.76
3	1.1.2 meadows, grasslands	431	BMP-2	16	0:01:10	0:01:10	0
					0:01:09	0:01:07	−2.89
			Dismounted	3.42	0:05:38	0:06:13	+10.35
					0:05:14	0:04:43	−9.87
4	1.3.5 other roads	886	BMP-2	26	0:01:57	0:01:57	0
					0:01:51	0:01:55	+3.6
5	1.3.5 other roads	1,194	Lorry (T-815)	33	0:01:52	0:01:52	0
					0:01:42	0:01:47	+4.9
6	1.3.4 third-class roads	1,803	Lorry (T-815)	57	0:01:38	0:01:38	0
					0:01:50	0:01:56	+5.45
7	1.3.7 unpaved roads	1,812	Lorry (T-815)	28.7	0:05:40	0:05:40	0
					0:07:22	0:03:37	−50.9
			Dismounted	5.01	0:21:39	0:21:38	−0.07
					0:19:27	0:18:21	−5.65
8	1.1.1 forests, trees	637	Dismounted	3.01	0:09:35	0:11:16	+17.56
					0:09:27	0:07:47	−17.63
9	1.1.4 forests, scrub pines	569	Dismounted	1.24	0:19:21	0:27:33	+42.37
					0:15:42	0:15:44	−0.21
10a	More ls	2,352	Dismounted	More ls $\left(V_i^h\right)$	0:22:41	0:21:52	−3.60
10b	More ls.	2,459	Dismounted	More ls $\left(V_i^h\right)$	0:31:06	0:29:00	−6.75

4 Simulation of Enemy Activity

An irreplaceable role in the decision-making process of the commander and the crew in a military operation is the process of evaluating the intent of the enemy. During this process, the strength, weaponry and equipment of the enemy, their deployment and readiness, as well as the possibilities of reinforcement and target objects of interest must be evaluated. By analyzing many heterogeneous input variables, we try to estimate and spatially specify the activity or maneuver the enemy intends to implement. Similar to the overall rating of the variance of an enemy's activity, it varies depending on the experience of the particular analyst who performs it. A number of such comprehensively evaluated conclusions are then created on the basis of intuition, based on the life experiences and character of the author. For this reason, it is very difficult to create information software that would allow this multi-criteria and highly intuitive analysis with an acceptable degree of probability to implement. To do this, it is important to realize that the realization of each evaluated option is most likely influenced by any change in the situation in the operation area. This may be a significant change in weather, a change in attitudes and the mood of the local population, or an increase in the depletion of maneuvering units and a decline in their morale. This makes it very difficult to produce a universal algorithm of estimating the intent and activity of the enemy.

However, the optimized maneuver route model implemented in TDSS can be used for partially automated but quick simulation of enemy activity. A prerequisite of its function is the specification of the position of the deployment of the enemy's forces and means in the operation area, their equipment and the technique used, and the intended target objects or directions of interest. By determining the starting point of the enemy's forces and means and the method of further movement, based on the type of the technique used, the model calculates the "surface" of the difficulty of moving the surrounding spaces. The basis for the calculation will be unambiguous information on terrain, relief, and current weather forecasts, as well as the likely position of the enemy's own deployment of resources and resources. By specifying the maneuver target point, based on the evaluator's intuitive estimate, the implemented model suggests the safest and fastest moving route of an enemy's move, if any. Calculations of model was verified by means of tests in the field, see Table 1. The results of tests number 7, 8 and 9 in the field were affected by unpredictable human factor of driver and micro-relief forms significantly.

An intuitive estimate of the maneuver target will also play a significant role here. The basis for the estimate is the actual strength and composition of the enemy, their deployment, their known abilities and the morale of the fighters to carry out the planned activity, the supply of persons and the needed material. However, it is still necessary to estimate the long-term intentions and objectives of the enemy, derived from their combat activities in the past. This means knowing the places of repeated attacks, consistently selected static objects for attack, long-term shelters and material hideouts, safe havens, and dislocation of reinforcements. Subsequently, the user or intelligence analyst determines target objects or enemy maneuver spaces.

The default space for the deployment of enemy forces and resources and the target maneuver space can be subsequently defined in the implemented model, including the specification of the mode of movement. As a result of the model's mathematical calculations, an enemy maneuver route, optimized for terrain, safety and time, will be created. Displaying the route makes it possible to estimate the anticipated movement of the enemy's forces and means and implement it into our own decision-making process. During this route, it is possible to estimate spaces and times, the execution of a trap or a planned artillery attack, an attack of air forces, the locations of explosive baits or barriers. The calculated time of reaching the target space, or various spatial parts of the movement, allows us to prepare an attack of our own forces in the form of a maneuver with motion, together with gunfire. Implementation of anticipated routes of enemy forces maneuver into projection of optimal route of own forces maneuver is going to be next step of model development. Together with the more detailed description of the enemy forces situation it has been analyzing by author of the article since finishing of his doctoral study.

5 Conclusion

The environment of a networked digitized battlefield provides all commanders with an important unified knowledge base for command and control in the form of a common picture of the battlefield situation. It delivers information updated in real-time on the laptop screen. Thus, a single overview of the situation allows independent coordination of the joint efforts of all units in the military operation. Adding automated maneuvering route planning software will also greatly increase the speed of command and control process in the operation and efficiency of the use of forces and resources. By speeding up decision-making, the ability to instantly respond to sudden changes on the battlefield will be greatly enhanced, creating a vital prerequisite for gaining operational superiority over the enemy.

The above described model of the optimal maneuver route, using raster representation of tactical geographic data, partially optional criteria inputs and map algebra links, is a flexible tool for command and control information support. Variant settings of each model variable can simulate the effects of the most important aspects of the battlefield on the creation of the maneuver route, which gives the system a great deal of adaptability. The results of input data modeling and mathematical-algorithmic calculations are then shown by TDSS to the user in seconds from the start of their processing. To a large extent, it replaces the analysis of the situation on the commander's battlefield, and enables quick creation of variants of their own activities, optimized in terms of safety and time. The implemented model, therefore, significantly increases the efficiency of using our own resources and resources.

The model also has a legitimate use in the intelligence analysis of variants of enemy activities. In this case, it allows you to quickly generate the movement paths of its forces and means and times of reaching the pre-selected targets. Through the TDSS, gather and release points of enemy units, as well as places with limited passage capacity, can be predicted. This simulation of the enemy's activity finds its use even when assessing the options of maneuvering our own forces and means, in performing a war game, as well as coordinating the cooperation of all units in a military operation. Generated enemy units' transfer routes and calculated times of reaching any point on the route allow for planning a focused artillery fire or air strikes. And last but not least, qualified simulation of the enemy's activity can be further used during training of the planning and decision-making process of the commanders and staffs. Implementation of anticipated influence of enemy forces maneuver is going to be further development of the model.

References

1. FM 6-0 Mission Command: Command and Control of Army Forces, Headquarters Department of the Army, Washington (2003). PIN: 080933-000, A1p
2. FM 3-0 Operations: Headquarters Department of the Army, Washington, 4–3p. (2011). PIN: 079091-000, C1
3. Source: author of the article (OTS VŘ PozS)
4. Group of authors: NEC Understanding network enabled capability, 12p. Publishing Newsdesk Communications Ltd., Londýn (2009). ISBN 978-1-905435-94-4
5. Hodicky, J., Frantis, P.: Decision support system for a commander at the operational level. In: 2009 Proceedings 1st International Conference on Knowledge Engineering and Ontology Development KEOD, pp. 359–362 (2009)
6. Nohel, J.: The possibilities of information support in the planning of the maneuver of units Dissertation thesis, University of Defence, Brno (2015)
7. Source: author of the article (TDSS)
8. Mezníková, L.: Analysis of shortest routes on maps for orientational running, Diploma paper, ČVUT University, 57p. Prag (2011). http://geo.fsv.cvut.cz/proj/dp/2011/lenka-meznikova-dp-2011.pdf
9. Talhofer, V., et al.: Military topography, military publication, Odbor doktrín VeV – VA, Vyškov, 186–187pp. (2011). Pub-28-68-01
10. Military geographic institute Prag, ÚDP PŮDY, handbook. PRAHA, 2000, 6p. http:www.is.muni.cz/el/1431/podzim2005/Z8105/Databaze_pudy.doc

Automation in Experimentation with Constructive Simulation

Jan Hodicky, Dalibor Prochazka$^{(\boxtimes)}$, and Josef Prochazka

Centre for the Security and Military Strategic Studies,
University of Defense, Brno, Czech Republic
{jan.hodicky,dalibor.prochazka,
josef.prochazka}@unob.cz

Abstract. Today constructive simulations are used mainly to support training and education subdomain from the modelling and simulation applications portfolio as defined in the NATO Modelling and Simulation (M&S) Master Plan. Other M&S application areas, namely Support to Operations, Capability Development, Mission Rehearsal and Procurement can benefit from already implemented constructive simulations. The recommended approach is to use constructive simulation to design, execute and analyze an experiment to get insights in problems being solved in the previously mentioned M&S application areas. The first part of the article descripts the value of experimentation for the military and explains fidelity, cost and automation factors in Live, Virtual and Constructive simulation if used for experimentation purposes. Further basic building blocks of an experiment with constructive simulation are described. Starting from the scenario development block up to the analytical and customized visualization block to better fit a need of the customer of the experiment results. The second part describes the current architecture of constructive simulation and its challenges when trying to cover all the building blocks of an experiment. The common denominator of all challenges is automation. Therefore the role of human being and automata will be discussed in the context of an experiment. The last part covers the Test Case when the constructive simulation, MASA Sword, is used to demonstrate current state of the art and limitation of automation in the experimentation field.

Keywords: Automation · Constructive simulation · Experiment

1 Introduction

NATO Science and Technology Organization governs all collaborative activities under six panels and one group. Panels are represented by Applied Vehicle Technology, System Analysis and Studies, Human Factor and Medicine, System Concepts and Integration, Information Systems Technology, Sensors and Electronics Technology. The only group at this level is named NATO Modelling and Simulation Group (NMSG). The mission of NMSG is to promote co-operation among Alliance bodies, NATO member nations and partner nations to maximize the effective utilisation of Modelling and Simulation (M&S) and to foster the MS interoperability through the related standardization activities. The strategic guidance in the M&S domain is scoped

© Springer Nature Switzerland AG 2019
J. Mazal (Ed.): MESAS 2018, LNCS 11472, pp. 566–576, 2019.
https://doi.org/10.1007/978-3-030-14984-0_42

by the NATO Modelling and Simulation Master Plan version 2 [1]. This document, in its second part called NATO M&S Implementation Plan, identifies the following application areas that can capitalize on M&S:

- Support to Operations;
- Capability Development;
- Mission Rehearsal;
- Training and Education;
- Procurement.

Training and Education is the most widespread application area in NATO and national military environment [2]. Almost all NATO nations have their own National Simulation Centre with defined collective training objectives. However the other application areas are not very well supported and still stand by to prove its value.

Experimentation is a tool to get insights of a problem domain, to test specified hypothesis or validate proposed solution [3]. In fact experiment approach is used in all previously defined M&S application areas. It is only way to increase the knowledge of a problem domain in rigorous way. Experimentation can reduce uncertainty in decision making process at all levels of command introduced there by the current complex and dynamic security and operational environment. Experimentation can be applied at different levels of abstraction. As an example at strategic level, it can reduce of uncertainty related to the description of future development in the security environment [4] or at very tactical level it can search for optimum reconnaissance procedure for unmanned aerial systems [5–7], or looking for optimal manoeuvre of tactical units [8]. The operational level example might be represented by the surface-to-air missile systems optimization [9] or training coordination with the air force simulator [10].

Analysis and synthesis were the main methods applied in the article. Analysis revealed a need to automate elements of constructive simulation and synthesis showed the building blocks of an experiment to be automated. The Use Case approach was used to demonstrate the current state of the art in the automation in constructive simulation employed for experimentation.

2 Fidelity, Cost and Automation in Simulation for Experimentation

Many M&S practitioners describe three types of simulation. [11–13]. The first one, Live simulation, contains real people operating real system, like soldier on a field training. Effects on real people are simulated. The second one, Virtual simulation, contains real players operating in simulated environment with simulated effects. As an example a pilot in a flight simulator can be taken. The third one, Constructive simulation, contains simulated people decision making and simulated environment with simulated effects. Strategic games might be taken as an example from this category. The fourth category that implies combination of simulated people in real environment hasn't been yet agreed in the community, even if there are attempts to explicitly define it. With specific attention to the training, the last category might be called Cognitive simulation [14], where real operators supervising the real systems in which the simulated people; the code replicating the human behaviour; make their own decisions and give orders or released synchronization tasks based on the autonomy level to the system and/or to simulated human over the real terrain with simulated effects.

Three main factors differ based on the previously defined types of simulation. It is simulation fidelity, automation and cost. Simulation fidelity is the extent to which the appearance and behaviour of the simulation matches the appearance and behaviour of the simulated system [15]. Or it might be rephrased be the question: How close is the simulation to the replicated reality? Automation is the level of automated activities related to the execution and analysis of simulation based event that needn't to have human support. Automation is the prerequisite for cooperation and collaboration of entities being simulated [16, 17]. Finally the cost in this context is strictly related to the development, maintenance and reuse of simulation.

Figure 1 shows difference between current status of fidelity, automation and cost of live, virtual and constructive simulation and expected status of these factors in close future. Regardless the level of implemented automation live simulation will stay the type of simulation with the highest fidelity and the highest cost. Constructive simulation is and will be the type of simulation with lowest fidelity and lowest cost. Virtual simulation fidelity and cost is and will stay between levels of fidelity and cost of constructive and live simulation. However in the light of artificial intelligence domain development the constructive simulation fidelity will be much closer to the real environment that is to be replicated then nowadays. Moreover the reuse and manage of constructive simulation with higher level of automation will be less expensive.

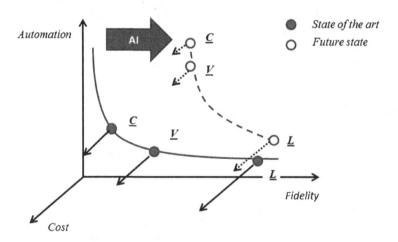

Fig. 1. Fidelity, cost and automation factors in live, virtual and constructive simulation

Because of low fidelity of current constructive simulation rooted in very low ability to automate human behaviour, it is still requested to have combined experimental environment employing all three categories of simulation. However tendency there is very clear. Constructive simulation with higher level of automation will be the key type of simulation used for experimentation purposes and it is the main subject of the following chapters.

3 Experiment Lifecycle

An experimental approach used to increase knowledge of a studied domain is composed of five key phases that compos a process described by the experiment life cycle. Figure 2 shows experiment life cycle with its phases and their relation. Some authors simplified the experiment life cycle like composition of four phases [18], however the approach used in the article better fits to the purpose of constructive simulation involvement.

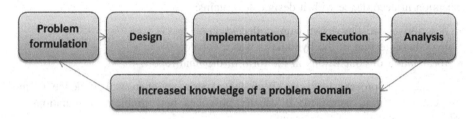

Fig. 2. Experiment lifecycle and its phases

The first phase, problem formulation, is aimed at making explicit statements in a problem domain that deserves to be approached by the experimentation. The statement has to be easy understandable by the problem domain community and finding a solution is required by the community as well. This phase defines among others object of the experiment, goal and scope of the experiment, constrains and expected validity of results of the experiments, customer of the experiment results, important measures from customer point of view. It answers question why the experiment should be performed.

The second phase, experiment design, puts the overall planning activity of the experiment together. This phase contains activities like:

- context definition in the sense of scenario to be used for experimentation;
- hypothesis formulation (if it is needed for a test hypothesis type of experiment);
- definition of metrics to be used to set the environment of the experiment and to evaluate the results of the experiment (independent metrics/inputs and dependent metrics/outputs; outputs are usually constructed as Measure of Performance (MoP) or Measures of Effectiveness (MoE); these are further composed of Indicators with weights towards a selected MoE or MoP);
- design of experiment for the selection of experiment runs with defined metrics [19];
- selection of data from scenario to be used as inputs for the experiment run;
- design of data collection process;
- validity evaluation composed of conclusion validity, internal validity, construct validity and external validity [19].

This phase answers question how the experiment should be performed.

The third phase is aimed at the experiment implementation. The environment of the experiment in which it is going to be performed must be modified or created to fit the purpose of the experiment, if it doesn't exist.

The fourth phase, experiment execution, carries out designed runs of the experiment with specified inputs simultaneously measuring the outputs. It contains:

- training on environment of the experiment with involved staff;
- executions of all runs and collection of output data;
- validation of output data.

The last phase, experimentation analysis, brings the conclusion of the experiment execution in accordance with it design. It contains:

- descriptive statistics drawn from collected data;
- hypothesis testing, if the hypothesis was involved;
- formulation of conclusions in the form of the final report.

In fact, the process is cyclic. With the experiment analysis knowledge of the problem domain is increased. It usually provokes new problem formulations or enforces modification of the current one.

4   Experiment with Constructive Simulation

Uncertainty is introduced by complex operational and security environment where military is performing its missions. If experimentation is used the validity of experiment results and conclusion can be assured only if the experiment is carried out in many runs and can be repeated. One run of experiment in live simulation is very complicated and required enormous resources. Therefore natural way is to use constructive simulation for experimentation. Current main drawback is that military constructive simulations were designed exclusively for training and education purposes and doesn't support any, or in very limited scope, automation.

Further part of the chapter describes constructive simulation building blocks that are used for the training purposes and might be used for experimentation as well; see Fig. 3. The missing features of these building blocks are named from the experimentation point of view.

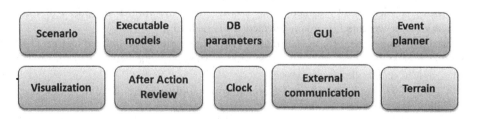

Fig. 3. General building blocks of constructive simulation

Scenario building block is used to design and implement scenario that is composed of Order of Battle (ORBAT), structure and forces of enemy. The missing experimentation features are:

- validation of ORBAT and enemy forces in respect of real world data;
- real time scenario generation based on the experiment objective, MoPs and MoEs.

Executable models block is responsible for making calculation of all entities behaviour represented in the simulation. Models are designed as highly aggregated, the JTLS can be taken as an example, or detailed ones, JCATS can be understood as an example. The hybrid mixture of these two categories is not available. The missing experimentation features is hybrid mixture of highly aggregated and detailed simulation.

DB parameters block contains all parameters that are inputs for executable models block. The probabilities of hit and kill or weapon characteristic are examples of these parameters. The missing experimentation features is validation mechanism of DB parameters in respect of real world data.

GUI block is in charge of providing interface between operator of simulation and simulation executable. The missing experimentation feature is standardized GUI to manage all experimentation phases.

Event planner block is used to plan activities with entities represented in simulation. The own and enemy forces are instructed to follow prebuild orders. It has been implemented at the system level; however the missing experimentation feature is an implementation of event planners at the aggregated level.

Visualisation block shows the current situation of all entities represented in simulation. It might be used in real time or in connection with after action review block.

After action review block is used to demonstrate the flow of events and status of all entities represented in time in simulation. The missing feature from the automation point of view is the automation of displaying relation of causes and effects in the executed scenario.

Clock block is responsible to execute all models in accordance of the logical time of simulation. The missing feature from the automation point of view is ability to change the speed of models execution to the maximum to reduce time of overall experimentation life cycle.

External communication block ensures the interoperability among federated simulations. There is no identified lack of experimentation features.

Terrain block contains all geographical data used to create the environment over which entities are represented. The missing feature from the automation point of view is validation of scenario in respect of real world data.

5   Constructive Simulation for Experimentation - Test Case

Test Case with MASA that support Sword was constructed in the way to reveal if the product is able to automate some activity or behaviour that might help to better design, execute and analyse experiment.

MASA Sword is a constructive simulation tool consisting of several applications. The tool functionality covers:

- command post training, from battalion to division levels;
- course of action analysis;
- officer training;
- doctrine and equipment analysis;
- public safety, emergency management;
- decision support for C2 systems.

The tool promises flexibility in all these areas. It is given by possibility to create or modify entities, units, crowds and their behaviour including logistics, human factors, information exchange etc. It is empowered by artificial intelligence functions, which give entities behaviour complying doctrines without necessity to be controlled by humans.

Sword is composed of application given in Table 1.

Table 1. Sword applications

Application	Description
Adapt	Gives access to adaptation tools used to customize Sword
Authoring	Create and customize physical databases
Terrain gen	Create new Sword terrains from various data sources (vector, elevation, raster)
Prepare	Create and modify exercises
Play	Start or join an exercise
Start	Start and automatically join an exercise
Join	Join an exercise
Replay	Replay an exercise and analyze the course of action of different recorded sessions
Admin	Configure and manage general Sword parameters and data

Test case was designed as an experimentation related to areas of effects from equipment and weapon system level to units composition, behaviour and doctrines.

All physical models and their parameters are stored in a relational database, accessible through the Adapt application GUI: A user can modify or create from scratch models and their parameters. They are (not complete list) supplies, resource networks, launchers, weapon systems, sensors, transmitters, active protection, objects (kind of dynamic terrain features), missions, doctrine models, equipment, units, crowd, populations, knowledge groups, human factors and automata.

An automaton is a group of units that can be controlled through a virtual commanding unit, simulating a dynamic chain of command. Missions are composed of scripts creating intelligent behaviour (Direct IA) and other parameters, doctrine models consist of missions and their parameters. The artificial intelligence modules and automata controlling simulated entities enables to prepare exercises without human intervention during simulation. Knowledge groups enable to build information exchange models that enable decision making processes.

Having physical models and a terrain database (preparation of terrain databases is out of scope presented paper), next step is to set up an exercise according to an experiment goal(s). The Preparation application is used to:

- select terrain;
- create objects (obstacles, dikes, fires, jamming areas, minefields, roads, crossing points, bridges etc.);
- create urban areas (cities, districts, building blocks and their properties);
- create, position and initialize forces;
- set up logistic;
- create course of actions by tasking them;
- set up simulation environment, (global and local weather conditions), available light.

The crucial step of an experimental exercise preparation is configuring performance indicators. These indicators are used in the experimentation as the main inputs to the MoE and MoP. It can be done in the Preparation application performing four operations, namely:

- data extraction;
- data selection;
- variable definition;
- operations on data.

Setting up these indicators is performed by creating calculations called Scores in Sword - formulas that aim to provide an indication of the state of forces as the exercise progresses. Scores are created during the preparation phase, stored in scripts and calculated during the exercise execution. Scores can be viewed as the exercise progresses and the results can be used for analysis later on.

There are three options to create scores: automatically, generating Sword's predefined scores whose syntax is delivered in a file with the product, semi-manually, by making selections using an assistant to construct a customized score or manually, by building the formula either by hand or insertion in the Score Edition window. User defined variables allow to specify the entities to which the score formula applies. This approach can provide any required data concerning simulation outcomes for post-exercise analysis. Scores data can be exported, but only manually either during simulation or during post-exercise analysis. running the Replay application.

The Play application provides environment for user to observe simulation, make decisions, interact with simulation and other players and see outcomes. To reduce cost of experiment, make use of simulation faster than real time and minimal user intervention are required, but these options depend on experiment goals.

The Replay application is used for exercise evaluation. It provides access to simulation results and data through application GUI (the same as Play) and trough Scores created in Preparation or Play applications.

One trial of the simulation experiment, i.e. one run of prepared exercise with specific parameters can be executed without human intervention. It requires:

- well defined scenario including course of actions implemented by entities tasking (orders and missions);
- valid doctrine models to ensure realistic behaviour of entities;
- defined Scores to collect data for further analysis.

  Sword provides these basic capabilities.

  Changing physical parameters of models, environment parameters or human factors must be done manually, as well as export of Scores data, so they are limits to automation of series of trials and a whole experiment.

6 Missing Experimental Constructive Simulation Features Based on the Test Case

Tested constructive simulation, Sword, as other constructive simulation tools, was primarily designed for training, so developers were not focused on automation of preparation, execution and analysis. What is missing in implementation, dealing with an experiment automation process is following:

- Automated execution of runs. To automate execution of a series of trials would require capability to run automatically the exercise with the same parameters (to be able perform statistical analysis of global outcome, filtering out random factors built in models).
- Automation of Design of Experiment that enables selection of inputs and outputs of an experiment and defines number of simulation runs and combination of inputs values that creates single run to reduce the risk originated in the complexity of operational environment.
- Automation of simulation execution based on Design of Experiment. Design of Experiment creates all combination of inputs parameters and runs that must be then automatically run; without human intervention. It corresponds to the automation of a series of exercises with changing parameters (physical models, doctrines, environment) according to experiment goals.
- Capability to generate required data automatically. The first step is to automatic export of selected Scores.
- Automated collection and processing of output data. The data as exported – time series of Scores values - will need automated procession to facilitate analysis. Some relational database could be an output of this step.
- Automation of analysis. Having the database, a tool to facilitate its analysis, using statistical methods and visualisation would support outcomes evaluation.
- Automation of indicators construction – in this case represented by Scores. Now scores are created in the Preparation application and they relate to a specific exercise. Construction might be done automatically based on the experimental goals and scenario available.

These requirements are formulated to facilitate a Sword supported experiment design, execution and analysis. To specify them more accurately, further discussion between developer and customer is necessary.

7 Conclusion

Described Test Case with MASA demonstrated current state of the art in experiment automation. Automation reduces manpower need to run design, run and analyse experiment and is very foundation of fast experimentation execution resulting in risk reduction related to the decision taken from experiment results. Most of the constructive simulation allows:

- Automation of Human behaviour own and enemy. Simulation is able to automate own and enemy forces and systems acting autonomously without any operator intervention.
- Automation of Doctrine. Simulation is able at defined level of unit aggregation to implement national doctrine, e.g. how the subordinates unit are behaving in the case of defence activities.
- Automation of Umpire. Simulation evaluates automatically all effects resulting from entities interactions.

The missing building blocks of constructive simulation that anchor all defined missing experimentation features are:

- Measures Construction block allowing automation of MoE, MoP and Indicators construction based on the experiment goal and scenario.
- Design of Experiment block allowing composition of inputs and outputs values and specification of these inputs values and number of simulation runs.
- Automated Execution block allowing execution of all designed runs without any human intervention.
- Automated Data Collection block allowing collecting outputs in the manner to execute analytical phase automatically based on the analytical tool available.
- Automation of Analysis block allowing taking results from all experiment runs and showing it in the way native for decision making based on the scenario and experiment goals.

Acknowledgments. This work is sponsored by the Czech MoD project called STRATAL (2016-2020).

References

1. NATO ACT: NATO Modelling and Simulation Master Plan, 2nd edn (2012). AC/323/NMSG(2012)-015
2. Hodicky, J., Prochazka, D.: NATO modelling and simulation education and training curriculum- do we need a new NATO military discipline?. In: Proceedings of the 2017 International Conference on Military Technologies (ICMT), University of Defence, Brno, pp. 700–704 (2017). ISBN 978-1-5386-1988-9
3. NATO ACT: Bi-SC 75-4. Experimentation Directive. HQ SACT, Norfolk (2010)

4. Fucik, J., Melichar, J., Kolkus, J., Prochazka, J.: Military technology evolution assessment under growing uncertainty and complexity: methodo-logical framework for alternative futures. In: Proceedings of the 2017 International Conference on Military Technologies, pp. 682–689. Institute of Electrical and Electronics Engineers Inc., Piscataway (2017). ISBN 978-1-5386-1988-9

5. Drozd, J., Stodola, P., Křišťálová, D., Kozůbek, J.: Experiments with the UAS reconnaissance model in the real environment. In: Mazal, J. (ed.) MESAS 2017. LNCS, vol. 10756, pp. 340–349. Springer, Cham (2018). https://doi.org/10.1007/978-3-319-76072-8_24

6. Stodola, P., Mazal, J.: Model of optimal cooperative reconnaissance and its solution using metaheuristic methods. Defence Sci. J. **67**(5), 529–535 (2017). ISSN 0011-748X

7. Stodola, P.: Improvement in the model of cooperative aerial reconnaissance used in the tactical decision support system. J. Defense Model. Simul. **14**(4), 483–492 (2017). ISSN 1548-5129

8. Stodola, P., Nohel, J., Mazal, J.: Model of optimal maneuver used in tactical decision support system. In: Methods and Models in Automation & Robotics (MMAR 2016), pp. 1240–1245. West Pomeranian University of Technology in Szczecin, Mezizdroje (2016). ISBN 978-150901866-6.1

9. Farlik, J., Tesar, F.: Aspects of the surface-to-air missile systems modelling and simulation. In: Mazal, J. (ed.) MESAS 2017. LNCS, vol. 10756, pp. 324–339. Springer, Cham (2018). https://doi.org/10.1007/978-3-319-76072-8_23

10. Farlik, J.: Conceptual operational architecture of the air force simulator: simulation of air defense operations. In: 2015 International Conference on Military Technologies ICMT (2015). Article no. 7153723

11. Hurt, T., McKelvy, T., McDonnell, J.: The modeling architecture for technology, research, and experimentation. In: Proceedings Winter Simulation Conference, pp. 1261–1265 (2006). Article no. 4117746

12. Hodson, D., Baldwin, R.: Characterizing, measuring, and validating the temporal consistency of Live—Virtual constructive environments. Simulation **85**(10), 671–682 (2009)

13. Lutz, R., Drake, D.: Gateway concepts for enhanced LVC interoperability. In: 2011 Spring Simulation Interoperability Workshop 2011, Spring SIW, pp. 113–119 (2011)

14. Hodicky, J.: Autonomous systems operationalization gaps overcome by modelling and simulation. In: Hodicky, J. (ed.) MESAS 2016. LNCS, vol. 9991, pp. 40–47. Springer, Cham (2016). https://doi.org/10.1007/978-3-319-47605-6_4

15. Maran, N.J., Glavin, R.J.: Low- to high-fidelity simulation - a continuum of medical education? Med. Educ. Suppl. **37**(1), 22–28 (2003)

16. Vysocky, A., Novak, P.: Human – robot collaboration in industry. MM Sci. J. **2016**(02), 903–906 (2016). https://doi.org/10.17973/mmsj.2016_06_201611. ISSN 18031269. Accessed 24 May 2018

17. Lazna, T., Gabrlik, P., Jilek, T., Zalud, L.: Cooperation between an unmanned aerial vehicle and an unmanned ground vehicle in highly accurate localization of gamma radiation hotspots. Int. J. Adv. Robot. Syst. **15**(1) (2018). https://doi.org/10.1177/1729881417750787

18. Guide for understanding and implementing defense experimentation (GUIDEx). The Technical Cooperation Programme (TTCP), pp. 1–388 (2006)

19. Wohlin, C., et al.: Experimentation in Software Engineering. Springer, London (2012). ISBN 978-3-642-29043-5

Autonomous Air Defense Effectors Deployment Algorithms for Modeling and Simulation Purposes

Jan Farlik$^{(\boxtimes)}$, Miroslav Kratky, and Simona Simkova

Department of Air Defence, University of Defence, Brno, Czech Republic
{jan.farlik, miroslav.kratky, simona.simkova}@unob.cz

Abstract. At present, the development of information technologies leads to a significant development of simulation technologies and modeling possibilities. In the area of air defense, it is mainly about the algorithmization of the command and control processes together with the development of the possibilities of modeling and simulation of an air targets engagement process, e.g. the process of guiding the missile against a target. The process of destroying the target itself, however, is preceded by many activities, of which the modeling has been very difficult for a long time before. One of these activities is the process of deploying air defense systems to fulfill their purpose, i.e. to ensure the continuous coverage of a territory by their weapon systems. Ability to describe this process in simulations gives the possibility for future implementation not only in the complex simulators but also for use in real command and control systems for the needs of subordinate elements autonomous tasking. This article describes one option of such process algorithmization. This method has been used in the Department of Air Defence Systems tactical ground air defense simulator for the autonomous deployment of available effectors on the battlefield.

Keywords: Air defense · Deployment modeling · Decision making modeling · Simulator

1 Introduction

Algorithmization of command and control processes and operational tactical tasks has long been one of the activities developed by automated command and control systems. The implementation of these algorithms enhances the ability of command and control systems to successfully solve the tasks associated with combat activities. Due to the computing power parameters, it has been impossible for many years to develop these algorithms above the level that was determined by the computing speed of the computers at that time. With the evolution of computing, it is increasingly possible to implement algorithms that require big data [1]. These data are today mainly represented by maps and fusion information of the digital battlefield [2]. The autonomy of data processing operations enables the commander's decision-making process to be more effective, with more insight into the battlefield without having to deal with the calculations themselves [3]. Before the combat itself, however, it is necessary to carry out a number of activities, such as the deployment of units. Previously, this set of tasks was

© Springer Nature Switzerland AG 2019
J. Mazal (Ed.): MESAS 2018, LNCS 11472, pp. 577–587, 2019.
https://doi.org/10.1007/978-3-030-14984-0_43

exclusively the domain of the expert's estimation and knowledge of the commander. At present, a number of autonomous applications can be created to offer the commander an optimized deployment of units on the battlefield [4] or their better visualization [5]. These applications, or rather algorithms, must first be implemented in tactical simulators to ensure their thorough testing and verification. At the same time it is possible to train commanders in the given issue (in the case of this article it is the algorithmization of the deployment of fire units), because tactical simulators enable mainly the training of acquired knowledge and skills.

One of the important tasks currently dealt with by the commander's own judgment and tactical knowledge is the issue of deploying two or more firing units to provide mutual coverage. This means that once the first firing unit is deployed, the fire coverage efficiency is calculated, and then the other units are deployed so that the first unit gaps are eliminated. This process secures a continuous area of defense of fire units to defend the designated object (in the form of a point or area defense). Algorithmization of this task significantly speeds up the commander's decision process. This algorithm can be used not only for implementation in automated command and control systems or tactical simulators, but may also be one of partial algorithms for future autonomous weapon defense systems. In this paper, the developed algorithm is demonstrated on two ground air defense units' deployment that have the task of defending a designated object. Described algorithm is implemented for optical visibility in this paper. For calculations of radar visibility, it would be necessary to apply the relevant equations (radiolocation equations, beacon equation) [6].

2   Input Conditions of the Algorithm

To construct the algorithm for calculating the insertion of firing elements into the terrain within the simulator, only the optimization of the two firing elements will be considered within this article. The proposed algorithm can be extended to multiple firing elements in the future. The algorithm will include the design of a combat assembly of two firing elements in the "Mutual support" configuration, which assumes the deployment of the firing elements so that one element covers the deaf gaps (places where one of the firing elements cannot operate).

The input data of the algorithm is the digital model of the defended area. For a thorough exploration of the terrain it is possible to use a net with a density of $12 \cdot 12$ m. The use of a coarser grid would cause inaccurate and erroneous calculations because some terrain elements such as steep hills, artificial obstacles, etc. could be neglected. For the algorithm purposes, a digital surface model (DSM) is better than the digital terrain model (DTM) because the DTM does not include artificial barriers (while DSM does), which is key for ground based air defense (GBAD) units' deployment planning.

The specific input conditions of the algorithm are:

- It is assumed that the target fly at speeds of up to 300 m.s^{-1}.
- A defended object coordinates are known.
- There are two firing elements with defined range.
- To simplify the calculation, it will be assumed that the effective range of the air defense unit is cylindrical with a maximum range of detection D_{max}.

- The DSM for the defended area is given.
- For the purposes of the algorithm, point defense will be considered, not area defense.
- The first firing unit is placed within the terrain (simulated terrain).

3 Algorithm for Placing Two Firing Elements

Before introducing the algorithm itself, it is necessary to state the relationships that will be used during the algorithm calculations for the target detection range. The target is assumed to copy the terrain during its flight. If the attacking target is to fly at a certain altitude above the terrain and try to copy the terrain to be undetected for as long as possible, it will appear above the horizon at a certain distance D from the position of the firing unit sensors.

$$h_{hor} = \frac{D^2}{12.745} \ [\text{m}] \tag{1}$$

Where h_{hor} is the altitude, where the target will appear above the horizon at the distance D [7, 8].

The above stated equation must be expanded because the terrain is not flat but contains a number of terrain inequalities such as hills, mountains, valleys, etc. It is therefore necessary to extend the calculation to convert the discovery of the target above the mountainous horizon.

3.1 Terrain Obstacles Calculation

Figure 1 shows an example of mountainous terrain for which the DSM is known. The resulting calculation will obtain the so-called concealment angle ε_c, which will be the input parameter to calculate the distance at which the target is detected. The algorithm is simplified here by counting the closest distance from the firing position, where a continuous tracking of the target, which appears above the terrain, is available.

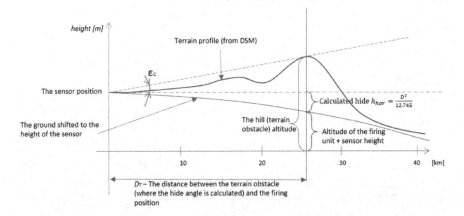

Fig. 1. An example of terrain profile and concealment angle calculated

3.2   Concealment Angle Calculation

The algorithm will always calculate the largest concealment angle from the firing element (the antenna), that means if the angle of concealment is less distant, the terrain is inclined further, considering the curvature of the earth. At each step of optical visibility calculation (see below), the concealment angle for each DSM cell will be calculated in a given direction from zero (firing position) to the maximum distance D_{max} (given by the firing unit boundary). Figure 1 gives details for the following calculation.

$$\varepsilon_c = arctan \frac{terrainheight(h_T) - sensorheight(h_A) - hideheight(h_{hor})}{D_T} \qquad (2)$$

For example, if a firing position is at a height of 195 m above sea level (ASL) and its antenna height is 10 m, then the imaginary displaced earth surface for calculation will be $h_A = 195 + 5 = 200$ m. The height of the hill for which the angle of concealment is calculated will be, for example, $h_T = 400$ m ASL. The distance of the hill is for example $D_T = 26$ km from the firing position. At this distance, the h_s is the concealment altitude according to (1) $- h_s = 26^2/12.745 = 53$ m. Then, after fitting into Eq. (2):

$$\varepsilon_c = arctan \frac{400 - 200 - 53}{26000} \cong 0.32° \qquad (3)$$

The algorithm proceeds from the firing position to the maximum required distance (e.g., the given visibility or radar range) after a step equal to the network resolution available to the DSM (for example, if the DSM network is 12 m, then the calculation step will be 12 m). In each next step, the algorithm must determine whether the hidden angle increases. If so, the old entry will be overwritten with this new, bigger concealment angle. At the end of the series of these calculations we get a distance from fire position where the

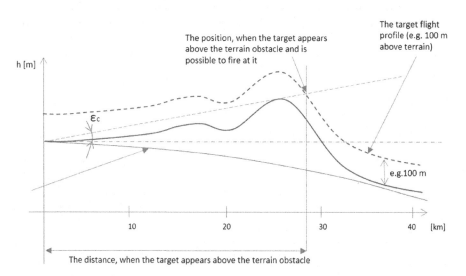

Fig. 2. An example of flight profile of the target

angle of concealment is greatest. This (for example, the peak of the hill) will then be used to calculate the distance to the defined flight height when it appears above the horizon (picture below).

The Fig. 2 shows the moment when the target appears above the horizon and can be engaged. The next part of the algorithm is designed as follows: after we have the maximum concealment angle ε_c and the height of the target flight over the terrain in the given direction (azimuth), step by step (corresponding to the DSM step, e.g. 12 m) from the largest distance to the nearest, we count the height at which the target could be seen at the given distance.

For example, for a given azimuth it was calculated that the maximum concealment angle (for 26 km from firing unit) is $0.32°$ (as an example above). We are now approaching the distance 26 km from the distance of 40 km (sensor range) in steps of 12 m. In these steps, we always calculate the height of H_{cmin} at which the target appears above the horizon. The calculation is according to the modified Eq. (2) only by the fact that we now have a concealment angle (in this case $0.32°$) and we need to obtain the height value at which the target will arise above the horizon (see figure below). In the figure below, we will calculate at what height the target would have to fly above the horizon to be detected (above the horizon). The target altitude is in this case counted at the distance of 35 km. At this distance, according to DSM, the ground height is 220 m. Then by Eq. (2) we get the equation:

$$tg\varepsilon_c = \frac{(h_T + H_{cmin}) - h_A - h_{hor}}{D_T}, \tag{4}$$

where h_{hor} is a concealment altitude at the calculated D_T distance. Here $D_T = 35$ km, and after fitting to Eq. (1) $h_{hor} = 96.12$ m. Then, from Eq. (3) we get the required height H_{cmin} as:

$$H_{cmin} = h_A + h_{hor} - h_T + (D_T \cdot tg\varepsilon_c), \tag{5}$$

in this case:

$$H_{cmin} = 200 + 96.12 - 220 + (35000 \cdot tg(0, 32°)) = 271.6\,\text{m} \tag{6}$$

This indicates that if the target is at the altitude of 100 m, 35 km far from firing position, it will not be detected and fireable. For the given distance, "uncovered" or "undetected" attribute is recorded and continues until the first detection (emerging above the horizon) (Fig. 3) .

Now we will summarize the coverage diagram that will be used in the next part of the algorithm to calculate the weak points of the firing position.

To obtain a full-circle coverage diagram, we have to do the above stated calculation several times, always for the desired azimuth. We propose a calculation step of $5°$, which can be reduced if necessary. For a full circle diagram, $360°/5° = 72$ calculations will be needed. Therefore, for each direction it is necessary to obtain a table of heights in the defined distances obtained from the DSM, here for example after 12 m intervals.

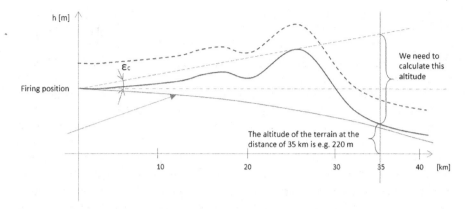

Fig. 3. Calculation if the target is at given distance visible or not

Then, for each direction, it is necessary to perform the above calculations to tell, at what distance the target flying at a constant altitude above the terrain is detected. The result will be a diagram and a table (see picture and table below) (Fig. 4).

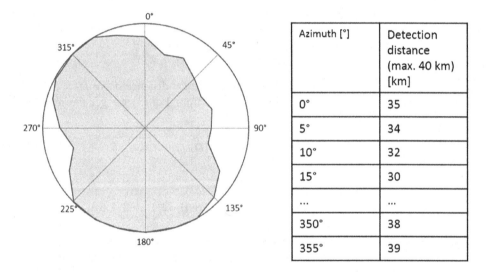

Azimuth [°]	Detection distance (max. 40 km) [km]
0°	35
5°	34
10°	32
15°	30
...	...
350°	38
355°	39

Fig. 4. An example of the calculation output

3.3 Calculation in a Certain Direction

For further steps, it is necessary to calculate the weak points in a certain direction. When knowing the assumed direction of the attack, we calculate the percentage of the ability to cover the given space. If the direction of attack is not known, we suggest to calculate the rounded percentage of the ability to cover the given space in steps of 5° with the width of the observed segment of 90°.

calculation of the coverage (visibility) of the given space compared to the ideal state.

Taking into account the 90° segment of the radius r (radar range), its ideal area is full coverage (visibility) given by the content of the circular sector. The area of ideal coverage S_{Ideal} is then given by the quarter circle with the radius D_{max}.

$$S_{Ideal} = \frac{\pi D_{max}^2}{4} \tag{7}$$

For example, if the radar range is 40 km, then the ideal coverage area would be 1256 km².

Now we take the desired segment, for example, if the assumed direction of attack is in the direction of 45°, we will calculate the area ±45°, i.e. from 0° to 90°. The desired area can be obtained by successive calculations from the obtained target detection values at a certain altitude above the terrain at a certain distance. If we obtained a set of values for targets flying at a altitude of 100 m above terrain in the form of 5° segments table (total of 72 values for a 360°), we have to calculate an approximate coverage area for each 5th segment. The 90° segment coverage will then consist of the sum of 18 values (90°/5° = 18). Below are the steps to calculate the actual coverage area of one 5° segment from the values obtained from the terrain analysis (from DSM). Then we calculate the distance the target reaches above the horizon. The figure below is an example of the segment that was obtained from the calculations of the visibility (Fig. 5).

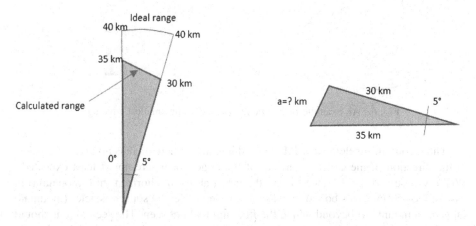

Fig. 5. Calculation of triangle area obtained from previous calculations

It is clear from the figure that for the next calculation we need to obtain a third triangle edge. This dimension is obtained through a cosine sentence, α = 5° in this proposed case.

The D_0 and D_5 are calculated visibilities in the $0°$, $5°$ directions from the simulated firing unit. For next calculation, there would be D_5 and D_{10}, D_{10} and D_{15}, etc. Based on this data, it is possible to calculate the area of the triangle area using the Heron formula.

Ideal coverage would be on the area of $S_{Ideal}/18$, which is approximately 69.78 km^2 (1256/18). However, the coverage of the given $5°$ space (segment) is about 65% (45.63/69.78 = 0.65). In this way, calculations of all $5°$ segments in the $90°$ segment have to be made and the sum to obtain the total coverage area (target visibility). Subsequently, the sum of the covered areas of the individual $5°$ segments is computed and the total coverage area is obtained. By the ratio of the ideal coverage area, we obtain a percentage coverage of the given $90°$ segment.

If we know the assumed direction of attack, we can calculate the percentage coverage in this particular space only, while the others not. If we do not know the assumed direction of attack, and we have a total of 72 rectangles, where we know the percentage of coverage, we propose to select the segment with the smallest percentage of ability to cover the given space and place the second firing element in this segment.

Placing the second firing unit to cover the gaps of first one.

Now, the $90°$ space with the lowest ability to fire on a target flying at a certain altitude (as shown below) was obtained (Fig. 6).

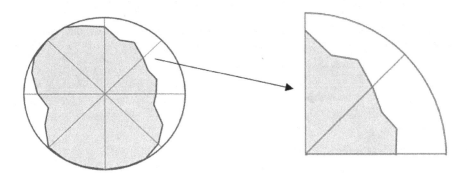

Fig. 6. An example of the gained area with the smallest coverage

The second firing element will be placed in places where there is no greater obstacle in the direction of the expected arrival of the target (or where not at least expected). When we counted the distance to find the target above the horizon (hill, mountain) in case of the first unit, this boundary was more or less given by some obstacles (mountain range or a mountain) beyond which the first unit had not seen. The second unit should therefore be located in the axis of the selected $90°$ sector at the "foot" of the obstacle that limits the visibility limit of the first unit, see the Fig. 7.

The algorithm proposes a procedure whereby the second unit is first placed at the edge of the range of the first unit. Here, the coverage calculation is performed by the same algorithm as for the first unit. Given that before this boundary, the first unit have the largest concealment angle, we suppose that the terrain would be sloping or being lower behind this terrain obstacle. The other unit should thus cover the rest of the required range of the first unit, which is the purpose of the whole algorithm. If this is not possible for any reason, we suggest continuing to test the percent coverage of the required space in steps up to the required coverage level of the first unit.

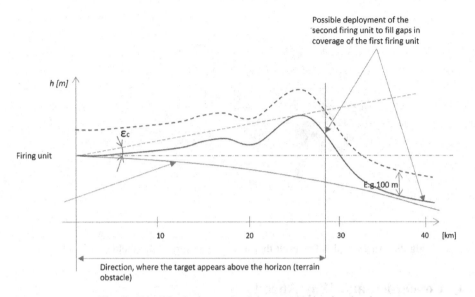

Fig. 7. Possible deployment of the second simulated unit

If there would not be any improvement (for example due to the existence of valleys), we suggest moving with the second unit always at 5° segments (in the case of faster computers, perhaps 1° segments) to the left and right of the original position at the visibility limit of the first unit. In these positions, the visibility and overall calculation of the penetration of the first and second units would again be calculated (Fig. 8).

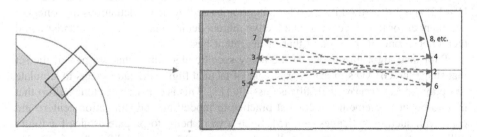

Fig. 8. Algorithm procedure to scan and calculate the appropriate location of the second unit from the boundary of the first unit

Subject Matter Excerpts would have to set the limit percentage of the required coverage of the first and second units. This would set a threshold, e.g. 90% coverage, when it would be possible to terminate the deployment of the second unit at a time when coverage would be greater than this threshold (Fig. 9).

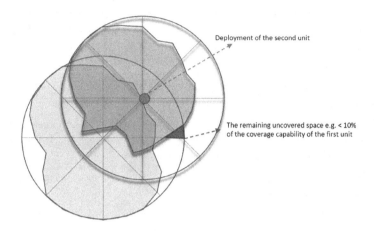

Fig. 9. An example of a result that meets the coverage threshold e.g. $\geq 90\%$

4   Conclusion and Way Ahead

Algorithm for deployment of firing units described in this paper offers a simple way, how to calculate the effective coverage of two or more effectors. The method used is not limited for ground based air defense purposes, but could be used also for generic deployment of simulated units in cases, when we have some range of interest (firing range, visual range, area of interest, etc.) and we need to cover certain gaps of other acting units or simply elements. This algorithm or method could be upgraded to calculate more variables, describing the effect of some element in some area [9]. Authors' research covered modelling and simulation of GBAD units for tactical flight simulator [10–12] and this method is tested for autonomous deployment calculations that offers to commander (or to trainee) several effective mutual deployments. The commander could then choose one of the variant that fit his intent best.

Published algorithm is one of the many specific algorithms that are part of the set that shape the GBAD simulator, or in case of tactical flight simulator, helps to simulate an opponent forces as realistically as possible [13]. This is consistent to today effort that tries to better implement the combat practice to modelling and simulation centers and cope with such a challenges [14, 15]. Such a work helps to prepare armies for future military autonomous systems as well as to overcome the gaps and difficulties for such a process [16].

References

1. Frantis, P.: Big data in the air force - process, use and understand for safety. In: 33rd Digital Avionics Systems Conference, pp. 8C2:1–8C2:6. Institute of Electrical and Electronics Engineers Inc., Colorado Springs (2014)
2. Kristalova, D., et al.: Geographical data and algorithms usable for decision-making process. In: Hodicky, J. (ed.) MESAS 2016. LNCS, vol. 9991, pp. 226–241. Springer, Cham (2016). https://doi.org/10.1007/978-3-319-47605-6_19
3. Stodola, P., Mazal, J.: Tactical decision support system to aid commanders in their decision-making. In: Hodicky, J. (ed.) MESAS 2016. LNCS, vol. 9991, pp. 396–406. Springer, Cham (2016). https://doi.org/10.1007/978-3-319-47605-6_32
4. Stodola, P., Nohel, J., Mazal, J.: Model of optimal maneuver used in tactical decision support system. In: Methods and Models in Automation & Robotics (MMAR 2016), pp. 1240–1245. West Pomeranian University of Technology in Szczecin, Szczecin (2016)
5. Frantis, P.: Visualization of common operational picture. In: WIT Transactions on Engineering Sciences, pp. 347–354, WITPress, United States (2014)
6. Hamtil, I., Sebela, M., Stefek, A.: Radar information creation with use of a simulation environment. IET Radar Sonar Navig. 7(4), 333–341 (2013). ISSN 1751-8784
7. Horizon. In Wikipedia: the free encyclopedia, Wikimedia Foundation, San Francisco (2001). https://en.wikipedia.org/wiki/Horizon. 31 July 2018
8. Young, A.: Distance to the Horizon. Distance to the Horizon (2018). https://aty.sdsu.edu/explain/atmos_refr/horizon.html. 31 July 2018
9. Farlik, J.: Simulation of surface-to-air missile units - cluster design. In: IEEE International Conference on Military Technology Proceeding, ICMT 2015, pp. 647–652. University of Defence, Brno (2015)
10. Farlik, J., Kratky, M., Hamtil, I.: The Air defence missile system effective coverage determination using computer simulation. In: International Conference on Military Technology Proceeding, ICMT 2015, pp. 669–673. Brno, University of Defence (2015)
11. Farlik, J., Stary, V., Casar, J.: Simplification of missile effective coverage zone in air defence simulations. In: Proceedings of the 2017 International Conference on Military Technologies (ICMT), pp. 733–737. Institute of Electrical and Electronics Engineers Inc., Piscataway (2017). ISBN 978-1-5386-1988-9
12. Farlik, J., Tesar, F.: Aspects of the surface-to-air missile systems modelling and simulation. In: Mazal, J. (ed.) MESAS 2017. LNCS, vol. 10756, pp. 324–339. Springer, Cham (2018). https://doi.org/10.1007/978-3-319-76072-8_23
13. Farlik, J.: Conceptual operational architecture of the air force simulator: simulation of air defense operations. In: International Conference on Military Technology Proceeding, ICMT 2015, pp. 675–679. University of Defence, Brno (2015). ISBN 978-80-7231-976-3
14. Hodicky, J.: Modelling and simulation in the autonomous systems' domain – current status and way ahead. In: Hodicky, J. (ed.) MESAS 2015. LNCS, vol. 9055, pp. 17–23. Springer, Cham (2015). https://doi.org/10.1007/978-3-319-22383-4_2
15. Hodicky, J., Prochazka, D.: Challenges in the implementation of autonomous systems into the battlefield. In: Proceedings of the 2017 International Conference on Military Technologies (ICMT), pp. 743–747. Institute of Electrical and Electronics Engineers Inc., Piscataway (2017). ISBN 978-1-5386-1988-9
16. Hodicky, J.: Autonomous systems operationalization gaps overcome by modelling and simulation. In: Hodicky, J. (ed.) MESAS 2016. LNCS, vol. 9991, pp. 40–47. Springer, Cham (2016). https://doi.org/10.1007/978-3-319-47605-6_4

Autonomous Systems and Chinese Strategic Thinking

Jakub Fučík$^{(\boxtimes)}$ ⓘD, Libor Frank ⓘD, and Richard Stojar ⓘD

Centre for Security and Military Strategic Studies, University of Defence,
Tučkova 23, 602 00 Brno, Czech Republic
{jakub.fucik,libor.frank,richard.stojar}@unob.cz

Abstract. Since the 1980s, it has been possible to identify changes in the military-strategic dimension, which are associated with the concept of the Revolution in Military Affairs (RMA). The development of new (weapon) technologies (e.g. precision-guided munitions, unmanned aerial vehicles) represents opportunity (and challenges) especially for states to enhance their military power and change their status in the system of international relations. Nowadays, the U.S. still represent the main actor of this system, however, we can identify several rivals on the global or regional levels. People's Republic of China (PRC) is one of these challengers. From this point of view, the development of autonomous systems (AxS) could strengthen the Chinese military capabilities and enhance the threat for the position of the U.S. and their allies (NATO). However, this presumption depends on the transformation of the Chinese strategic thinking that could address such opportunity. The aim of this paper is to analyse how the Chinese strategic thinking reflects the development of AxS and identify the changes which are connected with this issue. This paper will provide the necessary understanding of the PRC's approach to the AxS, which, from this point of view, represents implications for military planning and strategy development not only of the U.S., but also NATO and its member states.

Keywords: Autonomous systems · Chinese strategic thinking ·
Military planning · Modern military technologies · Strategy development

1 Introduction

Nowadays, the armed forces of more than sixty countries of the world are using remotely controlled Unmanned Systems (UxS) for reconnaissance, survey or monitoring purposes. The number of states which employ armed UxS is also gradually growing. It can be assumed that this general trend, i.e., the growing number of states which operate UxS of various categories, will only intensify in all "traditional" military domains (land, naval, air). Compared to remotely controlled systems, the Autonomous Systems (AxS) require no or only minimal involvement of the human operator [1]. Individual systems should be able not only to obtain information about the environment but also to process (evaluate) this information and take appropriate decisions on their own. The motivation to establish those systems is directly based on their increased

© Springer Nature Switzerland AG 2019
J. Mazal (Ed.): MESAS 2018, LNCS 11472, pp. 588–598, 2019.
https://doi.org/10.1007/978-3-030-14984-0_44

effectiveness in combat. Similar to remotely controlled systems, the idea of minimising the human losses on the part of the operator's own armed forces plays the key role [2, 3]. Moreover, AxS enable to reduce (or completely remove) the cognitive charge of their operators.

The development of AxS is strongly connected with the changes in the military-strategic dimension, which are associated with the concept of the Revolution in Military Affairs (RMA) [4]. Since the 1980s, People's Republic of China (PRC) has been one of the states which started to implement the premises of the RMA into its strategic thinking. From this point of view, the development of autonomous systems (AxS) could strengthen the Chinese military capabilities and enhance the threat for the position of the U.S. and their allies (NATO).

The aim of this paper is to analyse how the Chinese strategic thinking reflects the development of AxS and identify the changes which are connected with this issue. This paper will provide the necessary understanding of the PRC's approach to the AxS, which, from this point of view, represents implications for military planning and strategy development not only of the U.S., but also NATO and its member states. This paper is a qualitative case study of the Chinese strategic thinking and its relation·to the AxS.

2 Revolution in Military Affairs

For the purpose of this paper, RMA is defined as the process and condition of revolutionary changes in the nature or method of warfare based on the external manifestations (actions) which employ the threat of force or the use of force to achieve political aims [5]. The "revolutionary" then refers to the radical nature of these changes, which, in relation to the original system and its elements, must occur abruptly de facto preserving just a minimum similarity (e.g. in features by which the system is identified). We cannot, therefore, speak of a progressive (gradual) transition and the establishment of new elements into the existing framework and its evolutionary transition. With regard to the military dimension of this revolution, we can use the modified characteristics defined by Jeffrey R. Cooper, who speaks about: "… discontinuous increase in military capability and effectiveness" [6].

However, this paper does not focus on the RMA in the context of the Chinese strategic thinking in the aforementioned "general" form, but rather on its expression through specific processes or changes. This is marked by the period of about the 1980s, still continuing today. Relevant changes in the method of warfare are founded on a technological level with the introduction and use of advanced weapons and information systems (e.g. precision-guided munitions - PGM, unmanned aircraft, and remote sensing devices/sensors).

Changes in the doctrinal dimension are represented by the establishment of the concepts of so-called System of Systems (SoS) and Network Centric Warfare (NCW). The former is based on two fundamental elements - information and integration (co-operation). The prerequisite is an amalgamation of particular systems and components, such as command, control, computers, communications and information (C4I), into one coherent functional framework [7]. Essentially, the aim is to provide situational awareness on the battlefield in real time for all relevant components of the armed forces.

The second concept is associated with the very existence and use of functional links among the units on the battlefield, which are integrated into the aforementioned framework. Their interdependence allows to maximize their combat skills and, on the other hand, to compensate for weaknesses (e.g., through an almost perfect fire support, information about the intentions of the enemy). Full use of this potential is connected, e.g., to the implementation of the so-called "swarming" tactic, which in itself implies synchronized and highly flexible combat deployment of a large number of small clusters (military units) [8].

In practical terms, the army, which fully applies both concepts, is allowed to interfere (invade) the opponent accurately at his most vulnerable areas, to prevent his possible attempts to initiate counterattacks or enact countermeasures and therefore completely take over the combat initiative and paralyze the opponent.

3   Chinese Strategic Thinking

3.1   Strategic Culture

The overall form of the Chinese strategic thinking is inherently influenced by the interactions between two basic subtypes of the strategic culture. These subtypes are defined by sets of Chinese values, believes, norms, etc. With regard to a time scale, their importance (for example, for the formulation of the Chinese international politics) in history (and also in present) was changing, i.e., in some cases one of the motives prevailed, but always both subtypes were/are shaping the Chinese strategic thinking [9, 10]. At the same time, the above-mentioned link points to the continuity of the Chinese strategic culture. Undoubtedly, for example, the "victory" of the communist regime influenced its character, but only in the context of strengthening or weakening the importance of one of the subtypes or adding new elements (e.g., deepening the ideological dimension). By no means all these changes have led to the emergence of a completely new strategic culture.

According to the terminology of Alastair I. Johnston, the first subtype can be described by the term *Parabellistic Strategic Culture* – from the realpolitical axiom "Si vis pacem, parabellum" [11]. In general, it is an offensive (aggressive) approach to, e.g., the creation of a "grand strategy" of the state, which is supported by the Clausewitz's thesis about war and the use of force ("War is a continuation of politics by other means."); in the case of China, regardless of weaker material capabilities [12–15]. The second subtype is referred to as the *Conflict Strategic Culture* [9]. The general basis of this direction is the Confucian philosophy. In response to security threats and other incentives, diplomatic and non-military means are preferred. In this respect, the use of force is seen as an additional element, which is used mainly in defensive intentions [9, 16]. The current concept of the "scientific development" (in the perspective of the RMA), which has been accentuated in Chinese politics and strategic thinking basically since the beginning of the new century (see Sect. 3.2), expresses the values described above [17, 18].

3.2   Strategic Thinking

The development of the Chinese strategic thinking into the current character was initiated after the Sino-Vietnamese war in 1979. In principle, China failed to fulfil almost any of its objectives (such as the land gains or the withdrawal of Vietnamese troops from Cambodia). Chinese ground forces struggled in the conflict with low efficiency of their direct and indirect fire, logistical obstacles, or failure of communication [19]. The original strategy of *People's War*, which was set by Mao Zedong in 1935, was replaced by *People's War under Modern Conditions*. This strategy, compared to the pre-existing Mao's one, puts less emphasis on the quantitative superiority of the armed forces and rather focuses on the existence and use of a professional army (a combination of various types of weapon systems, especially ground forces). At the same time, it has changed the character of the supposed defence of the territory, which no longer has to lure the adversary deep in the territory of China, but instead to stop him at the external borders. From the point of view of the theoretical definition of RMA, the deviation from the belief of the necessity of nuclear conflict (the use of nuclear weapons) to the gradual emphasis of modern conventional technologies is also significant [20].

Further changes were being implemented after 1985 with the new strategy - *Limited War under Modern Conditions*. The basic premise consisted of the belief that the likelihood of a massive foreign invasion with the aim of a total defeat of China was greatly reduced. This assumption was further strengthened with the end of the Cold War and the collapse of the USSR. On the contrary, the likelihood of a limited armed conflict has increased, due to the escalation of the dispute over a particular maritime or terrestrial geographic area (South China Sea). This characteristic corresponds to the relative shortness of the alleged conflict and the remoteness of the territory from the central regions of the PRC, but at the same time located near the external borders [21]. As a result, increasing demands (and consequently emphasis) on mobility and the rapid deployment of relevant military units have been made. Compared to the previous strategy, there is a noticeable shift from the defensive concept, to the pre-emptive strike against the opponent in the context of "active defence". Similarly, the role of purely professional (elite) Rapid Response Forces and up-to-date weapon and support systems are emphasized. In this sense, the ground component of the armed forces is no longer preferred, but the importance of joint operations is still not accentuated [22].

Between 1993 and 1996, this strategy was replaced by *Limited War under Hi-Tech Conditions*. Similar to the previous strategy, the area of the supposed deployment of the Chinese armed forces was limited to several provinces covering all dimensions of military operations (land, naval, airborne, space, information/cyberspace operations) under the generic "War Zone Campaign" [23]. However, compared with previous assumptions, these campaigns should not be conducted only in the immediate vicinity of the Chinese frontiers (border disputes), but in accordance with power aspirations within the global environment (international system) [24]. The operational doctrine was newly oriented to joint operations of all branches of the armed forces, reflecting the above-specified environment of their deployment. Emphasis was put on fast deployability, high mobility, and the ability to reach (temporary) local superiority as needed to meet the assigned tasks. In relation to ground forces, the need for overall mechanization is emphasized in this sense, which results in the achievement of the established capabilities. For the naval and air forces (later also for space and information) there are key capabilities to deny the enemy the ability to deploy his armed forces effectively or to project his military power on a given battlefield (anti-access/area denial - A2/AD) [22].

In terms of institutionalization, the People's Liberation Army (PLA) started reforms, which outlined the new basic (general) assumptions/characteristics through the establishment of the ideas and tools of the RMA. In the first place, there had to be an organizational link among the capabilities of active and forward defence, which involved the incorporation of the possibility of projection of Chinese military power outside the Chinese territory. In this sense, from the point of view of the Chinese strategic planning, there is a radical expansion of the strategic depth of both, the conflict zone and the relevant (security/military) interests, which are no longer limited by the factual boundaries of the state [25].

Second, in Chinese terms, the identified RMA elements maximize the offensive direction of the armed forces. In an environment where the opponents have the ability to destroy their targets precisely in conjunction with a digitized battlefield, from this point of view, the party that takes the initiative (retains it) and strikes before the counterparty, gains the edge. This element is even more emphasized on the condition that Chinese armed forces do not contain such qualitative levels (in the terms of weapon and support systems) as the adversary. In general, during the 1990s, PLA shaped itself in such a way, which was to be balanced by proactive and pre-emptive actions [26].

Third, the Chinese strategic planning and the organization of the armed forces had to be able to adapt and absorb new technological innovations and respond to a changing international environment. In the same sense, the PRC tried to focus on a possible conflict with the major powers of the international system, but at the same time to dispose of forces and resources flexibly deployable in smaller and (more time-intensive) clashes. Also, the approach of the Chinese institutions seems to be rather attempting to prepare/transform the military power with a long-term view to ensure the competitiveness or ideally superiority in the future rather than to focus on the current gain of a short-term benefit [27].

Other reforms were started between 2002 and 2004 with the strategy *Local War under Informationized Conditions*. A fundamental change is the direct anchoring of not only the need for mechanized units but also the digitization. The transformation of the armed forces itself took place under the simultaneous digitization and mechanization processes, with further emphasis on information warfare. The aim was to ensure that the Chinese transformation of the armed forces proceeds in line with contemporary trends of similar activities of other world powers (Russian Federation, the United States of America) [28]. In the context of military operations, a new concept of "integrated joint operations" was applied. These operations differ primarily from the original "joint operations" by another type of actor who is in conflict. Previously, there were relatively separate branches (elements) of the armed forces that had their own information systems, and the joint operation was based on the creation of ad hoc connections. In contrast, under integrated operations, the main player is the system that incorporates all components of the armed forces and the necessary operational elements (C4ISR, destruction capabilities and logistics). At the same time, in these operations, the boundaries between the various components of the armed forces may be blurred due to close mutual co-operation and combat deployment. Coordination is ensured through relevant information systems immediately following developments on the battlefield. Integration itself is defined as a permanent structure that fulfils the conditions of a fast and flexible response [22].

In the follow-up strategic documents (from 2006, 2009, 2011, and 2013), this orientation is not only confirmed, but also gradually intensified in the context of importance for the fulfilment of Chinese national interests, defence, or improvement of the power position in the international system and the successful conduct of military operations (also in the space domain). The latest military strategy of 2015 then contains the essential provisions that shift the orientation to information warfare to a new "level". Under the auspices of the current leader of the People's Republic of China - Xi Jinping, the original strategy has been modified into *Winning Informationized Local Wars*. The title itself refers to the future ambitions of the PRC and their linkages to information warfare. In this context, it is also interesting to emphasize the development of maritime and air forces and their role in the effective implementation of integrated joint operations. Last but not least, there is the premise of the transformation of strategic/missile forces, which emphasizes the combination of both conventional and nuclear components and their use in precision strikes against the opponent [29–34].

4   The Role of AxS in the Chinese Strategic Thinking

The general direction of the transformation of the Chinese armed forces stems from the concept of information warfare, which both represents the "target" domain of new combat capabilities (e.g., cyberspace) and serves as a necessary basis for establishing other (modern) weapon and support systems that use elements associated with this kind of warfare. In this sense, PLA distinguishes six sub-sets/sets of capabilities and relevant (information) technologies, which include: (1) operational safety; (2) deception; (3) psychological operations; (4) electronic warfare; (5) operations in cyberspace (cyberwar); (6) physical destruction (enemy information systems - e.g., via electro-magnetic pulse). All of these capabilities represent strong connection among "traditional" (land, naval, air) and "new" (space, cyberspace) operational domains.

In this context, autonomous systems are perceived as an opportunity to strengthen PLA's power and also create new possibilities in terms of its projection and operability. From the PLA point of view, in the armed conflict, the potential use of AxS capabilities is envisaged in the combination with the aim of achieving the greatest possible success in achieving the stated goals. From a practical point of view, of course, it will always depend on variables such as the nature of the tasks the armed forces have to conduct, the conditions and environments in which they operate, the character of the adversary, and which specific capabilities will be used [35]. AxS should also ensure general requirement for flexibility of the armed forces and directly fulfil the established strategic framework (see Sect. 3). Focal point of this assumption is based not only on capability to analyse and process vast amount of information and almost-real-time decision-making. Important advantage is also compensation or removal of the human "weaknesses" (e.g. demands for rest and meals, variable reaction time, or other physical and psychical issues). At the same time, an increased emphasis on the management of operations directly related to cyberspace is interlinked not only with effective usage of own (Chinese) unmanned and/or autonomous systems (active and passive defence), but also with the capabilities to suppress/counter similar assets on the side of adversary.

In addition to this role, the second approach known as *Elements of Excellence* or *Assassin's Mace* (Shashoujian) was established during the 1990s. The naming refers to an old Chinese legend of a hero who managed to overcome a much stronger (more powerful) opponent thanks to such a weapon. Similarly, the technologies and weapon systems developed under this heading would have the ability to generate future overwhelming dominance over the adversary (also in terms of deterrence) and in the possible armed conflict (for example, the Taiwan Strait or the South China Sea) to ensure his defeat [36]. In general, there is an attempt by the PRC to conceal these projects as much as possible. In this respect, the exploitation of the AxS is also associated with a surprise element to increase their effectiveness. At the same time, their use is directed against the adversary's weaknesses, which should help achieve a quick and convincing victory. Generally, this assumption implies the basic elements of "asymmetric strategies" that focus on building and developing such assets that directly target/utilize the weakest characteristics of the adversary. In this context, the development of the AxS based on both - already known and/or exotic technologies, which are currently in the conceptual stage [37], can be identified as the analogy to the imaginary search of the "silver bullet".

A significant impetus for the development of the AxS came in 1997 from the already mentioned establishment of the principles of the *Limited War under Hi-Tech Conditions* strategy. A new research and development program was launched under the code name Program 973 (National Program of Basic Research). The program also includes other multidisciplinary projects and links the AxS, e.g., with information technologies, nanotechnologies, or biotechnologies [38]. Simultaneously, the role of the AxS was/is also highlighted by the prevailing *School of Revolution in Military Affairs* in the Chinese strategic thinking, which reflects the premises of the RMA mentioned in Sect. 2. AxS (interconnected with other modern weapon systems) should enable military strikes over a long distance, which is directly related to the introduction of advanced guidance systems and precision guided munition. At the tactical level it is reflected, e.g., in the concentrated fire of dispersed units [39]. Secondly, the AxS should support the creation of small (mobile) combat formations without reducing their combat capabilities. This premise is connected to integrated C4ISR systems and thus conveying information as a key element not only for ensuring the functioning of these systems but also in denying these benefits to the adversary [40]. Thirdly, the AxS should help establish an interconnection between information superiority and operation effectiveness [41].

Further integration of the AxS into the PLA should be established through two new branches of the armed forces. These are the Strategic Support Forces, which were created at the end of 2015, and the Joint Logistic Support Forces, which were created in September 2016. Within the scope of the Strategic Support Force, there are, in particular, the cyber-space/space-related operations, and also, in the needs of the rest of the armed forces, tasks related to information warfare and managing and using Chinese (military) capabilities (for example, the AxS). The Joint Logistic Support Forces are directly subordinated to the Central Military Commissions. Their purpose is to provide comprehensive logistical support for the full range of integrated joint operations, for example, also with regard to the projection of the AxS (with a planned global reach) [42].

5   Conclusion

The general character of the Chinese strategic thinking is based on the Chinese strategic culture. In this context we can distinguish between two sub-types - *Parabellistic Strategic Culture* and *Conflict Strategic Culture*. Interactions of these sub-types create the framework for certain elements of the strategic thinking and its development. Such a process can be identified since the 1970s with the transformation of the Chinese strategies. The progress from *People's War*, through *People's War under Modern Conditions, Limited War under Modern Conditions, Limited War under Hi-Tech Conditions, Local War under Informationized Conditions,* to the up-to-date *Winning Informationized Local Wars* incorporates the thesis and the premises of the (current) Revolution in Military Affairs [43, 44].

In this context, AxS represent key elements and tools in this process. AxS are perceived as the component linked to the information domain and information warfare. They provide the opportunity to strengthen the PLA's power in terms of its projection and operability. From the PLA's point of view, in the armed conflict, the potential use of AxS capabilities is envisaged in combination with the aim of achieving (if possible) the greatest possible success in meeting the stated goals. AxS should also ensure the general requirements for the flexibility of the armed forces and directly fulfil the established strategic framework. Moreover, AxS should provide the ability to generate future overwhelming dominance over the adversary in terms of "asymmetric strategies" that focus on building and developing such assets that directly target/utilize the weakest characteristics of the adversary. On the other hand, their role is not perceived as the sole one. Only in connection with other "hi-tech" systems and technologies (PGMs, space capabilities, etc.) they should create the desired synergic effect and ensure the strengthening of the PRC's military power.

Acknowledgements. The work presented in this paper has been supported by the Ministry of Defence of the Czech Republic (Research Project "STRATAL" No. 907930101023).

References

1. Hodicky, J., Prochazka, D.: Challenges in the implementation of autonomous systems into the battlefield. In: ICMT 2017 - 6th International Conference on Military Technologies, Brno, pp. 743–744 (2017)
2. Stojar, R.: Bezpilotní prostředky a problematika jejich nasazení v soudobých konfliktech (The Unmanned Aerial Vehicles and Issues Connected with Their Use in Contemporary Conflicts). Obrana a strategie, vol. 16, no. 2, pp. 5–18 (2016)

3. Stojar, R., Frank, L.: Changes in armed forces and their significance for the regular armed forces. In: The 18th International Conference. The Knowledge-Based Organization: Conference Proceedings 1 - Management and Military Sciences, vol. 1, pp. 142–145 (2012)
4. Fučík, J., Kříž, Z.: Informační revoluce, vojensko-technická revoluce, nebo revoluce ve vojenských záležitostech (Information Revolution, Military-Technical Revolution or Revolution in Military Affairs). Obrana a strategie, vol. 13, no. 2, pp. 15–24 (2013)
5. Gray, C.S.: Strategy for Chaos - Revolution in Military Affairs and The Evidence of History. Frank Cass, London (2005)
6. Cooper, J.R.: Another View of the Revolution in Military Affairs, p. 13. Strategic Studies Institute (1994). http://www.strategicstudiesinstitute.army.mil/pdffiles/pub240.pdf
7. Owens, W.A., Offley, E.: Lifting the Fog of War. Farrar Straus Giroux, New York (2000)
8. Alberts, D.S., Gartska, J.J., Stein, F.P.: Network Centric Warfare: Developing and Leveraging Information Superiority. CCRP, Washington D.C. (1999)
9. Feng, H.: A dragon on defense: explaining China's strategic culture. In: Johnson, J.L., Kartchner, K.M., Larsen, J.A. (eds.) Strategic Culture and Weapons of Mass Destruction: Culturally Based Insights into Comparative National Security Policymaking, pp. 171–187. Palgrave Macmillan, New York (2009). https://doi.org/10.1057/9780230618305_11
10. Sondhaus, L.: Strategic Culture and Ways of War, pp. 98–105. Routledge, New York (2006)
11. Johnston, A.I.: Cultural Realism: Strategic Culture and Grand Strategy in Chinese History, pp. 61–68. Princeton University Press, Princeton (1995)
12. Johnston, A.I.: Cultural Realism: Strategic Culture and Grand Strategy in Chinese History. Princeton University Press, Princeton (1995)
13. Johnston, A.I.: Cultural realism and strategy in maoist China. In: Katzenstein, P. (ed.) The Culture of National Security: Norms and Identity in World Politics, pp. 216–270. Columbia University Press, New York (1996)
14. Scobell, A.: China and Strategic Culture. Strategic Studies Institute, US Army War College, Carlisle Barracks (2002)
15. Zhang, T.: Chinese strategic culture: traditional and present features. Comp. Strat. 21(2), 73–90 (2002)
16. Feng, H.: Chinese Strategic Culture and Foreign Policy Decision-Making: Confucianism, Leadership and War, pp. 19–35. Routledge, New York (2007)
17. Fewsmith, J.: Promoting the scientific development concept. China Leadersh. Monit. (11) (2004). http://media.hover.org/sites/default/files/documents/clm11_jf.pdf
18. Hooper, W. The Scientific Development Concept (2010). http://www.theoligarch.com/scientific_development_concept_china_political_philosophy.htm
19. Feigenbaum, E.A.: China's Techno-Warriors: National Security and Strategic Competition from the Nuclear to the Information Age. Stanford University Press, Stanford (2003)
20. Fisher Jr., R.D.: China's Military Modernization: Building for Regional and Global Reach, pp. 69–70. Praeger Security International, Westport (2008)
21. Ibid. p. 69
22. Li, N.: New developments in PLAs operational doctrine and strategies. In: Li, N., McVadon, E., Wang, Q. (eds.) China's Evolving Military Doctrine: Issues and Insights, pp. 5–12. Pacific Forum CSIS (2006). http://www.comw.org/cmp/fulltext/0612li.pdf
23. Bajwa, J.S.: Chinese military strategy: war zone campaign concept. Indian Defence Review (2015). http://www.indiandefencereview.com/news/chinese-military-strategy-war-zone-campaign-concept/
24. Ji, Y.: Learning and catching up: china's revolution in military affairs initiative. In: Goldman, E.O., Mahnken, T.G. (eds.) The Information Revolution in Military Affairs in Asia, pp. 97–123. Palgrave MacMillan, New York (2004). https://doi.org/10.1057/9781403980441_5

25. Yongjun, G.: Fangkong zuozhan ying shuli quanquyu zhengti fangkong de sixian (Air defence should be guided by the theory of area and integrated defense), pp. 47–49. Junshi xueshu, no. 11 (1995)

26. Zhigang, S.: Jiji fangyu zhanlie sixiang zhai xinshiqi junshi douzheng de tixian (The application of active defense strategy in the military preparation in the new era). The Journal of PLA National Defence University, p. 100 (1998)

27. Bojun, T.: Dangde sandai lingdao jiti yu keji qianjun (The Party's three generation leadership and strengthening the armed forces through technological breakthroughs). China Military Science, no. 3, pp. 65–73 (1997)

28. Lai, D.: Introduction. In: Kamphausen, R., Lai, D., Scobell, A. (eds.) The PLA at Home and Abroad: Assessing the Operational Capabilities of China's Military, pp. 1–44. Strategic Studies Institute, Carlisle Barracks (2010)

29. China's National Defense in 2004. http://english.gov.cn/official/2005-07/28/content_18078.htm

30. China's National Defense in 2006. http://www.china.org.cn/english/features/book/194421.htm

31. China's National Defense in 2008 (2009). http://www.china.org.cn/government/whitepaper/node_7060059.htm

32. China's National Defense in 2010 (2011). http://news.xinhuanet.com/english2010/china/2011-03/31/c_13806851.htm

33. Diversified Employment of China's Armed Forces (2013). http://www.nti.org/media/pdfs/China_Defense_White_Paper_2013.pdf

34. China's Military Strategy (2015). http://eng.mod.gov.cn/Press/2015-05/26/content_4586805.htm

35. Yoshihara, T.: Chinese Information Warfare: A Phantom Menace or Emerging Threat? Strategic Studies Institute, Carlisle (2001)

36. Yang, A.: China's revolution in military affairs: rattling Mao's army. In: Goldman, E.O., Mahnken, T.G. (eds.) The Information Revolution in Military Affairs in Asia, pp. 125–138. Palgrave MacMillan, New York (2004). https://doi.org/10.1057/9781403980441_6

37. Fisher Jr., R.D.: China's Military Modernization: Building for Regional and Global Reach, p. 81. Praeger Security International, Westport (2008)

38. Springut, M., Schlaikjer, S., Chen, D.: China's Program for Science and Technology Modernization: Implications for American Competitiveness, p. 28. CENTRA Technology, Inc., Arlington (2011). https://www.uscc.gov/sites/default/files/Research/USCC_REPORT_China%27s_Program_forScience_and_Technology_Modernization.pdf

39. Youyuan, Ch.: Junshi jishu gemin yu zhanyi lilun de fazhan (RMA and the development of campaign theory), pp. 37–38. The Journal of PLA National Defence University (1999)

40. Yongfeng, H., et al.: Shuzhihua budui yu zhanchang (Digitalized troops and battlefield). Junyiwen chubanshe, Beijing (1998)

41. Qingshan, L.: Xinjunshi gemin yu gaojishu zhanzhen (New revolution in military affairs and hi-tech warfare), chapters 5 and 6. The PLA Academy of Military Science Press, Beijing (1995)

42. Wuthnow, J., Saunders, P.C.: Chinese Military Reform in the Age of Xi Jinping: Drivers, Challenges, and Implications, pp. 15-17. National Defense University Press, Washington, D.C. (2017). http://www.css.ethz.ch/content/dam/ethz/special-interest/gess/cis/center-for-securities-studies/resources/docs/INSS_US-ChinaPerspectives-10.pdf

43. Fucik, J., Kolkus, J., Melichar, J., Prochazka, J.: Military technology evolution assessment under growing uncertainty and complexity: methodological framework for alternative futures. In: ICMT 2017 - The 6th International Conference on Military Technologies, Brno, pp. 682–689 (2017). Analysis and prediction of (probable) future development of this issue could be based. Alternative Futures Methodological Framework
44. Frank, L.: Creation of scenarios and other methods as a tool for predicting the future security and operating environment. In: The 17th International Conference. The Knowledge-Based Organization: Conference Proceedings 1 - Management and Military Sciences, pp. 418–422 (2011)

Modelling of the Force Protection Process Automation in Military Engineering

Jaroslav Záleský[✉] and Tibor Palasiewicz[✉]

University of Defence, Brno, Czech Republic
{jaroslav.zalesky,tibor.palasiewicz}@unob.cz

Abstract. Article deals with a set of problems linked to a Engineer Force Protection Provision algorithm design and evaluation of input factors series. This algorithm is generally compatible with The NATO Force Protection Process Model adjusting it to a part of engineer forces' decision making process. The base of the algorithm is an application of repeated numerical matrix pattern and its word interpretation. The article provides a thought content being a possible key idea for the suitable software development.

Keywords: Force protection · Engineer support · Protective measures · Threat · Risk · Risk analyses · Risk management

1 Introduction

One of key success conditions of any military activity is own forces casualties reduction to such level that enable them keeping at disposal personal and material resources sufficiency therefore having preponderance over an adversary. Force protection presents a sectional field reflecting the demand mentioned above being multidisciplinary domain implicating all of military branches during a fulfilling of their tasks resulting from their predetermination. General abilities of forces necessary for successful force protection support are illustrated on Fig. 1. Force protection engineer measures are underlined on Fig. 2. The planning and execution force protection philosophy is based on the general force protection model (Fig. 3), presenting a force protection measures projection algorithm including an engineer provisions design. The algorithm is based on a thought model encompassing processes enabling to prevent potential incidents or to react to them by force protection measures adoption. Engineer measures act as possible means for a risk avoidance or it´s reduction. Their content and scope design followed by their planning and execution essentially belong to the risk management acting as a backbone activity of a planning and execution process of force protection measures.

Analyses procedures of processes leading to particular engineer force protection measure design had demonstrated that a specific engineer risk management algorithm based on above mentioned general force protection model has not been still exist. It has created an opportunity to develop such algorithm therefore to fill an "empty area" in the mentioned sphere.

© Springer Nature Switzerland AG 2019
J. Mazal (Ed.): MESAS 2018, LNCS 11472, pp. 599–613, 2019.
https://doi.org/10.1007/978-3-030-14984-0_45

2 Appropriate Engineer Force Protection Measures Design Process Based on Risk Assessment of Critical Resources Damage

Design process of engineer provisions acting as risk reduction means is illustrated in the chapter. The risk is based on an impact of a particular event reaching from a particular threat occurring and resulting to the loss of particular resource. There are suggested following steps being parts of the process:

- resources criticality assessment,
- resources vulnerability assessment,
- risk assessment,
- appropriate engineer measures design leading to a risk reduction.

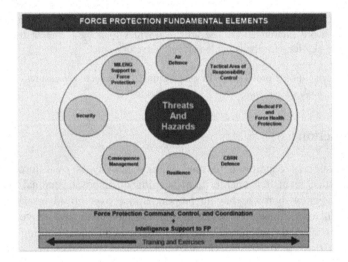

Fig. 1. General capabilities of forces required for a force protection support. (Source: STANAG 2528, p. 20)

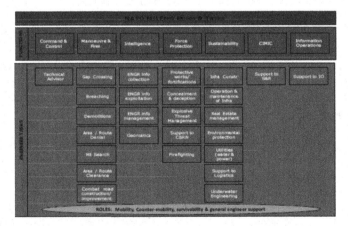

Fig. 2. System base of engineer roles and tasks. (Source: STANAG 2394, p. 18)

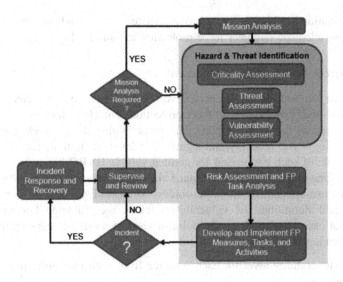

Fig. 3. Force protection model (Source: STANAG 2528, p. 35)

2.1 Resources Criticality Assessment

Critical assessment can be based on two main attributes necessary for a source to be critical. Its significance for particular task accomplishment can be considered as the first attribute while its restorability in case of a loss presents the second one. Assessment scales of source's significance and restorability have been formulated in Tables 1 and 2. Each degree has been determined as based on possible consequences' alternatives of resources' loss and their restoration possibilities.

Value of criticality H_k can be evaluated using a formula

$$Hk = STv * STo, \tag{1}$$

where H_k = criticality value of particular source, ST_v = significance degree of source and ST_o = recovery degree of source.

Possible criticality values determined using the formula mentioned above while taking into account different combinations of significance and recovery degrees have been expressed in Table 3. Each sources' evaluation using scale of criticality values can be formulated as:

- extremely critical source (criticality value 20 to 15),
- highly critical source (criticality value 12 to 9),
- moderately critical source (criticality value 8 to 5),
- low critical source (criticality value 4 to 1).

The evaluation enables assessors to prioritize sources and suggest which ones would require adequate engineer force protection measures adoption.

2.2 Resources Vulnerability Assessment

Each resource (without additional protective provision) can be characterized by vulnerability from the point of view of force protection. The character describes its ability to be eliminated or damaged by particular threat. Vulnerability levels can be determined by analogy with recovery and significance levels definitions (see Tables 1 and 2). An example of such expression for combat vehicle BVP-2 has been illustrated in Table 4 and Diagram 1. Based on similar vulnerability evaluations of each asset it is possible to develop and maintain resources´ vulnerability records. For numerical expression and definitions of vulnerability levels see Table 5. For vulnerability assessment of each resource from the point of view of all threats causing its potential loss the following procedure can be used:

- specific vulnerability level for each resource from each particular threat identified can be assigned,
- all vulnerability levels can be expressed with a table,
- the table will be then transformed to a diagram,
- based on an interpretation of data from a diagram it is possible to state resource´s level of vulnerability from each threat to consider which potential risk will be significant enough to evaluate its level.

Table 1. Significance degree of a resource (Source: Záleský J, p. 77)

Significance degree of a resource ST_v	Numerical quantification	Definition
Indispensable resource	5	Resource which loss or damage makes a task accomplishment impossible
Highly significant resource	4	Resource which loss or damage will require significant change of a task accomplishment course of action
Moderate significant resource	3	Resource which loss or damage will cause acceptable delay of task accomplishment or immediate resource restoration necessity
Low significant resource	2	Resource which loss or damage will cause task accomplishment constraint without affecting a final success
Insignificant resource	1	Resource which loss or damage will not affect task accomplishment

Table 2. Recovery degree of a resource (Source: Záleský J, p. 78)

Recovery degree of a resource STo	Numerical quantification	Definition
Unrecoverable resource	4	Source which availability for task accomplishment is significantly limited and it is impossible to share it with other forces. Restoration or recovery of such resource requires conduct such activities which are impracticable under particular conditions or the time required for its conduction exceeds the time of task accomplishment
Hardly recoverable resource	3	Source which availability for task accomplishment is limited but it is possible to share it with other forces however the accomplishment itself has to be modified under such conditions. Resource recovery or restoration will require course of action change or required accomplishment time reevaluation
Resource recoverable with difficulties	2	Source which availability for task accomplishment is limited but it is possible to share it with other forces until it is recovered or restored
Easy recoverable resource	1	Source which availability for task accomplishment is unlimited or which recovery or restoration will not affect course of action

Table 3. Resource criticality value (Source: Záleský J, p. 78)

Recovery degree of a resource ST_o		Significance degree of a resource ST_v				
		Indispensable resource	Highly significant resource	Moderate significant resource	Low significant resource	Insignificant resource
		5	4	3	2	1
Unrecoverable resource	4	20	16	12	8	4
Hardly recoverable resource	3	15	12	9	6	3
Resource recoverable with difficulties	2	10	8	6	4	2
Easy recoverable resource	1	5	4	3	2	1

Table 4. Vulnerability level assignment from point of view of particular threats (Source: Záleský J, p. 80)

Threat	Vulnerability level
Anti-tank mine	4
100 kg explosive VB - IED	4
Machine gun direct fire	1
RPG	5
Molotov cocktail	1
Artillery shell 155 mm	3

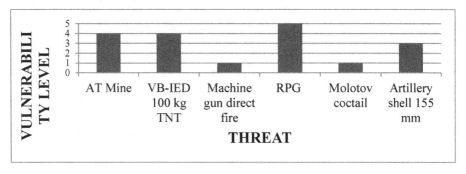

Diagram 1. Vulnerability level assignment to particular threats (Source: Záleský J, p. 82)

2.3 Risk Assessment

A purpose of a risk assessment is an incident occurrence probability estimation and expected impact forecast [4, 5]. Based on results of such sub assessments it is possible to access specific level of risk of task accomplishing hamper or limitation. Risk assessment process consists of following steps:

- likehood assessment of incident occurrence,
- assessment of incident´s expected impact on a task accomplishing,
- level of risk assessment based on sub assessments of incident´s impact and probability,
- risk prioritization. [4, 5].

Likehood Assessment of Incident Occurrence
Occurrence likehood of particular incident has to be estimated in case of each threat. To express it by quantitative way the Table 6 can be used.

Each incident occurrence likehood category can be expressed by percent using rectangular probability distribution as the most useful mathematic tool. Then a numerical expression can be used. Particular incident likehood can be estimated based on its occurrence in a particular operation during directly determined period of time. The time range will depend on particular conditions.

Severity Assessment of Particular Threat Impact on Particular Resource and Its Effect on Particular Task Accomplishment

Particular threat impact of particular resource assessment can be expressed by a level degree describing consequences of such impact on fighting power, combat task accomplishments or for combat readiness.

Following scale of potential consequences can be used for each threat impact severity assessment:

- catastrophic impact,
- critical impact,
- marginal impact,
- negligible impact.

Severity assessment of an incident caused by threat exploiting a vulnerability of particular resource can be based on following two key factors:

- particular resource vulnerability to a particular threat – expressed with vulnerability level STzr,
- particular resource criticality for particular task accomplishment – expressed with criticality value Hk.

Based on those factors severity level can be expressed using following equation:

$$Dz = STzr * Hk, \tag{2}$$

where D_z = severity level, ST_{zr} = vulnerability level of particular resource and H_k = criticality value of particular resource.

Equation mentioned above is analogical to critical value (H_k) assessment mathematical formula. If particular numerical values expressed in Tables 3 and 5 are put down to the equation the level of severity will assume values expressed in the Table 7.

Apart from numerical expression severity value can be described using definitions (see Table 8).

Specific Risk Level Assessment

Value of particular risk level of particular task nonfulfillment caused by particular incident impact resulting to a particular resource loss due to particular threat application can be expressed with following equation:

$$Ru = Dz * Pv, \tag{3}$$

where R_u = incident appearance risk level, D_z = severity level and P_v = likehood category.

Putting down numerical values of likehood category (see Table 6) and severity level (see Table 7) to the equation shown above the value of appearance risk level assumes values expressed in the Table 9. Levels of risk can be classified into five categories (see Table 10). Based on such classification and prioritization force protection measures can be designed and implemented to reduce risk level assessed above.

Table 5. Vulnerability levels of personnel, equipment, material and structures (Source: Záleský J, p. 81)

Vulnerability level ST_{zr}	Definition			
	Personnel	Equipment	Material	Structure
5	Threat causes fatal injuries to personnel	Threat causes total destruction of equipment or damages requiring more 1000 man-hours to be repaired	Threat causes permanent loss of 60% available material or more	Total damage of structure. Building has been unable to fulfill its purpose
4	Threat causes major to fatal injuries to personnel	Threat causes damages requiring 400 to1000 man-hours to be repaired	Threat causes permanent loss of more than 30% of available material and damage of more than 30% of available material limits its function	Strong damage. Structure lost its essential characteristics and it has been already disabled to fulfil its purpose
3	Threat causes minor to major injuries to personnel	Threat causes damages requiring 100 to 400 man-hours to be repaired	Threat causes damage of more than 30% of available material limiting its function	Middle damage. Structure has still fulfilled its purpose however loosing significant part of its essential characteristics it requires conduction of extensive repairs to be fully useful. During the time of repairs the structure will be untuneful for its purpose
2	Threat causes no injury to major injuries to personnel	Threat causes damages requiring 60 to100 man-hours to be repaired	Threat causes damage of less than 30% of available material limiting its function	Small damages. It will be necessary to conduct repairs. Structure will be useful during a time of repairs
1	Threat causes no injury to personnel	Threat causes no damages or damages requiring 40 to 60 man-hours to be repaired	Threat causes no or small damage of a material	Minor damages. No or small repairs will be conducted. Structure will be fully useful

Table 6. Incident likehood categories (Source: Záleský J, p. 83)

Likehood category P_v	Numerical expression	Probability scope (%)
Frequent	5	81–100
Likely	4	61–80
Occasional	3	41–60
Seldom	2	21–40
Unlikely	1	0–20

Table 7. Severity level values (Source: Záleský J, p. 84)

Resource vulnerability level ST_{zr}	Resource criticality value H_k												
	20	16	15	12	10	9	8	6	5	4	3	2	1
5	100	80	75	60	50	45	40	30	25	20	15	10	5
4	80	64	60	48	40	36	32	24	20	16	12	8	4
3	60	48	45	36	30	27	24	18	15	12	9	6	3
2	40	32	30	24	20	18	16	12	10	8	6	4	2
1	20	16	15	12	10	9	8	6	5	4	3	2	1

2.4 Appropriate Engineer Force Protection Measure Determination as a Mean to Reduce Value of Risk

Particular engineer force protection measure (set of measures) draft suitable for a particular resource in a logical sequence of risk level evaluated before seems to be a key issue. The measure has to be designed in detail in logical sequence of draft mentioned above.

Measure effectivity presumption has to be based on the fact that the vulnerability level of particular resource achieved after the measure adoption will be lower than the former one existing before the measure adoption. The proper measure design requires applying the resource destruction or damaging risk reduction rate. The rate may be based on the comparison of resource vulnerability before and after the adoption of a particular measure.

For easy mathematical expression of facts mentioned above the vulnerability mitigation coefficient Ksz can be established using an equation

$$Ksz = \frac{STzrpre}{STzrpos}, \tag{4}$$

where K_{zs} = vulnerability mitigation coefficient, ST_{zrpre} = resource vulnerability level before force protection measure design and ST_{zrpos} = resource vulnerability level after force protection measure design.

If the numerical value of the coefficient is 1 or more the resource vulnerability level will be the same or higher after measure adoption. If it is lower than 1 the level will be lower too and the measure will be effective.

Reduced risk level of incident causing the loss of particular resource therefore the task completion failure or limitation can be evaluated using an equation:

$$R\mathrm{mod} = Ksz * Ru, \tag{5}$$

where R_{mod} = reduced value of incident risk, K_{sz} = vulnerability mitigation coefficient and R_u = initial value of incident risk before the measure was designed.

The risk can be also reduced by decreasing of a likehood and engineer camouflage and deception measures can be usefull means to reach it but the rate of likehood reduction measure assignment requires more difficult method using mathematical probability models. Therefore, it may be the topic of individual article.

Table 8. Severity level values definitions (Source: Záleský J, p. 85)

Severity level	Numerical expression	Consequences
Catastrophic	**100 − 50**	Full mission failure or the loss of ability to accomplish it, death or permanent personnel disability to accomplish the task, loss of main systems equipment or material critical for mission success, significant equipment material or stallations damage
Critical	**49 − 25**	Significantly limited ability to fulfill the task or unit readiness, permanent partial disability or temporary (more than three months) full disability of personnel to fulfil a task, extensive significant damage of systems, equipment and installations
Marginal	**24 − 15**	Limited ability to fulfill the task or unit readiness, small systems 'equipment's and installations 'damages, several days' waste of time caused by personnel wounds or diseases (healing less than three mounts)
Negligible	**12 − 1**	Small or no loss of ability to fulfill the task first aid or small rate medical care necessary, bare damages of equipment, systems or installations (remain fully operational or useful)

Characteristics of Particular Engineer Measures Supporting Force Protection

Following measures illustrated on Fig. 2 represent the real and tangible part of force protection engineer measures design. If adopted, the elimination risk of personnel, equipment, material, installation, or breakdown of activity critical for mission accomplishment will be eliminated or limited.

Their combination creates the synergic effect increasing the efficiency of force protection more than, if they would have adopted sequentially:

- **protective works and field fortifications** (chicanes or route access control points, fences, screens, or bunkers surrounding a facility or vehicle, equipment or troop concentration, preparation of sites for tactical air and aviation units, advice/assistance with the construction of protective barriers, perimeter protection

Table 9. Potential values of incident risk (Source: Záleský J, p. 86)

Severity level Dz		Likehood category Pv				
		Frequent	Likely	Occasional	Seldom	Unlikely
		5	4	3	2	1
Catastrophic	100	500	400	300	200	100
	80	400	320	240	160	80
	75	375	300	225	150	75
	64	320	256	192	128	64
	60	300	240	180	120	60
	50	250	200	150	100	50
Critical	48	240	192	144	96	48
	45	225	180	135	90	45
	40	200	160	120	80	40
	36	180	144	108	72	36
	32	160	128	96	64	32
	30	150	120	90	60	30
	27	135	108	81	54	27
	25	125	100	75	50	25
Marginal	24	120	96	72	48	24
	20	100	80	60	40	20
	18	90	72	54	36	18
	16	80	64	48	32	16
	15	75	60	45	30	15
Negligible	12	60	48	36	24	12
	10	50	40	30	20	10
	9	45	36	27	18	9
	8	40	32	24	16	8
	6	30	24	18	12	6
	5	25	20	15	10	5
	4	20	16	12	8	4
	3	15	12	9	6	3
	2	10	8	6	4	2
	1	5	4	3	2	1

systems, support to CBRN collective protection, advice on the construction of field fortifications, construction of command posts, construction of artillery gun positions, tank scrapes and weapon pits, preparation of alternate positions, preparation of sites for tactical air and aviation units, strengthening field fortifications and building reinforcement),

- **concealment and deception** (terrain camouflage capacity exploitation assessment, assistance with natural camouflage measures design, assistance with artificial camouflage measures implementation, dummy objects building, decoy installation, anti-radar camouflage measures, thermal camouflage measures, explosives usage for the purpose of deception),

- **explosive threat management** (planning, command, control and training of activities connected in with explosive hazards, EOD activities, engineer part of C-IED),
- **support to CBRN** (field fortifications building and collective protection means installation, mobility support within contaminated areas and around them, assistance with decontamination points building, assistance with industrial disasters consequences disposal),
- **Firefighting** (fire protection means installation, assistance with fire extinguishing and localization, fire-fighting equipment building) [2].

Table 10. Risk level categories (Source: Záleský J, p. 88)

Category	Numerical expression	Definition
Extremely high risk	**500 − 180**	The loss of ability to accomplish the mission, if threats occur during its accomplishment. Frequent catastrophic casualties or their high likehood level, frequent critical casualties. It means that the risk of incident can cause serious consequences related to mission accomplishment. Decision of mission accomplishment continuation has to be properly evaluated in consideration of potential benefit achieved if task is fulfilled by the way suggested before risk was assessed
High risk	**160 − 75**	Serious loss of ability to accomplish the mission it scheduled time, disability to fulfill partial tasks or disability to fulfill the task in compliance with requirements if threat occurs during the task accomplishment. Occasional or seldom occurrence of catastrophic casualties. Likely to occasional occurrence of critical casualties. Frequent marginal casualties. It implies if the dangerous incident occurs, it will cause significant consequences. Decision of mission accomplishment continuation will have to be properly evaluated in consideration of potential benefit achieved if task is fulfilled by the way suggested before risk was assessed
Moderate risk	**72 − 30**	If the threat occurs during the mission accomplishment, the deterioration of ability to fulfill a task in accordance with requirements may be expected. The final impairment of task result quality would be a consequence. Unlikely catastrophic casualties. Seldom-critical casualties. Frequent to likely marginal casualties. Frequent negligible casualties
Low risk	**27 − 1**	Expected casualties cause marginal or negligible consequences to task accomplishment. Unlikely critical casualties, the probability of marginal casualties can be classified as seldom or unlikely. Negligible casualties seldom or unlikely. Severity of expected casualties cause no or limited consequences to the task accomplishing. Wounds, diseases or damages are not expected, or their impact on mission accomplishment will be not significant or long lasting

The Development Procedure of Engineer Force Protection Measures as Means to Reduce Risk

Rules of suitable engineer force protection measures design mentioned above can be formulated as a procedure that is a result of their applicability research. The procedure consists of following steps:

1. Significance degree quantification of particular resource based on particular task analyses (see Table 1),
2. Recovery degree quantification of the resource based on its availability and capabilities to distribute it to particular unit or troop (see Table 2),
3. Criticality value calculation via Eq. 1,
4. Arrangement of all resources necessary for particular mission accomplishment in compliance with criticality value,
5. Assignment of all identified threats to each resource that can be threaten by such hazards,
6. Vulnerability level assignment of each resource from each threat relevant for it (see Table 5),
7. Severity level calculation for each relationship threat-resource via Eq. 2,
8. Likehood evaluation of each threat occuring for each resource (see Table 6),
9. The risk calculation of event capable to limit or harm the usage of particular resource critical for particular mission accomplishment due to particular threat exploring particular vulnerability. Using of Eq. 3,
10. Acceptability evaluation of each risk calculated,
11. Prioritization of all risks in compliance with their value,
12. Particular engineer measures adoption and their impact evaluation on risk reduction. The evaluation is based on equation vulnerability levels before and after the adoption comparison (see step 6) with usage of Table 5,
13. Vulnerability mitigation coefficient calculation using the Eq. 4
14. Reevaluation of risk level after the particular measure adoption for each relationship threat-resource. Using Eq. 5,
15. Repeated arrangement of all risks in compliance with their value and their acceptance decision or next possible measure adoption.

The process illustrated above even though it seems to be difficult, can be routinely repeated. If some resource is then recognized as low critical and generally available in terms of a price and a quantity it will not be necessary to continue the risk assessment process to protect it. Likewise, if the threat although generally perceived does not affect the resource in particular situation or if the resource is invulnerable by the hazard, it will be void to access the potential risk.

Tables 3, 7, and 9 containing data calculated with an application of particular equations after data from Tables 1, 2, 5, and 6 had been inserted can be used for calculation advance.

Data ranges in Tables 8 and 10 specifying severity and risk levels reach from singular numerical values reaching from insertion of numerical expression of vulnerability levels, significance and recovery degrees to particular equations. The generally accepted axiom has been taken in account, that catastrophical and critical severity levels of risks represent the highest necessity of force protection measures adoption including engineer ones. Therefore, it is the reason why the scale of these severity levels has been developer so wide (Fig. 4).

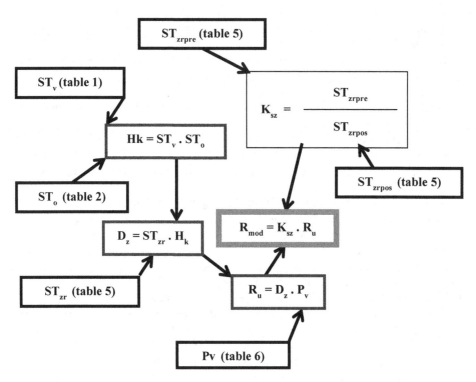

Fig. 4. Design process of engineer force protection measures as means for risk reduction (Source: Author)

3   Conclusion

Process suggested in the article would be useful for each engineer measure development and assessment based on risk level of incident occurring that can cause the particular resource loss or damage originating its usability thwarting for particular mission accomplishment. The process ca be applied also for more measures designed together to accomplish their synergic effect. The area for additional research has been opened therefore. Each step of the process can be used separately too as a mean for decision making process. The potentials to develop software based on mathematic equations and data tables expressed in the article can be taken in account as well together with the usage of existing software having mathematical functions. The example of such software can be MS Office Excel, MS Office Project or MATLAB.

References

1. Allied Joint Doctrine for Force Protection. STANAG 2528. NATO Standardization Office, Bruxels, pp. 20, 35. AJP-3.14 (2015)
2. Allied Tactical Doctrine for Military Engineering. STANAG 2394. NATO Standardization Office, Bruxels, pp. 18, 49–50. ATP-3.12.1 (2016)
3. Záleský J.: Ženijní opatření ochrany vojsk v operacích mimo území České republiky. Doctoral thesis. Univerzita obrany v Brně, Fakulta ekonomiky a management, Brno, pp. 77–86, 88 (2012)
4. Composite Risk Management: Headquarters department of the army, Washington, DC, pp. 7–8, 9. FM 5-19 (2006)
5. Comparative study: Australian defence risk management framework: DSTO systems sciences laboratory, Adelaide, pp. 10–11. DSTO-GD-0427 (2005)

Approaches to Realize the Potential of Autonomous Underwater Systems in Concept Development and Experimentation

Thomas Mansfield$^{(\boxtimes)}$, Pilar Caamaño Sobrino, Arnau Carrera Viñas,
Giovanni Luca Maglione, Robert Been, and Alberto Tremori

NATO STO Centre for Maritime Research and Experimentation,
Viale San Bartolomeo, 400, La Spezia, Italy
{thomas.mansfield,pilar.caamano,arnau.carrera,
giovanni.maglione,robert.been,
alberto.tremori}@cmre.nato.int

Abstract. Recent NATO reports highlight the rapid progress being made in the development of autonomous underwater systems. In contrast, national reports indicate that their benefits are not being fully realized in a timely manner in operational scenarios. One approach to improve NATO's adoption of these systems is to provide guidance in the NATO concept development and experimentation process specially aimed at articulating autonomous system behaviors and allowing efficient experimentation with their capabilities. This position paper reviews the latest techniques and approaches for articulating and testing autonomous system capabilities in industry, academia and within NATOs national militaries. Discussed techniques focus on encouraging and developing understanding and trust in the commander and operator stakeholder communities as well improving the efficiency of autonomous system testing. Potential future guidance and the structure of these activities within the existing NATO CD&E framework are presented for further discussion.

Keywords: Autonomous underwater vehicles ·
Concept development and experimentation ·
Design of experiments · Virtualization · Mine countermeasure

1 Introduction

Advances in sensors, robotics and computing are allowing the development of advanced autonomous systems [1]. These systems offer a wide range of military benefits including the ability to conduct missions in remote and hostile environments without placing personnel in harm's way as well as new human-machine teaming concepts [2]. The benefits of autonomous systems are particularly apparent in the underwater domain, where hazardous activities such as mine counter measure missions must be conducted in uncertain environments with limited or no in-mission Command and Control (C2) infrastructure [3].

To develop new operational capabilities, NATO typically uses its Concept Development and Experimentation (CD&E) process [4]. While this process has been

© NATO 2019
J. Mazal (Ed.): MESAS 2018, LNCS 11472, pp. 614–626, 2019.
https://doi.org/10.1007/978-3-030-14984-0_46

designed to incorporate a wide range of technologies, the increased technical and conceptual complexity of operations involving autonomous systems has not allowed the rapid advances in autonomous system capabilities to be considered in a timely manner [5].

This position paper reviews recent advances from industry, academia and national defense to provide a summary of the latest techniques and approaches that may compliment and bolster the existing NATO CD&E process. The paper continues to propose a possible update to the NATO CD&E toolset to enable the efficient consideration of autonomous systems in future operations.

Section 2 of this paper summarizes the key components of the NATO CD&E process. Section 3 discusses the key concepts and challenges presented by autonomous systems in the underwater domain. A review of the latest developments in autonomous system design and test are presented in Sect. 4. Section 5 provides conclusions and links between the review findings and the CD&E process steps.

2 The NATO CD&E Process

NATO defines its existing CD&E process as a technology agnostic approach that allows the proposal and test of a range of potential future concepts of operation in all military fields [6].

Key to the concept of CD&E is the use of a spiraling approach, with iterating and separate concept development and experimentation stages. The iterative, spiraling approach is managed in a series of increasing capability maturity levels (CMLs). An overview of the approach is provided in Fig. 1.

The CML groupings represent stages of maturity from the low maturity CML 1, where up to five novel concepts of operation are selected to the more mature concepts discussed at CML 6 where the selected and refined operational scenario is demonstrated and validated implementation requirements are obtained.

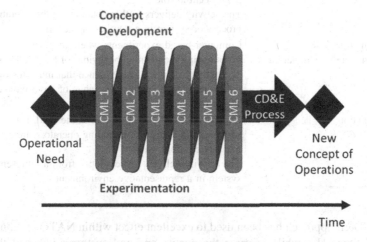

Fig. 1. Summary of NATO's CD&E process.

While the CD&E process does not mandate the use of any specific tools or techniques for the concept development stage, techniques commonly used by NATO are summarized in Table 1 [7].

Table 1. Summary of existing and commonly used NATO concept development techniques

Existing concept development techniques	
Category	Description
Analysis	The definition of the problem statement by a team of end-users and experts in the problem field The activity delivers clear problem statement
Brainstorming	Whiteboard based group sessions in which new knowledge and ideas are generated, discussed and linked. This methodology may apply approaches from the NATO Alternatives Analysis (AltA) handbook [8] The activity delivers a list of potential solutions
Evaluation	Workshops where the brainstorming output is translated to the concept and evaluated by a team of experts in the problem field The activity delivers a shortlist of potential solutions and rationale for their selection

Further, the NATO tools commonly used in the experimentation phase of the NATO CD&E process are summarized in Table 2 [7].

Table 2. Summary of existing and commonly used NATO experimentation techniques

Existing experimentation techniques	
Category	Description
Table-top gaming	Paper or computer supported games in which the users play a central role The activity delivers further detail about the operation of a process or system in a number of scenarios
Experimentation with virtual/constructive simulation	Computer-based experiments where the focus lies on investigating the detailed behaviour of a modelled system The activity delivers test evidence that indicates the performance of a system in a range of environments or scenarios
Live simulation	Experiments with a real system in the field, using real software and hardware, including operators, in a suitable live test environment The activity delivers test evidence from a representative system in a representative environment

The CD&E approach has been used to excellent effect within NATO to reduce both costs and timescales while increase the quality and end customer value of the final solutions [9]. Specific issues that prevent the timely adoption of autonomous systems are encountered when CD&E techniques are used to discuss and analyze the additional technical and behavioral complexity of autonomous systems [10–12]. Further description of the challenges presented by autonomous systems and the specific challenges in underwater operation are discussed in the following section of this paper.

3 Autonomy in the Underwater Domain

The term 'autonomous system' generally refers to a system that is required to operate with some degree of human independence [13]. Differing from 'unmanned systems', the more complex autonomous systems are able to interact and respond to their environment without the involvement of a human in the loop. This leads to a new level of technical challenge, where not only the systems function but also its behavior needs to be understood and proven [14]. Challenges specific to the underwater domain stem from both the lack of in-mission communications, requiring the operation of the system for long periods without human control and intervention, and the difficulty in sensing and understanding the complex underwater environment.

The key performance parameters for autonomous systems include elements such as how precisely it can observe the environment through its sensors, how effectively it can combine sensor data sources and whether the resulting behavior is the most appropriate given the environment and the required mission objectives.

The evaluation of these key performance parameters must be made from the viewpoint of two key CD&E process stakeholders; operators and commanders [15].

For **commanders**, a key challenge presented by the use of autonomy is in understanding and developing trust in the mission specific concept of operations and the associated trade-offs across a wide range of conflicting parameters of interest. At the level of the commander, concepts are currently both difficult to articulate during concept development and difficult to test, analyze and present following the experimentation stage of the process.

Operators must be able to understand and operate the human-machine interfaces needed to conduct in their mission. Again, the behavior of the system must be understandable and predicable to the operator to build trust and confidence in the system.

3.1 Barriers to the Inclusion of Autonomous Systems in the CD&E Process

The complexity of autonomous systems, combined with the lack of in-mission human supervision to detect and act upon unexpected failures has led to a lack of trust by both commander and operator stakeholder groups and is limiting the adoption of autonomous underwater systems in operations [16].

Further, the additional complexity of autonomous systems requires efficiency improvements in the experimentation and test steps of the CD&E process. The need for further efficiency is driven by the non-deterministic nature of autonomous systems. The number of required test cases, the amount of data generated, and the complexity of the analysis process provide additional barriers to the adoption of autonomous systems in the CD&E process [17, 18].

A range of research activities are currently underway to communicate the potential of complex concepts and technologies to a range of stakeholders. A review of emerging approaches aimed at communicating capabilities, building trust and efficiently testing autonomous systems is presented in the following section of this paper.

4   Techniques for Assessing Autonomous System Capabilities

This section of the paper reviews recent advances from academia, industry and the NATO nations aimed at assessing autonomous system capabilities that may be used as best practice case studies for future NATO CD&E techniques.

4.1   Understanding the States and Interfaces of Autonomous Systems

Due to the non-deterministic nature of autonomous systems and the large number of potential operational unknowns the systems may encounter, guidance is required in the CD&E process to effectively test the system in a representative range of environments. The lack of guidance in this area leads to both complex test phases that does not result in a clear comprehension of the tests coverage in relation to problem space. Solutions to reducing the time and complexity of the test stages for autonomous systems have been pioneered by a range of national programs [19]. One example which will be used to demonstrate work typical to this area is the work carried out by the US Army Robotic Intelligence Evaluation Program [20]. Their work has investigated the use of a design of experiments (DoE) based methodology to both comprehensively test the intelligence of autonomous systems while limiting the number of required test cases.

The first step of the approach uses 'parameter effect propagation' to limit the number of tests that are required by running only those tests that provide a unique situation to the autonomous system.

Parameter effect propagation is the process of recognizing each of the individual sets of parameter values (i.e., all the possible scenarios) and estimating the effect on the autonomous system. Central to the identification of parameter values is the recognition that the autonomous underwater vehicle (AUV) decision system can only be affected by its sensor inputs and that the senor inputs are often only limited in their scope.

This initial stage of the analysis results in the population of a critical test matrix where the function of the system (e.g. communication packet loss, battery level, sonar received signal strength) are identified. A pictorial summary of the methodology is shown in Fig. 2.

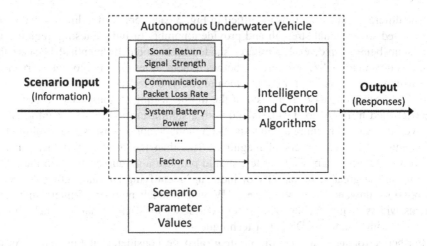

Fig. 2. Factors affected by environmental parameters

Following the identification of the AUV inputs factors, an assessment of the impact of the scenarios on that factor should be made for each scenario.

Before testers can determine how the parameters of a scenario will uniquely affect an AUV, they need to establish the parameters. Using the list of customer defined, testers can produce a list of scenario parameters for each. The list of scenario parameters is determined by examining the variables in a scenario's mission along with all the ways the environment can interact with the system. This approach limits the number of test cases as the environment can only interact with the system in a few ways. Like humans, AUVs can only base decisions on what they detect. The published method may been applied to aid the understanding of AUVs. An example critical test matrix for an example parameter, sonar reflected signal strength, has been populated and is shown in Fig. 3.

Sonar Reflected Signal Strength

		Water Salinity (ppt)							
		0	10	20	30	40	50	60	
Sea Floor Type	Clutter								A
	Gravel								
	Mud								
	Sand								B
	Ripples								
	Posidonia								
Fault Conditions	Transmitter Failure								C
	Receiver Failure								

Fig. 3. Critical test matrix test zones

The different shaded regions identified in Fig. 3 each represent a different value for the reflected sonar signal strength and provide a number of unique testing regions. To the system being tested, each similarly shaded case would be identical because the system is unable to realize a difference between them. If testers were to run every case, there would be 56 tests just for environmental effects on reflected sonar strength alone. Instead, choosing one case from each of the unique testing regions leaves testers with four (Labelled from 'A' to 'D') tests and with a significant portion of the information they would have had running all 56 tests. With more complex sensors, the resulting test matrix might be many orders of magnitude less than the original full factorial test design. In addition, each resulting test should propose a unique problem to the AUV resulting in the greatest probability of inducing, identifying and attributing emergent and possibly unwanted behavior. An AUV with a small number of interacting subsystems will typically end up with a limited test set for each scenario, which can be addressed with standard CD&E test techniques.

Further, error and fault conditions may also be considered at this stage. As an example, a high frequency sonar may return much less information when surveying from a posidonia covered sea floor. This scenario is likely to return the same information as broken sonar scenarios, and the AUV will treat them as the same. This is identified as an extension to existing matrix test zones in Fig. 3.

For many of the identified parameter, such as communication packet loss rate, the test plan must specify which values to use. If the parameter is continuous or there is a multitude of discreet values, a traditional approach is for testers to choose values at the 95 ($\approx 2\sigma$) and 99 ($\approx 3\sigma$) percent extremes on the probability distribution function of a parameter. However, with a system containing so many parameters, testing only the extremes would leave possible many common parameter interactions untested. A more comprehensive approach is to select parameter values along a probability distribution function at a given percentage step size. This allows the tester to not only select the extremes but also test more of the most common values. The percentage step size is determined by running a sensitivity analysis on the AUV factors by this particular parameter.

This sensitivity analysis forms the next step of the DoE process. Effects of subsystem factors can be predicted through analytical calculations and then verified through field testing or through empirical experimentation. Testers should not be concerned with the multitude of parameter value combinations, but rather concerned with the sets of factor effects derived from the parameter permutations. A key concept in a number of autonomous system test approaches [17, 20] is that running 100 different tests that induce the same factor effect set in the system will not tell the tester nearly as much as running 100 different tests where each of which induces a unique factor effect set.

A test team should step through each parameter value for each scenario and calculate a corresponding factor effect set. It is important to take the set of parameter values that define a scenario together so that the factor effect set can accurately represent interaction between multiple parameters. For example, range and sea floor type can each affect sonar sensors, but the combination of different values of these two parameters can produce drastic effects on a system. Further, if the sea floor is out of range, it makes no difference to the sonar sensor on an AUV what the sea floor covering is because it may not be able to "see" at all. If all the effects of all the values of one parameter dominate all the effects of all the values of another parameter, there is no reason to have the second parameter and it can be removed from the test scenario.

4.2 Analyzing and Understanding the Experimentation Phase

Once the DoE test environments have been identified and tests conducted, the next stage of the CD&E process mandates the analysis of the results. While the DoE approach limits the number of tests that need to be performed, it is likely that there are still too many test results to review in full with the system commanders. Further work is required to identify approaches to effectively communicate the findings of the experimentation phase, allowing the progression to the next CML level.

A current area of work that may allow this capability is in the development of 3D visualization environments that allow system behavior to be demonstrated to end users [21]. Simulation capabilities have been provided by several available tools [22, 23] for autonomous ground vehicles. These tools typically show, on one screen, the motion and actions of the autonomous system along with a user interface that allows the audience to 'play' with the environment around the autonomous system. An example of this approach can be seen applied to autonomous ground vehicles in Fig. 4.

The addition of an input, also shown in Fig. 4, allows the users trust to be reported. Areas where the user does not trust the system, for example if the user notices that the autonomous system is moving toward home with an explosive still loaded, a flag can be raised and the behavior investigated further in the next CML concept development stage of the process.

Fig. 4. A virtualized environment to demonstrate autonomous system behaviour

Without the users' inputs from the 3D visualization, this potentially dangerous behavior would be much more difficult to detect.

Further to watching and witnessing the behavior of the system, work has also been carried out that allows the system commander and operators to interact in the creation of scenarios. Their involvement in creating the possible scenarios is vital in the concept development phase of the CD&E process. Using Event Sequence Charts [21], Inputs and scenario development can be managed even in complex scenarios.

An example of an Event Sequence Chart for an underwater autonomous system mine detection system can be seen in Fig. 5.

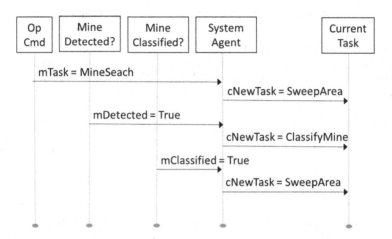

Fig. 5. Event sequence chard for underwater mine identification and classification

Based on UML sequence diagrams, these charts allow the development of sequenced events by the user that can form the basis of either the concept development of experimentation phased of the CD&E process.

4.3   Aiding Communication and Understanding in Concept Development

Following the efficient integration of autonomous system testing into the experimentation analysis of the results by all relevant stakeholders the complexity of autonomous system behavior may benefit from additional tools to clarify new concepts and ways of working. Pioneered by the education sector, a large body of work has been carried out, into allowing the sharing of ideas and concept development aided by AR [24] and VR [25].

Utilizing and building on the models and approaches already discussed in this paper, this approach allows the articulation of complex ideas in an initiative manner to a range of stakeholders. Further, commonly used systems also allow distributed collaborations of specialized personal, encouraging the involvement of the most appropriate personnel efficiently within the CD&E process and driving further improvements to the CD&E process.

5 Conclusions

Recent developments in autonomous system technology have provided an opportunity for new concepts of operation to be developed in a range of NATOs undersea activities. NATOs existing CD&E provides an excellent and adaptable framework for allowing the efficient development integration of emerging technologies. Despite this, the additional complexity of autonomous systems, along with the removal of in-mission humans in the loop, presents a series of challenges that limit the rate of adoption of this improved technology in the existing CD&E framework.

This paper has reviewed the creation of a range of processes and techniques from industry, academia and NATO nations to identify those that may enable autonomous systems to integrate better into the NATO CD&E process.

To enable the experimentation phase, methods for simplifying and clarifying the relevant test cases while increasing the rigor and robustness of testing by identifying common system boundaries has been presented. Building up this with a DoE approach, supported by clear reporting metric matrices, allows both the efficient and fast testing of the system and, more importantly, the results of the testing to be understood in terms of their coverage of the problem space.

Once tested, methods for demonstrating the capabilities and behaviors of the system have been described. These approaches aim to build trust with both system operators and commanders by allowing them to intuitively witness and interact with the system. Areas of the scenario where trust is lost are recorded, highlighted and used to update subsequent CD&E CML stages.

Further, advances spearheaded by the education sector can be used to enhance and streamline the concept development stages. The VR and AR in concept development allow the articulation of complex autonomous system behaviors and can be used to allow subject matter experts to interact with system users as required.

The application of these approaches can be used to enhance the tools available in the existing CD&E process. A summary of the possible improvements identified by this paper and their links to the existing CD&E process is presented Tables 3 and 4.

Table 3. Enhanced approaches and toolsets to aid the CD&E process

Concept development techniques		
Category	Existing techniques	Potential additional techniques
Analysis	The definition of the problem statement by a team of end-users and experts in the problem field. The activity delivers clear problem statement	The generation of a virtualised environment that shows and describes the current operational experiment. The virtualised environment may be combined with VR or AR as required to allow all stakeholders to develop a detailed understanding of the key issues and challenges to be solved

(continued)

Table 3. (*continued*)

Concept development techniques		
Category	Existing techniques	Potential additional techniques
Brainstorming	Whiteboard based group sessions in which new knowledge and ideas are generated, discussed and linked. This methodology may apply approaches from the NATO Alternatives Analysis (AltA) handbook [8]	Distributed group sessions with VR and AR tools that involve both subject matter experts and system operators and commanders to collaboratively develop potential solutions
Evaluation	Workshops where the brainstorming output is translated to the concept and evaluated by a team of experts in the problem field. The activity delivers a shortlist of potential solutions and rationale for their selection	Interactive use of a virtualised 3D environment to develop an improved understanding of autonomous system behaviours. This has recently been demonstrated in the development of new anti-submarine warfare concepts that involve multiple autonomous systems in the maritime domain at CMRE

Table 4. Enhanced approaches and toolsets to aid the CD&E process

Experimentation techniques		
Category	Existing techniques	Potential additional techniques
Table-top gaming	Paper or computer supported games in which the users play a central role. The activity delivers further detail about the operation of a process or system in a number of scenarios	Interactive use of virtualised models with end user inputs to highlight areas that require further development to provide a suitable solution
Experimentation with virtual/constructive simulation	Computer-based experiments where the focus lies on investigating the detailed behaviour of a modelled system. The activity delivers test evidence that indicates the performance of a system in a range of environments or scenarios	Targeted testing based on DoE principles to highlight unique areas of operations. The outputs of the experimentation should result in the population of a metric matrix to identify the strengths and weaknesses of the system as well as an idea of the problem space covered
Live simulation	Experiments with a real system in the field, using real software and hardware, including operators, in a suitable live test environment. The activity delivers test evidence from a representative system in a representative environment	Live events carried out when identified by the critical test matrix, maximising the value and efficiency each test

References

1. Dyndal, G.L., Berntsen, T.A., Redse-Johansen, S.: Autonomous military drones: no longer science fiction. NATO Review Magazine, Oslo, Norway (2017)
2. Williams, A.P., Scharre, P.D.: Autonomous Systems - Issues for Defence Policy Makers. NATO Headquarters SACT, Norfolk (2015)
3. Yuh, J.: Design and control of autonomous underwater robots: a survey. Auton. Robot. 8(1), 7–24 (2000)
4. de Nijs, H.: Concept development and experimentation policy and process. HQ SACT, Norfolk, USA (2010)
5. Boulanin, V., Verbruggen, M.: Mapping the Development of Autonomy in Weapon Systems. Stockholm International Peace Research Institute, Stockholm (2017)
6. NATO: NATO concept development and experimentation (CD&E) process. North Atlantic Military Committee (2010)
7. van der Wiel, W., et. al.: Concept maturity levels bringing structure to the CD&E process. In: Interservice/Industry Training, Simulation and Education Conference, Orlando, USA (2010)
8. NATO: The NATO Alternative Analysis Handbook, 2nd edn. NATO, Brussels (2017)
9. Software V&V Working Group: IEEE STD 1012-2012 - IEEE Standard for System and Software Verification and Validation. IEEE (2012)
10. Pecheur, C.: Verification and Validation of Autonomy Software at NASA. NASA Ames Research Center, USA (2000)
11. Schumann, J., Visser, W.: Autonomy software: V&V challenges and charecteristics. In: IEEE Aerospace Conference, Big Sky, USA (2006)
12. Hodicky, J., Prochazka, D.: Challenges in the implementation of autonomous systems into the battlefield. In: 6th International Conference on Military Technologies, Brno, CZE (2017)
13. Callow, G., Watson, G., Kalawsky, R.: System modelling for run-time verification and validation of autonomous systems. In: Conference in Systems of Systems Engineering, Loughborough, UK (2010)
14. Defence Science Board: The Role of Autonomy in DoD Systems. Office of the Under Secretary of Defence for Acquisition, Technology and Logistics, Washington, USA (2012)
15. Tremori, A., et al.: A verification, validation and accreditation process for autonomous interoperable systems. In: Mazal, J. (ed.) MESAS 2017. LNCS, vol. 10756, pp. 314–323. Springer, Cham (2018). https://doi.org/10.1007/978-3-319-76072-8_22
16. Palmer, G., Selwyn, A., Zwillinger, D.: The "Trust V": building and measuring trust in autonomous systems. In: Mittu, R., Sofge, D., Wagner, A., Lawless, W.F. (eds.) Robust Intelligence and Trust in Autonomous Systems, pp. 55–77. Springer, Boston, MA (2016). https://doi.org/10.1007/978-1-4899-7668-0_4
17. Helle, P., Schamai, W., Strobel, C.: Testing of autonomous systems - challenges and current state-of-the-art. In: INCOSE International Symposium, Edinburg, UK (2016)
18. Hodicky, J.: Autonomous systems operationalization gaps overcome by modelling and simulation. In: Hodicky, J. (ed.) MESAS 2016. LNCS, vol. 9991, pp. 40–47. Springer, Cham (2016). https://doi.org/10.1007/978-3-319-47605-6_4
19. Thompson, M.: Testing the intelligence of unmanned autonomous systems. Int. Test Eval. Assoc. 29, 380–387 (2008)
20. Ahner, D.K., Parson, C.R.: Workshop report: test and evaluation of autonomous systems. USA Department of Defense, Washington D.C., USA (2016)

21. Heitmeyer, C.K., Leonard, E.I.: Obtaining trust on autonomous systems: tool for formal model synthesis and validation. In: Workshop on Formal Methods in Software Engineering, Florence, Italy (2015)
22. Heitmeyer, C.L., Archer, M., Bharadwaj, R., Jeffords, R.D.: Tools for constructing requirements specifications: the SCR toolset at the age of 10. Comput. Syst. Sci. Eng. **20**(1), 19–35 (2005)
23. Knexus: eBotworks. http://www.knexusresearch.com
24. Phon, D.N.E., Ali, M.B., Halim, N.D.A.: Collaborative augmented reality in education: a review. In: International Conference in Teaching and Learning in Computing and Engineering, Kuching, Malaysia (2014)
25. Carruth, D.W.: Virtual reality for education and workforce training. In: International Conference on Emerging eLearning Technologies and Applications, Stary Smokovec, Slovakia (2017)

Author Index

Printed in the United States
By Bookmasters